C. Remigius Fresenius

A System of Instruction in Qualitative Chemical Analysis

C. Remigius Fresenius

A System of Instruction in Qualitative Chemical Analysis

ISBN/EAN: 9783337140328

Printed in Europe, USA, Canada, Australia, Japan

Cover: Foto ©berggeist007 / pixelio.de

More available books at **www.hansebooks.com**

A

SYSTEM OF INSTRUCTION

IN

QUALITATIVE CHEMICAL ANALYSIS.

BY

DR. C. REMIGIUS FRESENIUS,

PRIVY AULIC COUNCELLOR OF ───── OF NASSAU;
DIRECTOR OF THE CHEMICAL LABORATORY AT WIESBADEN;
PROFESSOR OF CHEMISTRY, NATURAL PHILOSOPHY, AND TECHNOLOGY AT THE
WIESBADEN AGRICULTURAL INSTITUTE.

Sixth Edition.

EDITED BY

J. LLOYD BULLOCK, F.C.S.

LONDON:
JOHN CHURCHILL & SONS, NEW BURLINGTON STREET.
MDCCCLXIV.

EDITOR'S PREFACE TO THE SIXTH EDITION.

A STEADY and ever-increasing demand for this work in Germany has led to the issue of successive editions until an eleventh has been reached. In every new edition the Author has introduced whatever improvement in the processes described, or in the distribution of his material, the progress of the science has suggested. The present is characterized by two important additions :—

1st. The original plan of the work excluded all the rarer elements and their compounds; but, inasmuch as many of these have recently acquired importance, as chemical reagents or in the arts, it has been deemed advisable to embrace the whole. Now, therefore, all the known elements are treated of, and processes given for their preparation and detection; but in order to avoid increasing too much the size of the volume, or to embarrass the beginner in the study of analysis, many of these are printed in smaller type.

2nd. The other addition is that of SPECTRUM ANALYSIS, the most interesting, beautiful, and important acquisition which Analytical Chemistry has ever received. This is treated of fully, as its intrinsic importance and the requirements of students of such a work as this demanded. The reader will also observe that the new process known as Dialysis has not been overlooked.

J. LLOYD BULLOCK.

3, HANOVER-STREET, *Dec.*, 1863.

CONTENTS.

PART I.

INTRODUCTORY PART.

PRELIMINARY REMARKS. — PAGE

Definition, general principles, objects, utility, and importance of qualitative chemical analysis. Conditions and requirements for a successful study of that science 1

SECTION I.
OPERATION, § 1 3
1. Solution, § 2 3
2. Crystallization, § 3 . . . 5
3. Precipitation, § 4 . . . 6
4. Filtration, § 5 7
5. Decantation, § 6 . . . 8
6. Evaporation, § 7 . . . 9
7. Distillation, § 8 . . . 10
8. Ignition, § 9 10
9. Sublimation, § 10 . . . 11
10. Fusion and Fluxing, § 11 . . 11
11. Deflagration, § 12 . . . 12
12. The use of the blowpipe, § 13 . 13
13. The use of lamps, particularly of gas lamps, § 14 . . . 17
14. Observation of the coloration of flame by certain bodies, and spectral analysis, § 15 . . 21

Appendix to the First Section.

Apparatus and utensils, § 16 . . 25

SECTION II.
Reagents § 17 27

A. REAGENTS IN THE HUMID WAY.

I. SIMPLE SOLVENTS 29
1. Water, § 18 29
2. Alcohol, § 19 30
3. Ether,
4. Chloroform, } § 20 . . 30
5. Sulphide of carbon,

II. ACIDS AND HALOGENS, § 21 . . 31
 a. Oxygen acids.
 1. Sulphuric acid, § 22 . . 31
 2. Nitric Acid, § 23 . . . 33
 3. Acetic acid, § 24 . . . 33
 4. Tartaric acid, § 25 . . . 34

 b. Hydrogen acids and halogens.
 1. Hydrochloric acid, § 26 . . 34
 2. Chlorine and chlorine water, § 27 35
 3. Nitrohydrochloric acid, § 28 . 36
 4. Hydrofluosilicic acid, § 29 . . 37

 c. Sulphur acids.
 1. Hydrosulphuric acid, § 30 . . 37

III. BASES AND METALS, § 31 . . 42
 a. Oxygen bases.
  α. Alkalies.
  1. Potassa and soda, § 32 . . 42
  2. Ammonia, § 33 . . . 44

  β. Alkaline earths.
  1. Baryta, § 34 45
  2. Lime, § 35 46

  γ. Heavy metals and their oxides.
  1. Zinc, § 36 46
  2. Iron, § 37 47
  3. Copper, § 38 47
  4. Hydrate of teroxide of bismuth, § 39 47

 b. Sulphur bases.
 1. Sulphide of ammonium, § 40 . 48
 2. Sulphide of sodium, § 41 . . 49

IV. SALTS.
 a. Salts of the alkalies.
 1. Sulphate of potassa, § 42 . . 50
 2. Phosphate of soda, § 43 . . 50
 3. Oxalate of ammonia, § 44 . . 50
 4. Acetate of soda, § 45 . . 51
 5. Carbonate of soda, § 46 . . 51
 6. Carbonate of ammonia, § 47 . 52
 7. Bisulphite of soda, § 48 . . 52
 8. Nitrite of potassa, § 49 . . 53
 9. Bichromate of potassa, § 50 . 53
 10. Granular antimonate of potassa, § 51 54
 11. Molybdate of ammonia, dissolved in nitric acid, § 52 . . 54
 12. Chloride of ammonium, § 53 . 55
 13. Cyanide of potassium, § 54 . 55
 14. Ferrocyanide of potassium, § 55 . 56

CONTENTS.

	PAGE
15. Ferricyanide of potassium, § 56	57
16. Sulphocyanide of potassium, § 57	57

b. Salts of the alkaline earths.

1. Chloride of barium, § 58	58
2. Nitrate of baryta, § 59	58
3. Carbonate of baryta, § 60	59
4. Sulphate of lime, § 61	59
5. Chloride of calcium, § 62	59
6. Sulphate of magnesia, § 63	60

c. Salts of the oxides of the heavy metals.

1. Sulphate of protoxide of iron, § 64	60
2. Sesquichloride of iron, § 65	61
3. Nitrate of silver, § 66	62
4. Acetate of lead, § 67	62
5. Nitrate of suboxide of mercury, § 68	62
6. Chloride of mercury, § 69	63
7. Sulphate of copper, § 70	63
8. Protochloride of tin, § 71	64
9. Bichloride of plantinum, § 72	64
10. Sodio-protochloride of palladium, § 73	65
11. Terchloride of gold, § 74	65

V. COLORING MATTERS AND INDIFFERENT VEGETABLE SUBSTANCES, § 75 . . 65

1. *Test papers.*

α. Blue litmus-paper	65
β. Reddened litmus-paper	66
γ. Georgina paper	66
δ. Turmeric-paper	66
2. Indigo solution, § 76	67

B. REAGENTS IN THE DRY WAY.

I. *Fluxes and decomposing agents.*

1. Mixture of carbonate of soda and carbonate of potassa, § 77	67
2. Hydrate of baryta, § 78	68
3. Fluoride of calcium, § 79	69
4. Nitrate of soda, § 80	69

II. *Blowpipe reagents.*

1. Carbonate of soda, § 81	69
2. Cyanide of potassium, § 82	70
3. Biborate of soda, § 83	71
4. Phosphate of soda and ammonia, § 84	71
5. Nitrate of protoxide of cobalt, § 85	72

SECTION III.

REACTIONS, OR DEPORTMENT OF BODIES WITH REAGENTS, § 86 . . 73

A. REACTIONS, OR DEPORTMENT AND PROPERTIES OF THE METALLIC OXIDES AND OF THEIR RADICALS, § 87 . . 74

FIRST GROUP.

Potassa, soda, ammonia; oxide of cæsium, oxide of rubidium, lithia, § 88 74

	PAGE
Special reactions of the more common oxides of the first group.	
a. Potassa, § 89	75
b. Soda, § 90	76
c. Ammonia, § 91	78
Recapitulation and remarks, § 92	79

Special reactions of the rarer oxides of the first group.

1. Oxide of cæsium	80
2. Oxide of rubidium § 93	80
3. Lithia	80

SECOND GROUP.

Baryta, strontia, lime, magnesia, § 94 81
Special reactions.

a. Baryta, § 95	82
b. Strontia, § 96	83
c. Lime, § 97	85
d. Magnesia, § 98	86
Recapitulation and remarks, § 99	88

THIRD GROUP.

More common oxides of the third group: alumina, sesquioxide of chromium. Rarer oxides of the third group: berylla, thoria, zirconia, yttria, oxide of terbium, oxide of erbium, oxide of cerium, oxide of lanthanium, oxide of didymium, titanic acid, tantalic acid, hyponiobic acid, § 100 . . 90

Special reactions of the more common oxides of the third group.

a. Alumina, § 101	90
b. Sesquioxide of chromium, § 102	92
Recapitulation and remarks, § 103	93

Special reactions of the rarer oxides of the third group, § 104 . . 94

1. Berylla	94
2. Thoria	94
3. Zirconia	94
4. Yttria	95
5. Oxide of terbium	95
6. Oxide of erbium	95
7. Oxides of cerium	95
8. Oxide of lanthanium	96
9. Oxide of didymium	96
10. Titanic acid	97
11. Tautalic acid	98
12. Hyponiobic acid	98

FOURTH GROUP.

More common oxides of the fourth group: oxide of zinc, protoxide of manganese, protoxide of nickel, protoxide of cobalt, protoxide of iron, sesquioxide of iron. Rarer oxides of the fourth group: sesquioxides of uranium, oxides of vanadium, oxides of thallium, § 105 99

Special reactions.

a. Oxide of zinc, § 106	99
b. Protoxide of manganese, § 107	101

CONTENTS.

	PAGE
c. Protoxide of nickel, § 108	102
d. Protoxide of cobalt, § 109	104
e. Protoxide of iron, § 110	105
f. Sesquioxide of iron, § 111	107
Recapitulation and remarks, § 112	108

Special reactions of the rarer oxides of the fourth group, § 113 . . . 110
 a. Oxides of uranium . . . 110
 b. Oxides of vanadium . . . 110
 c. Oxides of thallium . . . 111

FIFTH GROUP.

More common oxides of the fifth group: oxide of silver, suboxide of mercury, oxide of mercury, oxide of lead, teroxide of bismuth, oxide of copper, oxide of cadmium. Rarer oxides of the fifth group: oxides of palladium, rhodium, osmium, ruthenium, § 114 . . . 111

Special reactions of the more common oxides of the fifth group.

First division of the fifth group: oxides which are precipitated by hydrochloric acid.
 a. Oxide of silver, § 115 . . . 112
 b. Suboxide of mercury, § 116 . . . 113
 c. Oxide of lead, § 117 . . . 114
Recapitulation and remarks, § 118 . 116

Second division of the fifth group: oxides which are not precipitated by hydrochloric acid.

Special reactions.
 a. Oxide of mercury, § 119 . . . 116
 b. Oxide of copper, § 120 . . . 117
 c. Teroxide of bismuth, § 121 . . 119
 d. Oxide of cadmium, § 122 . . 121
Recapitulation and remarks, § 123 . 121

Special reactions of the rarer oxides of the fifth group, § 124 . . 123
 a. Protoxide of palladium . . 123
 b. Sesquioxide of rhodium . . 123
 c. Oxides of osmium . . . 123
 d. Oxides of ruthenium . . . 124

SIXTH GROUP.

More common oxides of the sixth group; teroxide of gold, binoxide of platinum, protoxide of tin, binoxide of tin, teroxide of antimony, arsenious acid and arsenic acid. Rarer oxides of the sixth group: oxides of iridium, molybdenum, tungsten, tellurium, selenium, § 125 . . . 125

First division of the sixth group. Special reactions.
 a. Teroxide of gold, § 126 . . 125
 b. Binoxide of platinum, § 127 . 126
Recapitulation and remarks, § 128 . 127

Second division of the sixth group. Special reactions.
 a. Protoxide of tin, § 129 . . 128
 b. Binoxide of tin, § 130 . . 129
 c. Teroxide of antimony, § 131 . 131
 d. Arsenious acid, § 132 . . 134
 e. Arsenic acid, § 133 . . . 143
Recapitulation and remarks, § 134 . 144

Special reactions of the rarer oxides of the sixth group, § 135 . . 147
 a. Oxide of iridium . . . 147
 b. Oxides of molybdenum . . 148
 c. Oxides of wolframium or tungsten 148
 d. Oxides of tellurium . . . 149
 e. Oxide of selenium . . . 149

B. REACTIONS OR DEPORTMENT OF THE ACIDS AND THEIR RADICALS WITH REAGENTS, § 136 . . . 150

Classification of acids in groups, § 136 151

I. INORGANIC ACIDS.

FIRST GROUP.

Acids which are precipitated from neutral solutions by chloride of barium: chromic acid (sulphurous acid, hyposulphurous acid, iodic acid), sulphuric acid (hydrofluosilicic acid), phosphoric acid (bibasic, monobasic phosphoric acid), boracic acid, oxalic acid, hydrofluoric acid (phosphorous acid), carbonic acid, silicic acid, § 137 151

First division of the first group of the inorganic acids, § 138 . . . 152
 Chromic acid 152
 Remarks 153

Rarer acids of the first division of the first group, § 139 . . . 154
 a. Sulphurous acid . . . 154
 b. Hyposulphurous acid . . 154
 c. Iodic acid 154

Second division of the first group of the inorganic acids.
 Sulphuric acid, § 140 . . . 155
 Remarks 156
 Hydrofluosilicic acid, § 141 . . 156

Third division of the first group of the inorganic acids.
 a. Phosphoric acid, § 142 . . 156
  α. Bibasic phosphoric acid, ⎫
  β. Monobasic phosphoric ⎬ § 143 160
   acid, ⎭
 b. Boracic acid, § 144 . . . 160
 c. Oxalic acid, § 145 . . . 162
 d. Hydrofluoric acid, § 146 . . 163
Recapitulation and remarks, § 147 . 166
 Phosphorous acid, 148 . . . 167

CONTENTS.

Fourth division of the first group of the inorganic acids.

	PAGE
a. Carbonic acid, § 149	167
b. Silicic acid, § 150	169
Recapitulation and remarks, § 151	170

SECOND GROUP OF INORGANIC ACIDS.

Acids which are precipitated by nitrate of silver, but not by chloride of barium: hydrochloric acid, hydrobromic acid, hydriodic acid, hydrocyanic acid (hydroferro— and hydroferricyanic acid), hydrosulphuric acid (nitrous acid, hypochlorous acid, chlorous acid, hypophosphorous acid).

a. Hydrochloric acid, § 152	171
b. Hydrobromic acid, § 153	172
c. Hydriodic acid, § 154	174
d. Hydrocyanic acid, § 155	176

Appendix.

a. Hydroferrocyanic acid	178
b. Hydroferricyanic acid	178
c. Hydrosulphuric acid, § 156	178
Recapitulation and remarks, § 157	180
Rarer acids of the second group, § 158	181
1. Nitrous acid	181
2. Hypochlorous acid	182
3. Chlorous acid	182
4. Hypophosphorous acid	182

THIRD GROUP OF THE INORGANIC ACIDS.

Acids which are not precipitated by salts of baryta nor by salts of silver: nitric acid, chloric acid (perchloric acid).

a. Nitric acid, § 159	183
b. Chloric acid, § 160	184
Recapitulation and remarks, § 161	185
Perchloric acid, § 162	186

II. ORGANIC ACIDS.

FIRST GROUP.

Acids which are invariably precipitated by chloride of calcium: oxalic acid, tartaric acid (paratartaric or racemic acid), citric acid, malic acid.

	PAGE
a. Oxalic acid, § 163	186
b. Tartaric acid, § 163	186
c. Citric acid, § 164	187
d. Malic acid, § 165	188
Recapitulation and remarks, § 166	189
Racemic or paratartaric acid, § 167	190

SECOND GROUP OF THE ORGANIC ACIDS.

Acids which chloride of calcium fails to precipitate under any circumstances, but which are precipitated from neutral solutions by sesquichloride of iron: succinic acid, benzoic acid.

a. Succinic acid, § 168	191
b. Benzoic acid, § 169	192
Recapitulation and remarks, § 170	192

THIRD GROUP OF THE ORGANIC ACIDS.

Acids which are not precipitated by chloride of calcium nor by sesquichloride of iron: acetic acid, formic acid (lactic acid, propionic acid, butyric acid).

a. Acetic acid, § 171	193
b. Formic acid, § 172	194
Recapitulation and remarks, § 173	195
Rarer acids of the third group of organic acids, § 174	196
1. Lactic acid	196
2. Propionic acid and butyric acid	196

PART II.

SYSTEMATIC COURSE OF QUALITATIVE CHEMICAL ANALYSIS.

	PAGE
Preliminary remarks on the course of qualitative analysis in general, and on the plan of this part of the present work in particular	201

SECTION I.

PRACTICAL PROCESS FOR THE ANALYSIS OF COMPOUNDS AND MIXTURES IN GENERAL.

I. *Preliminary examination*, § 175	203
A. The body under examination is solid.	
I. It is neither a pure metal nor an alloy, § 176	204
II. The substance is a metal or an alloy, § 177	209
B. The substance under examination is a fluid, § 178	209
II. Solution of bodies, or classification of substances according to their deportment with certain solvents, § 179	210
A. The substance under examination is neither a metal nor an alloy, § 180	211
B. The substance under examination is a metal or an alloy, § 181	213
III. *Actual examination.*	
SIMPLE COMPOUNDS.	
A. *Substances soluble in water.*	
Detection of the base, § 182	214
Detection of the acid, § 183	219

CONTENTS.

	PAGE
I. Detection of inorganic acids, § 183	219
II. Detection of organic acids, § 184	221
B. *Substances insoluble or sparingly soluble in water, but soluble in hydrochloric acid, nitric acid, or nitrohydrochloric acid.*	
Detection of the base, § 185	223
Detection of the acid, § 185	225
I. Detection of inorganic acids, § 186	225
II. Detection of organic acids, § 187	226
C. *Substances insoluble or sparingly soluble in water, hydrochloric acid, nitric acid, or nitrohydrochloric acid.*	
Detection of the base and the acid, § 188	226

COMPLEX COMPOUNDS.

A. *Substances soluble in water, and also such as are insoluble in water, but dissolve in hydrochloric acid, nitric acid, or nitrohydrochloric acid.*	
Detection of the bases	228
Treatment with hydrochloric acid: detection of silver, suboxide of mercury (lead), § 189	228
Treatment with hydrosulphuric acid, precipitation of the metallic oxides of Group V., 2nd division, and of Group VI., § 190	231
Treatment of the precipitate produced by hydrosulphuric acid with sulphide of ammonium; separation of the 2nd division of Group V. from Group VI., § 191	232
Detection of the metals of Group VI. Arsenic, antimony, tin, gold, platinum, § 192	233
Detection of the metallic oxides of Group V., 2nd division. Oxide of lead, teroxide of bismuth, oxide of copper, oxide of cadmium, oxide of mercury, § 193	236
Precipitation with sulphide of ammonium, detection and separation of the oxides of Groups III. and IV. Alumina, sesquioxide of chromium, oxide of zinc, protoxide of manganese, protoxide of nickel, protoxide of cobalt, proto- and sesquioxide of iron, and also of those salts of the alkaline earths which are precipitated by ammonia from their solution in hydrochloric acid: phosphates, borates, oxalates, silicates, and fluorides, § 194	238
Separation and detection of the oxides of Group II., which are precipitated by carbonate of ammonia in presence of chloride of ammonium, viz., baryta, strontia, lime, § 195	244
Examination for magnesia, § 196	245

	PAGE
Examination for potassa and soda, § 197	246
Examination for ammonia, § 198	246

DETECTION OF THE ACIDS.

1. *Substances soluble in water.*	
I. In the absence of organic acids, § 199	247
II. In presence of organic acids, § 200	249
2. *Substances insoluble in water, but soluble in hydrochloric acid, nitric acid, or nitrohydrochloric acid.*	
I. In the absence of organic acids, § 201	251
II. In presence of organic acids, § 202	252
B. *Substances insoluble or sparingly soluble both in water and in hydrochloric acid, nitric acid, or nitrohydrochloric acid.*	
Detection of the bases, acids, and non-metallic elements, § 203	252

SECTION II.

PRACTICAL COURSE IN PARTICULAR CASES.

I. Special method of effecting the analysis of cyanides, ferrocyanides, &c., insoluble in water, and also of insoluble mixed substances containing such compounds, § 204	256
II. Analysis of silicates, § 205	258
A. Silicates decomposable by acids, § 206	258
B. Silicates which are not decomposed by acids, § 207	260
C. Silicates which are partially decomposed by acids, § 208	262
III. Analysis of natural waters, § 209	262
A. Analysis of the fresh waters (spring-water, well-water, brook-water, river-water, &c.), § 210	263
B. Analysis of mineral waters, § 211	266
1. Examination of the water, § 212	266
a. Operations at the spring, § 212	266
b. Operations in the laboratory, § 213	267
2. Examination of the sinter-deposit, § 214	271
IV. Analysis of soils, § 215	273
1. Preparation and examination of the aqueous extract, § 216	274
2. Preparation and examination of the acid extract, § 217	276
3. Examination of the inorganic constituents insoluble in water and acids, § 218	277
4. Examination of the organic constituents of the soil, § 219	277
a. Examination of the organic substances soluble in water	277

CONTENTS.

  b. Treatment with an alkaline carbonate 277
  c. Treatment with caustic alkali 278
 V. Detection of inorganic substances in presence of organic substances, § 220 278
  1. General rules for the detection of inorganic substances in presence of organic matters, which by their color, consistence, &c., impede the application of the reagents, or obscure the reactions produced, § 221 . . . 278
  2. Detection of inorganic poisons in articles of food, in dead bodies, &c., in chemico-legal cases, § 222 279
 I. Method for the detection of arsenic, § 223 280
  A. Method for the detection of undissolved arsenious acid . . 281
  B. Method of detecting soluble arsenical and other metallic compounds, by means of Dialysis, § 224 281
  C. Method for the detection of arsenic in whatever form of combination it may exist, which allows also a quantitative determination of that poison, and permits at the same time the detection of other metallic poisons which may be present, § 225 283
   1. Decoloration and solution 283
   2. Treatment of the solution with hydrosulphuric acid . . 284
   3. Purification of the precipitate produced by hydrosulphuric acid 284
   4. Preliminary examination for arsenic and other metallic poisons of the fifth and sixth groups . 285
   5. Treatment of the purified precipitate produced by hydrosulphuric acid in cases where arsenic alone is assumed to be present . 286
   6. Treatment of the purified precipitate produced by hydrosulphuric acid in cases where there is reason to suppose that another metal is present, perhaps with arsenic 286
   7. Reduction of the sulphide of arsenic 287
   8. Examination of the reserved residues for other metals of the fifth and sixth groups . . 288
    a. Residue I. . . . 288
    b. Residue II. . . . 289
    c. Residue III. . . . 289
    d. Residue IV. . . . 289
   9. Examination of the reserved filtrate for metals of the third and fourth groups . . . 289
 II. Method for the detection of hydrocyanic acid, § 226 . . . 290
 III. Method for the detection of phosphorus, § 227 . . . 292

  A. Detection of unoxidized phosphorus 292
  B. Detection of phosphorous acid . 296
 3. Examination of the inorganic constituents of plants, animals, or parts of the same, of manures, &c. (analysis of ashes). . 297
  A. Preparation of the ash, § 228 . 297
  B. Examination of the ash . . 297
   a. Examination of the part soluble in water . . . 297
   b. Examination of the part soluble in hydrochloric acid . 298
   c. Examination of the residue insoluble in hydrochloric acid 299

SECTION III.

EXPLANATORY NOTES AND ADDITIONS TO THE SYSTEMATIC COURSE OF ANALYSIS.

 I. Additional remarks to the preliminary examination, To §§ 175—178 300
 II. Additional remarks to the solution of substances, &c., To §§ 179—181 301
 III. Additional remarks to the actual examination, To §§ 182—204 . 302
  A. General review and explanation of the analytical course . . 302
   a. Detection of the bases . . 302
   b. Detection of the acids . . 305
  B. Special remarks and additions to the systematic course of analysis. To § 189 308
   §§ 190 and 191 . . . 309
   § 192 310
   § 193 310
   § 194 311
   §§ 195—198 312
   § 203 312
   § 204 313

APPENDIX.

 I. Deportment of the most important medicinal alkaloids with reagents, and systematic method of effecting the detection of these substances, § 229 . . . 315
 I. *Volatile alkaloids.*
  1. Nicotina, § 230 . . . 316
  2. Conia, § 231 . . . 317
 II. *Non-volatile alkaloids.*

FIRST GROUP.

Non-volatile alkaloids which are precipitated by potassa or soda from the solution of their salts, and redissolve readily in an excess of the precipitant . . 318
 Morphia, § 232 318

SECOND GROUP.

Non-volatile alkaloids which are

CONTENTS.

precipitated by potassa from the solutions of their salts, but do not redissolve to a perceptible extent in an excess of the precipitant, and are precipitated by bicarbonate of soda even from acid solutions . . . 320
  a. Narcotina, § 223 . . . 320
  b. Quina, § 234 . . . 321
  c. Cinchonia, § 235 . . . 322
Recapitulation and remarks, § 236   323

THIRD GROUP.

Non-volatile alkaloids which are precipitated by potassa from the solutions of their salts, and do not redissolve to a perceptible extent in an excess of precipitant, but are not precipitated from acid solutions by the bicarbonates of the fixed alkalies . 324
  a. Strychnia, § 237 . . . 324
  b. Brucia, § 238 . . . 326
  c. Veratria, § 239 . . . 327
Recapitulation and remarks, § 240   328
Salicine, § 241 . . . 329
Systematic course for the detection of the alkaloids treated of in the preceding paragraphs, and of salicine . . . 329
I. Detection of the alkaloids, and of salicine, in solutions supposed to contain only one of these substances, § 242 . . . 329
II. Detection of the alkaloids, and of salicine, in solutions supposed to contain several or all of these substances, § 243 . . . 331
Detection of the alkaloids, in presence of coloring and extractive vegetable or animal matters, § 244   332
  1. Stas's method of detecting poisonous alkaloids . . . 333
  2. Otto's modifications of Stas's method . . . 335
  3. Uslar and Erdmann's method . 335
  4. Method of detecting strychnia, based upon the use of chloroform   337
    a. Rodgers and Girdwood's method   337
    b. Prollin's method . . . 337
  5. Method of effecting the detection of strychnia in beer, by Graham and Hofmann . . . 337
  6. Separation by dyalisis . . 338
II. General plan of the order and succession in which substances should be analyzed for practice, § 245 . . . 338
III. Arrangement of the results of the analysis performed for practice, § 246 . . . 339
IV. Table of the more frequently occurring forms and combinations of the substances treated of in the present work; arranged with special regard to the class to which they respectively belong, according to their solubility in water, in hydrochloric acid, nitric acid, or nitrohydrochloric acid . . . 342
Preliminary remarks, § 247 . . 342
Table . . . 344
Notes . . . 344
V. Table of weights and measures . 347

INDEX 349

PART I.

INTRODUCTORY.

PRELIMINARY REMARKS.

DEFINITION, GENERAL PRINCIPLES, OBJECTS, UTILITY, AND IMPORTANCE OF QUALITATIVE CHEMICAL ANALYSIS—CONDITIONS AND REQUIREMENTS FOR A SUCCESSFUL STUDY OF THAT SCIENCE.

CHEMISTRY is the science which treats of the various materials entering into the structure of the earth, their composition and decomposition, their mutual relations and their deportment in general. A special branch of this science is designated *Analytical Chemistry*, inasmuch as it pursues a distinct and definite object—viz. the analysis of compound bodies, and the determination of their component elements. Analytical chemistry, again, is subdivided into two branches—viz. *qualitative analysis*, which simply studies the *nature* and properties of the component parts of bodies; and *quantitative analysis*, which ascertains the *quantity* of every individual constituent present. The office of qualitative analysis, therefore, is to exhibit the constituent parts of a substance of *unknown* composition in forms of *known* composition, from which the constitution of the body examined and the presence of its several component elements may be positively inferred. The efficiency of its method depends upon two conditions—viz, 1st, it must attain the object in view with unerring certainty, and 2nd, it must attain it in the most expeditious manner. The object of quantitative analysis, on the other hand, is to exhibit the elements revealed by the qualitative investigation in forms which will permit the most accurate estimate of their weight, or to effect by other means the determination of their quantity.

These different ends are, of course, attained respectively by very different ways and means. The study of qualitative analysis must, therefore, be pursued separately from that of quantitative analysis, and must naturally precede it.

Having thus generally defined the meaning and scope of qualitative analysis, we have now still to consider, in the first place, the preliminary information required to qualify students for a successful cultivation of this branch of science, the rank which it holds in the domain of chemistry, the bodies that fall within the sphere of its operations, and its utility and importance; and, in the second place, the principal parts into which its study is divided.

It is, above all, absolutely indispensable for a successful pursuit of qualitative investigations that the student should possess some knowledge of the chemical *elements*, and of their most important combinations, as well as of the principles of chemistry in general; and that he should combine with this knowledge some readiness in the apprehension of chemical processes. The practical part of this science demands moreover strict order, great cleanness and neatness, and a certain skill in manipulation. If the student joins to these qualifications the habit of invariably ascribing the failures with which he may happen to meet to some error or defect in his operations, or, in other words, to the absence of some condition or other indispensable to the success of the experiment—and a firm reliance on the immutability of the laws of nature cannot fail to create this habit—he possesses every requisite to render his study of analytical chemistry successful.

Now, although chemical analysis is based on general chemistry, and cannot be cultivated without some previous knowledge of the latter, yet, on the other hand, we have to look upon it also as one of the main pillars upon which the entire structure of the science rests, since it is of almost equal importance for all branches of theoretical as well as of practical chemistry; and I need not expatiate here on the advantages which the physician, the pharmaceutist, the mineralogist, the rational farmer, the manufacturer, the artisan, and many others derive from it.

This consideration would surely in itself be sufficient reason to recommend a thorough and diligent study of this branch of science, even if its cultivation lacked those attractions which yet it unquestionably possesses for every one who devotes himself zealously and ardently to it. The human mind is constantly striving for the attainment of truth; it delights in the solution of problems; and where do we meet with a greater variety of them, more or less difficult of solution, than in the province of chemistry? But as a problem to which, after long pondering, we fail to discover the key, wearies and discourages the mind; so, in like manner, do chemical investigations, if the object in view is not attained—if the results do not bear the stamp of truth, of unerring certainty. A *half-knowledge* is therefore, as indeed in every department of science, but more especially *here*, to be considered worse than no knowledge at all; and a mere *superficial* cultivation of chemical analysis is consequently to be particularly guarded against.

A qualitative investigation may be made with a twofold view—viz., either, 1st, to prove that a certain body is or is not contained in a substance, *e.g.*, lime in spring-water; or, 2nd, to ascertain *all* the constituents of a chemical compound or mixture. Any substance may of course become the object of a chemical analysis.

But all elements are not equally important for the purposes of practical chemistry, a certain number of them only being found more widely disseminated in nature, and more generally employed in pharmacy, in the arts and manufactures, and in agriculture, whilst the others are met with only as constituents of rarely occurring minerals; the elements of the former class alone, therefore, and the more important of their compounds, will be considered more fully in the present work, whilst those of the latter class will be discussed more briefly and in a manner to enable the learner to separate, without difficulty, the study of the former from that of the latter. This arrangement will serve to render the study of the science more easy to beginners, and to lighten the labors of practical chemists.

The study of qualitative analysis is most properly divided into four principal parts—viz.,

1. CHEMICAL OPERATIONS.
2. REAGENTS AND THEIR USES.
3. REACTIONS, OR DEPORTMENT OF THE VARIOUS BODIES WITH REAGENTS.
4. SYSTEMATIC COURSE OF QUALITATIVE ANALYSIS.

It will now be readily understood that the pursuit of chemical analysis requires *practical skill and ability* as well as *theoretical knowledge ;* and that, consequently, a mere speculative study of that science can be as little expected to lead to success as purely empirical experiments. To attain the desired end, theory and practice must be combined.

SECTION I.

OPERATIONS.

§ 1.

THE operations of analytical chemistry are essentially the same as those of synthetical chemistry, though modified to a certain extent to adapt them to the different object in view, and to the small quantities operated upon in analytical investigations.

The following are the principal operations in qualitative analysis.

§ 2.

1. SOLUTION.

The term "*solution*," in its widest sense, denotes the perfect union of a body, no matter whether gaseous, liquid, or solid, with a fluid, resulting in a homogeneous liquid. However, where the substance dissolved is *gaseous*, the term "*absorption*" is more properly made use of; and the solution of one fluid in another is more generally called a *mixture*. The application of the term solution, in its usual and more restricted sense, is confined to the perfect union of a *solid* body with a fluid.

A solution is the more readily effected the more minutely the body to be dissolved is divided. The fluid by means of which the solution is effected, is called the *solvent*. We call the solution *chemical*, where the solvent enters into chemical combination with the substance dissolved; *simple*, where no definite combination takes place.

In a *simple* solution the dissolved body exists in the free state, and retains all its original properties, except those dependent on its form and cohesion; it separates unaltered when the solvent is withdrawn. Common salt dissolved in water is a familiar instance of a simple solution. The salt in this case imparts its peculiar taste to the fluid. On evaporating the water, the salt is left behind in its original form. A simple solution is called *saturated* if the solvent has received as much as it can retain of the dissolved substance. But as fluids dissolve generally larger quantities of a substance the higher their temperature, the term *saturated*, as applied to *simple* solutions, is only relative, and refers invariably to a certain temperature. It may be laid down as a general rule that elevation of temperature facilitates and accelerates simple solution.

A *chemical* solution contains the dissolved substance not in the same state nor possessed of the same properties as before; the dissolved body is no longer free, but intimately combined with the solvent, which latter also has lost its original properties; a new substance has thus been produced, and the solution manifests therefore now the properties of this new substance. A chemical solution also may be *accelerated* by elevation of temperature; and this is indeed usually the case, since heat generally promotes the action of bodies upon each other. But the *quantity* of the dissolved body remains always the same in proportion to a given quantity of the solvent, whatever may be the difference of temperature—the combining proportions of substances being invariable and altogether independent of the gradations of temperature.

The reason of this is, that in a chemical solution the solvent and the body upon which it acts have invariably opposite properties, which they strive mutually to neutralize. Further solution ceases as soon as this tendency of mutual neutralization is satisfied. The solution is in this case also said to be saturated or, more properly, *neutralized*, and the point which denotes it to be so is termed the point of saturation or neutralization.

The substances which produce chemical solutions are, in most cases, either acids or alkalies. With few exceptions, they have first to be converted to the fluid state by means of a simple solvent. When the opposite properties of acid and base are mutually neutralized, and the new compound is formed, the actual transition to the fluid state will ensue only if the new compound possesses the property of forming a simple solution with the liquid present; *e.g.*, if solution of acetic acid in water is brought into contact with oxide of lead, there ensues, first, a chemical combination of the acid with the oxide, and then a simple solution of the new-formed acetate of lead in the water of the menstruum.

In pharmacy solutions are often made in a porcelain mortar, by triturating the body to be dissolved with the solvent added gradually in small quantities at a time; in chemical laboratories solutions are rarely made in this manner, but generally by digesting or heating the substance to be dissolved with the fluid in beaker-glasses, flasks, test-tubes, or dishes. In the preparation of chemical solutions the best way generally is to mix the body to be dissolved in the first place with water (or with whatever other indifferent fluid may happen to be used), and then gradually add the chemical agent. By this course of proceeding a large excess of the latter is avoided, an over-energetic action guarded against, the process greatly facilitated, and complete solution ensured, which is a matter of some importance, as it will not seldom happen in chemical combinations that the product formed refuses to dissolve if an excess of the chemical solvent is present; in which case the molecules first formed of the new salt, being insoluble in the menstruum present, gather round and enclose the particles still unacted on, weakening thereby or preventing altogether further chemical action upon them. Thus, for instance, Witherite (carbonate of baryta) dissolves readily if water is poured upon the pulverised mineral and hydrochloric acid gradually added; but it dissolves with difficulty and imperfectly if it is projected into a concentrated solution of hydrochloric acid in water; since chloride of barium will indeed dissolve in water, but not in hydrochloric acid.

CRYSTALLIZATION and PRECIPITATION are the reverse of solution, since they have for their object the conversion of a fluid or dissolved substance to the solid state. As both generally depend on the same cause, viz., on the absence of a solvent, it is impossible to assign exact limits to either; in many cases they merge into one another. We must, however, consider them separately here, as they differ essentially in their extreme forms, and as the special objects which we purpose to attain by their application are generally very different.

§ 3.

2. CRYSTALLIZATION.

We understand by the term crystallization, in a more general sense, every operation or process whereby bodies are made to pass from the fluid to the solid state, and to assume certain fixed, mathematically definable, regular forms. But as these forms, which we call *crystals*, are the more regular, and consequently the more perfect, the more slowly the operation is carried on, we always connect with the term "crystallization" the accessory idea of a *slow* separation—of a *gradual* conversion to the solid state. The formation of crystals depends on the regular arrangement of the ultimate constituent particles of bodies (*molecules* or *atoms*); it can only take place, therefore, if these atoms possess perfect freedom of motion, and thus in general only when a substance passes from the fluid or gaseous to the solid state. Those instances in which the mere ignition, or the softening or moistening of a solid body, suffices to make the tendency of the molecules to a regular arrangement (crystallization) prevail over the diminished force of cohesion—such as, for instance, the turning white and opaque of moistened barley-sugar—are to be regarded as exceptional cases.

To induce crystallization, the causes of the fluid or gaseous form of a substance must be removed. These causes are either *heat alone*, *e.g.*, in the case of fused metals; or *solvents alone*, as in the case of an aqueous solution of common salt; or *both combined*, as in the case of a hot saturated solution of nitrate of potassa in water. In the first case we accordingly obtain crystals by cooling the fused mass; in the second by evaporating the menstruum; and in the third by either of these means. The most frequently occurring case is that of crystallization by cooling hot saturated solutions. The liquors which remain after the separation of the crystals are called *mother-waters* or *mother-liquors*. The term *amorphous* is applied to such solid bodies as have no crystalline form.

We have recourse to crystallization mostly either to obtain the crystallized substance in a solid form, or to separate it from other substances dissolved in the same menstruum. In many cases also the form of the crystals or their deportment in the air, viz., whether they remain unaltered or effloresce or deliquesce upon exposure to the air, will afford an excellent means of distinguishing between bodies otherwise resembling each other; for instance, between sulphate of soda and sulphate of potassa. The process of crystallization is usually effected in evaporating dishes or, for very small quantities, in watch-glasses.

In cases where the quantity of fluid to be operated upon is only small, the surest way of getting well-formed crystals is to let the fluid evaporate in the air or, better still, under a bell-glass, under which is also placed an open vessel half-filled with concentrated sulphuric acid. Minute crystals are examined best with a lens or under the microscope.

§ 4.

3. PRECIPITATION.

This operation differs from the preceding in this, that the dissolved body is converted to the solid state, not slowly and gradually, but *suddenly*, no matter whether the substance separating is crystalline or amorphous, whether it sinks to the bottom of the vessel or ascends or remains suspended in the liquid. Precipitation is either caused by a modification of the solvent—thus sulphate of lime (gypsum) separates immediately from its solution in water upon the addition of alcohol; or it ensues in consequence of the separation of an educt insoluble in the menstruum—thus metallic copper precipitates if a solution of chloride of copper is brought into contact with zinc, as this separates the copper from the chlorine, and the eliminated metal is insoluble in the water of the menstruum. Precipitation, lastly, takes place also where, by the action of simple or double chemical affinity, new compounds are formed which are insoluble in the menstruum; thus oxalate of lime precipitates upon addition of oxalic acid to a solution of acetate of lime; chromate of lead if chromate of potassa in solution is mixed with solution of nitrate of lead. In decompositions of this kind, induced by simple or double affinity, one of the new compounds remains generally in solution, and the same is sometimes the case also with the educt; thus in the instances just mentioned the chloride of zinc, the acetic acid, and the nitrate of potassa remain in solution. It may, however, happen also that both the product and the educt, or two products, precipitate, and that nothing remains in solution; this is the case, for instance, when a solution of sulphate of magnesia is mixed with water of baryta, or when a solution of sulphate of silver is precipitated with chloride of barium.

Precipitation is resorted to for the same purposes as crystallization, viz., either to obtain a substance in the solid form, or to separate it from other substances dissolved in the same menstruum. But in qualitative analysis we have recourse to this operation more particularly for the purpose of detecting and distinguishing substances by the color, properties, and general deportment which they exhibit when precipitated either in an isolated state or in combination with other substances. The solid body separated by this process is called the *precipitate*, and the substance which acts as the immediate cause of the separation is termed the *precipitant*. Various terms are applied to precipitates by way of particularizing them according to their different nature; thus we distinguish crystalline, pulverulent, flocculent, curdy, gelatinous precipitates, &c.

The terms *turbid* and *turbidity*, or *cloudy* and *cloudiness*, are made use of to designate the state of a fluid which contains a precipitate so finely divided and so inconsiderable in amount, that the suspended particles, although impairing the transparency of the fluid, yet cannot be clearly distinguished. The separation of flocculent precipitates may generally be promoted by a vigorous shake of the vessel; that of crystalline precipitates, by stirring the fluid and rubbing the inside of the vessel with a glass rod; lastly, elevation of temperature is also an effective means of promoting the separation of most precipitates. The process is therefore conducted, according to circumstances, either in test-tubes, flasks, or beakers.

The two operations described respectively in §§ 5 and 6, viz. *filtration* and *decantation*, serve to effect the mechanical separation of fluids from matter suspended therein.

§ 5.

4. Filtration.

This operation consists simply in passing the fluid from which we wish to remove the solid particles mechanically suspended therein through a filtering apparatus, formed usually by a properly folded piece of unsized paper placed in a funnel. An apparatus of this description allows the fluid to trickle through with ease, whilst it completely retains the solid particles. We employ smooth filters and plaited filters; the former in cases where the separated solid substance is to be made use of, the latter in cases where it is simply intended to clear the solution. Smooth filters are prepared by double-folding a circular piece of paper, with the folds at right angles; they must in every part fit close to the funnel. The preparation of plaited filters is more properly a matter for ocular demonstration than for description. In cases where the contents of the filter require washing, the paper must not project over the rim of the funnel. It is in most cases advisable to moisten the filter previously to passing the fluid through it; since this not only tends to accelerate the process, but also to prevent the solid particles being carried through the pores of the filter. The paper selected for filters must be as free as possible from inorganic substances, especially such as are dissolved by acids, as sesquioxide of iron, lime, &c. The common filtering paper of commerce seldom comes up to our requirements in this respect, and I would therefore always recommend to wash it carefully with acid and water whenever it is intended for use in *accurate analyses*. With the stronger sorts of paper this may be done by placing the paper, cut into pieces of suitable size, in a layer of moderate thickness, in a shallow porcelain dish, pouring over it a mixture of one part of hydrochloric acid or nitric acid with about nine parts of water, and letting it digest for several hours at a moderate heat. The fluid is then poured off, and the paper repeatedly washed with water (finally with distilled water), until litmus paper is no longer reddened by the washings: the water is then drained off, and the entire layer is carefully transferred to a quire of blotting-paper, and left there until the filters can be taken off singly without injury; they are then finally dried by exposing them to a gentle heat, either singly or placed in thin layers between sheets of blotting-paper. With the finer sorts of paper (Swedish) I prefer washing the filters in the funnel. To this end they are first sprinkled with a little moderately diluted hydrochloric or nitric acid, and then thoroughly washed with water, finally with distilled water.

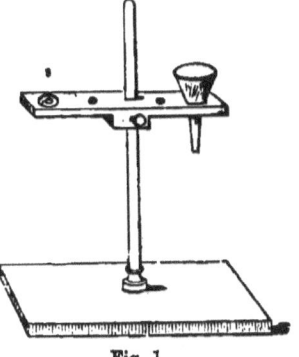

Fig. 1.

Filtering paper, to be considered good, must, besides being pure, also let fluids pass readily through, yet completely retain even the finest pulverulent precipitates, such as sulphate of baryta, oxalate of lime, &c. Where a paper satisfying these requirements cannot be readily procured it is advisable to keep two sorts, one of closer texture for the separation of very finely divided precipitates, and one of greater porosity for the speedy separation of grosser particles. The funnels must be of glass or porcelain (§ 16, 10); they are usually placed on an appropriate stand, to keep them in a fixed position. The stand shown in Fig. 1 is particularly well adapted for the reception of the small-sized funnels used in qualitative analyses.

§ 6.

5. DECANTATION.

This operation is frequently resorted to instead of filtration, in cases where the solid particles to be removed are of considerably greater specific gravity than the liquid in which they are suspended; as they will in such cases speedily subside to the bottom, thereby rendering it easy either to decant the supernatant fluid by simply inclining the vessel, or to draw it off by means of a syphon or pipette.

In cases where filtration or decantation are resorted to for the purpose of obtaining the solid substance, the latter has to be freed afterwards by repeated washing from the liquid still adhering to it. This operation is termed *washing* or *edulcoration*. The washing of precipitates collected on a filter is usually effected by means of a washing-bottle, such as is shown in Fig. 2.

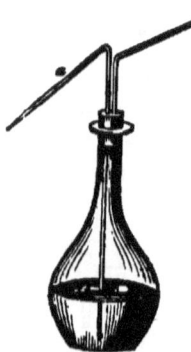

Fig. 2.

The drawing needs no elaborate explanation. The outer end of the tube *a* is drawn out to a fine point. By blowing air into the flask through the other tube, a fine jet of water is expelled through *a*, with a certain degree of force, which is particularly well suited for washing precipitates. Washing-bottles of this construction afford also the advantage that they do equally well for washing with hot water. They are, for this purpose, and to enble the operator to use them with the greater ease, either furnished with a handle, or with a double or treble coil of twine bound tight round the neck.

As the success of an analytical operation often depends absolutely upon the proper washing of a precipitate, it may as well be mentioned at once that the operation ought never to be considered completed before the object of it has been fully attained. And this is the case only when the precipitate has been absolutely freed from the fluid adhering to it. The operator should, in this respect, never trust to mere belief or guessing, but should always make quite sure by properly testing the last washings. With fixed bodies it generally suffices to slowly evaporate a drop of the last washings on platinum-foil, when complete volatilisation will show that the end in view has been fully attained.

There are four operations which serve to separate volatile substances from less volatile or from fixed bodies, viz., *evaporation*, *distillation*, *ignition*, and *sublimation*. The two former of these operations refer exclusively to fluids, the two latter exclusively to solids.

§ 7.

6. EVAPORATION.

This is one of the most common operations in analytical chemistry. It serves to separate volatile fluids from less volatile or from fixed bodies (no matter whether solid or fluid), in cases where the residuary substance alone is of importance, whilst the evaporating matter is entirely disregarded;—thus, for instance, we have recourse to evaporation for the purpose of removing from a saline solution part of the water, in order to bring about crystallization of the salt; we resort to this process also for the purpose of removing the whole of the water of the menstruum from the solution of a non-crystallizable substance, so as to obtain the latter in a solid form, &c. The evaporated water is entirely disregarded in either of these cases, the only object in view being to obtain, in the former case a more concentrated fluid, and in the latter a dry substance. These objects are invariably attained by converting the fluid which is to be removed to the gaseous state. This is generally done by the application of heat; sometimes also by leaving the fluid for a certain time in contact with the atmosphere, or with an enclosed volume of air constantly kept dry by hygroscopic substances, such as concentrated sulphuric acid, chloride of calcium, &c.; or, lastly, in many cases, by placing the fluid in rarefied air, with simultaneous application of hygroscopic substances. As it is of the utmost importance in qualitative analyses to guard against the least contamination, and as an evaporating fluid is the more liable to this the longer the operation lasts, the process is usually conducted with proper expedition, in porcelain or platinum dishes, over the flame of a spirit or gas-lamp, in a separate place free from dust and not exposed to draughts of air. If the operator has no place of the kind, he must have recourse to the much less suitable proceeding of covering the dish; the best way of doing this is to place over the dish a large glass funnel secured by a retort-holder, in a manner to leave sufficient space between the rim of the funnel and the border of the dish; the funnel is placed slightly aslant, that the drops running down its sides may be received in a beaker. Or the dish may also be covered with a sheet of filter-paper previously freed from inorganic substances by washing with dilute hydrochloric or nitric acid (see § 5); were common and unwashed filter-paper used for the purpose, the sesquioxide of iron, lime, &c., contained in it would dissolve in the vapors evolved (more especially if acid), and the solution dripping down into the evaporating fluid would speedily contaminate it. These precautions are necessary of course only in accurate analyses. Larger quantities of fluid are evaporated best in glass flasks standing aslant, covered with a cap of pure filtering paper, over a charcoal fire or gas; or also in tubular retorts with neck rising obliquely upward, and open tubulature. Evaporating processes at 212° are conducted in a suitable steam apparatus, or

Fig. 3.

in the water-bath shown in Fig. 3. Evaporation to dryness is not usually conducted over an open fire, but generally either on the water-bath or the sand-bath, or on a heated iron plate.

§ 8.

7. DISTILLATION.

This operation serves to separate a volatile liquid from a less volatile or a fixed substance (no matter whether solid or fluid) where the object is to recover the evaporating fluid. In order to attain this end, it is necessary to reconvert the liquid from the gaseous form in which it evaporates into the fluid state. A distilling apparatus consists consequently always of three parts, no matter whether separable or not. These three parts are—1st, a vessel in which the liquid to be distilled is heated, and thus converted into vapor; 2nd, an apparatus in which this vapor is cooled again or *condensed*, and thus reconverted to the fluid state; and 3rd, a vessel to receive the fluid thus reproduced by the condensation of the vapor (the distillate). For the distillation of large quantities

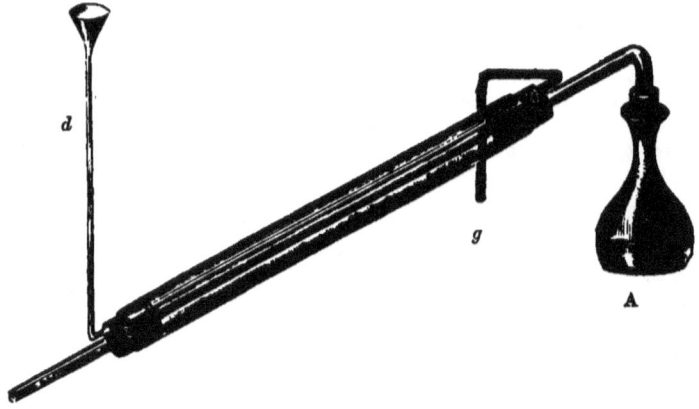

Fig. 4.

metallic apparatus are used (also copper stills with head and condenser of tin), or large glass retorts; in analytical investigations we generally employ the apparatus shown in Fig. 4.

§ 9.

8. IGNITION.

Ignition is, in a certain manner, for solid bodies what evaporation is with regard to fluids; since it serves (at least generally) to separate volatile substances from less volatile or from fixed bodies in cases where the residuary substance alone is of importance. The process of ignition always presupposes the application of a high temperature, in which respect it differs from drying or exsiccation. The form or state which the eliminated substance assumes on cooling—whether it remains gaseous, as in the ignition of carbonate of lime; or assumes the liquid state, as in the ignition of hydrate of lime; or solidifies, as in the ignition of a mixture containing chloride of ammonium—is a matter of perfect indifference as regards the name given to the operation.

The process of ignition is mostly employed, as has just been said, to effect the elimination of a volatile body. In some instances, however, substances are ignited simply for the purpose of modifying their state, without any volatilization taking place; thus the sesquioxide of chromium is converted by ignition into the so-called insoluble modification, &c. In analytical investigations substances under examination are often ignited also, that the operator may from their deportment at a red heat draw a conclusion as to their nature in general, their fixity, their fusibility, the presence or absence of organic matter, &c.

Crucibles are the vessels generally made use of in ignition. In operations on a large scale Hessian or black-lead crucibles are used, heated by charcoal or coke; in analytical experiments small-sized crucibles or dishes are selected, of porcelain, platinum, silver, or iron, or glass tubes sealed at one end, according to the nature of the substances to be ignited; these crucibles, dishes, or tubes are heated over a *Berzelius* spirit-lamp or a properly constructed gas-lamp.

§ 10.

9. SUBLIMATION.

The term *sublimation* designates the process which serves to convert solid bodies into vapor by the application of heat, and subsequently to recondense the vapor to the solid state by refrigeration;—the substance volatilized and recondensed is called a *sublimate*. Sublimation is consequently a *distillation of solid bodies*. We have recourse to this process mostly to effect the separation of substances possessed of different degrees of volatility. Its application is of the highest importance in analysis for the detection of certain substances, *e. g.* of arsenic. The vessels used in sublimation are of various shapes, according to the different degrees of volatility of the substances operated upon. In sublimations for analytical purposes we generally employ sealed glass tubes.

§ 11.

10. FUSION AND FLUXING.

We designate by the term "fusion" the conversion of a solid substance into the fluid form by the application of heat; fusion is most frequently resorted to for the purpose of effecting the combination or the decomposition of bodies. The term "fluxing" is applied to this process in cases where substances insoluble or difficult of solution in water and acids are by fusion in conjunction with some other body modified or decomposed in such a manner that they or the new-formed compounds will subsequently dissolve in water or acids. Fusion and fluxing are conducted either in porcelain, silver, or platinum crucibles, according to the nature of the compound. The crucible is supported on a triangle of moderately stout platinum wire, resting on, or attached to, the iron ring of the *Berzelius* spirit-lamp or the gas-lamp. Triangles of thick iron wire, especially when laid upon the stouter brass ring of the lamp, carry off too much heat to allow of the production of very high temperatures.

Small quantities of matter are also often fused in glass tubes sealed at one end.

Resort to fluxing is especially required for the analysis of the sulphates of the alkaline earths, and also for that of many silicates. The flux most commonly used is carbonate of soda or carbonate of potassa, or, better still, a mixture of both in equal atomic proportions (see § 77). In certain cases hydrate of baryta is used instead of the alkaline carbonates. But in either case the operation is conducted in a platinum crucible.

I have to add here a few precautionary rules for the prevention of damage to the platinum vessels used in these operations. No substance evolving chlorine ought to be treated in platinum vessels; no nitrates of the alkalies, hydrate of potassa and soda, metals, or sulphides of metals or cyanides of the alkalies should be fused in such vessels; nor should readily deoxidizable metallic oxides, or salts of the heavy metals with organic acids be ignited in them, or phosphates in presence of organic compounds. It is also detrimental to platinum crucibles, and especially to their covers, to expose them direct to an intense charcoal fire, as the action of the ash is likely to lead to the formation of silicide of platinum, which renders the vessel brittle. It is always advisable to support platinum crucibles used in ignition or fusion on triangles of platinum wire. Soiled platinum crucibles are cleaned by rubbing with wet sea-sand, the round grains of which do not scratch the metal. Where this fails to remove the stains the desired object may be attained by fusing bisulphate of potassa or borax in the crucible, boiling subsequently with water, and polishing finally with sea-sand.

We have still to speak here of another operation which bears some affinity to fusion, viz.—

§ 12.
11. DEFLAGRATION.

We understand by the term "*deflagration*," in a more general sense, every process of decomposition attended with noise or detonation—(the *cause* of the decomposition is a matter of perfect indifference as regards the application of the term in this sense).

We use the same term, however, in a more restricted sense, to designate the oxidation of a substance in the dry way, at the expense of the oxygen of another substance mixed with it (usually a nitrate or a chlorate), and connect with it the idea of a sudden and violent combustion attended with vivid incandescence and noise or detonation. Deflagration is resorted to either to produce the desired oxide—thus sulphide of arsenic is deflagrated with nitrate of potassa to obtain arsenate of potassa; —or it is applied as a means to prove the presence or absence of a certain substance—thus salts are tested for nitric or chloric acid by fusing them in conjunction with cyanide of potassium, and observing whether this process will cause deflagration or not, &c.

To attain the former object the perfectly dry mixture of the substance under examination and of the deflagrating agent is projected in small portions at a time into a red-hot crucible. Experiments of the latter description are invariably made with very minute quantities; the process is in such cases best conducted on a piece of thin platinum foil, or in a small spoon.

§ 13.
12. THE USE OF THE BLOWPIPE.

This operation belongs exclusively to the province of analytical chemistry, and is of paramount importance in many analytical processes. We have to examine here, 1, the apparatus required; 2, the mode of its application; and, 3, the results of the operation.

The blowpipe (Fig. 5) is a small instrument, usually made of brass or German silver. It was originally used by metallurgists for the purpose of soldering, whence it derived the name of "soldering pipe" (*Löthrohr*) by which the Germans designate it. It consists of three distinct parts; viz. 1st, a tube (*a b*), fitted, for greater convenience, with a horn or ivory mouthpiece, through which air is blown from the mouth; 2nd, a small cylindrical vessel (*c d*), into which *a b* is screwed airtight, and which serves as an air-chamber and to retain the moisture of the air blown into the tube; and 3rd, a smaller tube (*f g*), also fitted into the vessel (*c d*). This small tube, which forms a right angle with the larger one, is fitted at its aperture either simply with a finely perforated platinum plate, or more conveniently with a finely perforated platinum cap (*h*), screwed in air-tight. The construction of the cap is shown in Fig. 6. It is, indeed, a little dearer than a simple plate, but it is also much more durable. If the opening of the cap gets stopped up, the obstruction may generally be removed by heating it to redness before the blowpipe.

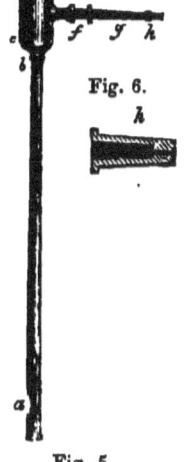

Fig. 6.

Fig. 5.

The proper length of the blowpipe depends upon the distance to which the operator can see with distinctness; it is usually from twenty to twenty-five centimetres. The form of the mouthpieces varies. Some chemists like them of a shape to be encircled by the lips; others prefer the form of a trumpet mouthpiece, which is only pressed against the lips. The latter require less exertion on the part of the operator, and are accordingly generally chosen by those who have a great deal of blowpipe work.

The blowpipe serves to conduct a continuous fine current of air into a gas-flame, or into the flame of a candle or lamp, or sometimes into a spirit-of-wine flame. The flame of a candle (and equally so that of gas or of an oil lamp), burning under ordinary circumstances, is seen to consist of three distinct parts, as shown in Fig. 7, viz., 1st, a dark nucleus in the centre (*a*); 2nd, a luminous cone surrounding this nucleus (*e f g*); and, 3rd, a feebly luminous mantle encircling the whole flame (*b c d*). The dark nucleus is formed by the gases which the heat evolves from the wax or fat, and which cannot burn here for want of oxygen. In the luminous cone these gases come in contact with a certain amount of air insufficient for their complete

Fig. 7.

combustion. In this part, therefore, it is principally the hydrogen of the carbides of hydrogen evolved which burns, whilst the carbon separates in a state of intense ignition, which imparts to the flame the luminous appearance observed in this part. In the outer coat the access of air is no longer limited, and all the gases not yet burned are consumed here. This part of the flame is the hottest, and the extreme apex is the hottest point of it. Oxidizable bodies oxidize therefore with the greatest possible rapidity when placed in it, since all the conditions of oxidation are here united, viz. high temperature and an unlimited supply of oxygen. This outer part of the flame is therefore called the *oxidizing flame*.

On the other hand, oxides having a tendency to yield up their oxygen suffer *reduction* when placed within the *luminous* part of the flame, the oxygen being withdrawn from them by the carbon and the still unconsumed carbide of hydrogen present in this sphere. The luminous part of the flame is therefore called the *reducing flame*.

Now the effect of blowing a fine stream of air across a flame is, first, to alter the shape of the flame, as, from tending upward, it is now driven sideways in the direction of the blast, being at the same time lengthened and narrowed; and, in the second place, to extend the sphere of combustion from the outer to the inner part. As the latter circumstance causes an extraordinary increase of the heat of the flame, and the former a concentration of that heat within narrower limits, it is easy to understand the exceedingly energetic action of the blowpipe flame. The way of holding the blowpipe and the nature of the blast will always depend upon the precise object in view, viz., whether the operator wants a *reducing* or an *oxidizing* flame. The easiest way of producing most efficient flames of both kinds is by means of coal-gas delivered from a tube terminating in a flat top with a somewhat slantingly downward-turned slit 1 centimetre long and $1\frac{1}{4}$ to 2 millimetres wide; as with the use of gas the operator is enabled to control and regulate not only the blowpipe flame, but the gas stream also. The task of keeping the blowpipe steadily in the proper position may be greatly facilitated by firmly resting that instrument upon some moveable metallic support, such as, for instance, the ring of *Bunsen's* gas-lamp for supporting dishes, &c.

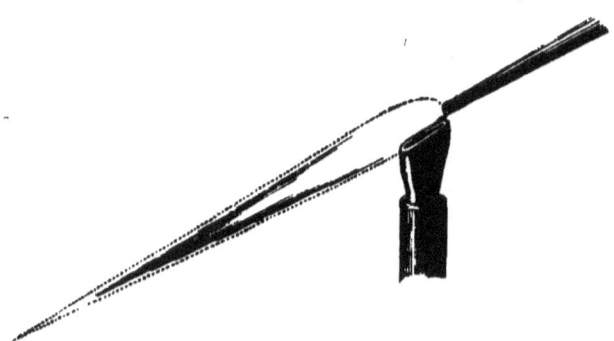

Fig. 8.

Fig. 8 shows the flame for reducing, Fig. 9 the flame for oxidizing. The luminous parts are shaded.

The *reducing* flame is produced by keeping the jet of the blowpipe just on the border of a tolerably strong gas flame, and driving a moderate blast across it. The resulting mixture of the air with the gas is only imperfect, and there remains between the inner bluish part of the flame and the outer barely visible part a luminous and reducing zone, of which the hottest point lies somewhat beyond the apex of the inner cone. To produce the oxidizing flame, the gas is lowered, the jet of the blowpipe pushed a little further into the flame, and the strength of the current somewhat increased. This serves to effect an intimate mixture of the air and gas, and an inner pointed, bluish cone, slightly luminous towards the apex is formed, and surrounded by a thin, pointed, light-bluish, barely visible mantle. The hottest part of the flame is at the apex of the inner cone. Difficultly fusible bodies are exposed to this part to effect their fusion; but bodies to be oxidized are held a little beyond the apex, that there may be no want of air for their combustion. An oil-lamp with broad wick of proper thickness may be used instead of gas; a thick wax-candle also will do. For an oxidizing flame a small spirit-lamp will in most cases answer the purpose.

Fig. 9.

The *current* is produced with the cheek muscles alone, and not with the lungs. The way of doing this may be easily acquired by practising for some time to breathe quietly with puffed-up cheeks and with the blowpipe between the lips; with practice and patience the student will soon be able to produce an even and uninterrupted current.

The *supports* on which substances are exposed to the blowpipe flame are generally either wood charcoal, or platinum wire or foil.

Charcoal supports are used principally in the reduction of metallic oxides, &c., or in trying the fusibility of bodies. The substances to be operated upon are put into small conical cavities scooped out with a penknife or with a little tin tube. Metals that are volatile at the heat of the reducing flame evaporate wholly or in part upon the reduction of their oxides; in passing through the outer flame the metallic fumes are re-oxidized, and the oxide formed is deposited around the portion of matter upon the support. Such deposits are called incrustations. Many of these exhibit characteristic colors leading to the detection of the metals. Thoroughly-burnt pieces of charcoal only should be selected for supports in blowpipe experiments, as imperfectly-burnt pieces are apt to spirt and throw off the matter placed on them. The charcoal of the wood of the pine, linden, or willow, is greatly preferable for supports to that of harder and denser woods. Smooth pieces ought to be selected for supports, as knotty pieces are apt to spirt when heated, and to throw off the matter placed on them. The most convenient way is to saw the charcoal of well-seasoned and straight-split pinewood into parallelopipedic pieces, and to blow or brush off the dust; they may then be handled without fear of soiling the hands. Those sides alone are used on which the annual rings are visible on the edge, as on the other sides the fused matters are apt to spread over the surface of the charcoal (*Berzelius*).

The properties which make charcoal so valuable as a material for supports in blowpipe experiments are—1st, its infusibility; 2nd, its low conducting power for heat, which permits substances being heated more strongly upon a charcoal than upon any other support; 3rd, its porosity, which makes it imbibe readily fusible substances, such as borax, carbonate of soda, &c., whilst infusible bodies remain on the surface; 4th, its power of reducing oxides, which greatly contributes to effecting the reduction of oxides in the inner blowpipe flame.

We use *platinum wire*, and occasionally also *platinum foil*, in all oxidizing processes before the blowpipe, and also when fusing substances with fluxes, with a view to try their solubility in them and to watch the phenomena attending the solution and mark the color of the bead; lastly, also to introduce substances into the flame, to see whether they will color it.

The wire, which should be about the thickness of lute-strings, is cut into lengths of 8 centimetres, and each length twisted at both ends into a small loop (Fig. 10).

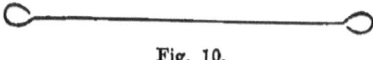

Fig. 10.

When required for use, the loop is moistened with a drop of water, then dipped into the powdered flux (where a flux is used), and the portion adhering exposed to the flame of a gas- or spirit-lamp. When the bead produced, which sticks to the loop, is cold it is moistened again, and a small portion of the substance to be examined put on and made to adhere to it by the action of a gentle heat. The loop is then finally exposed, according to circumstances, to the inner or to the outer blowpipe flame.

What renders the application of the blowpipe particularly useful in chemical experiments is the great expedition with which the intended results are attained. These results are of a twofold kind, viz., either they afford us simply an insight into the general properties of the examined body, and enable us accordingly only to determine the *class* to which it belongs, *i.e.*, whether it is fixed, volatile, fusible, &c.; or the phenomena which we observe enable us at once to recognise the particular body which we have before us. We shall have occasion to describe these phenomena when treating of the deportment of the different substances with reagents.

As in the use of the blowpipe one hand is always necessarily engaged, and the production of a continued blast requires practice and some slight exertion, and as, lastly, it is not very easy to maintain the blowpipe flame always steadfastly, so that the substances exposed to it are invariably and always undeviatingly kept in the desired parts of the flame, many chemists have long been endeavouring to devise some self-acting blowpipe apparatus, and many contrivances of the kind have been proposed and have found favor. In some of them the air-current is produced by means of a gasometer, in others by means of a caoutchouc balloon, in others again by a species of hydrostatic blast, &c. But the simplest self-acting blowpipe apparatus, by which most of the objects attainable with the blowpipe may be most suitably and conveniently accomplished, is the flame of a *Bunsen's* gas-lamp, which burns without luminosity and without soot. A description of this lamp follows in the next paragraph.

§ 14.

13. THE USE OF LAMPS, PARTICULARLY OF GAS-LAMPS.

As we have to deal in analytical chemistry mostly with smaller quantities of matter only, we use in processes of qualitative analysis requiring the application of heat, such as evaporation, distillation, ignition, &c., generally lamps, either spirit-lamps or, where coal-gas is obtainable, most advantageously, gas-lamps.

Of *spirit-lamps* there are two kinds in use, viz., the simple spirit-lamp, as shown in Fig. 13, and the *Berzelius* lamp with double draught (Fig. 11). In the construction of the latter lamp it should be borne in mind that the part containing the wick and the vessel with the spirit must be in separate pieces, connected only by means of a narrow tube; otherwise troublesome explosions are apt to occur in lighting the lamp. Nor should the chimney be too narrow, or the stopper fit air-tight on the mouth through which the spirit of wine is poured in. A lamp should be selected that may be readily moved up and down the pillar of the

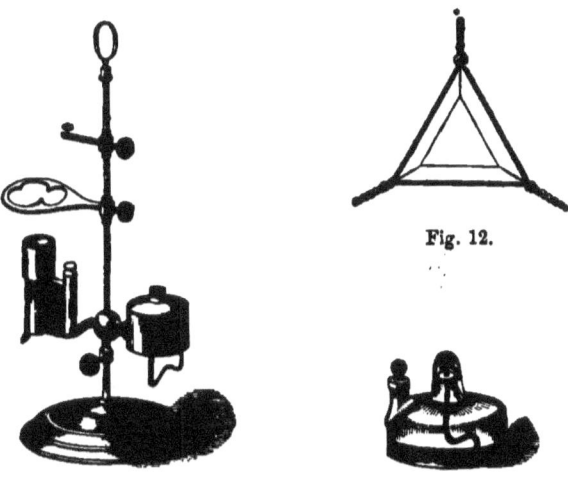

Fig. 12.

Fig. 11.   Fig. 13.

stand, which must be fitted with proper brackets, and also with a moveable brass ring to support dishes and flasks in processes of ebullition, and a ring of moderately stout iron wire to support the triangle for holding the crucibles in the processes of ignition and fusion.

Of the various forms of lamps in use, the one shown in Fig. 11 is the most suitable and elegant. Fig. 12 shows a triangle of platinum wire fixed within an iron wire triangle; this serves to support the crucible in processes of ignition. Glass vessels, more particularly beakers, which it is intended to heat over the lamp, are most conveniently rested on a circular piece of gauze made of fine iron wire such as is used in making sieves of medium fineness.

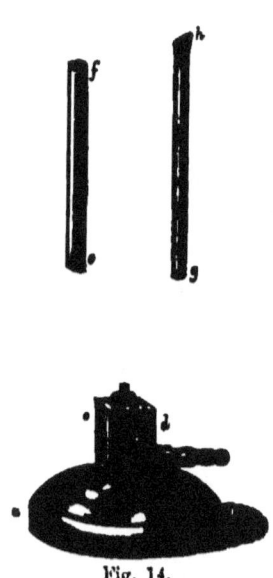

Fig. 14.

Of the many *gas-lamps* proposed, *Bunsen's*, as shown in its simplest form in Figs. 14 and 15, is the most convenient. $a\,b$ is a foot of cast iron measuring 7 centimetres in diameter. In the centre of this is fixed a square brass box, $c\,d$, which slightly slants towards the top; the sides of this box are 25 millimetres high and 16 millimetres wide; it has a cylindric cavity of 12 millimetres deep and 10 millimetres in diameter. Each side of the box has, 4 millimetres from the upper rim, a circular aperture of 8 millimetres diameter, leading to the inner cavity. One of the sides has fitted into it, 1 millimetre below the circular aperture, a tube over which is drawn vulcanized india-rubber which serves to convey the gas to the apparatus. This tube is turned in the shape shown in Fig. 14; it has a bore of 4 millimetres diameter. The gas conveyed into it through the india-rubber re-issues from a tube placed in the centre of the cavity of the box. This tube, which is 4 millimetres thick at the top, thicker at the lower end, projects 3 millimetre above the rim of the box; the gas issues from a narrow opening which appears formed of 3 radii of a circle, inclined to each other at an angle of 120°. The length of each radius is 1 millimetre; the opening of the slit is ⅛ millimetre wide; $e\,f$ is a brass tube 90 millimetres long, open at both ends, and having an inner diameter of 9 millimetres; the screw at the lower end of this tube fits into a nut in the upper part of the cavity of the box. With this tube screwed in, the lamp is completed. On opening the stop-cock the gas rushes from the trifid slit into the tube $e\,f$, where it mixes with the air coming in through the circular apertures (*c*). When this mixture is kindled at f, it burns with a straight, upright, bluish flame, entirely free from soot, which may be regulated at will by opening the stop-cock more or less; a partial opening of the cock suffices to give a flame fully answering the purpose of the common simple spirit-lamp; whilst with the full stream of gas turned on, the flame, which will now rise up to 2 decimetres in height, burning with a roaring noise, affords a most excellent substitute for the Berzelius lamp. If the flame is made to burn very low, it will often occur that it recedes; in other words, that instead of the mixture of gas and air burning at the mouth of the tube $e\,f$, the gas takes fire on issuing from the slit, and burns below in the tube. This defect may be perfectly obviated by covering the tube ef at the top with a little wire cap. Flasks, &c., which it is intended to heat over the gas-lamp, are most conveniently supported on wire gauze. If it is wished to use the gas-lamp for blowpipe operations, the tube gh must be inserted into ef; this tube terminates in a flattened top slanting at an angle of 68° to the axis, and having an opening in it 1 centimetre long and 1½ to 2 millimetres wide. The insertion of gh into ef serves to close up the air-holes in the box, and pure gas, burning with a luminous flame, issues accordingly now from the top of the tube. Fig. 15 shows the apparatus complete, fixed in the fork of an iron stand; this arrangement permits the lamp being moved backward and forward between the prongs of the fork, and up and down the pillar of the stand. The moveable ring on the same pillar serves to support the objects to be operated upon.

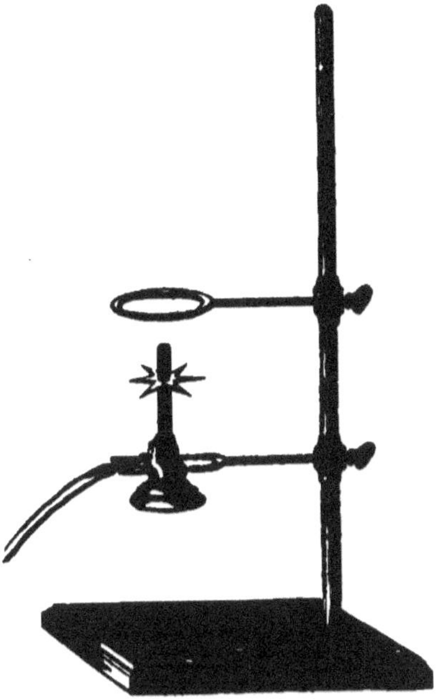

Fig. 15.

The 6 radii round the tube of the lamp serve to support an iron-plate chimney (see Fig. 16), or a porcelain plate used in quantitative analyses.

To heat crucibles to the brightest red heat, or to a white heat, the gas-blast is resorted to. But even without this the action of the gas-lamp may be considerably heightened by heating the crucible within a small clay furnace, as recommended by O. L. ERDMANN. Fig. 16 shows the simple contrivance by which this is effected. The furnaces are 115 millimetres high, and measure 70 millimetres diameter in the clear. The thickness of material is 8 millimetres.

Fig. 16.

Bunsen[*] has devised also a somewhat improved form of his lamp, to fit it for processes of reduction, oxidation, fusion, and volatilization, and also to serve as a substitute for the blowpipe-blast. (See Fig. 17).

The illustration shows the part a, which is fitted for screwing on and off; also the conical iron-plate chimney, b, which is 30 millimetres

---

* Ann. d. Chem. u. Pharm., iii. 257.

Fig. 17.

wide at the top, and 55 millimetres at the bottom, and rests on the supporters $c\ c\ c\ c$ in such a manner that the burner-tube d is placed in the axis of the chimney and ends 45 millimetres below the upper mouth of the latter. As this construction, on the one hand, permits an easy regulation of the access of air, the chimney, on the other hand, ensures a tall, steady, and evenly-burning flame of the shape shown in the illustration. Looking attentively at the flame in the illustration, we distinguish in it an inner part and two mantles surrounding it. The inner part corresponds to the dark nucleus of the common candle, oil or gas flame, and contains the mixture of gas and air issuing from the burner. If the gas-cock is so adjusted that the apex of the inner part of the flame is on an exact level with the upper mouth of the chimney, a flame is obtained of perfectly constant dimensions, which remains quite steady, and is sharply defined in its parts, and may also, at all times, be reproduced in exactly the same condition. The mantle immediately surrounding the inner part contains still some unconsumed carbide of hydrogen; the outer mantle, which looks bluer and less luminous, consists of the last products of combustion. The hottest part of the flame has, according to Bunsen's calculation, a temperature of 2300° centigrade (4172° Fahrenheit.) This hottest part lies in the mantles surrounding the inner part of the flame, in a zone extending a few millimetres upwards and downwards from the transverse section of the flame across the apex of the inner part. We will term this region the *zone of fusion*. BUNSEN calls it the "*Schmelzraum*." It serves to try the deportment of bodies at a temperature of about 2300° centigrade (4172° Fahrenheit.) The outer margin of this zone of fusion acts as *oxidizing flame*, the inner part of it as *reducing flame*. The spot where the reducing action is the most powerful and energetic lies immediately above the apex of the inner part of the flame. The flame of this lamp is most admirably suited to bring out the coloration which many substances impart to flames, and by which the attentive observer is enabled to detect many bodies, even though present in such exceedingly minute traces that all other known means of detection fail to discover them. The subject of the coloration of flames will be discussed more fully in the next paragraph. Here we will simply state, in addition, that the substitution of the gas-flame in lieu of the blowpipe affords the great advantage that the samples for examination may be placed, by means of a holder, in any desired part of the flame, and kept there quite fixed and immoveable. A holder of the kind required is shown in Fig. 18.

The arm a is easily moved along the pillar c by means of the spring slide b. The glass tube d, which bears, fused into its sealed end, a platinum wire about 0.145 millimet. thick, with the outer end twisted into a small loop, is pushed over the horizontal arm a. If this loop is moistened, and then dipped into the powder of the substance to be examined, a portion of the powder adheres to it. If this loop is now held near the flame, the powder agglutinates or fuses, and sticks fast to the loop, which is then thrust into the desired part of the flame. Decrepitating substances must previously be ignited in a covered platinum crucible. If fluids are to be examined, with a view to ascertain whether they hold flame-coloring substances in solution, the round loop of the platinum wire is flattened by a few blows with the hammer into the shape of a platinum ring. If this is dipped into the fluid to be examined, and taken out again, there remains adhering to the inner circle a drop of the liquid. This is evaporated by holding the loop near the flame, taking care, however, to keep the liquid from boiling, and the residue is then examined by thrusting the loop into the zone of fusion. (BUNSEN).

Fig. 18.

§ 15.

14. OBSERVATION OF THE COLORATION OF FLAME BY CERTAIN BODIES AND SPECTRUM ANALYSIS.

Many substances have the property of coloring a colorless flame in a very remarkable manner. As most of these substances impart each of them a different and distinct, and accordingly characteristic, tint to the flame, the observation of this colorization of flame affords an excellent, easy, and safe means of detecting many of these bodies. Thus, for instance, salts of soda impart to flame a yellow, salts of potassa a violet, salts of lithia a carmine tint, and may thus be easily distinguished from each other.

The flame of BUNSEN's gas-lamp, with chimney, described in § 14, and shown in Fig. 17, is more particularly suited for observations of the kind. The substances to be examined are put on the small loop of a fine platinum wire, and thus, by means of the holder shown in Fig. 18, placed within the zone of fusion of the gas-flame. A particularly striking coloration is imparted to the flame by the salts of the alkalies and alkaline earths. If different salts of one and the same base are compared in this way, it is found that every one of them, if at all volatile at high temperatures, or permitting at least the volatilization of the base, imparts the same color to the flame, only with different degrees of intensity, the most volatile of the salts producing also the most intense colorization; thus, for instance, chloride of potassium gives a more intense coloration than carbonate of potassa, and this latter again a more intense one than silicate of potassa. In the case of difficultly volatile compounds, the coloration of the flame may often be brought about, or made more apparent, by adding some other body which has the power of decomposing the compound under examination. Thus, for instance, in silicates containing only a few per cents of potassa, the latter body cannot be directly detected by coloration of flame ; but this detection may be accomplished by adding to the silicate a little pure gypsum, as this will cause formation of silicate of lime and of sufficiently volatile sulphate of potassa.

But however decisive a test the mere coloration of flame affords for the detection of certain metallic compounds, when present unmixed with others, this test becomes apparently quite useless in the case of mixtures of compounds of several metals. Thus, for instance, mixtures of salts of potassa and soda show only the soda flame, mixtures of salts of baryta and strontia only the baryta flame, &c. This defect may be remedied, however, in two ways, with the most surprising success. Both ways have only quite recently been discovered.

The one way, started first by CARTMELL,[*] and perfected afterwards by BUNSEN[†] and by MERZ,[‡] consists in looking at the colored flame through some colored medium (colored glasses, indigo solution, &c.). Such colored media, in effacing the flame coloration of the one metal, bring out that of the other metal mixed with it. For instance, if a mixture of a salt of potassa and a salt of soda is exposed to the flame, the latter will only show the yellow soda coloration ; but if the flame be now looked at through a deep-blue-tinted cobalt glass, or through solution of indigo, the yellow soda coloration will disappear and will be replaced by the violet potassa tint. A simple apparatus suffices for all observations and experiments of the kind ; all that is required for the purpose being,—

1. A hollow prism (see Fig. 19) composed of mirror plates, the chief section of which forms a triangle with two sides of 150 millimetres, and one side of 35 millimetres length.

Fig. 19.

The indigo solution required to fill this prism is prepared by dissolving 1 part of indigo in 8 parts of fuming sulphuric acid, adding to the solution 1800–2000 parts of water, and filtering the fluid. When using this apparatus, the prism is moved in a horizontal direction close before the eyes in such a way that the rays of the flame are made to penetrate successively thicker and thicker layers of the effacing medium.

2. A blue, a violet, a red, and a green glass. The blue glass is tinted with protoxide of cobalt ; the violet glass with sesquioxide of manganese ; the red glass (partly colored, partly uncolored) with suboxide of copper ; and the green glass with sesquioxide of iron and protoxide of

---

[*] Phil. Mag., xvi. 328. [†] Annal. d. Chem. u. Pharm., iii. 257.
[‡] Journ. f. prakt. Chem., 80, 487.

copper. The common colored glasses sold in the shops for ornamenting windows, will generally be found to answer the purpose. As regards the tints imparted to the flame by the different bodies, when viewed through the aforesaid media, and their combinations, by which those bodies are severally identified, the information required will be found in Section III., in the paragraphs treating of the several bases and acids.

The other method, which is called *Spectrum Analysis*, was discovered by KIRCHHOFF and BUNSEN. It consists in letting the rays of the colored flame pass first through a narrow slit, then through a prism, and observing the so refracted rays through a telescope. A distinct spectrum is thus obtained for every flame-coloring metal: this spectrum consists either, as in the case of baryta, of a number of colored lines lying side by side; or, as in the case of lithia, of two separate, differently-colored lines; or, as in the case of soda, of a single yellow line. These spectra are characteristic in a double sense—viz., the spectrum lines have a distinct color, and they occupy also a fixed position.

It is this latter circumstance which enables us to identify without difficulty, in the spectrum observation of mixtures of flame-coloring metals, every individual metal. Thus, for instance, a flame in which a mixture of potassa, soda, and lithia salts is evaporated, will give, side by side, the spectra of the several metals in the most perfect purity.

KIRCHHOFF and BUNSEN have constructed two kinds of apparatus, which are both of them suited for spectrum observation, and enable the operator to determine by measure the spots in which the spectrum lines make their appearance. Both are constructed upon the same principle. A description, with illustration, of the larger of the two, which is also the most perfect one, has been published in *Poggendorff's* "Annalen," 113, 374, and in the "Zeitschrift für Analytische Chemie," 1862, 49.

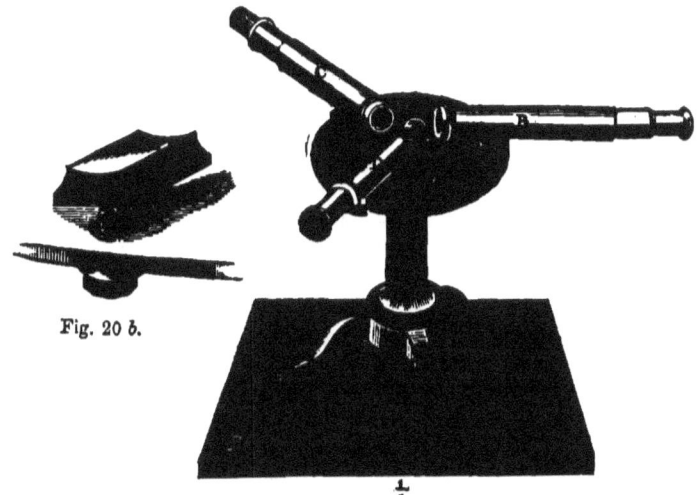

Fig. 20 b.

$\frac{1}{5}$

Fig. 20 a.

The smaller, more simple, and accordingly cheaper apparatus, which suffices for all common purposes, and will probably be used most in chemical laboratories, we will describe here. It is shown in Fig. 20 *a*.

A is an iron disk, in the centre of which a prism, with circular refracting faces of about 25 millimetres diameter, is fastened by a bow, which presses upon the upper face of the prism, and is secured below to the iron plate by a screw. The same disk has also firmly fastened to it the three tubes B, C, and D. Each of these tubes is soldered to a metal block, of which Fig. 20 *b* gives an enlarged representation. This block contains the nuts for two screws, which pass through wider openings in the iron plate, and are firmly secured beneath when the tube has been adjusted in the proper position. B is the observation telescope; it has a magnifying power of about 6, with an object-glass of 20 millimetres diameter. The tube C is closed at one end by a tin-foil disk, into which the perpendicular slit is cut through which the light is admitted. The tube D carries a photographic copy of a millimetre-scale, produced in the camera obscura on a glass plate of about one-fifteenth the original dimensions. This scale is covered with tin-foil, with the exception of the narrow strip upon which the divisional lines and the numbers are engraved. It is lighted by the flame of a taper or candle placed close behind it. The axes of the tubes B and D are directed, at the same inclination, to the centre of one face of the prism, whilst the axis of the tube C is directed to the centre of the other face of the prism. This arrangement makes the spectra produced by the refraction of the colored light passing through C, and the image of the scale in D produced by total reflection appear in one and the same spot, so that the positions occupied by the spectrum lines may be read off on the scale. The prism is placed in about that position in which there is a minimum divergence of the rays of the sodium line; and the telescope is set in that direction in which the red and the violet potassium lines are about equidistant from the middle of the field of view. The colorless flame into which the flame-coloring bodies are to be introduced, is placed 10 centimetres from the slit. BUNSEN's lamp described page 19, and shown in Fig. 17, gives the best flame. The lamp is adjusted so as to place the upper border of the chimney about 20 millimetres below the lower end of the slit. When this lamp has been lighted, and a bead of substance—say of sulphate of potassa, for instance—introduced into the zone of fusion by means of the holder described page 20, and shown in Fig. 18, the iron disk of the spectrum apparatus, which, with all it carries, is moveable round its vertical axis, is turned until the point is reached where the luminosity of the spectrum is the most intense.

To cut off foreign light in all spectrum observations, a black cloth, with three circular openings in it for the three tubes, is thrown over the prism and the tubes. The spectra produced by the alkalies and the alkaline earths are shown in Table I., Fig. 1. The solar spectrum has been added simply as a guide to the position and bearings of the lines. The spectra are represented as they appear in the apparatus furnished with an astronomic telescope. In the third section, in the chapters treating of the several bodies, attention will be called to the lines which are most characteristic for each metal. Here I will simply state the manner in which the highest degree of certainty is imparted to spectrum analysis. This is done by exposing the beads of the pure and unmixed metallic compounds to the flame, and marking on copied scales the position which the most striking spectrum lines occupy on the scale of the apparatus, in the manner shown, by way of illustration, in Table I., Fig. 2, with regard to the strontium spectrum. It is self-evident that the spectrum of an unknown substance can only pass for the strontium spectrum, if the characteristic lines not only agree with those of the latter in point of color, but appear also in exactly the same position where they are marked on the strontium scale.

The drawings of such scales every operator must, of course, make for his own apparatus; and they become useless for the intended purpose if any alteration is made in the position of the prism or the scale. It is therefore always advisable to set the apparatus so that it can be easily readjusted to its original position, which is most readily done by making the left border of the sodium line coincide with the number 50 of the scale.

With the introduction of spectrum analysis a new era has, in many respects, begun for chemical analysis, as by means of this discovery we can detect such minute quantities of bodies as by no other method. Spectrum analysis is marked moreover by a certainty above all doubt, and gives results in a few seconds, which could formerly be obtained only, if at all, in hours or days.

APPENDIX TO SECTION I.

§ 16.

APPARATUS AND UTENSILS.

As many students of chemical analysis might find some difficulty in the selection of the proper apparatus, &c., I append here a list of the articles which are required for the performance of simple experiments and investigations, together with instructions to guide the student in the purchase or making of them.

1. A BERZELIUS SPIRIT LAMP (§ 14, Fig. 11).
2. A GLASS SPIRIT LAMP (§ 14, Fig. 13). Or, instead of these two, where coal-gas is procurable, a *Bunsen's* Gas-lamp, best one with chaplet and chimney (§ 14, Figs. 14, 15, and 17).
3. A BLOWPIPE (see § 13).
4. A PLATINUM CRUCIBLE. Select a crucible which will contain about a quarter of an ounce of water, with a cover shaped like a shallow dish; it must not be too deep in proportion to its breadth.
5. PLATINUM FOIL, as smooth and clean as possible, and not very thin: length about 40 millimetres; width about 25 millimetres.
6. PLATINUM WIRE (see § 13 and § 14, Figs. 10 and 18). Three stronger wires and three finer wires are amply sufficient. They are kept most conveniently in a glass filled with water, most of the beads being dissolved by that fluid when left in contact with it for some time; the wires may thus be kept always clean.

7. A STAND WITH TWELVE TEST TUBES—16 to 18 centimetres is about the proper length of the tubes, from 1 to 2 centimetres the proper width. The tubes must be made of thin white glass, and so well annealed that they do not crack even though boiling water be poured into them. The rim must be quite round, and slightly turned over; it ought not to have a lip, as this is useless and simply prevents the tube being closely stopped with the finger, and also shaking the contents. The stand shown in Fig. 21 will be found most suitable. The pegs on the upper shelf serve for the clean tubes, which may thus be always kept dry and ready for use.

Fig. 21.

8. SEVERAL BEAKER GLASSES AND SMALL RETORTS of thin, well annealed glass.

9. SEVERAL PORCELAIN EVAPORATING DISHES, AND A VARIETY OF SMALL PORCELAIN CRUCIBLES. Those of the royal manufacture of Berlin are unexceptionable, both in shape and durability. Meissen porcelain will also answer the purpose.

10. SEVERAL GLASS FUNNELS of various sizes. They must be inclined at an angle of 60°, and merge into the neck at a definite angle.

11. A WASHING BOTTLE of a capacity of from 300 to 400 cubic centimetres (see § 6).

12. SEVERAL GLASS RODS AND GLASS TUBES. The latter are bent, drawn out, &c., over a Berzelius spirit-lamp; the former are rounded at the ends by fusion.

13. A selection of WATCH-GLASSES.

14. A small AGATE MORTAR.

15. A pair of small STEEL or BRASS PINCERS, about four or five inches long.

16. A WOODEN FILTERING STAND (see § 5).

17. A TRIPOD of thin iron, to support the dishes, &c., which is intended to heat over the small spirit or gas lamp:

18. The Colored Glasses described in § 15, especially blue and green.

SECTION II.

REAGENTS.

§ 17.

A VARIETY of phenomena may manifest themselves upon the decomposition or combination of bodies. In some cases liquids change their color, in others precipitates are formed; sometimes effervescence takes place, and sometimes deflagration, &c. Now if these phenomena are very striking, and attend only upon the action of two definite bodies upon one another, it is obvious that the presence of one of these bodies may be detected by means of the other : if we know, for instance, that a white precipitate of certain definite properties is formed upon mixing baryta with sulphuric acid, it is clear that, if upon adding baryta to any liquid, we obtain a precipitate exhibiting these properties, we may conclude that this liquid contains sulphuric acid.

Those substances which indicate the presence of others by any striking phenomena are called *reagents*.

According to the different objects attained by the application of these bodies, we make a distinction between *general* and *special* reagents. By *general* reagents we understand those which serve to determine the *class* or *group* to which a substance belongs; and by *special* reagents those which serve to detect and determine bodies individually. That the line between the two divisions cannot be drawn with any degree of precision, and that one and the same substance is often made to serve both as a general and a special reagent, cannot well be held a valid objection to this classification, which is in fact simply intended to induce a habit of employing reagents always for a settled purpose—viz., either simply to find out the *group* to which the substance under examination belongs, or to determine the latter *individually*.

Now whilst the usefulness of *general* reagents depends principally upon their efficiency in strictly characterizing groups of bodies, and often effecting a complete separation of the bodies belonging to one group from those belonging to another, that of *special* reagents depends upon their being *characteristic* and *sensitive*. We call a reagent *characteristic*, if the alteration produced by it, in the event of the body tested for being present, is so distinctly marked as to admit of no mistake. Thus iron is a characteristic reagent for copper, protochloride of tin for mercury, because the phenomena produced by these reagents—viz., the separation of metallic copper and of globules of mercury—admit of no mistake. We call a reagent *sensitive* or *delicate*, if its action is distinctly perceptible, even though a very minute quantity only of the substance tested for be present ; such is, for instance, the action of starch upon iodine.

Very many reagents are *both* characteristic and delicate ; thus, for instance, terchloride of gold for protoxide of tin ; ferrocyanide of potassium for sesquioxide of iron and oxide of copper, &c.

I need hardly mention that, as a general rule, reagents must be chemically pure—*i.e.*, they must consist purely and simply of their essential constituents, and must contain no admixture of foreign substances. We must therefore make it an invariable rule *to test the purity of our reagents before we use them*, no matter whether they be articles of our own production or purchased. Although the *necessity* of this is fully admitted on all hands, yet we find that in *practice* it is too often neglected; thus it is by no means uncommon to see alumina entered among the substances detected in an analysis, simply because the solution of potassa used as one of the reagents happened to contain that earth; or iron, because the chloride of ammonium used was not free from that metal. The directions given in this section for testing the purity of the several reagents refer, of course, only to the presence of foreign matter resulting from the mode of their preparation, and not to mere accidental admixtures.

One of the most common sources of error in qualitative analysis proceeds from missing the proper measure—the right quantity—in the application of reagents. Such terms as *"addition in excess," "supersaturation,"* &c., often induce novices to suppose that they cannot add *too* much of the reagent, and thus some will *fill* a test tube with acid, simply to *supersaturate* a few drops of an alkaline fluid, whereas *every drop* of acid added, after the neutralization point has once been reached, is to be looked upon as an *excess* of acid. On the other hand, the addition of an insufficient amount is to be equally avoided, since a reagent added in insufficient quantity often produces phenomena quite different from those which will appear if the same reagent be added in excess: *e.g.*, a solution of chloride of mercury yields a *white* precipitate if tested with a *small* quantity of hydrosulphuric acid; but if treated with the same reagent *in excess*, the precipitate is *black*. Experience has, however, proved that the most common mistake beginners make, is to add the reagents too copiously. The reason why this over-addition must impair the accuracy of the results is obvious; we need simply bear in mind that the changes effected by reagents are perceptible within certain limits only, and that therefore they may be the more readily overlooked the nearer we approach these limits by diluting the fluid.

No special and definite rules can be given for avoiding this source of error; a general rule may, however, be laid down, which will be found to answer the purpose, if not in all, at least in the great majority of cases. It is simply this: *let the student always reflect before the addition of a reagent for what purpose he applies it, and what are the phenomena he intends to produce.*

We divide reagents into two classes, according to whether the state of fluidity which is indispensable for the manifestation of the action of reagents upon the various bodies, is brought about by the application of heat, or by means of liquid solvents; we have consequently, 1, *Reagents in the humid way;* and 2, *Reagents in the dry way.* For greater clearness we subdivide these two principal classes as follows:—

A. REAGENTS IN THE HUMID WAY.

    I. SIMPLE SOLVENTS.

    II. ACIDS (and HALOGENS).

        a. Oxygen acids.
        b. Hydrogen acids and halogens.
        c. Sulphur acids.

III. BASES (and METALS).
  a. *Oxygen bases.*
  b. *Sulphur bases.*
IV. SALTS.
  a. *Of the alkalies.*
  b. *Of the alkaline earths.*
  c. *Of the oxides of the heavy metals.*
V. COLORING MATTERS AND INDIFFERENT VEGETABLE SUBSTANCES.

B. REAGENTS IN THE DRY WAY.
  I. FLUXES.
  II. BLOWPIPE REAGENTS.

A. REAGENTS IN THE HUMID WAY.
I. SIMPLE SOLVENTS.

Simple solvents are fluids which do not enter into chemical combination with the bodies dissolved in them; they will accordingly dissolve any quantity of matter up to a certain limit, which is called the point of saturation, and is in a measure dependent upon the temperature of the solvent. The essential and characteristic properties of the dissolved substances (taste, reaction, color, &c.) are not destroyed by the solvent. (See § 2.)

§ 18.
1. WATER (HO).

Preparation.—Pure water is obtained by distilling spring water from a copper still with head and condenser made of pure tin, or from a glass retort; which latter apparatus, however, is less suitable for the purpose. The distillation is carried to about three-fourths of the quantity operated upon. If it is desired to have the distilled water perfectly free from carbonic acid and carbonate of ammonia, the portions passing over first must be thrown away. In the larger chemical and in most pharmaceutical laboratories, the distilled water required is obtained from the steam apparatus which serves for drying, heating, boiling, &c. Rain water collected in the open air may in many cases be substituted for distilled water.

Tests.—Pure distilled water must be colorless, inodorous and tasteless, and must leave no residue upon evaporation in a platinum vessel. Sulphide of ammonium must not alter it (copper, lead, iron); its transparency must not be in the least impaired by basic acetate of lead (carbonic acid, carbonate of ammonia), nor, even after long standing, by oxalate of ammonia (lime), chloride of barium (sulphates), or nitrate of silver (metallic chlorides).

Uses.—We use water* principally as a simple solvent for a great variety of substances; the most convenient way of using it is with the washing bottle (see § 6, Fig. 2), by which means a stronger or finer stream may be obtained. It serves also to effect the conversion of several neutral metallic salts (more particularly terchloride of antimony and the salts of bismuth) into soluble acid and insoluble basic compounds.

* In analytical experiments we use only *distilled* water; whenever, therefore, the term "water" occurs in the present work, distilled water is meant.

§ 19.
2. Alcohol ($C_4H_6O_2$).

Preparation.—Two sorts of alcohol are used in chemical analyses: viz., 1st, spirit of wine of 0·83 or 0·84 spec. gr. = 91 to 88 per cent. by volume (*spiritus vini rectificatissimus* of the pharmaceutist); and 2nd, absolute alcohol. The latter may be prepared most conveniently by mixing, in a distilling vessel, 1 part of fused chloride of calcium with two parts of rectified spirit of wine of about 90 per cent. by volume, digesting the mixture 2 or 3 days, until the chloride of calcium is dissolved, and then distilling slowly and in fractional portions. So long as the distillate shows a specific gravity below 0·810 = 96·5 per cent. by volume), it may pass for absolute alcohol. The portions coming over after are received in a separate vessel.

Tests.—Pure alcohol must completely volatilize, and ought not to leave the least smell of fusel oil when rubbed between the hands; nor should it alter the color of moist blue or red litmus paper. When kindled, it must burn with a faint bluish barely perceptible flame.

Uses.—Alcohol serves, (*a*) to effect the separation of bodies soluble in this fluid from others which do not dissolve in it, *e.g.* of chloride of strontium from chloride of barium; (*b*) to precipitate from aqueous solutions many substances which are insoluble in dilute alcohol, *e.g.* gypsum, malate of lime; (*c*) to produce various kinds of ether, *e.g.* acetic ether, which is characterized by its peculiar and agreeable smell; (*d*) to reduce, mostly with the co-operation of an acid, certain peroxides and metallic acids, *e.g.*, binoxide of lead, chromic acid, &c.; (*e*) to detect certain substances which impart a characteristic tint to its flame, especially boracic acid, strontia, potassa, soda, and lithia.

§ 20.
3. Ether (C_4H_5O).
4. Chloroform (C_2HCl_3).
5. Sulphide of Carbon (CS_2).

These solvents find but limited application in the qualitative analysis of inorganic bodies. They serve indeed almost exclusively to detect and isolate bromine and iodine. Chloroform and sulphide of carbon are preferable to ether in this respect.

These preparations are made much better on a large than on a small scale, and the best way therefore is to procure them by purchase.

Tests.—*Ether* must have a specific gravity of 0·725, and require 9 parts of water for solution. The solution must not alter the color of test papers. Ether must, even at the common temperature, rapidly and completely evaporate on a watch-glass. *Chloroform* must be colorless and transparent and have a specific gravity of 1·49. It must have no acid reaction, nor impair the transparency of solution of nitrate of silver. Mixed with 2 vols. of water, and shaken, its volume must not appear perceptibly diminished. It must even at the common temperature readily and completely evaporate on a watch-glass. *Sulphide of carbon* should be colorless, readily and completely volatile even at the common temperature, and exercise no action upon carbonate of lead.

II. ACIDS AND HALOGENS.

§ 21.

The acids—at least those of more strongly pronounced character—are soluble in water. The solutions taste acid and redden litmus paper. Acids are divided into oxygen acids, sulphur acids, and hydrogen acids.

The *oxygen acids*, produced generally by the combination of a non-metallic element with oxygen, combine with water in definite proportions to hydrated acids. It is with these hydrates that we have usually to do in analytical processes; they are contained in the aqueous solutions of the acids, and are commonly designated by the simple name of the free acid, as the accession of water does not destroy their acid properties. In the action of hydrated acids upon oxides of metals, the oxide takes the place of the water of hydration, and an oxygen salt is formed ($HO, SO_3 + KO = KO, SO_3 + HO$). Where these salts are the product of the combination of an acid with a strong base, their reaction (supposing the combining acid also to be a strong acid) is neutral; the salts formed with weaker bases, for instance with the oxide of a heavy metal, generally show acid reaction, but are nevertheless called neutral salts if the oxygen of the base bears the same proportion to that of the acid in which it is found in the distinctly neutral salts of the same acid, or, in other terms, if it corresponds with the saturation capacity of the acid. Sulphate of potassa (KO, SO_3) has a neutral reaction, whilst the reaction of sulphate of copper (CuO, SO_3) is acid; yet the latter is nevertheless called neutral sulphate of copper, because the oxygen of the oxide of copper in it bears a proportion of 1 : 3 to that of the sulphuric acid, which is the same proportion as the oxygen of the potassa bears to that of the sulphuric acid in the confessedly neutral sulphate of potassa.

The *hydrogen acids* are formed by the combination of the salt radicals with hydrogen. Most of these possess the characteristic properties of acids in a high degree. They neutralize oxygen bases, with formation of haloid salts and water; HCl and $NaO = NaCl$ and HO,—$3HCl$ and $Fe_2O_3 = Fe_2Cl_3$ and $3HO$. The haloid salts produced by the action of powerful hydrogen acids upon strong bases have a neutral reaction; whilst the solutions of those haloid salts that have been produced by the action of powerful hydrogen acid upon weak bases (such as the oxides of the heavy metals) have an acid reaction.

The *sulphur acids* are more frequently the result of the combination of metallic than of non-metallic elements with sulphur; they combine with sulphur bases to sulphur salts; $HS + KS = KS, HS$,—$SbS_3 + 3NaS = 3NaS, SbS_3$. The sulphur acids being weak acids, the soluble sulphur salts have all of them alkaline reaction.

a. OXYGEN ACIDS.

§ 22.

1. SULPHURIC ACID (HO, SO_3).

We use—
 a. Concentrated sulphuric acid of commerce, generally known in Germany as *English sulphuric acid.*
 b. Concentrated pure sulphuric acid.

The following is the best method of preparing pure sulphuric acid: Pour into 4 parts of water 1 part of concentrated sulphuric acid, and conduct into the mixture for some time a slow stream of hydrosulphuric acid. Let the mixture stand at rest for several days, then decant the clear supernatant fluid from the precipitate, which consists of sulphur, sulphide of lead, perhaps also sulphide of arsenic, and heat the decanted fluid in a tubulated retort with obliquely upturned neck and open tubulature until sulphuric acid fumes escape with the aqueous vapor. The acid so purified is fit for many purposes of chemical analysis; if it is wished, however, to free it also from non-volatile substances, it may be distilled from a luted non-tubulated retort, heated directly over charcoal. To avoid bumping of the liquid, it is advisable to rest the bottom of the retort on a reversed crucible cover. The neck of the retort must reach so far into the receiver that the acid distilling over drops directly into the body. Refrigeration of the receiver by means of water is unnecessary and even dangerous. To prevent the flask coming into direct contact with the hot neck of the retort, some asbestos in long fibres is wrapped round that part of the neck where such contact might be apprehended. As soon as the drops in the neck of the retort become oily, the receiver is changed, and the concentrated acid which now passes over is kept in a separate vessel.

c. Common dilute sulphuric acid. This is prepared by adding to 5 parts of water in a leaden or porcelain dish gradually, and whilst stirring, 1 part of concentrated sulphuric acid. The sulphate of lead which separates is allowed to subside, and the clear fluid finally decanted from the precipitate.

Tests.—Pure sulphuric acid must be colorless; when colorless solution of sulphate of protoxide of iron is poured upon it in a test tube, no red tint must mark the line of contact of the two fluids (nitric acid, hyponitric acid); when diluted with twenty parts of water it must not impart a blue tint to a solution of iodide of potassium mixed with starch paste (hyponitric acid).

Mixed with pure zinc and water, it must yield hydrogen gas which, on being passed through a red-hot tube, must not deposit the slightest trace of arsenic. It must leave no residue upon evaporation on platinum, and must remain perfectly clear upon dilution with four or five parts of spirit of wine (oxide of lead, sesquioxide of iron, lime). The presence of small quantities of lead is detected most easily by adding some hydrochloric acid to the sulphuric acid in a test tube. If the point of contact is marked by turbidity (chloride of lead), lead is present.

Uses.—Sulphuric acid has for most bases a greater affinity than almost any other acid; it is therefore used principally for the liberation and expulsion of other acids, especially of phosphoric, boracic, hydrochloric, nitric, and acetic acids. Several substances which cannot exist in an anhydrous state (*e.g.* oxalic acid), are decomposed when brought into contact with concentrated sulphuric acid; this decomposition is owing to the great affinity which sulphuric acid possesses for water. The nature of the decomposed body may in such cases be inferred from the liberated products of the decomposition. Sulphuric acid is also frequently used for the evolution of certain gases, more particularly of hydrogen and hydrosulphuric acid. It serves also as a *special* reagent for the detection and precipitation of baryta, strontia, and lead. What kind of sulphuric acid is to be used, whether the pure or the purified acid, or the ordinary acid of commerce, whether concentrated or dilute, depends upon what the circumstances in each case may require. It will, however, be found that the necessary directions on this point are generally given in the present work.

§ 23.

2. NITRIC ACID (HO, NO_5).

Preparation.—*a.* Heat crude nitric acid of commerce, as free as possible from chlorine, and of a specific gravity of *at least* 1·31*, in a glass retort to boiling, with addition of some nitrate of potassa; let the distillate run into a receiver kept cool, and try from time to time whether it still continues to precipitate or cloud solution of nitrate of silver. As soon as this ceases to be the case, change the receiver, and distil until a trifling quantity only remains in the retort. Dilute the distillate with water until the specific gravity of the diluted acid is 1·2.

b. Dilute crude nitric acid of commerce of about 1·38 specific gravity with two-fifths of its weight of water, and add solution of nitrate of silver as long as a precipitate of chloride of silver continues to form; then add a further slight excess of solution of nitrate of silver, let the precipitate subside, decant the perfectly clear supernatant acid into a retort or an alembic with ground head; add some nitrate of potassa free from chlorine, and distil until only a small quantity remains, taking care to attend to the proper cooling of the fumes distilling over. Dilute the distillate, if necessary, with water until it has a specific gravity of 1·2.

Tests.—Pure nitric acid must be colorless and leave no residue upon evaporation on platinum foil. Addition of solution of nitrate of silver or of nitrate of baryta must not cause the slightest turbidity in it. It is advisable to dilute the acid with water before adding these reagents, as otherwise nitrates will precipitate.

Uses.—Nitric acid serves as a chemical solvent for metals, oxides, sulphides, oxygen salts, &c. With metals and sulphides of metals the acid first oxidizes the metal present, at the expense of part of its own oxygen, and then dissolves the oxide to a nitrate. Most oxides are dissolved by nitric acid at once as nitrates; and so are also most of the insoluble salts with weaker acids, the latter being expelled in the process by the nitric acid. Nitric acid dissolves also salts with soluble non-volatile acids, as *e.g.* phosphate of lime, with which it forms nitrate of lime and acid phosphate of lime. Nitric acid is used also as an oxidizing agent: for instance, to convert protoxide of iron into sesquioxide, protoxide of tin into binoxide, &c.

§ 24.

3. ACETIC ACID ($HO, C_4H_3O_3 = HO, \bar{A}$).

A highly concentrated acetic acid is not required in qualitative analytical processes; the common acetic acid of commerce, which contains 25 per cent. of anhydrous acid, and has a specific gravity of 1·04, fully answers the purpose.

Tests.—Pure acetic acid must leave no residue upon evaporation, and —after saturation with carbonate of soda—emit no empyreumatic odor.

---

* A weaker acid will not answer the purpose.

Hydrosulphuric acid, solution of nitrate of silver, and solution of nitrate of baryta must not color or cloud the dilute acid, nor must sulphide of ammonium after neutralization of the acid by oxide of ammonium. Solution of indigo must not lose its color when heated with the acid.

If the acid is not pure, add some acetate of soda and redistil from a glass retort not quite to dryness; if it contains sulphurous acid (in which case hydrosulphuric acid will produce a white turbidity in it), digest it first with some binoxide of lead or finely pulverized binoxide of manganese, and then distil with acetate of soda.

Uses.—Acetic acid possesses a greater solvent power for some substances than for others; it is used therefore to distinguish the former from the latter; thus it serves, for instance, to distinguish oxalate of lime from phosphate of lime. Acetic acid is occasionally used also to acidulate fluids where it is wished to avoid the employment of mineral acids.

§ 25.

4. TARTARIC ACID $(2 HO, C_8 H_4 O_{10} = 2 HO, T)$.

The tartaric acid of commerce is sufficiently pure for the purposes of chemical analysis. It is kept best in powder, as its solution suffers decomposition after a time, a white film forming upon its surface. For use it is dissolved in a little water with the aid of heat.

Uses.—The addition of tartaric acid to solutions of sesquioxide of iron, protoxide of manganese, alumina, and various other oxides of metals, prevents the usual precipitation of these metals by an alkali; this non-precipitation is owing to the formation of double tartrates, which are not decomposed by alkalies.

Tartaric acid may therefore be employed to effect the separation of these metals from others the precipitation of which it does not prevent. Tartaric acid forms a difficultly soluble salt with potassa, but not so with soda; it is therefore one of our best reagents to distinguish between the two alkalies. *Bitartrate of soda* answers this latter purpose still better than the free acid. This reagent is prepared by dissolving one of two equal portions of tartaric acid in water, neutralizing the solution with carbonate of soda, then adding the other portion of the acid, and evaporating the solution to the crystallization point. For use one part of the salt is dissolved in 10 parts of water.

b. HYDROGEN ACIDS AND HALOGENS.

§ 26.

1. HYDROCHLORIC ACID (HCl).

Preparation.—Pour a cooled mixture of seven parts of concentrated sulphuric acid and two parts of water over four parts of chloride of sodium in a retort; expose the retort, with slightly raised neck, to the heat of a sand-bath until the evolution of gas ceases; conduct the evolved gas, by means of a double-limbed tube, into a flask containing six parts of water, and take care to keep this vessel constantly cool. To prevent the gas from receding the tube ought only to dip about one line into the water of the flask. When the operation is terminated, try the specific gravity of the acid produced, and dilute with water until it marks from 1·11 to

1·12. If you wish to ensure the absolute purity of the acid, and its perfect freedom from every trace of arsenic and chlorine, you must take care to free the sulphuric acid intended to be used in the process from arsenic and the oxygen compounds of nitrogen, according to the directions of § 22. A pure acid may also be prepared cheaply from the crude hydrochloric acid of commerce by diluting the latter to a specific gravity of 1·12, and distilling the fluid, with addition of some chloride of sodium. If the crude acid contains chlorine this should be removed first by cautious addition of solution of sulphurous acid, before proceeding to the distillation; if, on the other hand, it contains sulphurous acid, this is removed in the same way by cautious addition of some chlorine water. Hydrochloric acid not unfrequently contains an exceedingly minute trace of chloride of arsenic, owing to the presence of arsenic in the sulphuric acid employed. To free it from this impurity, hydrosulphuric acid is conducted into it, the mixture allowed to stand at rest for some time, the clear fluid then decanted from the sulphur and sulphide of arsenic, and the decanted fluid heated, to expel the sulphuretted hydrogen.

Tests.—Hydrochloric acid intended for the purposes of chemical analysis must be perfectly colorless and leave no residue upon evaporation. If it turns yellow on evaporation, sesquichloride of iron is present. It must not impart a blue tint to a solution of iodide of potassium mixed with starch paste (chlorine), nor discolor a fluid made faintly blue with iodide of starch (sulphurous acid). Chloride of barium ought not to produce a precipitate in the highly diluted acid (sulphuric acid). Hydrosulphuric acid must leave it unaltered.

Uses.—Hydrochloric acid serves as a solvent for a great many substances. It dissolves many metals and sulphides of metals as chlorides, with evolution of hydrogen or of hydrosulphuric acid. It dissolves lower and higher oxides in the form of chlorides, the solution being in the case of the higher oxides mostly attended with liberation of chlorine. Salts with insoluble or volatile acids are also converted by hydrochloric acid into chlorides, with separation of the original acid; thus carbonate of lime is converted into chloride of calcium, with liberation of carbonic acid. Hydrochloric acid dissolves salts with non-volatile and soluble acids *apparently* without decomposing them (*e.g.* phosphate of lime); but the fact is that in cases of this kind a metallic chloride and a soluble acid salt of the acid of the dissolved compound are formed; thus, for instance, in the case of phosphate of lime chloride of calcium and acid phosphate of lime are formed. With salts of acids forming no soluble acid compound with the base present hydrochloric acid forms metallic chlorides, the liberated acids remaining free in solution (borate of lime). Hydrochloric acid is also applied as a *special* reagent for the detection and separation of oxide of silver, suboxide of mercury, and lead (see Section III., §§ 115—118); and likewise for the detection of free ammonia, with which it produces in the air dense white fumes of chloride of ammonium.

§ 27.

2. CHLORINE (Cl) AND CHLORINE WATER.

Preparation.—Mix 18 parts of common salt with 15 parts of *finely pulverized* good binoxide of manganese; put the mixture in a flask, pour a *completely cooled* mixture of 45 parts of concentrated sulphuric acid and 21 parts of water upon it, and shake the flask: a uniform and continuous evolution of chlorine gas will soon begin, which, when slackening, may be easily increased again by the application of a *gentle* heat. This method of *Wiggers* is excellent, and can be highly recommended. Conduct the chlorine gas evolved first through a flask containing a little water, then into a bottle filled with cold water, and continue the process until the fluid is saturated. Where it is desired to obtain chlorine water quite free from bromine, the flask into which the chlorine is first conducted is changed after about one-half of the chlorine has been expelled, and the gas which now passes over is conducted into a separate bottle filled with water. The chlorine water must be kept in a cellar and carefully protected from the action of light; since, if this precaution is neglected, it speedily suffers complete decomposition, being converted into dilute hydrochloric acid, with evolution of oxygen (resulting from the decomposition of water). Smaller quantities, intended for use in the laboratory, are best kept in a stoppered bottle protected from the influence of light by a case of pasteboard. Chlorine water which has lost its strong peculiar odor is unfit for use.

Uses.—Chlorine has a greater affinity than iodine and bromine for metals and for hydrogen. Chlorine water is therefore an efficient agent to effect the expulsion of iodine and bromine from their compounds. Chlorine serves moreover to effect the solution of certain metals (gold, platinum), to convert sulphurous acid into sulphuric acid, protoxide of iron into sesquioxide, &c.; and also to effect the destruction of organic substances, as in presence of these it withdraws hydrogen from the water, enabling thus the liberated oxygen to combine with the vegetable matters and to effect their decomposition. For this latter purpose it is most advisable to *evolve* the chlorine in the fluid which contains the organic substances; this is effected by adding hydrochloric acid to the fluid, heating the mixture, and then adding chlorate of potassa. This gives rise to the formation of chloride of potassium, water, free chlorine, and bichlorate of chlorous acid, which acts in a similar manner to chlorine.

§ 28.

3. NITRO-HYDROCHLORIC ACID. *Aqua regia.*

Preparation.—Mix one part of pure nitric acid with from three to four parts of pure hydrochloric acid.

Uses.—Nitric acid and hydrochloric acid decompose each other, the decomposition, mostly resulting, as *Gay-Lussac* has shown, in the formation of two compounds which are gaseous at the ordinary temperature, $N O_2 Cl_2$ and $N O_2 Cl$, and of free chlorine and water. If one equivalent of $N O_5$ is used to three equivalents of $H Cl$, it may be assumed that only chloro-hyponitric acid ($N O_2 Cl_2$), chlorine and water are formed ($N O_5 + 3 H Cl = N O_2 Cl_2 + Cl + 3 H O$).

This decomposition ceases as soon as the fluid is saturated with the gas; but it recommences the instant this state of saturation is disturbed by the application of heat or by decomposition of the acid. The presence of the free chlorine, and also, but in a very subordinate degree, that of the acids named, makes aqua regia our most powerful solvent for metals (with the exception of those which form insoluble compounds with chlorine). Nitro-hydrochloric acid serves principally to effect the solution of gold and platinum, which metals are insoluble both in hydrochloric and in nitric acid; and also to decompose various metallic sulphides, *e.g.*, cinnabar, pyrites, &c.

§ 29.

4. Hydrofluosilicic Acid ($SiFl_3$, HFl).

Preparation.—Take quartz sand, wash off every particle of dust, and dry thoroughly. Mix one part of the dry sand intimately with one part of perfectly dry fluor spar in powder; pour six parts of concentrated sulphuric acid over the mixture in a non-tubulated retort, which it is advisable to lute, and mix carefully by shaking the vessel. As the mixture swells up when getting warm, it must at first fill the retort only to one-third. The neck of the retort is connected air-tight with a small tubulated receiver, and the tubulus of the latter again, by means of vulcanized indiarubber, with a wide glass tube twice bent at a right angle. To the descending limb of the glass tube a funnel is attached by means of vulcanized indiarubber; this funnel is lowered into a beaker containing four parts of water. Promote the disengagement of fluosilicic gas, which commences even in the cold, by moderately heating the retort over red-hot charcoal. Towards the end of the process a pretty strong heat should be applied. Every gas bubble produces in the water a precipitate of hydrated silicic acid, with simultaneous formation of hydrofluosilicic acid, $3\ SiFl_3 + 2\ HO = 2\ (SiFl_3, HFl) + SiO_3$. The precipitated hydrate of silicic acid renders the liquid gelatinous, and it is for this reason that the aperture of the descending limb of the tube cannot be allowed to dip direct into the water, since it would in that case speedily be choked. It sometimes happens in the course, and especially towards the end of the operation, that complete channels of silica are formed in the gelatinous liquid, through which the gas gains the surface without undergoing decomposition if the liquid is not occasionally stirred. When the evolution of gas has completely ceased, throw the gelatinous paste upon a linen cloth, squeeze the fluid through, and filter it afterwards. Keep the filtrate for use.

Tests.—Hydrofluosilicic acid must produce no precipitate in solutions of salts of strontia (sulphate of strontia).

Uses.—Bases decompose with hydrofluosilicic acid, forming water and metallic silicofluorides. Many of these are insoluble, whilst others are soluble; the latter may therefore by means of this reagent be distinguished from the former. In the course of analysis hydrofluosilicic acid is applied simply for the detection and separation of baryta.

c. SULPHUR ACIDS.

§ 30.

1. Hydrosulphuric Acid (*Sulphuretted Hydrogen*) (HS).

Preparation.—Hydrosulphuric acid gas is evolved best from sulphide of iron, which is broken into small lumps and then treated with dilute sulphuric acid. Fused sulphide of iron may be procured so cheaply in commerce that it is hardly worth while to take the trouble of preparing it expressly. However, if you wish to prepare it yourself, this may be done by heating iron turnings, or 1 to 1½ inch long iron nails, in a Hessian crucible to a white heat, and then adding small lumps of roll-sulphur until the entire contents of the crucible are in fusion. As soon as this is the case, pour the fused mass out on sand, or into an old

Fig. 22.

Hessian crucible. Or make a hole in the bottom of the crucible in which you fuse the mass, when the sulphide of iron will as fast as it forms run through the hole in the bottom of the crucible, and may thus be easily received in a coal-shovel placed in the ash-pit. Or introduce an intimate mixture of thirty parts of iron filings and twenty-one parts of flowers of sulphur gradually and in small portions at a time into a red-hot crucible, awaiting always the incandescence of the portion last introduced before proceeding to the addition of a fresh one. When you have thus put the whole mixture into the crucible, cover the latter closely, and expose it to a more intense heat, sufficient to make the sulphide of iron fuse more or less.

The evolution of the gas is effected in the apparatus illustrated by Fig. 22.

Pour water over the sulphide of iron in a, add concentrated sulphuric acid, and shake the mixture; the evolved gas is washed in c. When a sufficient quantity of gas is evolved, pour the fluid off the still undecomposed sulphide of iron, rinse the bottle repeatedly with water, then fill it with that fluid, and keep it for the next operation. If you neglect this, the apparatus will speedily become incrusted with crystals of sulphate of protoxide of iron, which is apt to interfere injuriously with subsequent processes of evolution of gas.

For larger laboratories, or for chemists having to operate often and largely with hydrosulphuric acid, I can recommend the lead apparatus designed by myself, which I have now for several years employed with the most satisfactory results in my own laboratory* (see Figs. 23 and 24).

$a\,b\,c\,d$ and $e\,f\,g\,h$ (Fig. 24) are two cylindrical leaden vessels, soldered with pure lead. They are both of the same size (in my own apparatus 33 centimetres high and 30 centimetres in diameter). i is a false bottom of lead, perforated like a sieve, placed from 4 to 5 centimetres above the actual bottom of the vessel, and resting on leaden feet, which support it on the sides as well as also more particularly in the middle. The numerous holes in the sieve-like bottom have a diameter of $1\frac{1}{2}$ millimetre; k shows the opening through which the sulphide of iron is introduced into the vessel. In my apparatus this aperture has a diameter of 7 centimetres, and is closed by putting a greased leather ring on its broad smooth rim, and pressing down upon this, by means of three winged screws, the broad rim of the smooth-turned cover. l shows the opening through which the solution of sulphate of protoxide of iron is drawn off; it will be seen by the drawing that the bottom of the vessel

---

* The apparatus is made by *Mr. Stumpf*, of Wiesbaden, mechanist, and fully answers all reasonable demands, both as regards workmanship and price.

($g\,h$) slants towards the part where this opening is placed. The aperture has a diameter of 3 centimetres; it is closed by means of a smooth-turned broad and thick leaden cap, fitting on the smooth-turned broad rim, and pressed down upon it with a winged screw. The semi-elliptical bar or bow in which the nut is set is moveable, and hinged to the sides of l in a manner to admit of its being bent out of reach of the liquid on drawing off the latter. The construction of the filling tube m may be learned from the drawing, and equally so that of the tube $d\,h$, which is intended to convey the acid from the upper to the lower vessel and *vice versâ*. It will be seen from the drawing that this tube reaches down into the slanting and deepened part of the bottom $g\,h$, without, however, actually touching the latter. The tube $c\,e$ is closed at the top, and has therefore no

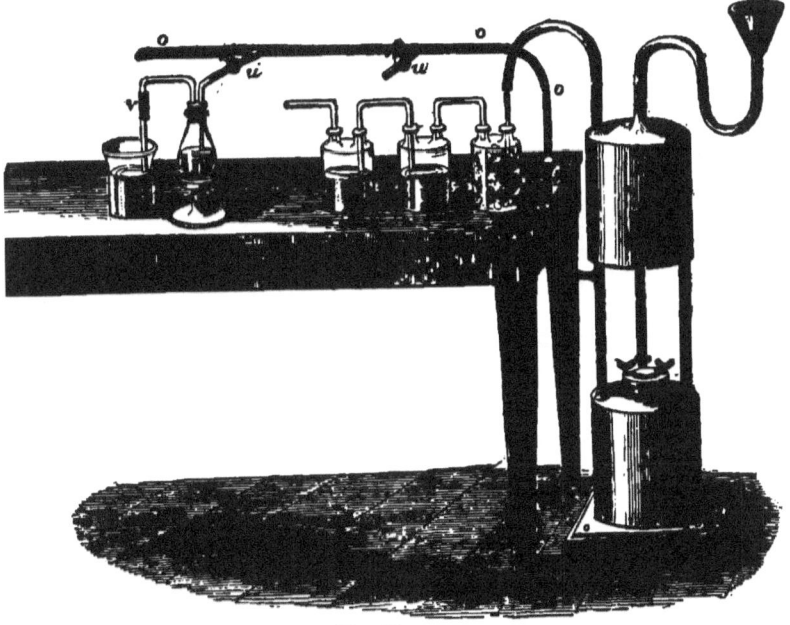

Fig. 23.

communication with the upper vessel, being simply intended to let off the gas evolved in $e\,f\,g\,h$; to which end it is connected laterally by a branch tube with the tube o; this latter tube is fitted with a stop-cock (n). The tube q is closed at both ends, and serves simply as an additional support for the upper vessel. The tubes in my apparatus have an inner diameter of 16 millimetres.

The process of filling is conducted as follows: put 3·3 kilogrammes* of fused sulphide of iron, broken in lumps, through the mouth k, upon the perforated bottom i; screw the covers properly down upon k and l, shut the cock n, and pour through the funnel of m first 7 litres of water, then 1 litre of concentrated sulphuric acid, and then again 7 litres of

* The quantities here given are calculated for an apparatus of the dimensions stated.

water. The air in $a\,b\,c\,d$ escapes in this operation through p, even after the latter tube is connected with the flasks r, s, t.

If the cock n is now opened, and one of the cocks u (Fig. 23), the acid will flow through the tube $d\,h$ into $e\,f\,g\,h$; and through o air will escape at first, followed by the hydrosulphuric acid evolved in $e\,f\,g\,h$. As is seen by the drawing, the tube o rises only to a certain elevation, where it makes a bend, running on thence in a horizontal direction. As many cocks $u\,u$ are added as is thought desirable; these are common brass gas stop-cocks, tight fitting and well ground in. They are connected with a small washing-bottle; a double-bent tube conveys the gas from the latter, with the co-operation of a straight tube connected with it at v by means of vulcanized indiarubber, into the fluid which it is intended to operate on; this arrangement greatly facilitates the cleansing of the straight tube dipping into the fluid. Upon now opening one of the cocks u, the cock n being of course also open, a current of gas of any desired strength is at once obtained, which will keep on for days in a continuous and steady stream. If all the cocks u are shut, the gas evolved in $e\,f\,g\,h$ forces the acid back to the upper vessel through the tube $h\,d$, and the evolution ceases.

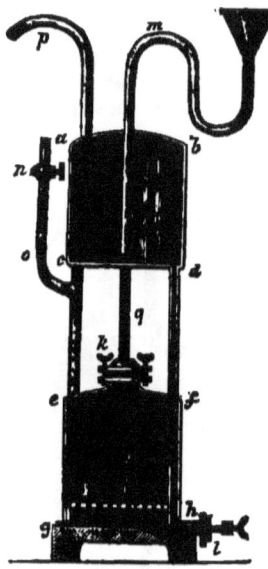

Fig. 24.

The cessation of the evolution of gas is not instantaneous, however, as the sulphide of iron in $e\,f\,g\,h$ remains still moistened with acid; moreover, small particles of the sulphides will always crumble off and, dropping through the sieve, come into contact with the rest of the acid covering the bottom $g\,h$. Now the gas which still continues to be evolved in $e\,f\,g\,h$, being no longer able to escape through o, forces the fluid up $h\,d$, and, passing through the acid in $a\,b\,c\,d$, makes its way out through p. To save this gas and keep it from poisoning the air, the flasks r, s, t, are connected with p. r contains cotton, and serves the purpose of a washing bottle;[*] s and t contain solution of ammonia; but the two flasks together should contain no more than either of them can conveniently hold, since, as the pressure of the gas increases or relaxes, the fluid is forced from s to t, or back from t to s. It will be readily understood that sulphide of ammonium is formed in these flasks.

The evolution of gas ceases completely when all the acid is consumed, but there remains still the one half of the sulphide of iron, as the quantity used of this is calculated for double the amount of acid. The solution of sulphate of protoxide of iron is therefore drawn off, and 1 litre of acid and 14 of water again poured in as before. This apparatus is now made also of much less dimensions to adapt it for smaller labora-

---

[*] A common washing-bottle filled with water could not well be used, as the water would very speedily recede.

tories. There are many other forms of apparatus used for the same purpose. Among others there is one devised by POHL, which is simple and convenient to use. It is shown in Fig. 25. The flask A, which contains dilute sulphuric acid, should hold from 2 to 2½ litres. The solid glass rod G, measuring at least 9 millimetres in diameter, and with the upper end ground, fits pretty tight into the indiarubber stopper B, so that it requires a certain degree of force to move it upwards or downwards. To the lower end of this rod is attached the perforated basket K, made of hard indiarubber. This basket is lined with coarse linen, and filled with lumps of sulphide of iron. If the glass rod G is pushed down sufficiently far just to dip into the dilute sulphuric acid in A, a slow stream of hydrosulphuric acid is evolved, which may be increased by lowering the basket, or stopped by drawing it up out of reach of the fluid in A. The tube R, which is interposed between the gas evolved and the exit tube, is filled with cotton, and serves the purpose of a washing-bottle.

Fig. 25.

Sulphuretted hydrogen water (solution of hydrosulphuric acid) is prepared by conducting the gas into very cold water, which has been previously freed from air by boiling. The operation is continued until the water is completely saturated with the gas, which may be readily ascertained by closing the mouth of the flask with the thumb, and shaking it a little: if upon this a pressure is felt from within, tending to push the thumb off the aperture of the flask, the operation may be considered at an end; but if, on the contrary, the thumb feels sucked into the mouth of the flask, this is a sure sign that the water is still capable of absorbing more gas.

Sulphuretted hydrogen water must be kept in well-closed vessels, otherwise it will soon suffer complete decomposition, the hydrogen being oxidized to water, and a small portion of the sulphur to sulphuric acid, the rest of the sulphur separating. The best way of preserving it unaltered for a very long time is to pour the freshly-prepared solution immediately into small phials, to cork these well, and to place them in an inverted position in small jars filled with water.

Tests.—Pure sulphuretted hydrogen water must be perfectly clear and strongly emit the peculiar odor of the gas; when treated with sesquichloride of iron, it must yield a copious precipitate of sulphur. Addition of ammonia must not impart a blackish appearance to it. It must leave no residue upon evaporation on platinum.

Uses.—Hydrosulphuric acid has a strong tendency to undergo decomposition with metallic oxides, forming water and metallic sulphides, which latter being mostly insoluble in water are usually precipitated in the process. The conditions under which the precipitation of certain sulphides ensues differ materially; by altering or modifying these conditions we may therefore divide the whole of the precipitable metals into groups, as will be found explained in Section III. Hydrosulphuric acid is therefore an invaluable agent to effect the separation of metals into principal groups. Some of the precipitated sulphides exhibit a characteristic color indicative of the individual metals which they respectively contain. The great facility with which hydrosulphuric acid is decomposed renders this substance also a useful reducing agent for many compounds; thus it serves, for instance, to reduce salts of sesquioxide of iron to salts of protoxide, chromic acid to the state of sesquioxide of chromium, &c. In these processes of reduction the sulphur separates in the form of a fine white powder. Whether the hydrosulphuric acid had better be applied in the gaseous form or in aqueous solution depends always upon the special circumstances of the case.

III. BASES AND METALS.

§ 31.

Bases are divided into oxygen bases and sulphur bases. The former result from the combination of metals or of compound radicals of similar character with oxygen, the latter from the combination of the same bodies with sulphur.

The *oxygen bases* are classified into alkalies, alkaline earths, earths proper, and oxides of the heavy metals. The alkalies are readily soluble in water; the alkaline earths dissolve with greater difficulty in that menstruum; and magnesia, the last member of the class, is only very sparingly soluble in it. The earths proper and the oxides of the heavy metals are insoluble in water or nearly so. The solutions of the alkalies and alkaline earths are caustic when sufficiently concentrated; they have an alkaline taste, change the yellow color of turmeric paper to brown, and restore the blue tint of reddened litmus paper; they saturate acids completely, so that even the salts which they form with strong acids do not change vegetable colors, whilst those with weak acids generally have an alkaline reaction. The earths proper and the oxides of the heavy metals combine likewise with acids to form salts, but, as a rule, they do not entirely take away the acid reaction of the latter.

The *sulphur bases* resulting from the combination of the metals of the alkalies and alkaline earths with sulphur are soluble in water. The solutions have a strong alkaline reaction. The other sulphur bases do not dissolve in water. All sulphur bases form with sulphur acids sulphur salts.

a. OXYGEN BASES.

a. ALKALIES.

§ 32.

1. POTASSA (K O) AND SODA (Na O).

The preparation of perfectly pure potassa or soda is a difficult operation. It is advisable therefore to prepare, besides perfectly pure caustic alkali, also some which is not quite pure, and some which being free from certain impurities may in many cases be safely substituted for the pure substance.

a. Common solution of soda.

Preparation.—Put into a clean cast-iron pan provided with a lid 3 parts of crystallized carbonate of soda of commerce and 15 parts of water, heat to boiling, and add, in small portions at a time, thick milk of lime prepared by pouring 3 parts of warm water over 1 part of quicklime, and letting the mixture stand in a covered vessel until the lime is reduced to a uniform pulpy mass. Keep the liquid in the pan boiling whilst adding the milk of lime, and for a quarter of an hour longer, then filter off a small portion, and try whether the filtrate still causes effervescence in hydrochloric acid. If this is the case, the boiling must be continued, and if necessary some more milk of lime must be added to the fluid. When the solution is perfectly free from carbonic acid, cover the pan, allow the fluid to cool a little, and then draw off the clear solution from the residuary sediment, by means of a siphon filled with water, and transfer it to a glass flask. Boil the residue a second and a third time with water, and draw off the fluid in the same way. Cover the flask close with a glass plate, and allow the lime suspended in the fluid to subside completely. Scour the iron pan clean, pour the clear solution back into it, and evaporate it to 6 or 7 parts. The solution so prepared contains from 9 to 10 per cent. of soda, and has a specific gravity of from 1·13 to 1·15. It must be clear, colorless, and as free as possible from carbonic acid; sulphide of ammonium must not impart a black color to it. Traces of silicic acid, alumina, and phosphoric acid are usually found in a solution of soda prepared in this manner; on which account it is unfit for use in accurate experiments.

Solution of soda is kept best in bottles closed with ground glass caps. In default of capped bottles, common ones with well-ground stoppers may be used, in which case the neck must be wiped perfectly dry and clean inside and the stopper coated with paraffin; since, if this precaution is neglected, it will be found impossible after a time to remove the stopper, particularly if the bottle is only rarely opened.

b. Hydrate of potassa purified with alcohol.

Preparation.—Dissolve some sticks of caustic potassa of commerce in rectified spirit of wine in a stoppered bottle by digestion and shaking; let the fluid stand, decant it, or filter it if necessary, and evaporate the clear fluid in a covered silver dish over the gas or spirit lamp until no more vapors escape; adding from time to time, during the evaporation, some water to prevent blackening of the mass. Place the silver dish in cold water until it has sufficiently cooled; remove the cake of caustic potassa from the dish, break it into coarse lumps in a hot mortar, and keep in a well-closed glass bottle. When required for use, dissolve a small lump in water.

The hydrate of potassa so prepared is sufficiently pure for most purposes; it contains, indeed, a minute trace of alumina, but is usually free from phosphoric acid, sulphuric acid, and silicic acid. The solution must remain clear upon addition of sulphide of ammonium; hydrochloric acid must only produce a barely perceptible effervescence in it. The solution acidified with hydrochloric acid must, upon evaporation to dryness, leave a residue which dissolves in water to a clear fluid; when boiled with molybdate of ammonia it must exhibit no yellow color; when treated with ammonia it ought not to deposit slight flakes of alumina immediately, but only after standing several hours in a warm place.

c. Hydrate of potassa prepared with baryta.

Preparation.—Dissolve pure crystals of baryta (§ 34) by heating with water, and add to the solution pure sulphate of potassa until a portion of the filtered fluid, acidified with hydrochloric acid and diluted, no longer gives a precipitate on addition of a further quantity of the sulphate (16 parts of crystals of baryta require 9 parts of sulphate of potassa). Let the turbid fluid clear, decant, and evaporate in a silver dish as in *b*. The hydrate of potassa so prepared is perfectly pure, except that it contains a trifling admixture of sulphate of potassa, which is left behind upon dissolving the hydrate in a little water. This hydrate is but rarely required, its use being in fact exclusively confined to the detection of minute traces of alumina.

Uses.—The great affinity which the fixed alkalies possess for acids renders these substances powerful agents to effect the decomposition of the salts of most bases, and consequently the precipitation of those bases which are insoluble in water. Many of the so precipitated oxides redissolve in an excess of the precipitant, as, for instance, alumina, sesquioxide of chromium, and oxide of lead; whilst others remain undissolved, *e.g.* sesquioxide of iron, teroxide of bismuth, &c. The fixed alkalies serve therefore also as a means to separate the former from the latter. Potassa and soda dissolve also many salts (*e.g.* chromate of lead), sulphur compounds, &c., and contribute thus to separate and distinguish them from other substances. Many of the oxides precipitated by the action of potassa or soda exhibit peculiar colors, or possess other characteristic properties that may serve to lead to the detection of the individual metal which they respectively contain; such are, for instance, the precipitate of protoxide of manganese, hydrate of protoxide of iron, suboxide of mercury, &c. The fixed alkalies expel ammonia from its salts, and enable us thus to detect that body by its smell, its action on vegetable colors, &c.

§ 33.

2. AMMONIA—*Oxide of Ammonium*—(NH_4O).

Preparation.—The apparatus illustrated by. Fig. 22 (§ 30) may also serve for the preparation of solution of ammonia, with this modification that, as no funnel tube is required in the process, the cork upon the flask *a* has only one perforation for the reception of the tube which serves to conduct the evolved ammonia into the washing bottle. Introduce into *a* 4 parts of chloride of ammonium in pieces about the size of a pea, and the dry hydrate of lime prepared from 5 parts of lime; mix by shaking the flask, and add cautiously a sufficient quantity of water to make the powder into lumps. Put a small quantity of water only into the washing bottle (which should be rather capacious); but have 10 parts of water in the flask which is intended for the final reception of the washed gas. Set the flask *a* now in a sand bath, connect it with the rest of the apparatus, place the flask *d* in a vessel of cold water, and apply heat. Evolution of gas speedily sets in. Continue to heat until no more bubbles appear. Open the cork of the flask *a*, to prevent the receding of the fluid. The solution of ammonia contained in the washing bottle is impure, but that contained in the receiver *d* is perfectly pure; dilute it with water until the specific gravity is about $0.96 = 10$ per cent. of ammonia. Keep the fluid in bottles closed with ground stoppers. This is the best way of preparing solution of ammonia in small quantities. That prepared on a large scale in cast-iron vessels is of course cheaper.

Tests.—Solution of ammonia must be colorless, and ought not to leave the least residue when evaporated on a watch-glass. When heated with an equal volume of lime water, it should cause no turbidity, at least not to a very marked extent (carbonic acid). When supersaturated with nitric acid, neither solution of nitrate of baryta nor of nitrate of silver must render it turbid, nor must sulphuretted hydrogen impart to it the slightest color.

Uses.—Solution of ammonia, although formed by conducting ammoniacal gas (NH_3) into water, and letting that gas escape upon exposure to the air, and much quicker when heated, may also be regarded as a solution of oxide of ammonium (NH_4O) in water, the first acceding equivalent of water (HO) being assumed to form NH_3O with NH_3. Upon this assumption solution of ammonia may accordingly be looked upon as an analogous fluid to solution of potassa and solution of soda, which greatly simplifies the explanation of all its reactions, the oxygen salts resulting from the neutralization of oxygen acids by solution of ammonia being also assumed to contain oxide of ammonium NH_4O, instead of NH_3. Ammonia is one of the most frequently used reagents. It is especially applied for the saturation of acid fluids, and also to effect the precipitation of a great many metallic oxides and earths; many of these precipitates redissolve in an excess of ammonia, as, for instance, the oxides of zinc, cadmium, silver, copper, &c., whilst others are insoluble in free ammonia. This reagent may therefore serve also to separate and distinguish the former from the latter. Some of these precipitates, as well as their solutions in ammonia, exhibit peculiar colors, which may at once lead to the detection of the individual metal which they respectively contain.

Many of the oxides which are precipitated by ammonia from neutral solutions are not precipitated by this reagent from acid solutions, their precipitation from the latter being prevented by the ammonia salt formed in the process. Compare § 53, chloride of ammonium.

β. ALKALINE EARTHS.

§ 34.

1. BARYTA (BaO).

Preparation.—There are a great many ways of preparing hydrate of baryta; but as Witherite is now easily and cheaply procurable, I prefer the following method to all others: mix intimately together 100 parts of finely pulverised Witherite, 10 parts of charcoal in powder, and 5 parts of resin, put the mixture in an earthenware pot, put on the lid and lute it on with clay or loam, and expose the pot so prepared to the heat of a brick-kiln. Break and triturate the baked mass, boil repeatedly with water in an iron pot, filter into vessels, stopper, and let them stand in the cold, when large quantities of crystals of hydrate of baryta ($BaO, HO + 8$ aq.) will make their appearance. Let the crystals drain in a properly covered funnel, dry rapidly between sheets of blotting paper, and keep them in well closed bottles. For use dissolve 1 part of the crystals in 20 parts of water, with the aid of heat, and filter the solution. The baryta water so prepared is purer than the mother liquor running off from the crystals, and is therefore preferable to it. The residue left in the pot, which is insoluble in water and consists of undecomposed Witherite and charcoal, is turned to account in the preparation of chloride of barium.

Tests.—Baryta water must, after precipitation of the baryta by pure sulphuric acid, give a filtrate remaining clear when mixed with spirit of wine, and leaving no fixed residue upon evaporation in a platinum crucible.

Uses.—Caustic baryta, being a strong base, precipitates the earths and metallic oxides insoluble in water from the solutions of their salts. In the course of analysis we use it simply to precipitate magnesia. Baryta water may also be used to precipitate those acids which form insoluble compounds with this base; it is applied with this view to effect the detection of carbonic acid, the removal of sulphuric acid, phosphoric acid, &c.

§ 35.

2. LIME (CaO).

We use—
a. Hydrate of lime.
b. Lime water.

The former is obtained by slacking pure calcined lime in lumps, in a porcelain dish, with half its weight of water. Hydrate of lime must be kept in a well-stoppered bottle.

To prepare lime water, digest hydrate of lime for some time with cold distilled water, shaking the mixture occasionally; let the undissolved portion of lime subside, decant, and keep the clear fluid in a well-stoppered bottle. If it is wished to have the lime water quite free from all traces of alkalies, which are almost invariably present in hydrate of lime prepared from calcined limestone, the liquids of the first two or three decantations must be removed, and the fluid decanted afterwards alone made use of.

Tests.—Lime water must impart a strongly-marked brown tint to turmeric paper, and give a not too inconsiderable precipitate with carbonate of soda. It speedily loses these properties upon exposure to the air, and is thereby rendered totally unfit for analytical purposes.

Uses.—Lime forms with many acids insoluble, with others soluble salts. Lime water may therefore serve to distinguish the former acids, which it precipitates from their solutions, from the latter, which it will of course fail to precipitate. Many of the precipitable acids are thrown down only under certain conditions, *e.g.* on boiling (citric acid), which affords a ready means of distinguishing between them by altering these conditions. We use lime water in analysis principally to effect the detection of carbonic acid, and also to distinguish between citric acid and tartaric acid. Hydrate of lime is chiefly used to liberate ammonia from ammonia salts.

γ. HEAVY METALS AND THEIR OXIDES.

§ 36.

1. ZINC (Zn).

Select zinc of good quality and, above all, perfectly free from arsenic. The method described § 132, 10 will serve to detect the presence of the slightest trace of this substance. Fuse the metal and pour the fused mass in a thin stream into a large vessel with water. Zinc which contains arsenic must be absolutely rejected, for no process of purification known to us that can in any way pretend to simplicity will ever succeed in removing every trace of that metal (Eliot and Storer).

Uses.—Zinc serves in qualitative analysis for the evolution of hydrogen, and also of arsenetted and antimonetted hydrogen gases (compare § 131, 10, and § 132, 10); it is occasionally used also to precipitate some metals from their solutions; in which process the zinc simply displaces the other metal ($CuO, SO_3 + Zn = ZnO, SO_3 + Cu$).

§ 37.

2. Iron (Fe).

Iron reduces many metals and precipitates them from their solutions in the metallic state. We use it especially for the detection of copper, which precipitates upon it with its characteristic color. Any clean surface of iron, such as a knife-blade, a needle, a piece of wire, &c., will serve for this purpose.

§ 38.

3. Copper (Cu).

We use copper exclusively to effect the reduction of mercury, which precipitates upon it as a white coating shining with silvery lustre when rubbed. A copper coin scoured with fine sand, or in fact any clean surface of copper, may be employed for this purpose.

§ 39.

4. Hydrate of Teroxide of Bismuth (BiO_3, HO).*

Preparation.—Dissolve bismuth, freed from arsenic by fusion with hepar sulphuris or nitrate of potassa, in dilute nitric acid; dilute the solution as much as is practicable without producing a permanent precipitate; filter, and evaporate the filtrate to crystallization. Wash the crystals with water containing nitric acid, triturate them with water, add ammonia in excess, and let the mixture digest for some time; then filter, wash, and dry the white precipitate, and keep it for use.

Tests.—Hydrosulphuric acid must throw down from a solution of this reagent in dilute nitric acid a precipitate insoluble in ammonia and sulphide of ammonium; the fluid filtered off from the precipitate treated with ammonia must therefore remain perfectly clear upon addition of hydrochloric acid, whilst in that filtered off from the precipitate treated with sulphide of ammonium hydrochloric acid must only produce a pure white turbidity (sulphur).

Uses.—Teroxide of bismuth when boiled with alkaline solutions of metallic sulphides decomposes with the latter, giving rise to the formation of metallic oxides and sulphide of bismuth. It is better adapted to effect decompositions of this kind than oxide of copper, since it enables

---

* The basic nitrate of teroxide of bismuth of commerce, if perfectly free from arsenic and antimony, may also be used instead of the hydrated teroxide.

the operator to judge immediately upon the addition of a fresh portion whether the decomposition is complete or not. It has still another advantage over oxide of copper, viz., it does not, like the latter, dissolve in the alkaline fluid in presence of organic substances; nor does it act as a reducing agent upon reducible oxygen compounds. We use it principally to convert tersulphide and pentasulphide of arsenic into arsenious and arsenic acids, for which purpose oxide of copper is altogether inapplicable, since it converts the arsenious acid immediately into arsenic acid, being itself reduced to the state of suboxide.

b. SULPHUR BASES.

§ 40.

1. Sulphide of Ammonium (NH_4S).

We use in analysis—
a. *Colorless proto-sulphide of ammonium.*
b. *Yellow bi-, ter-, &c., sulphide of ammonium.*

Preparation.—Transmit hydrosulphuric acid gas through 3 parts of solution of oxide of ammonium until no further absorption takes place; then add 2 parts more of the same solution of oxide of ammonium. The action of hydrosulphuric acid upon oxide of ammonium gives rise to the formation, first, of NH_4S (NH_4O and $HS = NH_4S$ and HO), then of NH_4S, HS; upon addition of the same quantity of solution of ammonia as has been saturated, the oxide of ammonium decomposes with the double sulphide of ammonium and hydrogen or, as it is commonly called, the hydrosulphate of sulphide of ammonium, and simple or proto-sulphide of ammonium is formed ($NH_4S, HS + NH_4O = 2(NH_4S) + HO$. The rule, however, is to add only two-thirds of the quantity of solution of ammonia, as it is better the preparation should contain a little hydrosulphate of sulphide of ammonium than that free ammonia should be present. To employ, as has usually been the case hitherto, hydrosulphate of sulphide of ammonium instead of the simple proto-sulphide is unnecessary, and simply tends to increase the smell of sulphuretted hydrogen in the laboratory, as the preparation allows that gas to escape when in contact with metallic sulphur acids.

Sulphide of ammonium should be kept in small well-stoppered bottles. It is colorless at first, and deposits no sulphur upon addition of acids. Upon exposure to the air, however, it acquires a yellow tint, owing to the formation of bisulphide of ammonium, which is attended also with formation of ammonia and water:

$$2(NH_4S) + O = NH_4S_2 + NH_3 + HO.$$

Continued action of the oxygen of the air upon the sulphide of ammonium tends at first to the formation of still higher sulphides; but afterwards the fluid deposits sulphur, and there remains in the end nothing in solution but pure ammonia, the whole of the sulphur being found deposited at the bottom of the vessel.

The sulphide of ammonium which has turned yellow by exposure to the air may be used for all purposes requiring the employment of yellow sulphide of ammonium. The yellow sulphide may also be expeditiously prepared by digesting the proto-sulphide with some sulphur. All kinds of yellow sulphide of ammonium deposit sulphur and look turbid and milky on being mixed with acids.

Tests.—Sulphide of ammonium must strongly emit the odor peculiar to it; with acids it must evolve abundance of sulphuretted hydrogen; the evolution of gas may be attended by the separation of a pure white deposit, but no other precipitate must be formed. Upon evaporation and exposure to a red heat on a platinum dish it must leave no residue. It must not even on heating precipitate or render turbid solution of magnesia or solution of lime (carbonate of ammonia or free ammonia).

Uses.—Sulphide of ammonium is one of the most frequently employed reagents. It serves (*a*) to effect the precipitation of those metals which hydrosulphuric acid fails to throw down from acid solutions, *e. g.* of iron, cobalt, &c. ($NH_4S + FeO, SO_3 = FeS + NH_4O, SO_3$); (*b*) to separate the metallic sulphides thrown down from acid solutions by hydrosulphuric acid, since it dissolves some of them to sulphur salts, as, for instance, the sulphides of arsenic and antimony, &c. (NH_4S, AsS_3, &c.), whilst leaving others undissolved—for instance, sulphide of lead, sulphide of cadmium, &c. The sulphide of ammonium used for this purpose must contain an excess of sulphur if the metallic sulphides to be dissolved will dissolve only as higher sulphides, as, for instance, SnS, which dissolves with ease only as SnS_2.

From solutions of salts of alumina and sesquioxide of chromium sulphide of ammonium precipitates hydrates of these oxides, with escape of sulphuretted hydrogen, as the sulphur compounds corresponding to these oxides cannot form in the humid way. [$Al_2O_3, 3SO_3 + 3NH_4S + 3HO = Al_2O_3, 3HO + 3(NH_4O, SO_3) + 3HS$]. Salts insoluble in water are thrown down by sulphide of ammonium unaltered from their solutions in acids; thus, for instance, phosphate of lime is precipitated unaltered from its solution in hydrochloric acid.

§ 41.

2. SULPHIDE OF SODIUM (Na S).

Preparation.—Same as sulphide of ammonium, except that solution of soda is substituted for solution of ammonia. Keep the fluid obtained in well-stoppered bottles. If required to contain some higher sulphide of sodium digest it with powdered sulphur.

Uses.—Sulphide of sodium must be substituted for sulphide of ammonium to effect the separation of sulphide of copper from sulphur compounds soluble in alkaline sulphides, *e. g.* from proto-sulphide of tin, as sulphide of copper is not quite insoluble in sulphide of ammonium.

IV. SALTS.

Of the many salts employed as reagents those of potassa, soda, and ammonia are used principally on account of their acids; salts of soda may therefore often be substituted for the corresponding potassa salts, &c. Thus it is almost always a matter of perfect indifference whether we use carbonate of soda or carbonate of potassa, ferrocyanide of potassium or ferrocyanide of sodium, &c. I have therefore here classified the salts of the alkalies *by their acids*. With the salts of the alkaline earths and those of the oxides of the heavy metals the case is different; these are not used for their acid, but for their base; we may therefore often substitute for one salt of a base another similar one, as *e. g.* nitrate or acetate of baryta for chloride of barium, &c. For this reason I have classified the salts of the alkaline earths and of the heavy metals *by their bases.*

a. SALTS OF THE ALKALIES.

§ 42.

1. Sulphate of Potassa (KO, SO_3).

Preparation.—Purify sulphate of potassa of commerce by recrystallization, and dissolve 1 part of the pure salt in 12 parts of water.

Uses.—Sulphate of potassa serves to detect and separate baryta and strontia. It is in many cases used in preference to dilute sulphuric acid, which is employed for the same purpose, as it does not, like the latter reagent, disturb the neutrality of the solution.

§ 43.

2. Phosphate of Soda ($2 NaO, HO, PO_5 + 24$ aq.).

Preparation.—Purify phosphate of soda of commerce by recrystallization, and dissolve 1 part of the pure salt in 10 parts of water for use.

Tests.—Solution of phosphate of soda must not become turbid when heated with ammonia. The precipitates which solution of nitrate of baryta and solution of nitrate of silver produce in it must completely, and without effervescence, redissolve upon addition of dilute nitric acid.

Uses.—Phosphate of soda precipitates the alkaline earths and all metallic oxides by double affinity. It serves in the course of analysis, after the separation of the oxides of the heavy metals, as a test for alkaline earths in general; and, after the separation of baryta, strontia, and lime, as a special test for the detection of magnesia; for which latter purpose it is used in conjunction with ammonia, the magnesia precipitating under these circumstances as basic phosphate of magnesia and ammonia.

§ 44.

3. Oxalate of Ammonia ($2 NH_4O, C_2O_3 + 2$ aq.).

Preparation.—Dissolve oxalic acid purified by recrystallization in 2 parts of distilled water, with the aid of heat, add solution of ammonia until the reaction is distinctly alkaline, and put the vessel in a cold place. Let the crystals drain. The mother liquor will, upon proper evaporation, give another crop of crystals. Purify all the crystals by recrystallization. Dissolve 1 part of the pure salt in 24 parts of water for use.

Tests.—The solution of oxalate of ammonia must not be precipitated nor rendered turbid by hydrosulphuric acid, nor by sulphide of ammonium. Ignited on platinum, the salt must volatilize without leaving a residue.

Uses.—Oxalic acid forms with lime, strontia, baryta, oxide of lead, and other metallic oxides, insoluble or very difficultly soluble compounds; oxalate of ammonia produces therefore in the aqueous solutions of the salts of these bases precipitates of the corresponding oxalates. In analysis it serves principally for the detection of lime.

§ 45.

4. ACETATE OF SODA (NaO, $C_4H_3O_3$ + 6 aq., or NaO, $\overline{A}$ + 6 aq.).

Preparation.—Dissolve crystallized carbonate of soda in a little water, add to the solution acetic acid to slight excess, evaporate to crystallization, and purify the salt by recrystallization. For use dissolve 1 part of the salt in 10 parts of water.

Tests.—Acetate of soda must be colorless and free from empyreumatic matter and inorganic acids.

Uses.—The stronger acids in the free state decompose acetate of soda, combining with the base, and setting the acetic acid free. In the course of analysis acetate of soda is used principally to precipitate phosphate of sesquioxide of iron (which is insoluble in acetic acid) from its solution in hydrochloric acid. It serves also to effect the separation of sesquioxide of iron and alumina, which it precipitates on boiling from the solutions of their salts.

§ 46.

5. CARBONATE OF SODA (NaO, CO_2 + 10 aq.).

Preparation.—Take bicarbonate of soda of commerce, put the powder into a funnel stopped loosely with some cotton, make the surface even, cover it with a disc of difficultly permeable paper with turned-up edges, and wash by pouring small quantities of water on the paper disc, until the filtrate, acidified with nitric acid, is not rendered turbid by solution of nitrate of silver, nor by solution of chloride of barium. Let the salt dry, and then convert it by gentle ignition into the simple carbonate. This is effected best in a crucible or dish of silver or platinum; but it may be done also in a perfectly clean vessel of cast iron, or, on a small scale, in a porcelain dish. Pure carbonate of soda may be obtained also by repeated recrystallization of carbonate of soda of commerce. For use dissolve 1 part of the anhydrous salt or 2·7 parts of the crystallized salt in 5 parts of water.

Tests.—Carbonate of soda intended for analytical purposes must be perfectly white. Its solution, after supersaturation with nitric acid, must not be rendered turbid by chloride of barium or nitrate of silver; nor must addition of sulphocyanide of potassium impart a red, or boiling with molybdate of ammonia a yellow tint to it, or give a yellow precipitate; the residue which remains upon evaporating its solution to dryness, after previous supersaturation with hydrochloric acid, must leave no residue (silicic acid) when redissolved in water.

Uses.—With the exception of the alkalies, carbonate of soda precipitates the whole of the bases, most of them as carbonates, but some also as hydrated oxides. Those bases which are soluble in water as bicarbonates require boiling for their complete precipitation from acid solutions. Many of the precipitates produced by the action of carbonate of soda exhibit a characteristic color, which may lead to the detection of the individual metals which they respectively contain. Solution of carbonate of soda serves also for the decomposition of many insoluble salts of the alkaline earths or of the metals, more particularly of those with organic acids. Upon boiling with carbonate of soda these salts are converted into insoluble carbonates, whilst the acids combine with the soda and are thus obtained in solution in the form of salts of soda. Carbonate of soda is often used also to saturate free acids.

§ 47.

6. Carbonate of Ammonia (NH_4O, CO_2).

Preparation.—We use for the purposes of chemical analysis purified sesquicarbonate of ammonia entirely free from any smell of animal oil, such as is prepared on a large scale from a mixture of chloride of ammonium and carbonate of lime by sublimation. The outer and the inner surface of the mass are carefully scraped. One part of the salt is dissolved by digestion with 4 parts of water to which one part of solution of caustic ammonia has been added.

Tests.—Pure carbonate of ammonia must completely volatilize. Neither solution of nitrate of baryta nor of nitrate of silver, nor sulphuretted hydrogen, must color or precipitate it, after supersaturation with nitric acid.

Uses.—Carbonate of ammonia precipitates, like carbonate of soda, most metallic oxides and earths; it is generally employed in preference to the latter reagent, because it introduces no non-volatile body into the solution. Complete precipitation of many of the oxides takes place also only on boiling. Several of the precipitates redissolve again in an excess of the precipitant. In like manner carbonate of ammonia dissolves many hydrated oxides and many sulphides, and thus enables us to distinguish and separate them from others which are insoluble in this reagent.

Carbonate of ammonia, like caustic ammonia and for the same reason, fails to precipitate from acid solutions many oxides which it precipitates from neutral solutions. (Compare § 53.) We use carbonate of ammonia in chemical analysis principally to effect the precipitation of baryta, strontia, and lime, and the separation of these substances from magnesia; also to separate sulphide of arsenic, which is soluble in it, from sulphide of antimony, which is insoluble in this reagent.

§ 48.

7. Bisulphite of Soda ($NaO, 2SO_2$).

Preparation.—Heat 5 parts of copper shreds with 20 parts of concentrated sulphuric acid in a flask, and conduct the sulphurous acid gas evolved, first through a washing bottle containing some water, then into a flask containing 4 parts of purified bicarbonate of soda (§ 46), or 7 parts of pure crystallized carbonate of soda, and from 20 to 30 parts of water, and which is not much more than half full; continue the transmission of the gas until the evolution of carbonic acid ceases. Keep the solution, which smells strongly of sulphurous acid, in a well-stoppered bottle.

Tests.—Sulphite of soda, when evaporated to dryness with pure sulphuric acid, must leave a residue,* the aqueous solution of which is not altered by hydrosulphuric acid, nor precipitated or colored yellow by heating with a solution of molybdate of ammonia mixed with nitric acid.

Uses.—Sulphurous acid has a great tendency to pass to the state of sulphuric acid by absorbing oxygen. It is therefore one of our most powerful reducing agents. Sulphite of soda, which has the advantage of being less readily decomposed than sulphurous acid, acts in an analogous manner upon addition of acid. We use it principally to reduce arsenic acid to arsenious acid, chromic acid to sesquioxide of chromium, and sesquioxide of iron to protoxide. It will serve also to effect the separation of tersulphide of arsenic, which is soluble in it, from the sulphides of antimony and tin, which are insoluble in this reagent.

§ 49.

8. Nitrite of Potassa (KO, NO_3).

Preparation.—Heat in a flask only half filled with the mixture 2 parts of starch, in pieces, with 8 parts of crude nitric acid of 1·4 specific gravity, and 8 parts of water, and conduct the nitrous fumes evolved, first through a larger empty flask, then into a flask containing solution of potassa, to complete saturation of the potassa. Where the solution of potassa contains silicic acid or alumina, as is mostly the case, the saturation point is marked by the separation of the said impurities. 5 parts of solution of potassa of 1·27 specific gravity is the quantity required in the process. As soon as the action begins, the flame under the evolution flask must be temporarily removed, or otherwise the action might become too energetic. Evaporate the filtered solution to dryness. Dissolve 1 part of the dry salt in about 2 parts of water when the reagent is required for use.

Tests.—Nitrite of potassa must upon addition of dilute sulphuric acid copiously evolve nitric oxide gas.

Uses.—Nitrite of potassa is an excellent means to effect the detection and separation of cobalt, in the solutions of which metal it produces a precipitate of nitrite of potassa and sesquioxide of cobalt (potassio-nitrite of sesquioxide of cobalt). It serves also in presence of free acid to liberate iodine from its compounds.

§ 50.

9. Bichromate of Potassa $(KO, 2CrO_3)$.

Preparation.—Purify the salt of commerce by recrystallization, and dissolve 1 part of the pure salt in 10 parts of water for use.

Uses.—Chromate of potassa decomposes most of the soluble salts of metallic oxides by double affinity. Most of the precipitated chromates are very sparingly soluble, and many of them exhibit characteristic colors which lead readily to the detection of the particular metal which they respectively contain. We use bichromate of potassa principally as a test for lead.

* The evaporation is attended with copious evolution of sulphurous acid.

§ 51.

10. GRANULAR ANTIMONATE OF POTASSA ($KO, SbO_5 + 7$ aq.).

Preparation.—Project a mixture of equal parts of pulverized tartar-emetic and nitrate of potassa in small portions at a time into a red-hot crucible. After the mass is deflagrated, keep it at a moderate red heat for a quarter of an hour longer, which will make it froth at first, but after some time it will be seen in a state of calm fusion. Remove the crucible now from the fire, let the mass get sufficiently cold, and then extract it with warm water. Transfer to a suitable vessel, which is easily done by rinsing, and decant the supernatant clear fluid from the heavy white powder deposit. Concentrate the decanted fluid by evaporation. After 1 or 2 days a doughy mass will separate. Treat this mass with three times its volume of cold water, working it at the same time with a spatula. This operation will serve to convert it into a fine granular powder, to which add the powder from which the fluid was decanted, wash slightly, and dry on blotting paper. 100 parts of tartar-emetic give about 36 parts of antimonate of potassa (BRUNNER).

Tests and *Uses.*—Granular antimonate of potassa is very sparingly soluble in water, requiring 90 parts of boiling and 250 parts of cold water for solution. The solution had always best be prepared immediately before required for use, by repeatedly shaking the salt with cold water, and filtering off the fluid from the undissolved portion. The solution must be clear and of neutral reaction; it must give no precipitate with solution of chloride of potassium, nor with solution of chloride of ammonium; but solution of chloride of sodium must produce a crystalline precipitate in it. Antimonate of potassa is a valuable reagent for soda, but only under certain conditions, for which see § 90.

§ 52.

11. MOLYBDATE OF AMMONIA (NH_4O, MoO_3), DISSOLVED IN NITRIC ACID.

Preparation.—Triturate sulphide of molybdenum with about an equal bulk of coarse quartz sand washed with hydrochloric acid, until the mass is reduced to a moderately fine powder; heat the powder to faint redness, with repeated stirring, until the mass has acquired a lemon-yellow color (which after cooling turns whitish). With small quantities this operation may be conducted in a flat platinum dish, with large quantities in a muffle. Extract the residuary mass with solution of ammonia, filter, evaporate the filtrate, heat the residue to faint redness until the mass appear yellow or white, and then digest this residuary mass for several days with nitric acid in the water bath, in order to convert the phosphoric acid which is almost invariably present in the ore to the tribasic state. When the nitric acid is evaporated dissolve the residue in 4 parts of solution of ammonia, filter rapidly, and pour the filtrate into 15 parts by weight of nitric acid of 1·20 specific gravity. Keep the mixture standing several days in a moderately warm place, which will cause the separation of any remaining traces of phosphoric acid or phospho-molybdate of ammonia. Decant the colorless fluid from the precipitate, and keep it for use. Heated to 104° Fahrenheit no white precipitate (molybdic acid or a molybdate) will separate; but if the temperature is raised beyond that point this will at once take place unless more nitric or hydrochloric acid be added (EGGERTZ).

Uses.—Phosphoric acid and arsenic acid form with molybdic acid and ammonia peculiar yellow compounds which are almost absolutely insoluble in the nitric acid solution of molybdate of ammonia. Molybdate of ammonia affords therefore an excellent means to detect these acids, and more especially very minute quantities of phosphoric acid in acid solutions containing sesquioxide of iron, alumina, and alkaline earths.

§ 53.

12. CHLORIDE OF AMMONIUM (NH_4, Cl).

Preparation.—Select sublimed white sal ammoniac of commerce. If it contains iron it must be purified. For that purpose add some sulphide of ammonium to the solution, let the precipitate which forms subside, and filter; add hydrochloric acid to the filtrate until the latter manifests a feebly acid reaction; boil the mixture some time, saturate with ammonia, filter if necessary, and crystallize. Dissolve 1 part of the salt in 8 parts of water for use.

Tests.—Solution of chloride of ammonium must upon evaporation on a platinum knife leave a residue which volatilizes completely upon continued application of heat. Sulphide of ammonium must leave it unaltered. Its reaction must be perfectly neutral.

Uses.—Chloride of ammonium serves principally to retain in solution certain oxides (*e.g.* protoxide of manganese, magnesia) or salts (*e.g.* tartrate of lime) upon the precipitation of other oxides or salts by ammonia or some other reagent. This application of chloride of ammonium is based upon the tendency of the ammonia salts to form double compounds with other salts. Chloride of ammonium serves also to distinguish between precipitates possessed of similar properties; for instance, to distinguish the *basic phosphate of magnesia and ammonia*, which is insoluble in chloride of ammonium, from other precipitates of magnesia. It is used also to precipitate from their solutions in potassa various substances which are soluble in that alkali, but insoluble in ammonia; *e.g.* alumina, sesquioxide of chromium, &c. In this process the elements of the chloride of ammonium transpose with those of the potassa, and chloride of potassium, water, and ammonia are formed. Chloride of ammonium is applied also as a *special* reagent to effect the precipitation of platinum as ammonio-bichloride of platinum.

§ 54.

13. CYANIDE OF POTASSIUM (K Cy).

Preparation.—Heat ferrocyanide of potassium of commerce (perfectly free from sulphate of potassa) gently, with stirring, until the crystallization water is completely expelled; triturate the anhydrous mass, and mix 8 parts of the dry powder with 3 parts of perfectly dry carbonate of potassa; fuse the mixture in a covered Hessian or, better still, in a covered iron crucible, until the mass is in a faint glow, appears clear, and a sample of it, taken out with a heated glass or small iron rod, looks perfectly white. Remove the crucible now from the fire, tap it gently, and let it cool a little until the evolution of gas has ceased; pour the fused cyanide of potassium into a heated tall, crucible-shaped vessel of clean scoured iron or silver, or into a moderately hot Hessian crucible, with proper care, to prevent the running out of any of the minute particles of iron which have separated in the process of fusion and have subsided to the bottom of the crucible. Let the mass now slowly cool in a somewhat warm place. The cyanide of potassium so prepared is exceedingly well adapted for analytical purposes, although it contains carbonate and cyanate of potassa; which latter is upon solution in water transformed into carbonate of ammonia and carbonate of potassa ($KO, C_2NO + 4HO = KO, CO_2 + NH_4O, CO_2$). Keep it in the solid form in a well-stoppered bottle, and dissolve 1 part of it in 4 parts of water, without application of heat, when required for use.

Tests.—Cyanide of potassium must be of a milk-white color and quite free from particles of iron or charcoal. It must completely dissolve in water to a clear fluid. It must contain neither silicic acid nor sulphide of potassium; the precipitate which salts of lead produce in its solution must accordingly be of a white color, and the residue which its solution leaves upon evaporation, after previous supersaturation with hydrochloric acid,* must completely dissolve in water to a clear fluid.

Uses.—Cyanide of potassium prepared in the manner described produces in the solutions of most of the salts with metallic oxides precipitates of cyanides of metals or of oxides or carbonates which are insoluble in water. The precipitated cyanides are soluble in cyanide of potassium, and may therefore by further addition of the reagent be separated from the oxides or carbonates which are insoluble in cyanide of potassium. Some of the metallic cyanides redissolve invariably in the cyanide of potassium as double cyanides, even in presence of free hydrocyanic acid and upon boiling; whilst others combine with cyanogen to new radicals, which remain in solution in combination with the potassium. The most common compounds of this nature are cobalticyanide of potassium and ferro- and ferricyanide of potassium. These differ from the double cyanides of the other kind particularly in this, that dilute acids fail to precipitate the metallic cyanides which they contain. Cyanide of potassium may accordingly serve also to separate the metals which form compounds of the latter description from others the cyanides of which are precipitated by acids from their solution in cyanide of potassium. In the course of analysis this reagent is of great importance, as it serves to effect the separation of cobalt from nickel; also that of copper, the sulphide of which metal is soluble in it, from cadmium, the sulphide of which is insoluble in this reagent.

§ 55.

14. FERROCYANIDE OF POTASSIUM ($2 K, C_6N_3Fe + 3$ aq. $= 2 K,$ Cfy $+ 3$ aq.).

Preparation.—The ferrocyanide of potassium is found in commerce sufficiently pure for the purposes of chemical analysis. 1 part of the salt is dissolved in 12 parts of water for use.

* This supersaturation with hydrochloric acid is attended with disengagement of hydrocyanic acid.

Uses.—Ferrocyanogen forms with most metals compounds insoluble in water, which frequently exhibit highly characteristic colors. These ferrocyanides are formed when ferrocyanide of potassium is brought into contact with soluble salts of metallic oxides, with chlorides, &c., the potassium changing places with the metals. Ferrocyanide of copper and ferrosesquicyanide of iron exhibit the most characteristic colors of all; ferrocyanide of potassium serves therefore particularly as a test for oxide of copper and sesquioxide of iron.

§ 56.

15. FERRICYANIDE OF POTASSIUM ($3 \text{ K}, \text{C}_{12}\text{N}_6\text{Fe}_2 = 3 \text{ K Cfdy}$).

Preparation.—Conduct chlorine gas slowly into a solution of 1 part of ferrocyanide of potassium in 10 parts of water, with frequent stirring, until the solution exhibits a fine deep red color by transmitted light (the light of a candle answers best), and a portion of the fluid produces no longer a blue precipitate in a solution of sesquichloride of iron, but imparts a brownish tint to it. Evaporate the fluid now in a dish to ¼ of its weight, and let crystallize. The mother liquor will upon further evaporation yield a second crop of crystals equally fit for use as the first. Dissolve the whole of the crystals obtained in 3 parts of water, filter if necessary; evaporate the solution briskly to half its volume, and let crystallize again. Whenever required for use, dissolve a few of the crystals, which are of a splendid red color, in a little water. The solution, as already remarked, must produce neither a blue precipitate nor a blue color in a solution of sesquichloride of iron.

Uses.—Ferricyanide of potassium decomposes with solutions of metallic oxides in the same manner as ferrocyanide of potassium. Of the metallic ferricyanides the ferriprotocyanide of iron is more particularly characterized by its color, and we apply ferricyanide of potassium therefore principally as a test for protoxide of iron.

§ 57.

16. SULPHOCYANIDE OF POTASSIUM ($\text{K}, \text{C}_2\text{N S}_2$ or $\text{K}, \text{Cy S}_2$).

Preparation.—Mix together 46 parts of anhydrous ferrocyanide of potassium, 17 parts of carbonate of potassa, and 32 parts of sulphur; introduce the mixture into an iron pan provided with a lid, and fuse over a gentle fire; maintain the same temperature until the swelling of the mass which ensues at first has completely subsided and given place to a state of tranquil and clear fusion; increase the temperature now, towards the end of the operation, to faint redness, in order to decompose the hyposulphite of potassa which has been formed in the process. Remove the half refrigerated and still soft mass from the pan, crush it, and boil repeatedly with alcohol of from 80 to 90 per cent. Upon cooling, part of the sulphocyanide of potassium will separate in colorless crystals; to obtain the remainder, distil the alcohol from the mother liquor. Dissolve 1 part of the salt in 10 parts of water for use.

Tests.—Solution of sulphocyanide of potassium must remain perfectly colorless when mixed with perfectly pure dilute hydrochloric acid.

Uses.—Sulphocyanide of potassium serves for the detection of sesquioxide of iron, for which substance it is at once the most characteristic and the most delicate test.

b. SALTS OF THE ALKALINE EARTHS.

§ 58.

1. CHLORIDE OF BARIUM (BaCl + 2 aq.).

Preparation.—*a. From heavy spar.* Mix together 8 parts of pulverized sulphate of baryta, 2 parts of charcoal in powder, and 1 part of common resin. Put the mixture in a crucible, and expose it in a wind furnace to a long-continued red heat; or put the mixture in an earthen pot, lute the lid on with clay, and expose to the heat of a brick-kiln (compare § 34). Triturate the crude sulphide of barium obtained, boil about $\frac{9}{10}$ of the powder with 4 times its quantity of water, and add hydrochloric acid until all effervescence of sulphuretted hydrogen has ceased, and the fluid manifests a slight acid reaction. Add now the remaining $\frac{1}{10}$ part of the sulphide of barium, boil some time longer, then filter, and let the alkaline fluid crystallize. Dry the crystals, redissolve them in water, and crystallize again.

b. From Witherite. Pour 10 parts of water upon 1 part of pulverized Witherite, and gradually add crude hydrochloric acid until the Witherite is almost completely dissolved. Add now a little more finely pulverized Witherite, and heat, with frequent stirring, until the fluid has entirely or very nearly lost its acid reaction; add solution of sulphide of barium as long as a precipitate forms; then filter, evaporate the filtrate to crystallization, and purify them by crystallizing again. For use dissolve 1 part of the chloride of barium in 10 parts of water.

Tests.—Pure chloride of barium must not alter vegetable colors; its solution must not be colored or precipitated by hydrosulphuric acid, nor by sulphide of ammonium. Pure sulphuric acid must precipitate every fixed particle from it, so that the fluid filtered from the precipitate formed upon the addition of that reagent leaves not the slightest residue when evaporated on platinum foil.

Uses.—Baryta forms with many acids soluble, with others insoluble compounds. This property of baryta affords us therefore a means of distinguishing the former acids, which are not precipitated by chloride of barium, from the latter, in the solution of the salts of which this reagent produces a precipitate. The precipitated salts of baryta severally show with other bodies (acids) a different deportment. By subjecting these salts to the action of such bodies we are therefore enabled to subdivide the group of precipitable acids and even to detect certain individual acids. This makes chloride of barium one of our most important reagents to distinguish between certain groups of acids, and more especially also to effect the detection of sulphuric acid.

§ 59.

2. NITRATE OF BARYTA (BaO, NO$_5$).

Preparation.—Treat carbonate of baryta, no matter whether Witherite or precipitated by carbonate of soda from solution of sulphide of barium, with dilute nitric acid free from chlorine, and proceed exactly as directed in the preparation of chloride of barium from Witherite. For use dissolve 1 part of the salt in 15 parts of water.

Tests.—Solution of nitrate of baryta must not be made turbid by solution of nitrate of silver. Other tests the same as for chloride of barium.

Uses.—Nitrate of baryta is used instead of chloride of barium in cases where it is desirable to avoid the presence of a metallic chloride in the fluid.

§ 60.

3. Carbonate of Baryta (BaO, CO_2).

Preparation.—Dissolve crystallized chloride of barium in water, heat to boiling, and add a solution of carbonate of ammonia mixed with some caustic ammonia, or of pure carbonate of soda, as long as a precipitate forms; let the precipitation subside, decant five or six times, transfer the precipitate to a filter, and wash until the washing water is no longer rendered turbid by solution of nitrate of silver. Stir the precipitate with water to the consistence of thick milk, and keep this mixture in a stoppered bottle. It must of course be shaken every time it is required for use.

Tests.—Pure sulphuric acid must precipitate every fixed particle from a solution of carbonate of baryta in hydrochloric acid (compare § 34, caustic baryta).

Uses.—Carbonate of baryta completely decomposes the solutions of many metallic oxides, *e.g.* sesquioxide of iron, alumina; precipitating from them the whole of the oxide as hydrate and basic salt, whilst some other metallic salts are not precipitated by it. It serves therefore to separate the former from the latter, and affords an excellent means of effecting the separation of sesquioxide of iron and alumina from protoxide of manganese, oxide of zinc, lime, magnesia, &c. It must be borne in mind, however, that the salts must not be sulphates, as carbonate of baryta equally precipitates the latter bases from these compounds.

§ 61.

4. Sulphate of Lime (CaO, SO_3, *crystallized* $CaO, SO_3 + 2$ aq.).

Preparation.—Digest and shake powdered crystallized gypsum for some time with water; let the undissolved portion subside, decant, and keep the clear fluid for use.

Uses.—Sulphate of lime, being a difficultly soluble salt, is a convenient agent in cases where it is wished to apply a solution of a lime salt or of a sulphate of a definite degree of dilution. As dilute solution of a lime salt it is used for the detection of oxalic acid; whilst as dilute solution of a sulphate it affords an excellent means of distinguishing between baryta, strontia, and lime.

§ 62.

5. Chloride of Calcium ($CaCl$, *crystallized* $CaCl + 6$ aq.).

Preparation.—Dilute 1 part of crude hydrochloric acid with 6 parts of water, and add to the fluid marble or chalk until the last portion added remains undissolved; add now some hydrate of lime, then sulphuretted hydrogen water until a filtered portion of the mixture is no longer altered by sulphide of ammonium. Then let the mixture stand 12 hours in a gentle heat; filter, exactly neutralize the filtrate, concentrate by evaporation, and crystallize. Let the crystals drain, and dissolve 1 part of the salt in 5 parts of water for use.

Tests.—Solution of chloride of calcium must be perfectly neutral, and neither be colored nor precipitated by sulphide of ammonium; nor ought it to evolve ammonia when mixed with hydrate of potassa or hydrate of lime.

Uses.—Chloride of calcium is in its action and application analogous to chloride of barium. For as the latter reagent is used to separate the *inorganic* acids into groups, so chloride of calcium serves in the same manner to effect the separation of the *organic* acids into groups, since it precipitates some of them, whilst it forms soluble compounds with others. And, as is the case with the baryta precipitates, the different conditions under which the various insoluble lime salts are thrown down enable us to subdivide the group of precipitable acids, and even to detect certain individual acids.

§ 63.

6. SULPHATE OF MAGNESIA (MgO, SO_3, *crystallized* MgO, SO_3, $HO + 6$ aq.).

Preparation.—Dissolve 1 part of sulphate of magnesia of commerce in 10 parts of water; if the salt is not perfectly pure, subject it to recrystallization.

Tests.—Sulphate of magnesia must have a neutral reaction. Its solution, when mixed with a sufficient quantity of chloride of ammonium, must, after the lapse of half an hour, not appear clouded or tinged by pure ammonia, or by carbonate or oxalate of ammonia, or by sulphide of ammonium.

Uses.—Sulphate of magnesia serves almost exclusively for the detection of phosphoric acid and arsenic acid, which it precipitates from aqueous solutions of phosphates and arsenates, in presence of ammonia and chloride of ammonium, in the form of almost absolutely insoluble highly characteristic double salts (basic phosphate or basic arsenate of magnesia and ammonia). Sulphate of magnesia is also employed to test the purity of sulphide of ammonium (see § 40).

c. SALTS OF THE OXIDES OF THE HEAVY METALS.

§ 64.

1. SULPHATE OF PROTOXIDE OF IRON (FeO, SO_3, *crystallized* $FeO, SO_3, HO + 6$ aq.).

Preparation.—Heat an excess of iron nails free from rust, or of clean iron wire, with dilute sulphuric acid until the evolution of hydrogen ceases; filter the sufficiently concentrated solution, add a few drops of dilute sulphuric acid to the filtrate, and allow it to cool. Wash the crystals with water very slightly acidulated with sulphuric acid, dry, and keep for use. The sulphate of protoxide of iron may also be prepared from the solution of sulphide of iron in dilute sulphuric acid which is obtained in the process of evolving hydrosulphuric acid.

Tests.—The crystals of sulphate of protoxide of iron must have a fine pale green color. Crystals that have been more or less oxidized by the action of the air, and give a brownish-yellow solution when treated with water, leaving undissolved basic sulphate of sesquioxide of iron behind, must be altogether rejected. Hydrosulphuric acid must not precipitate solution of sulphate of protoxide of iron after addition of some hydrochloric acid, nor even impart a blackish tint to it.

Uses.—Sulphate of protoxide of iron has a great disposition to absorb oxygen, and to be converted into the sulphate of the sesquioxide. It acts therefore as a powerful reducing agent. We employ it principally for the reduction of nitric acid, from which it separates nitric oxide by withdrawing three atoms of oxygen from it. The decomposition of the nitric acid being attended in this case with the formation of a very peculiar brownish-black compound of nitric oxide with an undecomposed portion of the salt of the protoxide of iron, this reaction affords a particularly characteristic and delicate test for the detection of nitric acid. Sulphate of protoxide of iron serves also for the detection of hydroferricyanic acid, with which it produces a kind of Prussian blue, and also to effect the precipitation of metallic gold from solutions of the salts of that metal.

§ 65.

2. Sesquichloride of Iron (Fe_2Cl_3).

Preparation.—Heat in a flask a mixture of 10 parts of water and 1 part of pure hydrochloric acid with small iron nails until no further evolution of hydrogen is observed, even after adding the nails in excess; filter the solution into another flask, and conduct into it chlorine gas, with frequent shaking, until the fluid no longer produces a blue precipitate in solution of ferricyanide of potassium. Heat until the excess of chlorine is expelled. Dilute until the fluid is twenty times the weight of the iron dissolved, and keep the dilute fluid for use.

Tests.—Solution of sesquichloride of iron must not contain an excess of acid; this may be readily ascertained by stirring a sample of it with a glass rod dipped in ammonia, when the absence of any excess of acid will be proved by the formation of a precipitate which shaking the vessel or agitating the fluid fails to redissolve. Ferricyanide of potassium must not impart a blue color to it.

Uses.—Sesquichloride of iron serves to subdivide the group of organic acids which chloride of calcium fails to precipitate, as it produces precipitates in solutions of benzoates and succinates, but not in solutions of acetates and formates. The aqueous solutions of the neutral acetate and formate of sesquioxide of iron exhibit an intensely red color; sesquichloride of iron is therefore a useful agent for detecting acetic acid and formic acid. Sesquichloride of iron is exceedingly well adapted to effect the decomposition of phosphates of the alkaline earths (see § 142). It serves also for the detection of hydroferrocyanic acid, with which it produces Prussian blue.

§ 66.

3. Nitrate of Silver (AgO, NO_5).

Preparation.—Dissolve pure silver in pure nitric acid, evaporate the solution to dryness, and dissolve 1 part of the salt in 20 parts of water.

Tests.—Dilute hydrochloric acid must completely precipitate all fixed particles from solution of nitrate of silver, which should have a neutral reaction; the fluid filtered from the precipitated chloride of silver must accordingly leave no residue when evaporated on a watch-glass, and must be neither precipitated nor colored by hydrosulphuric acid.

Uses.—Oxide of silver forms with many acids soluble, with others insoluble compounds. Nitrate of silver may therefore serve, like chloride of barium, to effect the separation and arrangement of acids into groups.

Most of the insoluble compounds of silver dissolve in dilute nitric acid; chloride, bromide, iodide, and cyanide, ferrocyanide, ferricyanide, and sulphide of silver are insoluble in that menstruum. Nitrate of silver is therefore a most excellent agent to distinguish and separate from all other acids the hydracids corresponding to the last enumerated compounds of silver. Many of the insoluble salts of silver exhibit a peculiar color (chromate of silver, arsenate of silver), or manifest a characteristic deportment with other reagents or upon the application of heat (formate of silver); nitrate of silver is therefore an important agent for the positive detection of certain acids.

§ 67.

4. Acetate of Lead ($Pb O, \overline{A}$, *crystallized* $Pb O, \overline{A} + 3$ aq.).

The best acetate of lead of commerce is sufficiently pure for the purpose of chemical analysis; for use dissolve 1 part of the salt in 10 parts of water.

Tests.—Sugar of lead must completely dissolve in water acidified with one or two drops of acetic acid; the solution must be quite clear and colorless; hydrosulphuric acid must throw down all fixed particles from it. On mixing the solution of sugar of lead with carbonate of ammonia in excess, and filtering the mixture, the filtrate must not show a bluish tint (copper).

Uses.—Oxide of lead forms with a great many acids compounds insoluble in water, which are marked either by peculiarity of color or characteristic deportment. The acetate of lead therefore produces precipitates in the solutions of these acids or of their salts, and essentially contributes to the detection of several of them. Thus chromate of lead, for instance, is characterized by its yellow color, phosphate of lead by its peculiar deportment before the blowpipe, and malate of lead by its ready fusibility.

§ 68.

5. Nitrate of Suboxide of Mercury ($Hg_2 O, NO_5$, *crystallized* $Hg_2 O, NO_5 + 2$ aq.).

Preparation.—Pour 1 part of pure nitric acid of 1·2 spec. gr. on 1 part of mercury in a porcelain dish, and let the vessel stand twenty-four hours in a cool place; separate the crystals formed from the undissolved mercury and the mother liquor, and dissolve them in water mixed with one-sixteenth part of nitric acid, by trituration in a mortar. Filter the solution, and keep the filtrate in a bottle with some metallic mercury covering the bottom of the vessel.

Tests.—The solution of nitrate of suboxide of mercury must give with dilute hydrochloric acid a copious white precipitate of subchloride of mercury; hydrosulphuric acid must produce no precipitate in the fluid filtered from this, or at all events only a trifling black precipitate (sulphide of mercury).

Uses.—Nitrate of suboxide of mercury acts in an analogous manner to the corresponding salt of silver. In the first place, it precipitates many acids, especially the hydracids; and, in the second place, it serves for the detection of several readily oxidizable bodies, *e.g.* of formic acid, as the oxidation of such bodies, which takes place at the expense of the oxygen of the suboxide of mercury, is attended with the highly characteristic separation of metallic mercury.

§ 69.

6. CHLORIDE OF MERCURY ($HgCl$).

The chloride of mercury of commerce is sufficiently pure for the purposes of chemical analysis. For use dissolve 1 part of the salt in 16 parts of water.

Uses.—Chloride of mercury gives with several acids, *e.g.* with hydriodic acid, peculiarly colored precipitates, and may accordingly be used for the detection of these acids. It is an important agent for the detection of tin where that metal is in solution in the state of protochloride; if only the smallest quantity of that compound is present the addition of chloride of mercury in excess to the solution is followed by separation of subchloride of mercury insoluble in water. In a similar manner chloride of mercury serves also for the detection of formic acid.

§ 70.

7. SULPHATE OF COPPER (CuO, SO_3, *crystallized* $CuO, SO_3, HO + 4$ aq.).

Preparation.—This reagent may be obtained in a state of great purity from the residue remaining in the retort in the process of preparing bisulphite of soda (§ 48), by treating that residue with water, applying heat, filtering, letting the filtrate crystallize, and purifying the salt by recrystallization. For use dissolve 1 part of the pure crystals in 10 parts of water.

Tests.—Pure sulphate of copper must be completely precipitated from its solutions by hydrosulphuric acid; ammonia and sulphide of ammonium must accordingly leave the filtrate unaltered.

Uses.—Sulphate of copper is employed in qualitative analysis to effect the precipitation of hydriodic acid in the form of subiodide of copper For this purpose it is necessary to mix the solution of 1 part of sulphate of copper with $2\frac{1}{2}$ parts of sulphate of protoxide of iron, otherwise half of the iodine will separate in the free state. The protoxide of iron changes in this process to sesquioxide, at the expense of the oxygen of the oxide of copper, which latter is thus reduced to the state of suboxide. Sulphate of copper is used also for the detection of arsenious and arsenic acids; it serves likewise as a test for the soluble ferrocyanides.

§ 71.

8. PROTOCHLORIDE OF TIN (SnCl, *crystallized* SnCl + 2 aq.).

Preparation.—Reduce English tin to powder by means of a file, or by fusing it in a small porcelain dish, removing from the fire, and triturating the fused liquid mass with a pestle until it has passed again to the solid state. Boil the powder for some time with concentrated hydrochloric acid in a flask (taking care always to have an excess of tin in the vessel) until no more hydrogen gas is evolved; dilute the solution with 4 times the quantity of water slightly acidulated with hydrochloric acid, and filter. Keep the filtrate for use in a well-stoppered bottle containing small pieces of metallic tin, or some pure tin-foil. If these precautions are neglected the protochloride will soon change to bichloride, with separation of white oxychloride, which will of course render the reagent totally unfit for the purpose for which it is intended.

Tests.—Solution of protochloride of tin must, when added to a solution of chloride of mercury, immediately produce a white precipitate of subchloride of mercury; when treated with hydrosulphuric acid it must give a dark brown precipitate; it must not be precipitated nor rendered turbid by sulphuric acid.

Uses.—The great tendency of protochloride of tin to absorb oxygen, and thus to form binoxide, or rather bichloride—as the binoxide in the moment of its formation decomposes with the free hydrochloric acid present—makes this substance one of our most powerful reducing agents. It is more particularly suited to withdraw part or the whole of the chlorine from chlorides. We employ it in the course of analysis as a test for mercury; also to effect the detection of gold.

§ 72.

9. BICHLORIDE OF PLATINUM (PtCl$_2$, *crystallized* PtCl$_2$ + 10 aq.).

Preparation.—Treat platinum filings, purified by boiling with nitric acid, with concentrated hydrochloric acid and some nitric acid in a narrow-necked flask, and apply a very gentle heat, adding occasionally fresh portions of nitric acid, until the platinum is completely dissolved. Evaporate the solution on the water-bath, with addition of hydrochloric acid, and dissolve the semifluid residue in 10 parts of water for use.

Tests.—Bichloride of platinum must, upon evaporation to dryness in the water-bath, leave a residue which dissolves completely in spirit of wine.

Uses.—Bichloride of platinum forms very sparingly soluble double salts with chloride of potassium and chloride of ammonium, but not so with chloride of sodium; it serves therefore to detect ammonia and potassa, and may, indeed, be looked upon as our most delicate reagent for the latter substance.

§ 73.

10. SODIO-PROTOCHLORIDE OF PALLADIUM (NaCl, PdCl).

Dissolve 5 parts of palladium in nitrohydrochloric acid (comp. § 72), add 6 parts of pure chloride of sodium, evaporate in the water-bath to dryness, and dissolve 1 part of the residuary double salt in 12 parts of water for use. The brownish solution affords an excellent means for detecting and separating iodine.

§ 74.

11. TERCHLORIDE OF GOLD $(AuCl_3)$.

Preparation.—Take fine shreds of gold, which may be alloyed with silver or copper, treat them in a flask with nitrohydrochloric acid in excess, and apply a gentle heat until no more of the metal dissolves. If the gold was alloyed with copper—which is known by the brownish-red precipitate produced by ferrocyanide of potassium in a portion of the solution diluted with water—mix it with solution of sulphate of protoxide of iron in excess. This will reduce the terchloride to metallic gold, which will separate in the form of a fine brownish-black powder; wash the powder in a small flask, and redissolve it in nitrohydrochloric acid; evaporate the solution to dryness on the water-bath, and dissolve the residue in 30 parts of water. If the gold was alloyed with silver, the latter metal remains as chloride upon treating the alloy with nitrohydrochloric acid. In that case evaporate the solution at once to dryness, and dissolve the residue in water for use.

Uses.—Terchloride of gold has a great tendency to yield up its chlorine; it therefore readily converts protochlorides into higher chlorides, protoxides, with the co-operation of water, into higher oxides. These peroxidations are usually indicated by the precipitation of pure metallic gold in the form of a brownish-black powder. In the course of analysis this reagent is used only for the detection of protoxide of tin, in the solutions of which it produces a purple color or a purple precipitate.

V. COLORING MATTERS AND INDIFFERENT VEGETABLE SUBSTANCES.

§ 75.

1. TEST PAPERS.

a. BLUE LITMUS PAPER.

Preparation.—Digest 1 part of litmus of commerce with 6 parts of water, and filter the solution; divide the intensely blue filtrate into 2 equal parts; saturate the free alkali in the one half by repeatedly stirring with a glass rod dipped in very dilute sulphuric acid, until the color of the fluid just appears red; add now the other half of the blue filtrate, pour the whole fluid into a dish, and draw slips of fine unsized paper through it; suspend these slips over threads, and leave them to dry. The color of litmus paper must be perfectly uniform and neither too light nor too dark. The paper must be readily wetted by aqueous fluids.

Uses.—Litmus paper serves to detect the presence of free acid in fluids, as acids change its blue color to red. It must be borne in mind, however, that the soluble neutral salts of most of the heavy metallic oxides produce the same effect.

β. Reddened Litmus Paper.

Preparation.—Stir blue solution of litmus with a glass rod dipped in dilute sulphuric acid, and repeat this process until the fluid has just turned distinctly red. Steep slips of paper in the solution, and dry them as in α. The dried slips must look distinctly red.

Uses.—Pure alkalies and alkaline earths, and also the sulphides of their metals, restore the blue color of reddened litmus paper; carbonates of the alkalies and the soluble salts of several other weak acids, especially of boracic acid, possess the same property. This reagent serves therefore for the detection of these bodies in general.*

γ. Georgina Paper (*Dahlia Paper*).

Preparation.—Boil the violet-colored petals of *Georgina purpurea* (purple dahlia) in water, or digest them with spirit of wine, and steep slips of paper in the tincture obtained. The latter should be neither more nor less concentrated than is necessary to make the paper when dry again appear of a fine and light violet blue color. Should the color too much incline to red this may be remedied by adding a very little ammonia to the tincture.

Uses.—Georgina paper is reddened by acids, whilst alkalies impart a beautiful green tint to it. It is therefore an extremely convenient substitute both for the blue and the reddened litmus paper. This reagent, if properly prepared, is a most delicate test both for acids and alkalies. Concentrated solutions of caustic alkalies turn Georgina paper yellow by destroying the coloring matter.

δ. Turmeric Paper.

Preparation.—Digest and heat 1 part of bruised turmeric root with 6 parts of weak spirit of wine, filter the tincture obtained, and steep slips of fine paper in the filtrate. The dried slips must exhibit a fine yellow tint; they must be readily wetted by aqueous fluids.

Uses.—Turmeric paper serves the same as reddened litmus paper and dahlia paper for the detection of free alkalies, &c., as they change its yellow color to brown. It is not quite so delicate a test as the other reagent papers; but the change of color which it produces is highly characteristic, and is very distinctly perceptible in many *colored* fluids; we therefore cannot well dispense with this paper. When testing with turmeric paper it is to be borne in mind that, besides the substances enumerated in β, several other bodies (boracic acid, for instance) possess the property of turning its yellow color to brown-red, more especially on drying. It affords an excellent means for the detection of boracic acid.

All test papers are cut into slips, which are kept in small well-closed boxes, or in bottles covered with black paper, as continued action of light destroys the color.

---

* Mr. A. S. Taylor has suggested that a very delicate test paper for detecting alkalies may be prepared by steeping slips of paper in an acid infusion of rose petals.

§ 76.

2. SOLUTION OF INDIGO.

Preparation.—Take from 4 to 6 parts of fuming sulphuric acid, add slowly and in small portions at a time 1 part of finely pulverized indigo, taking care to keep the mixture well stirred. The acid has at first imparted to it a brownish tint by the matter which the indigo contains in admixture, but it subsequently turns deep blue. Elevation of temperature to any considerable extent must be avoided, as part of the indigo blue is thereby destroyed; it is therefore advisable when dissolving larger quantities of the substance to place the vessel in cold water. When the whole of the indigo has been added to the acid, cover the vessel, let it stand forty-eight hours, then pour its contents into 20 times the quantity of water, mix, filter, and keep the filtrate for use.

Uses.—Indigo is decomposed by boiling with nitric acid, yellow-colored oxidation products being formed. It serves therefore for the detection of nitric acid. Solution of indigo is also well adapted to effect the detection of chloric acid and of free chlorine.

B. REAGENTS IN THE DRY WAY.

I. FLUXES AND DECOMPOSING AGENTS.

§ 77.

1. MIXTURE OF CARBONATE OF SODA AND CARBONATE OF POTASSA. ($NaO, CO_2 + KO, CO_2$).

Preparation.—Digest 10 parts of purified bitartrate of potassa in powder with 10 parts of water and 1 part of hydrochloric acid for several hours on the water-bath, with frequent stirring; put the mass into a funnel with a small filter inserted into the pointed end; let it drain; cover with a disc of rather difficultly permeable filtering paper with upturned edges, and wash by repeatedly pouring upon this small quantities of cold water; continue this washing process until the fluid running off is no longer rendered turbid by solution of nitrate of silver, after addition of nitric acid. Dry the bitartrate of potassa freed in this manner from lime (and phosphoric acid). It is now necessary to prepare pure nitrate of potassa. To effect this, dissolve nitrate of potassa of commerce in half its weight of boiling water, filter the solution into a porcelain or stoneware dish, using a hot funnel, and stir it well with a clean wooden or porcelain spatula until cold. Transfer the crystalline powder to a funnel loosely stopped with cotton, let it drain, press down tight, make even at the top, cover with a double disc of difficultly permeable filtering paper with upturned edges, and pour upon this at proper intervals small portions of water until the washings are no longer made turbid by solution of nitrate of silver. Empty now the contents of the funnel into a porcelain dish, dry in this vessel, and reduce the mass to a fine powder by trituration.

Mix now 2 parts of the pure bitartrate of potassa with 1 part of the pure nitrate of potassa; project the perfectly dry mixture in small portions at a time into a clean-scoured cast-iron pot heated to gentle redness; when the mixture has deflagrated heat strongly until a sample taken from the edges gives with water a perfectly colorless solution. Triturate the charred mass with water, filter, wash slightly, and evaporate the filtrate in a porcelain or, better still, in a silver dish, until the fluid is covered with a persistent pellicle. Let the mixture now cool, with constant stirring; put the crystals of carbonate of potassa on a funnel, let them well drain, wash slightly, dry thoroughly in a silver or porcelain dish, and keep the crystals in a well-stoppered bottle. The mother liquor leaves upon evaporation a salt which, though containing traces of alumina and silicic acid, may still be turned to account for many purposes.

Mix 13 parts of the pure carbonate of potassa prepared in the manner just now described with 10 parts of pure anhydrous carbonate of soda, and keep the mixture in a well-stoppered bottle. The mixture of carbonate of potassa and carbonate of soda may also be prepared by deflagrating 20 parts of pure bitartrate of potassa with 9 parts of pure nitrate of soda, treating with water, and evaporating the solution to dryness. Or by igniting pure tartrate of potassa and soda, extracting the carbonaceous mass with water, and evaporating the clear solution to dryness.

Tests.—The purity of the mixed salt is tested as directed § 46 (*carbonate of soda*). To detect any trace of cyanide of potassium that may happen to be present, add to the solution of the salt a little of a solution of proto-sesquioxide of iron, then hydrochloric acid in excess, when the bluish-green coloration of the fluid and the formation of a blue precipitate after a time will indicate the presence of cyanide of potassium.

Uses.—If silicic acid or a silicate is fused with about 4 parts (consequently with an excess) of carbonate of potassa or soda, carbonic acid escapes with effervescence, and a basic alkaline silicate is formed, which, being soluble in water, may be readily separated from such metallic oxides as it may contain in admixture; from this basic alkaline silicate hydrochloric acid separates the silicic acid as hydrate. If a fixed alkaline carbonate is fused together with sulphate of baryta, strontia, or lime, there are formed carbonates of the alkaline earths and sulphate of the alkali, in which new compounds both the base and the acid of the originally insoluble salt may now be readily detected. However, we do not use either carbonate of potassa or carbonate of soda separately to effect the decomposition of the insoluble silicates and sulphates; but we apply for this purpose the above-described mixture of both, because this mixture requires a far lower degree of heat for fusion than either of its two components, and thus enables us to conduct the operation over a *Berzelius* lamp, or over a simple gas-lamp. The fusion with alkaline carbonates is invariably effected in a platinum crucible, provided no reducible metallic oxides be present.

§ 78.

2. Hydrate of Baryta (BaO, HO).

Preparation.—The crystals of baryta prepared in the manner directed § 34 are heated gently in a silver or platinum dish, until the water of *crystallization* is completely expelled. The residuary white mass is pulverized, and kept for use in a well-closed bottle.

Uses.—Hydrate of baryta fuses at a gentle red heat without losing its water. Upon fusing silicates which resist decomposition by acids together with about 4 times their weight of hydrate of baryta, basic silicates are formed which acids will decompose. If, therefore, the fused mass is treated with water and hydrochloric acid, the solution evaporated to dryness, and the residue digested with dilute hydrochloric acid, the silicic acid is left behind, and the oxides are obtained in solution in the form of chlorides. We use hydrate of baryta as a flux when we wish to test silicates for alkalies. This reagent is preferable as a flux to the carbonate or nitrate of baryta, since it does not require a very high temperature for its fusion, as is the case with the carbonate, nor does it cause any spirting in the fusing mass, arising from disengagement of gas, as is the case with the nitrate. The operation of fluxing with hydrate of baryta is conducted in silver or platinum crucibles.

§ 79.

3. Fluoride of Calcium (CaFl).

Take fluor-spar as pure as can be procured and, more particularly, free from alkalies, reduce to fine powder, and keep this for use.

Uses.—Fluoride of calcium applied in conjunction with sulphuric acid serves to effect the decomposition of silicates insoluble in acids, and more especially to detect the alkalies which they contain. Compare Section III. *Silicic acid,* § 150.

§ 80.

4. Nitrate of Soda (NaO, NO_5).

Preparation.—Neutralize pure nitric acid with pure carbonate of soda exactly, and evaporate to crystallization. Dry the crystals thoroughly, triturate, and keep the powder for use.

Tests.—A solution of nitrate of soda must not be made turbid by solution of nitrate of silver or nitrate of baryta, nor precipitated by carbonate of soda.

Uses.—Nitrate of soda serves as a very powerful oxidizing agent, by yielding oxygen to combustible substances when heated with them. We use this reagent principally to convert several metallic sulphides, and more particularly the sulphides of tin, antimony, and arsenic, into oxides and acids; also to effect the rapid and complete combustion of organic substances; for the latter purpose, however, nitrate of ammonia is in many cases preferable; which latter reagent is prepared by saturating nitric acid with carbonate of ammonia.

II. Blowpipe Reagents.

§ 81.

1. Carbonate of Soda (NaO, CO_2).

Preparation.—See § 46.

Uses.—Carbonate of soda serves, in the first place, to promote the reduction of oxidized substances in the inner flame of the blowpipe. In fusing it brings the oxides into the most intimate contact with the charcoal support, and enables the flame to embrace every part of the substance under examination. It co-operates in this process also chemically by the transposition of its constituents (according to *R. Wagner,* in consequence of the formation of cyanide of sodium). Where the quantity operated upon was very minute the reduced metal is often found in the pores of the charcoal. In such cases the parts surrounding the little cavity which contained the sample are dug out with a knife, and triturated in a small mortar; the charcoal is then washed off from the metallic particles, which now become visible either in the form of powder or as small flat spangles, according to the nature of the particular metal or metals present.

Carbonate of soda serves, in the second place, as a solvent. Platinum wire is the most convenient support for testing the solubility of substances in fusing carbonate of soda. A few only of the bases dissolve in fusing carbonate of soda, but acids dissolve in it with facility. Carbonate of soda is also applied as a decomposing agent and flux, and more particularly to effect the decomposition of the insoluble sulphates, with which it exchanges acids, the newly formed sulphate of soda being reduced at the same time to sulphide of sodium; and to effect the decomposition of sulphide of arsenic, with which it forms a double sulphide of arsenic and sodium, and arsenite or arsenate of soda, thus converting it to a state which permits its subsequent reduction by hydrogen. Carbonate of soda also is the most sensitive reagent in the dry way for the detection of manganese, as it produces when fused in the outer flame of the blowpipe together with a substance containing manganese a green opaque bead, owing to the formation of manganate of soda.

§ 82.

2. CYANIDE OF POTASSIUM (K Cy).

Preparation.—See § 54.

Uses.—Cyanide of potassium is an exceedingly powerful reducing agent in the dry way; indeed it excels in its action almost all other reagents of the same class, and separates the metals not only from most oxygen compounds, but also from many sulphur compounds. This reduction is attended in the former case with formation of cyanate of potassa, by the absorption of oxygen, and in the latter case with formation of sulphocyanide of potassium, by the taking up of sulphur. By means of this reagent we may effect the reduction of metals from their compounds with the greatest possible facility; thus we may, for instance, produce metallic antimony from antimonious acid or from sulphide of antimony, metallic iron from sesquioxide of iron, &c. The readiness with which cyanide of potassium enters into fusion facilitates the reduction of the metals greatly; the process may usually be conducted even in a porcelain crucible over a spirit lamp. Cyanide of potassium is a most valuable and important agent to effect the reduction of binoxide of tin, antimonic acid, and more particularly of tersulphide of arsenic (see § 132). Cyanide of potassium is equally important as a blowpipe reagent. Its action is exceedingly energetic; substances like binoxide of tin, the reduction of which by means of carbonate of soda requires a tolerably strong flame, are reduced by cyanide of potassium with the greatest facility. In blowpipe experiments we invariably use a mixture of equal parts of carbonate of soda and cyanide of potassium; the admixture of carbonate of soda is intended here to check in some measure the excessive fusibility of the cyanide of potassium. This mixture of cyanide of potassium with carbonate of soda, besides being a far more powerful reducing agent than the simple carbonate of soda, has moreover this great advantage over the latter, that it is absorbed by the pores of the charcoal with extreme facility, and thus permits the production of the metallic globules in a state of the greatest purity.

§ 83.

3. BIBORATE OF SODA (*Borax*) ($NaO, 2BO_3$, *crystallized* + 10 aq.).

The purity of commercial borax may be tested by adding to its solution carbonate of soda or, after previous addition of nitric acid, solution of nitrate of baryta or of nitrate of silver. The borax may be considered pure if these reagents fail to produce any alteration in the solution; but if either of them causes the formation of a precipitate, or renders the fluid turbid, recrystallization is necessary. The pure crystallized borax is exposed to a gentle heat, in a platinum crucible, until it ceases to swell; it is then left to cool, and afterwards pulverized and kept for use.

Uses.—Boracic acid manifests a great affinity for oxides when brought into contact with them in a state of fusion. This affinity enables it, in the first place, to combine directly with oxides; secondly, to expel weaker acids from their salts; and, thirdly, to predispose metals, sulphides, and haloid compounds to oxidize in the outer flame of the blowpipe, that it may combine with the oxides. Most of the thus produced borates fuse readily, even without the aid of a flux, but far more so in conjunction with borate of soda; the latter salt acts in this operation either as a mere flux, or by the formation of double salts. Now in the biborate of soda we have both free boracic acid and borate of soda; the union of these two substances renders it one of our most important blowpipe reagents. In the process of fluxing with borax we usually select platinum wire for a support; the loop of the wire is moistened or heated to redness, then dipped into the powder, and exposed to the outer flame; a colorless bead of fused borax is thus produced. A small portion of the substance under examination is then attached to the bead, by bringing the latter into contact with it whilst still hot or having previously moistened it. The bead with the sample of the substance intended for analysis adhering to it is now exposed to the blowpipe flame, and the reactions to the manifestation of which this process gives rise are carefully observed and examined. The following points ought to be more particularly watched :—(1) Whether or not the sample under examination dissolves to a transparent bead, and whether or not the bead retains its transparency on cooling; (2) whether the bead exhibits a distinct color, which in many cases at once clearly indicates the individual metal which the analysed compound contains; as is the case, for instance, with cobalt; and (3) whether the bead manifests the same or a different deportment in the outer and in the inner flame. Reactions of the latter kind arise from the ensuing reduction of higher to lower oxides, or even to the metallic state, and are for some substances particularly characteristic.

§ 84.

4. PHOSPHATE OF SODA AND AMMONIA (*Microcosmic Salt*) (NaO, NH_4O, HO, PO_5, *crystallized* + 8 aq.).

Preparation.—*a.* Heat to boiling 6 parts of phosphate of soda and 1 part of pure chloride of ammonium with 2 parts of water, and let the solution cool. Free the crystals produced of the double phosphate of soda and ammonia from the chloride of sodium which adheres to them by recrystallization, with addition of some solution of ammonia. Dry the purified crystals, pulverize, and keep for use.

b. Take 2 equal parts of pure tribasic phosphoric acid, and add solution of soda to the one, solution of ammonia to the other, until both fluids have a distinct alkaline reaction; mix the two together, and let the mixture crystallize.

Tests.—Phosphate of soda and ammonia dissolves in water to a fluid with feebly alkaline reaction. The yellow precipitate produced in this fluid by nitrate of silver must completely dissolve in nitric acid. Upon fusion on a platinum wire, microcosmic salt must give a clear and colorless bead.

Uses.—On heating phosphate of soda and ammonia the ammonia escapes with the water of crystallization, leaving acid pyrophosphate of soda (NaO, HO, PO_5); upon heating more strongly the last equivalent of water escapes likewise, and readily fusible metaphosphate of soda (NaO, PO_5) is left behind. The action of microcosmic salt is quite analogous to that of biborate of soda. We prefer it, however, in some cases to borax as a solvent or flux, the beads which it forms with many substances being more beautifully and distinctly colored than those of borax. Platinum wire is also used for a support in the process of fluxing with microcosmic salt; the loop must be made small and narrow, otherwise the bead will not adhere to it. The operation is conducted as directed in the preceding paragraph.

§ 85.

5. NITRATE OF PROTOXIDE OF COBALT (CoO, NO_5, *crystallized* + 5 aq.).

Preparation.—Fuse in a Hessian crucible 3 parts of bisulphate of potassa, and add to the fused mass, in small portions at a time, 1 part of well-roasted cobalt ore (the purest zaffre you can procure) reduced to fine powder. The mass thickens, and acquires a pasty consistence. Heat now more strongly until it has become more fluid again, and continue to apply heat until the excess of sulphuric acid is *completely* expelled, and the mass accordingly no longer emits white fumes. Remove the fused mass now from the crucible with an iron spoon or spatula, let it cool, and reduce it to powder; boil this with water until the undissolved portion presents a soft mass; then filter the rose-red solution, which is free from arsenic and nickel, and mostly also from iron. Add to the filtrate a small quantity of carbonate of soda, so as to throw down a little carbonate of protoxide of cobalt, boil, and filter. Precipitate the solution, which is now free from iron, boiling with carbonate of soda, wash the precipitate well, and treat it still moist with oxalic acid in excess. Wash the rose-red oxalate of protoxide of cobalt thoroughly, dry, and heat to redness in a glass tube, in a current of hydrogen gas. This decomposes the oxalate into carbonic acid, which escapes, and metallic cobalt, which is left behind. Wash the metal, first with water containing acetic acid, then with pure water, dissolve in dilute nitric acid, treat—if necessary—with hydrosulphuric acid, filter the fluid from the sulphide of copper, &c., which may precipitate, evaporate the solution in the water-bath to dryness, and dissolve 1 part of the residue in 10 parts of water for use.

Tests.—Solution of nitrate of protoxide of cobalt must be free from other metals, and especially also from salts of the alkalies; when precipitated with sulphide of ammonium, and filtered, the filtrate must upon evaporation on platinum leave no fixed residue.

Uses.—Protoxide of cobalt forms upon ignition with certain infusible bodies peculiarly colored compounds, and may accordingly serve for the detection of these bodies (oxide of zinc, alumina, and magnesia; see Section III.).

SECTION III.

REACTIONS, OR DEPORTMENT OF BODIES WITH REAGENTS.

§ 86.

I STATED in my introductory remarks that the operations and experiments of qualitative analysis have for their object the conversion of the *unknown* constituents of any given compound into forms of which we *know* the deportment, relations, and properties, and which will accordingly permit us to draw correct inferences regarding the several constituents of which the analysed compound consists. The greater or less value of such analytical experiments, like that of all other inquiries and investigations, depends upon the greater or less degree of certainty with which they lead to definite results, no matter whether of a positive or negative nature. But as a question does not render us any the wiser if we do not know the language in which the answer is returned, so in like manner will analytical investigations prove unavailing if we do not understand the mode of expression in which the desired information is conveyed to us; in other words, if we do not know how to interpret the phenomena produced by the action of our reagents upon the substance examined.

Before we can therefore proceed to enter upon the practical investigations of analytical chemistry, it is indispensable that we should *really* possess the most perfect knowledge of the deportment, relations, and properties of the new forms into which we intend to convert the substances we wish to analyse. Now this perfect knowledge consists, in the first place, in a clear conception and comprehension of the conditions necessary for the formation of the new compounds and the manifestation of the various reactions; and, in the second place, in a distinct impression of the color, form, and physical properties which characterize the new compound. This section of the work demands therefore not only the most careful and attentive study, but requires moreover that the student should *examine and verify by actual experiment every fact asserted in it.*

The method usually adopted in elementary works on chemistry is to treat of the various substances and their deportment with reagents individually and separately, and to point out their characteristic reactions. I have, however, in the present work deemed it more judicious and better adapted to its elementary character, to arrange those substances which are in many respects analogous into groups, and thus, by comparing their analogies with their differences, to place the latter in the clearest possible light.

A.—Reactions, or Deportment and Properties of the Metallic Oxides and of their Radicals.

§ 87.

Before proceeding to the special study of the several metallic oxides, I give here a general view of the whole of them classified in groups—showing *which* oxides belong to each group. The *grounds* upon which the classification has been arranged will appear from the special consideration of the several groups.

First group—
Potassa, soda, ammonia (oxide of cæsium, oxide of rubidium, lithia).

Second group—
Baryta, strontia, lime, magnesia.

Third group—
Alumina, sesquioxide of chromium (berylla, thoria, zirconia, yttria-earths; oxide of terbium, oxide of erbium; oxides of cerium, lanthanium, didymium, titanium, tantalium, niobium).

Fourth group—
Oxides of *zinc, manganese, nickel, cobalt, iron* (uranium, vanadium, thallium).

Fifth group—
Oxides of *silver, mercury, lead, bismuth, copper, cadmium* (palladium, rhodium, osmium, ruthenium).

Sixth group—
Oxides and acids of *gold, platinum, tin, antimony, arsenic* (iridium, molybdenum, tellurium, tungsten, selenium).

Of these metallic oxides only those printed in *italics* are found distributed extensively and in large quantities in that portion of the earth's crust which is accessible to our investigations; these are therefore most important to chemistry, arts and manufactures, agriculture, pharmacy, &c. &c.; and we shall therefore dwell upon them at greater length. The remainder are more briefly considered in paragraphs printed in smaller type, which may be passed over by the younger class of students of analytical chemistry. The properties and reactions of the metals I have given only in the case of those that are more frequently met with in the metallic state in analytical operations.

§ 88.

FIRST GROUP.

More common oxides of the first group:—Potassa, Soda, Ammonia.
Rarer oxides of the first group:—Oxide of Cæsium, Oxide of Rubidium, Lithia.

Properties of the group.—The alkalies are readily soluble in water, as well in the pure or caustic state as in the form of sulphides, carbonates, and phosphates. (The salts of lithia, however, dissolve with difficulty.) Accordingly the alkalies do not precipitate one another in the pure state, nor as carbonates or phosphates (which in the case of lithia, however, presupposes a higher degree of dilution of the solutions), nor are they precipitated by hydrosulphuric acid under any condition whatever. The solutions of the pure alkalies as well as of their sulphides and carbonates restore the blue color of reddened litmus-paper, and impart an intensely brown tint to turmeric paper.

Special Reactions of the more common Oxides of the first group.

§ 89.

a. POTASSA (KO).

1. POTASSA and its HYDRATE and SALTS are not volatile at a faint red-heat. Potassa and its hydrate deliquesce in the air; the oily liquids formed do not solidify by absorption of carbonic acid.

2. Nearly the whole of the SALTS OF POTASSA are readily soluble in water. Those with colorless acids are colorless. The neutral salts of potassa with strong acids do not alter vegetable colors. Carbonate of potassa crystallizes (in combination with 2 equivalents of water) with difficulty, and deliquesces in the air. Sulphate of potassa is anhydrous and suffers no alteration in the air.

3. *Bichloride of platinum* produces in the neutral and acid solutions of the salts of potassa a yellow crystalline heavy precipitate of BICHLORIDE OF PLATINUM AND CHLORIDE OF POTASSIUM (*potassio-bichloride of platinum*) ($KCl, PtCl_2$). In concentrated solutions this precipitate separates immediately upon the addition of the reagent: in dilute solutions it forms only after some time, often after a *considerable* time. Very dilute solutions are not precipitated by the reagent. The precipitate consists of octahedrons discernible under the microscope. Alkaline solutions must be acidified with hydrochloric acid before the bichloride of platinum is added. The precipitate is difficultly soluble in water; the presence of free acids does not greatly increase its solubility; it is insoluble in alcohol. Bichloride of platinum is therefore a particularly delicate test for salts of potassa dissolved in spirit of wine. The best method of applying this reagent is to evaporate the aqueous solution of the potassa salt with bichloride of platinum nearly to dryness on the water-bath, and to pour a little water over the residue (or, better still, some spirit of wine, provided no substances insoluble in that menstruum be present), when the potassio-bichloride of platinum will be left undissolved. Care must be taken not to confound this double salt with ammonio-bichloride of platinum, which greatly resembles it (see § 91, 4.)

4. *Tartaric acid* produces in neutral or alkaline* solutions of salts of potassa—a white, quickly subsiding, *granular* crystalline precipitate of ACID TARTRATE OF POTASSA ($KO, HOC_4H_4O_{10}$). In concentrated solutions this precipitate separates immediately; in dilute solutions often only after the lapse of some time. Vigorous shaking or stirring of the fluid greatly promotes its formation. Very dilute solutions are not precipitated by this reagent. Free alkalies and free mineral acids dissolve the precipitate; it is sparingly soluble in cold, but pretty readily soluble in hot water. In acid solutions the free acid must, if practicable, first be expelled by evaporation and ignition, or the solution must be neutralized with soda or carbonate of soda, before they can be tested for potassa with tartaric acid.

Acid tartrate of soda answers still better as a test for potassa than free tartaric acid. The reaction is the same in kind, but different in degree, being much more delicate with the salt than with the free acid, since where the former is used the soda salt of the acid combined with the potassa is formed, whereas where free tartaric acid is the test ap-

* To alkaline solutions the reagent must be added until the fluid shows a strongly acid reaction.

plied the hydrate of the acid combined with the potassa is formed, which tends to increase the dissolving action of the water of the menstruum upon the acid tartrate of potassa, and thus to check the separation of the latter ($KO, NO_6 + NaO, HO, C_8H_4O_{10} = KO, HO, C_8H_4O_{10} + NaO, NO_5$).

5. If a potassa salt which is volatile at an intense red heat is held on the loop of a platinum wire in the fusion zone of the flame of *Bunsen's gas-lamp* (§ 14, fig. 17), the salt volatilises, and imparts a *blue violet* tint to the part of the flame above the sample. Chloride of potassium and nitrate of potassa volatilize rapidly, the carbonate and sulphate less rapidly, and the phosphate still more slowly; but they all of them distinctly show the reaction, though decreasing in degree. If it is wished to obtain a more uniform manifestation of the reaction, *i. e.* a manifestation independent of the nature of the acid that may chance to be combined with the potassa, the sample need simply be moistened with sulphuric acid, dried at the border of the flame, and then introduced into the fusion zone. With silicates, and other compounds of potassa of difficult volatility, the reaction may be ensured by fluxing the sample first with pure gypsum, as this serves to form silicate of lime and sulphate of potassa, which latter salt then readily colors the flame. Decrepitating salts are ignited in a platinum spoon before they are attached to the loop.

The sample of the potassa salt under examination may also be held before the apex of the *inner blowpipe flame* produced with a spirit-lamp. Presence of a salt of soda completely obscures the potassa coloration of the flame.

The spectrum of the potassa flame produced by the *spectrum apparatus* (§ 15, fig. 20) is shown on Table I. It contains two characteristic lines, the red line α and the indigo blue line β. If the potassa flame is observed through the *indigo prism* (§ 15, fig. 19) the coloration appears sky-blue, violet, and at last intensely crimson, even through the thickest layers of the solution. Admixtures of lime-, soda-, and lithia-compounds do not alter this reaction, as the yellow rays cannot penetrate the indigo solution, and the rays of the lithia flame also are only able to pass through the thinner layers of that solution, but not through the thicker layers; the exact spot where the penetrating power of the rays of the lithia flame ceases has to be marked by the operator on his indigo prism. But organic substances which impart luminosity to the flame might lead to mistakes, and must therefore, if present, first be removed by combustion. Instead of the indigo prism a blue glass may be used; if lithia is present the glass must be sufficiently thick to keep out the red lithia rays.

6. If a salt of potassa (chloride of potassium answers best) is heated with a small quantity of water, alcohol (burning with colorless flame) added, heated, and then kindled, the flame appears VIOLET. The presence of soda obscures this reaction, which is altogether much less delicate than the one described in 5.

§ 90.

b. SODA (NaO).

1. SODA and its HYDRATE and SALTS present in general the same properties and reactions as potassa and its corresponding compounds. The

oily fluid which soda forms by deliquescing in the air resolidifies speedily by absorption of carbonic acid. Carbonate of soda crystallizes readily; the crystals ($NaO, CO_2 + 10$ aq.) effloresce rapidly when exposed to the air. The same applies to the crystals of sulphate of soda ($NaO, SO_3 + 10$ aq.).

2. If a sufficiently concentrated solution of a soda salt with neutral or alkaline reaction is mixed, for greater convenience, in a watch-glass, with a solution of granular *antimonate of potassa* prepared according to the directions of § 51, the mixture remains clear at first, or appears only slightly colored; but upon rubbing the part of the glass wetted by the fluid with a glass rod, a crystalline precipitate of ANTIMONATE OF SODA ($NaO, SbO_5 + 7$ aq.) speedily separates, which makes its appearance first along the lines rubbed with the rod, and subsides from the fluid as a heavy sandy precipitate. From dilute solutions of soda salts the precipitate separates only after some time, occasionally as much as twelve hours. From very dilute solutions it does not separate at all. The precipitated antimonate of soda is *invariably* crystalline. Where it has separated slowly it occasionally consists of well-formed microscopic cubic octahedrons, but more frequently of four-sided columns tapering pyramid fashion; where it has separated promptly, it appears in the form of small boat-shaped crystals. Presence of larger quantities of salts of potassa interferes very considerably with the reaction. Acid solutions cannot be tested with antimonate of potassa, as free acids will separate from the latter substance hydrated antimonic acid or acid antimonate of potassa. It is indispensable therefore, before adding the reagent, to remove, if possible, the free acid by evaporation or ignition, or where this is not practicable, by neutralizing the acid solution with a little carbonate of potassa until the reaction is feebly alkaline. It should also be borne in mind that only such solutions can be tested with antimonate of potassa which contain no other bases besides soda and potassa.

3. If salts of soda are held in the fusion zone of *Bunsen's* gas-lamp, or in the *inner spirit-blowpipe flame*, they show, with regard to their relative volatility and the action of decomposing agents upon them, a similar deportment to the salts of potassa; the soda salts are, however, a little less volatile than the corresponding potassa salts. But the most characteristic sign of the presence of soda salts is the *intense yellow coloration* which they impart to the flame. This reaction will effect the detection of even the minutest quantities of soda, and is not obscured even by the presence of larger quantities of potassa.

The *soda spectrum* (Table I.) shows only a single yellow line, *a*. The reaction is so exceedingly delicate that the chloride of sodium contained in atmospheric dust generally suffices to give a soda spectrum, although a faint one.

It is characteristic of the soda flame that a crystal of bichromate of potassa appears colorless in its light, and that a slip of paper coated with iodide of mercury appears white with a faint shade of yellow (BUNSEN); also that it looks orange yellow when observed through a *green glass* (MERZ). These reactions are not obscured by presence of salts of potassa, lithia, and lime.

4. If salts of soda (chloride of sodium answers best) are treated as stated in the preceding paragraph on potassa, sub. 6, the *alcohol flame* is colored intensely YELLOW. The presence of a potassa salt does not impair the distinctness of this reaction.

5. *Bichloride of platinum* produces no precipitate in solutions of soda salts. Sodio-bichloride of platinum dissolves readily both in water and in spirit of wine; it crystallizes in rosy prisms.

6. *Tartaric acid* and *acid tartrate of soda* fail to precipitate even concentrated neutral solutions of soda salts.

§ 91.

c. AMMONIA (NH_4O).

1. Anhydrous AMMONIA (NH_3) is gaseous at the common temperature; but we have most frequently to deal with it in its aqueous solution, in which it betrays its presence at once by its penetrating odor. It is expelled from this solution by the application of heat. It may be assumed that the solution contains it as oxide of ammonium (NH_4O) (see § 33).

2. All the SALTS OF AMMONIA are volatile at a high temperature, either with or without decomposition. Most of them are readily soluble in water. The solutions are colorless. The neutral compounds of ammonia with strong acids do not alter vegetable colors.

3. If salts of ammonia are triturated together with *hydrate of lime*, best with the addition of a few drops of water, or are, either in the solid state or in solution, heated with solution of potassa or of soda, the ammonia is liberated in the gaseous state, and betrays its presence—1, by its characteristic ODOR; 2, by its REACTION on moistened TEST-PAPERS; and 3, by giving rise to the formation of WHITE FUMES when any object (*e.g.* a glass rod) moistened with hydrochloric acid, nitric acid, acetic acid, or any of the volatile acids, is brought in contact with it. These fumes arise from the formation of solid ammoniacal salts produced by the contact of the gases in the air. Hydrochloric acid is the most delicate test in this respect; acetic acid, however, admits less readily of a mistake. If the expulsion of the ammonia is effected in a small beaker, best with hydrate of lime, with addition of a very little water, and the beaker is covered with a watch-glass having a slip of moistened turmeric or reddened litmus-paper attached to the centre of the convex side, the reaction will show the presence of even very minute quantities of ammonia; only it is not immediate in such cases, but requires some time for its manifestation. It is promoted and accelerated by application of a gentle heat.

4. *Bichloride of platinum* shows the same deportment with salts of ammonia as with salts of potassa; the yellow precipitate of BICHLORIDE OF PLATINUM AND CHLORIDE OF AMMONIUM ($NH_4Cl, PtCl_2$) is, however, of a somewhat lighter color than potassio-bichloride of platinum. It consists, like the corresponding potassium compound, of octahedrons discernible under the microscope.

5. *Tartaric acid* throws down from most highly concentrated ammonia salt solutions with neutral reaction part of the ammonia as acid tartrate of ammonia ($NH_4O, HO, C_8H_4O_{10}$). Less concentrated solutions are not precipitated. *Acid tartrate of soda* precipitates concentrated solutions more completely, and produces a precipitate even in more dilute solutions. The precipitated acid tartrate of ammonia is white and crystalline. Its separation may be promoted by shaking the glass, or rubbing it inside with a glass rod. By solvents it is acted upon the same as the corresponding potassa salt, only that it is a little more readily soluble in water and in acids.

§ 92.

Recapitulation and remarks.—The salts of potassa and soda are not volatile at a moderate red heat, whilst the salts of ammonia volatilize readily; the latter may therefore be easily separated from the former by ignition. The expulsion of *ammonia* from its compounds by hydrate of lime affords the surest means of ascertaining the presence of this substance. Salts of potassa can be detected in the humid way *positively* only after the removal of the ammoniacal salts which may be present, since both classes of salts manifest the same or a similar deportment with bichloride of platinum and tartaric acid. After the removal of the ammonia the *potassa* is clearly and positively characterized by either of these two reagents. Let it be borne in mind always that the reactions will only show in concentrated fluids, and that dilute solutions must therefore first be concentrated. A single drop of a *concentrated* solution will give a positive result, which cannot be obtained with a large quantity of a dilute fluid. The most simple way of detecting the potassa in the two sparingly soluble compounds that have come under our consideration here—viz., the potassio-bichloride of platinum and the acid tartrate of potassa—is to decompose these salts by ignition; the former thereupon yields the potassa in the form of chloride of potassium, the latter in the form of carbonate of potassa. As regards *soda*, this alkali may be detected with positive certainty in the humid way by antimonate of potassa, provided the reagent be properly prepared and freshly dissolved, and the soda salt solution be concentrated, neutral, or feebly alkaline, and free from other bases, and that it be borne in mind that antimonate of soda invariably separates in the crystalline form, and not in a flocculent state. To detect in this way very minute quantities of soda in presence of a large proportion of potassa, precipitate the latter alkali first with bichloride of platinum, filter, remove the platinum from the filtrate by hydrosulphuric acid (§ 127), filter, evaporate the filtrate to dryness, ignite gently, dissolve the residue in a very little water, and then test the solution finally with antimonate of potassa.

Potassa and soda may be detected much more readily and speedily than in the humid way, and also with far greater delicacy, by the flame coloration. We have seen, indeed, that the soda coloration completely obscures the potassa coloration, even though the potassa salt contains only a trifling admixture of soda salt. But with the aid of the spectrum apparatus the spectra of the two are obtained so distinct and beautiful that a mistake is altogether impossible. And even without a spectrum apparatus the potassa coloration can always be distinctly recognised through the indigo prism, or through a blue glass, even in a flame colored strongly yellow by soda; and the soda coloration again may be placed beyond doubt, if necessary, with the aid of iodide of mercury paper, or green glass, in the manner already described.

Exceedingly minute traces of *ammonia* may be detected by the following test, which was first recommended by J. NESSLER. Dissolve 2 grammes of iodide of potassium in 5 cubic centimetres of water, heat the solution, and add iodide of mercury until the portion last added remains undissolved. Let this mixture cool, then dilute with 20 cubic centimetres of water. Let the fluid stand some time, filter, and mix 20 cubic centimetres of the filtrate with 30 cubic centimetres of a concentrated solution of potassa. Should the fluid turn turbid, filter it once more.

Upon adding to this solution a little of a fluid containing ammonia, or an ammonia salt, a reddish-brown precipitate is formed if the ammonia is present in some quantity; but there is, at any rate, always a yellow coloration produced, even if only *most minute* traces of ammonia are present. The precipitate consists of tetrahydrargyro-iodide of ammonium $(N Hg_4 I, 2 HO): 4 (Hg I, K I) + 3 KO + N H_3 = (N Hg_4 I + 2 HO) + 7 KI + HO$. Application of heat promotes the separation of the precipitate. Presence of chlorides of the alkali metals, or of salts of the alkalies with oxygen acid, does not interfere with the reaction; but presence of cyanide of potassium, and of sulphide of potassium, will prevent it.

§ 93.

Special Reactions of the rarer Oxides of the first group.

1. OXIDE OF CÆSIUM (CsO), and 2. OXIDE OF RUBIDIUM (RbO).

The cæsium and rubidium compounds are, it would appear, found pretty widely disseminated in nature, but always in very minute quantities only. They have hitherto been found chiefly in the mother liquors of mineral waters, and in a few minerals (lepidolite, for instance). The cæsium and rubidium compounds bear in general great resemblance to the potassium compounds, more particularly in this, that their concentrated aqueous solutions are precipitated by *tartaric acid* and by *bichloride of platinum*, and also that those of them that are volatile at a red heat tinge the *flame* violet. The most notable characteristic differences, on the other hand are that the precipitates produced by bichloride of platinum are far more insoluble in water than the potassio-bichloride of platinum; 100 grammes of water will, at 50° Fahrenheit, dissolve 900 milligrammes of potassio-bichloride of platinum, but only 154 milligrammes of the rubidio-bichloride of platinum, and as little as 50 milligrammes of the cæsio-bichloride of platinum—and, above all, that the flames colored by cæsium and rubidium compounds give *spectra* quite different from the potassium spectrum (see Table I). The cæsium spectrum is especially characterized by the two blue lines a and β, which are remarkable for their wonderful intensity and sharp outline; also by the line γ, which, however, is less strongly marked. Amongst the lines in the rubidium spectrum, the splendid indigo-blue lines marked a and β strike the eye by their extreme brilliancy. Less brilliant, but still very characteristic, are the lines δ and γ. Lastly, we have still to mention that carbonate of oxide of cæsium is soluble in absolute alcohol, whilst carbonate of oxide of rubidium is insoluble in that menstruum. Still, a separation of the two oxides is effected only with difficulty by this means, as they seem to form a double salt which is not absolutely insoluble in alcohol.

3. LITHIA (LiO).

Lithia is also found pretty widely disseminated in nature, but in minute quantities only. It is often met with in the analysis of mineral waters and ashes of plants, less frequently in the analysis of minerals, and only rarely in that of technical and pharmaceutical products. Lithia forms the transition from the first to the second group. It dissolves with difficulty in water; it does not attract moisture from the air. Most of its salts are soluble in water; some of them are deliquescent (chloride of lithium). Carbonate of lithia is difficultly soluble, particularly in cold water. *Phosphate of soda* produces in not over dilute solutions of salts of lithia upon boiling, a white crystalline precipitate of tribasic phosphate of lithia $(3 LiO, PO_5)$, which quickly subsides to the bottom of the precipitating vessel. This reaction, which is characteristic of lithia, is rendered much more delicate by adding with the phosphate of soda a little solution of soda, just sufficient to leave the reaction alkaline, evaporating the mixture to dryness, treating the residue with water, and adding an equal volume of liquid ammonia. By this course of proceeding even very minute quantities of lithia will be separated as $3 LiO, PO_5$. The precipitate fuses before the blowpipe, and gives upon fusion with carbonate of soda a clear bead; when fused upon charcoal it is absorbed by the pores of the latter body. It dissolves in hydrochloric acid to a fluid which, supersaturated with ammonia, remains clear in the cold, but upon boiling gives a heavy crystalline precipitate of $3 LiO, PO_5$. (Reactions by which the phosphates of lithia differ from the phosphates of the alkaline earths). *Tartaric acid* and *bichloride of platinum* fail to precipitate even concentrated solutions of salts of lithia. If salts of lithia are exposed to the *gas* or *blowpipe flame*, in the manner described in the chapter on the potassa reactions (§ 89, 5), they tinge the flames carmine-red. Silicates containing lithia demand addition of gypsum to produce this reaction. Phosphate of lithia will tinge the flame carmine-red if the fused bead is moistened with hydrochloric acid. The soda coloration conceals the lithia coloration; in presence of soda, therefore, the lithia tint must be viewed through a blue glass, or through a thin layer of indigo solution. Presence of a small proportion of potassa will not conceal the lithia coloration. In presence of a large proportion of potassa, the lithia may be identified by placing the bead in the point of fusion, viewing the colored flame through the indigo prism (§ 15), and comparing it with a pure potassa flame produced in the opposite fusion mantle. Viewed through thin layers, the lithia-colored flame appears now redder than the pure potassa flame; viewed through somewhat thicker layers, the flames appear at last equally red, if the proportion of the lithia to the potassa is only trifling; but when lithia predominates in the examined sample the intensity of the red coloration imparted by lithia decreases perceptibly when viewed through thicker layers, whilst the pure potassa-flame is scarcely impaired thereby. By this means lithia may still be detected in potassa salts, even though present only in the proportion of one part in several thousand parts of the latter. Soda, unless present in over-large quantities, interferes but little with these reactions (CARTMELL, BUNSEN).

The *lithium spectrum* (Table I.) is most brilliantly characterized by the splendid carmine-red line a, and the orange-yellow very faint line β. If alcohol be poured over chloride of lithium, and then ignited, the flame shows also a carmine-red tint. Presence of salts of soda will mask this reaction.

To detect small quantities of cæsium, rubidium, and lithium in presence of very large quantities of soda or potassa, extract the dry chlorides, with addition of a few drops of hydrochloric acid, with alcohol of 90 per cent., which leaves behind the far larger portion of the chloride of sodium and chloride of potassium. Evaporate the solution to dryness, dissolve the residue in a little water, and precipitate with bichloride of platinum. Filter the fluid off, wash the precipitate repeatedly with boiling water, to remove the potassio-bichloride of platinum present, and examine in the course of this process repeatedly by the spectroscope. The potassa spectrum will now be found to grow fainter and fainter, whilst the spectra of rubidium and cæsium will become visible, if these metals are present. Evaporate the fluid filtered off from the platinum precipitate to dryness, heat the residue to slight redness in the hydrogen current, to decompose the sodio-bichloride of platinum and the excess of bichloride of platinum, moisten with hydrochloric acid, drive off the acid again, and extract the chloride of lithium finally with a mixture of absolute alcohol and ether. The evaporation of the solution obtained leaves the chloride of lithium behind in a state of almost perfect purity; it may then be further examined and tested. Before drawing from the simple coloration of the flame the conclusion that lithia is present, it is advisable, in order to guard against the chance of error, to test a portion of the residue, dissolved in water, with carbonate of ammonia, to make quite sure that strontia or lime is not present. The addition of hydrochloric acid, which is repeatedly prescribed in the above process to precede the extraction of the chloride of lithium with alcohol, is necessary for this reason, that chloride of lithium is, even at a moderate red heat, converted by the action of aqueous vapor into caustic lithia, which then attracts carbonic acid, forming carbonate of lithia, which is insoluble in alcohol.

§ 94.

SECOND GROUP.

BARYTA, STRONTIA, LIME, MAGNESIA.

Properties of the group.—The alkaline earths are soluble in water in the pure (caustic) state. Magnesia, however, dissolves but very sparingly in water. The solutions manifest alkaline reaction; the alkaline reaction of magnesia is most clearly apparent when that earth is laid upon moistened test-paper. The neutral carbonates and phosphates of the alkaline earths are insoluble in water. The solutions of the salts of the alkaline earths are therefore precipitated by carbonates and phosphates of the alkalies. This reaction distinguishes the oxides of the second group from those of the first. From the oxides of the other groups they are distinguished by the solutions being neither precipitated by hydrosulphuric acid, nor by sulphide of ammonium. The alkaline earths and their salts are not volatile at a moderate red heat; they are colorless. The solutions of their nitrates and chlorides are not precipitated by carbonate of baryta.

Special Reactions.

§ 95.

a. BARYTA (BaO).

1. CAUSTIC BARYTA is pretty readily soluble in hot water, but rather sparingly so in cold water; it dissolves freely in dilute hydrochloric or nitric acid. Hydrate of baryta fuses at a red heat, without losing its water.

2. Most of the SALTS OF BARYTA are insoluble in water. The soluble salts do not affect vegetable colors, and are decomposed upon ignition, with the exception of chloride of barium. The insoluble salts dissolve in dilute hydrochloric acid, except the sulphate of baryta and the silicofluoride of barium. Nitrate of baryta and chloride of barium are insoluble in alcohol, and do not deliquesce in the air. Concentrated solutions of baryta are precipitated by hydrochloric or nitric acid added in large proportions, as chloride of barium and nitrate of baryta are not soluble in the aqueous solutions of the said acids.

3. *Ammonia* produces no precipitate in the aqueous solutions of salts of baryta; *potassa* or *soda* (free from carbonic acid) only in highly concentrated solutions. Water redissolves the bulky precipitate of CRYSTALS OF BARYTA ($BaO, HO + 8$ aq.) produced by potassa or soda. With acid fluids the application of heat is required to effect complete precipitation.

4. *Carbonates of the alkalies* throw down from solutions of baryta CARBONATE OF BARYTA (BaO, CO_2) in the form of a white precipitate. When carbonate of ammonia is used as the precipitant, or if the solution was previously acid, complete precipitation takes place only upon heating the fluid. In chloride of ammonium the precipitate is soluble to a trifling yet clearly perceptible extent; carbonate of ammonia therefore produces no precipitate in very dilute solutions of baryta containing much chloride of ammonium.

5. *Sulphuric acid* and all the soluble *sulphates*, more particularly also solution of *sulphate of lime*, produce even in very dilute solutions of baryta, a heavy, finely pulverulent, white precipitate of SULPHATE OF BARYTA (BaO, SO_3), which is insoluble in alkalies, nearly so in dilute acids, but perceptibly soluble in boiling concentrated hydrochloric and nitric acids, as well as in concentrated solutions of ammonia salts; however, in these latter only if there is no excess of sulphuric acid present. This precipitate is generally formed immediately upon the addition of the reagent; from highly dilute solutions, however, especially when strongly acid, it separates only after some time.

6. *Hydrofluosilicic acid* throws down from solutions of baryta SILICOFLUORIDE OF BARIUM ($BaFl, SiFl_2$) in the form of a colorless crystalline quickly subsiding precipitate. In dilute solutions this precipitate is

formed only after the lapse of some time; it is perceptibly soluble in hydrochloric and nitric acids. Addition of an equal volume of alcohol hastens the precipitation and makes it so complete that the filtrate remains clear upon addition of sulphuric acid.

7. *Phosphate of soda* produces in neutral or alkaline solutions of baryta a white precipitate of PHOSPHATE OF BARYTA ($2\,BaO, HO, PO_5$), which is soluble in free acids. Addition of ammonia only slightly increases the quantity of this precipitate, a portion of which is in this process converted into basic phosphate of baryta ($3\,BaO, PO_5$). Chloride of ammonium dissolves the precipitate to a clearly perceptible extent.

8. *Oxalate of ammonia* produces in moderately dilute solutions of baryta a white pulverulent precipitate of OXALATE OF BARYTA ($2\,BaO, C_4O_6 + 2\,aq.$), which is soluble in hydrochloric and nitric acids. When recently thrown down, this precipitate dissolves also in oxalic and acetic acids; but the solutions speedily deposit binoxalate of baryta ($BaO, HO, C_4O_6 + 2\,aq.$) in the form of a crystalline powder.

9. If soluble salts of baryta in powder are heated with dilute *spirit of wine*, they impart to the flame a GREENISH-YELLOW color, which, however, is not very characteristic.

10. If salts of baryta are held on the loop of a platinum wire in the hottest part of *Bunsen's gas flame*, the part of the flame above the sample is colored YELLOWISH-GREEN; or, if the baryta salts are held in the *inner spirit-blowpipe flame*, the same coloration is imparted to the part of the flame before the sample. With the soluble baryta salts, and also with the carbonate and sulphate of baryta, the reaction is immediate or very soon; but the phosphate demands previous moistening of the sample with sulphuric acid or hydrochloric acid, by which means the baryta may be detected by the flame coloration also in silicates decomposable by acids. Silicates which hydrochloric acid fails to decompose must be fluxed with carbonate of soda, when the carbonate of baryta produced will show the reaction. It is characteristic of the yellowish-green baryta coloration of the flame that it appears bluish-green when viewed through the green glass. If the sulphates are selected for the experiment, presence of lime and strontia will not interfere with the reaction. The *baryta spectrum* is shown in Table I. The green lines, α and β, are the most intense; γ is lesser marked, but still characteristic.

11. Cold solutions of *bicarbonates of the alkalies* or of *carbonate of ammonia* fail to decompose sulphate of baryta, or, to speak more correctly, they decompose that salt only to a scarcely perceptible extent; the same applies to a boiling solution of 1 *part of carbonate and 3 parts of sulphate of potassa*. Repeated action of boiling solutions of simple or monocarbonates of the alkalies upon sulphate of baryta succeeds in the end completely in decomposing that salt. It is readily decomposed also by fusion in conjunction with carbonates of the alkalies, which results in the formation of a sulphate of the fluxing alkali, which is soluble in water, and of carbonate of baryta, which is insoluble in that menstruum

§ 96.

b. STRONTIA (SrO).

1. STRONTIA and its HYDRATE and SALTS have nearly the same general properties and reactions as baryta and its corresponding compounds.—

Hydrate of strontia is more sparingly soluble in water than hydrate of baryta.—Chloride of strontium dissolves in absolute alcohol and deliquesces in moist air. Nitrate of strontia is insoluble in absolute alcohol and does not deliquesce in the air.

2. The salts of strontia show with *ammonia, potassa,* and *soda*, and also with the *carbonates of the alkalies* and with *phosphate of soda*, nearly the same reactions as the salts of baryta. Carbonate of strontia dissolves somewhat more difficultly in chloride of ammonium than is the case with carbonate of baryta.

3. *Sulphuric acid* and *sulphates* precipitate from solutions of strontia SULPHATE OF STRONTIA (SrO, SO_3) in the form of a white powder. Sulphate of strontia is insoluble in spirit of wine; addition of alcohol will therefore promote the separation of the precipitate. Application of heat greatly promotes the precipitation. Sulphate of strontia is far more soluble in water than sulphate of baryta; owing to this readier solubility, the precipitated sulphate of strontia separates from rather dilute solutions in general only after the lapse of some time; and this is invariably the case (even in concentrated solutions) if *solution of sulphate of lime* is used as precipitant. In hydrochloric acid and in nitric acid sulphate of strontia dissolves perceptibly. Presence of larger quantities of these acids will accordingly most seriously impair the delicacy of the reaction. Solution of sulphate of strontia in hydrochloric acid is, after dilution with water, rendered turbid by chloride of barium.

4. *Hydrofluosilicic acid* fails to produce a precipitate even in concentrated solutions of strontia; even upon addition of an equal volume of alcohol no precipitation takes place, except in very highly concentrated solutions.

5. *Oxalate of ammonia* precipitates even from rather dilute solutions OXALATE OF STRONTIA ($2 S_2 O, \overline{O} + 5$ aq.) in the form of a white powder, which dissolves readily in hydrochloric and nitric acid, and perceptibly in salts of ammonia, but is only sparingly soluble in oxalic and acetic acid.

6. If salts of strontia soluble in water or alcohol are heated with dilute spirit of wine, and the spirit is kindled, the flame appears of a very intense CARMINE color, more particularly upon stirring the alcoholic mixture.

7. If a strontia salt is held in the fusion zone of *Bunsen's gas flame*, or in the *inner spirit-blowpipe flame*, an INTENSELY RED color is imparted to the flame. With chloride of strontium the reaction is the most distinct, less clear with strontia and carbonate of strontia, fainter still with sulphate of strontia, and scarcely at all with strontia compounds with fixed acids. The sample is therefore, after its first exposure to the flame, moistened with hydrochloric acid, and then again exposed to the flame. If sulphate of strontia is likely to be present, the sample is first exposed a short time to the reducing flame (to produce sulphide of strontium), before it is moistened with hydrochloric acid. Viewed through the *blue glass*, the strontia flame appears purple or rose (difference between strontia and lime, which latter body shows a faint greenish-gray color when treated in this manner); this reaction is the most clearly apparent if the sample moistened with hydrochloric acid is let spirt up in the flame. In presence of baryta the strontia reaction shows only upon the first introduction of the sample moistened with hydrochloric acid into the flame. The *strontia spectrum* is shown in Table I. It contains a number of characteristic lines, more especially

the orange line a, the red lines β and γ, and the blue line δ, which latter is more particularly suited for the detection of strontia in presence of lime.

8. Sulphate of strontia is completely decomposed by continued digestion with solutions of *carbonate of ammonia* or of *bicarbonates of the alkalies*, but much more rapidly by boiling with a solution of 1 part of *carbonate of potassa and 3 parts of sulphate of potassa* (essential difference between sulphate of strontia and sulphate of baryta).

§ 97.

c. LIME (CaO).

1. LIME and its HYDRATE and SALTS present in their general properties and reactions, a great similarity to baryta and strontia and their corresponding compounds. Hydrate of lime is far more difficultly soluble in water than the hydrates of baryta and strontia; it dissolves also more sparingly in hot than in cold water. Hydrate of lime loses its water upon ignition. Chloride of calcium and nitrate of lime are soluble in absolute alcohol and deliquesce in the air.

2. *Ammonia, potassa, carbonates of the alkalies*, and *phosphate of soda* show nearly the same reactions with salts of lime as with salts of baryta. Recently precipitated CARBONATE OF LIME (CaO, CO_2) is bulky and amorphous—after a time, and immediately upon application of heat, it falls down and assumes a crystalline form. Recently precipitated carbonate of lime dissolves pretty readily in solution of chloride of ammonium; but the solution speedily becomes turbid, and deposits the greater part of the dissolved salt in form of crystals.

3. *Sulphuric acid* and *sulphate of soda* produce immediately in highly concentrated solutions of lime white precipitates of SULPHATE OF LIME ($CaO, SO_3, HO + aq.$), which redissolve completely in a large proportion of water, and are still far more soluble in acids. In less concentrated solutions the precipitates are formed only after the lapse of some time; and no precipitation whatever takes place in dilute solutions. Solution of sulphate of lime of course cannot produce a precipitate in salts of lime; but even a cold saturated solution of sulphate of potassa, mixed with 3 parts of water, produces a precipitate only after standing from twelve to twenty-four hours. In solutions of lime which are so very dilute that sulphuric acid has no apparent action on them, a precipitate will immediately form upon addition of alcohol.

4. *Hydrofluosilicic acid* does not precipitate salts of lime.

5. *Oxalate of ammonia* produces in solutions of lime a white pulverulent precipitate of OXALATE OF LIME. If the fluids are in any degree concentrated or hot, the precipitate ($2 CaO, C_4 O_6 + 2$ aq.) forms at once; but if they are very dilute and cold, it forms only after some time, in which latter case it is more distinctly crystalline and consists of a mixture of the above salt with $2 CaO, C_4 O_6 + 6$ aq. Oxalate of lime dissolves readily in hydrochloric and nitric acids; but acetic and oxalic acids fail to dissolve it to any perceptible extent.

6. Soluble salts of lime when heated with dilute *spirit of wine* impart to the flame of the latter a YELLOWISH-RED color, which is liable to be confounded with that communicated to the flame of alcohol by salts of strontia.

7. If salts of lime are held in the hottest part of *Bunsen's gas flame*, or in the *inner spirit-blow-pipe flame*, they impart to the flame a YELLOWISH-RED color. This reaction is the most distinct with chloride of calcium; sulphate of lime shows it only after its incipient conversion into basic salt, and carbonate of lime also most distinctly after the escape of the carbonic acid. Compounds of lime with fixed acids do not color flame; those of them which are decomposed by hydrochloric acid will, however, show the reaction after moistening with that acid. The reaction is in such cases promoted by flattening the loop of the platinum wire, placing a small portion of the lime compound upon it, letting it frit, adding a drop of hydrochloric acid, which remains hanging to the loop, and then holding the latter in the hottest part of the flame. The reaction shows now the most distinct light immediately upon the disappearance of the drop, which in this process, as in LEIDENFROST's phenomenon, evaporates without boiling (BUNSEN). Viewed through the *green glass*, the lime coloration of the flame appears finch-green colored on letting the sample moistened with hydrochloric acid spirt in the flame (difference between lime and strontia, which latter substance under similar circumstances shows a very faint yellow) (MERZ). In presence of baryta the lime reaction shows only upon the first introduction of the sample into the flame. The *lime spectrum* is shown in Table I. The intensely green line β is more particularly characteristic, also the intensely orange line α. It requires a very good apparatus to show the indigo-blue line to the right of G in the solar spectrum, as this is much less luminous than the other lines.

8. With monocarbonates and bicarbonates of the alkalies, sulphate of lime shows the same reactions as sulphate of strontia.

§ 98.

d. MAGNESIA (MgO).

1. MAGNESIA and its HYDRATE are white powders of far greater bulk than the other alkaline earths and their hydrates. Magnesia and hydrate of magnesia are nearly insoluble both in cold and hot water. Hydrate of magnesia loses its water upon ignition.

2. Some of the SALTS OF MAGNESIA are soluble in water, others are insoluble in that fluid. The soluble salts of magnesia have a nauseous bitter taste; in the neutral state they do not alter vegetable colors; with the exception of sulphate of magnesia, they undergo decomposition when gently ignited, and the greater part of them even upon simple evaporation of their solutions. Nearly all the salts of magnesia which are insoluble in water dissolve readily in hydrochloric acid.

3. *Ammonia* throws down from the solutions of neutral salts of magnesia part of the magnesia as HYDRATE (MgO, HO) in the form of a white bulky precipitate. The rest of the magnesia remains in solution as a double salt, viz., in combination with the ammonia salt which forms upon the decomposition of the salt of magnesia; these double salts are not decomposed by ammonia. It is owing to this tendency of salts of magnesia to form such double salts with ammoniacal compounds that ammonia fails to precipitate them in presence of a sufficient proportion of an ammonia salt with neutral reaction; or, what comes to the same, that ammonia produces no precipitate in solutions of magnesia contain-

ing a sufficient quantity of free acid, and that precipitates produced by ammonia in neutral solutions of magnesia are redissolved upon the addition of chloride of ammonium.

4. *Potassa, soda, caustic baryta,* and *caustic lime,* throw down from solutions of magnesia HYDRATE OF MAGNESIA. The separation of this precipitate is greatly promoted by boiling the mixture. Chloride of ammonium and other similar salts of ammonia redissolve the washed precipitated hydrate of magnesia. If the salts of ammonia are added in sufficient quantity to the solution of magnesia before the addition of the precipitant, small quantities of the latter fail altogether to produce a precipitate. However, upon boiling the solution afterwards with an excess of potassa, the precipitate will of course make its appearance, since this process causes the decomposition of the ammonia salt, removing thus the agent which retains the hydrate of magnesia in solution.

5. *Carbonate of potassa* and *carbonate of soda* produce in neutral solutions of magnesia a white precipitate of BASIC CARBONATE OF MAGNESIA $4\,(MgO, CO_2) + MgO, HO + 10$ aq. One-fifth of the carbonic acid of the decomposed alkaline carbonate is liberated in the process, and combines with a portion of the carbonate of magnesia to bicarbonate, which remains in solution. This carbonic acid is expelled by boiling, and an additional precipitate formed $(MgO, CO_2 + 3\text{ aq.})$ Application of heat therefore promotes the separation and increases the quantity of the precipitate. Chloride of ammonium and other similar salts of ammonia prevent this precipitation also, and redissolve the precipitates already formed.

6. If solutions of magnesia are mixed with *carbonate of ammonia,* the fluid always remains clear at first; but after standing some time, it deposits, more or less quickly or slowly according to the greater or less concentration or dilution of the solution, a crystalline precipitate of CARBONATE OF MAGNESIA AND AMMONIA $(NH_4O, CO_2 + MgO, CO_2 + 4\text{ aq.})$ In rather highly dilute solutions this precipitate will not form. Addition of ammonia promotes its separation. Chloride of ammonium counteracts it, but it cannot prevent the formation of the precipitate in rather highly concentrated solutions.

7. *Phosphate of soda* precipitates from solutions of magnesia, if they are not too dilute, PHOSPHATE OF MAGNESIA $(2\,MgO, HO, PO_5, + 14\text{ aq.})$ as a white powder. Upon boiling, basic phosphate of magnesia $(3\,MgO, PO_5 + 7\text{ aq.})$ separates, even from rather dilute solutions. But if the addition of the precipitant is preceded by that of chloride of ammonium and ammonia, a white crystalline precipitate of BASIC PHOSPHATE OF MAGNESIA AND AMMONIA $(2\,MgO, NH_4O, PO_5 + 12\text{ aq.})$ will separate even from very dilute solutions of magnesia; its separation may be greatly promoted and accelerated by stirring with a glass rod; even should the solution be so extremely dilute as to forbid the formation of a precipitate, yet the lines of direction in which the glass rod has moved along the side of the vessel will after the lapse of some time appear distinctly as white streaks. Water and solutions of salts of ammonia dissolve the precipitate but very slightly; but it is readily soluble in acids, even in acetic acid. In water containing ammonia it may be considered insoluble.

8. *Oxalate of ammonia* produces no precipitate in highly dilute solutions of magnesia; in less dilute solutions no precipitate is formed at

first, but after standing some time crystalline crusts of various oxalates of ammonia and magnesia make their appearance. In highly concentrated solutions oxalate of ammonia very speedily produces precipitates of oxalate of magnesia ($2\,MgO, C_4O_6 + 4\,aq.$), which contain small quantities of the above-named double salts. Chloride of ammonium, especially in presence of free ammonia, interferes with the formation of these precipitates, but will not in general absolutely prevent it.

9. *Sulphuric acid* and *hydrofluosilicic acid* do not precipitate salts of magnesia.

10. Salts of magnesia do not color flame.

§ 99.

Recapitulation and remarks.—The difficult solubility of the hydrate of magnesia, the ready solubility of the sulphate, and the disposition of salts of magnesia to form double salts with ammonia compounds, are the three principal points in which magnesia differs from the other alkaline earths. To detect magnesia in solutions containing all the alkaline earths, we always first remove the baryta, strontia, and lime. This is effected most conveniently by means of carbonate of ammonia, with addition of some ammonia and of chloride of ammonium, and application of heat; since by this process the baryta, strontia, and lime are obtained in a form of combination suited for further examination. If the solutions are somewhat dilute, and the precipitated fluid is quickly filtered, the carbonate of baryta, strontia, and lime is obtained on the filter, whilst the whole of the magnesia is found in the filtrate. But as chloride of ammonium dissolves a little carbonate of baryta, and also a little carbonate of lime, though much less of the latter than of the former, trifling quantities of these bases are found in the filtrate; nay, where only traces of them are present, they may altogether remain in solution. In accurate experiments, therefore, the separation is effected in the following way: Divide the filtrate into three portions, test one portion with dilute sulphuric acid for the trace of baryta which it may contain in solution, and another portion with oxalate of ammonium for the minute trace of lime which may have remained in solution. If the two reagents produce no turbidity even after some time, test the third portion with phosphate of soda for magnesia. But if one of the reagents causes turbidity, filter the fluid from the gradually subsiding precipitate, and test the filtrate for magnesia. Should both reagents produce precipitates, mix the two first portions together, filter after some time, and then test the filtrate to make sure that the precipitate thrown down by oxalate of ammonia is actually oxalate of lime, and not, as it may be, oxalate of magnesia and ammonia, dissolve it in some hydrochloric acid, and add dilute sulphuric acid, and then spirit of wine.

To show the presence of the baryta, strontia, and lime in the precipitate produced by carbonate of ammonia, dissolve the precipitate in some dilute hydrochloric acid; add solution of gypsum to a small portion of this solution, when the immediate formation of a precipitate will prove the presence of baryta. Evaporate the remainder of the hydrochloric acid solution to dryness, and treat the residue with absolute alcohol, which will dissolve the chloride of strontium and the chloride of calcium, leaving the greater part of the chloride of barium undissolved. Mix the alcoholic solution with an equal volume of water and a few drops of hydrofluosilicic acid, and let the mixture stand several hours, when the last traces of the baryta present will be found precipitated as silicofluoride of barium. Filter, and add sulphuric acid to the alcoholic filtrate. This will throw down the strontia and the lime. Filter the fluid from the precipitate, wash with weak spirit of wine, and convert the sulphates into carbonates by boiling with solution of carbonate of soda. Wash the carbonates, dissolve them in a small quantity of hydrochloric acid, evaporate the solution to dryness, dissolve the residue in a very little water, and divide the solution into two portions, after previous filtration if necessary. Mix the one portion with solution of gypsum. The formation of a precipitate after some time, often only after a considerable time, shows the presence of STRONTIA. To remove the latter earth from the other portion, mix this with a solution of sulphate of potassa, boil, filter, and test the filtrate with oxalate of ammonia for LIME. If much lime is present, a portion of it may fall down with the sulphate of strontia precipitate produced by sulphate of potassa; but there always remains sufficient of it in solution to permit its positive detection in the filtrate without difficulty. The best and most convenient way of effecting the detection of the alkaline earths in their phosphates, is to decompose these latter by means of sesquichloride of iron, with addition of acetate of soda (§ 142). The oxalates of the alkaline earths are converted into carbonates by ignition, preparatory to the detection of the several earths which they may contain. The following method will serve to analyse mixtures of the sulphates of the alkaline earths. Extract the mixture under examination with small portions of boiling water. The solution contains the whole of the sulphate of magnesia, besides a trifling quantity of sulphate of lime. Digest the residue, according to H. ROSE's direction, in the cold for 12 hours, with a solution of carbonate of ammonia, or boil it 10 minutes with a solution of 1 part of carbonate and 3 parts of sulphate of potassa, filter, wash, then treat with dilute hydrochloric acid, which will dissolve the carbonates of strontia and lime formed, but always also a minute trace of baryta (FRESENIUS), leaving behind the undecomposed sulphate of baryta. The latter may then be decomposed by fusion with carbonates of the alkalies. The solutions obtained are to be examined further according to the above directions.

The detection of baryta, strontia, and lime in the moist way is very instructive, but also rather laborious and tedious. By means of the spectral apparatus these alkaline earths are much more readily detected, even when present all three together. According to the nature of the acid, the sample is either introduced at once into the flame, or after previous ignition and moistening with hydrochloric acid. To detect very minute quantities of baryta and strontia in presence of large quantities of lime, ignite a few grammes of the mixed carbonates a few minutes in a platinum crucible strongly over the blast,* extract the ignited mass by boiling with a little distilled water, evaporate with hydrochloric acid to dryness, and examine the residue by spectrum analysis (ENGELBACH). But even without a spectrum apparatus the three alkaline earths may be detected in mixtures containing all three of them by the different coloration which they severally impart to flame. To this end the sample under examination is repeatedly moistened with sulphuric acid, then cautiously

* The carbonates of baryta and strontia are much more readily reduced to the caustic state in this process than would be the case in the absence of carbonate of lime.

dried, and introduced into the fusion zone of the gas flame. After the evaporation of the alkalies that may chance to be present, the baryta coloration (§ 95, 10) will make its appearance alone. After this coloration has completely disappeared, and the sample moistened with hydrochloric acid gives on spirting no longer a flame coloration of a bluish-green tint when viewed through the green glass, the sample is moistened again with hydrochloric acid, and tested for lime by viewing it through the green glass when spirting (§ 97, 7), for strontia by viewing it under the same circumstances through the blue glass (§ 96, 7) (MERZ).

§ 100.

THIRD GROUP.

More common oxides of the third group:—ALUMINA, SESQUIOXIDE OF CHROMIUM.

Rarer oxides of the third group:—BERYLLA, THORIA, ZIRCONIA, YTTRIA, OXIDE OF TERBIUM and OXIDE OF ERBIUM, OXIDES OF CERIUM, OXIDE OF LANTHANIUM, OXIDE OF DIDYMIUM, TITANIC ACID, TANTALIC ACID, HYPONIOBIC ACID.

Properties of the group.—The oxides of the third group are insoluble in water, both in the pure state and as hydrates. Their sulphides cannot be produced in the moist way. Hydrosulphuric acid therefore fails to precipitate the solutions of their salts. Sulphide of ammonium throws down, from the solutions of the salts in which the oxides of the third group constitute the base,* in the same way as ammonia, the hydrated oxides. This reaction with sulphide of ammonium distinguishes the oxides of the third from those of the two preceding groups.

Special Reactions of the more common Oxides of the third group.

§ 101.

a. ALUMINA (Al_2O_3).

1. ALUMINIUM metal is nearly white. It is not oxidized by the action of the air, in compact masses not even upon ignition. It may be filed, and is very ductile; its specific gravity is only 2·56. It is fusible at a bright red heat. In the pulverulent form it slowly decomposes water at a boiling heat; the compact metal does not show this reaction. Aluminium dissolves readily in hydrochloric acid, as well as in hot solution of potassa, with evolution of hydrogen. Nitric acid dissolves it only slowly, even with the aid of heat.

2. ALUMINA is non-volatile and colorless; the hydrate is also colorless. Alumina dissolves in dilute acids slowly and with very great difficulty, but more readily in concentrated hot hydrochloric acid. In

---

* The oxides of the third group may nearly all of them combine to saline compounds with acids as well as with bases; alumina, for instance, combines with potassa to aluminate of potassa, with sulphuric acid to sulphate of alumina. The oxides of the third group stand, accordingly, partly on the verge between bases and acids. Those which incline more to the latter, as is the case with the three last members of the group, are therefore also called acids.

fusing bisulphate of potassa, it dissolves readily to a mass soluble in water. The hydrate in the amorphous condition is readily soluble in acids; in the crystalline state it dissolves in them with very great difficulty. After previous ignition with alkalies, the alumina, or, more correctly speaking, the alkaline aluminate formed, is readily dissolved by acids.

3. The SALTS OF ALUMINA are colorless and non-volatile; some of them are soluble, others insoluble. The soluble salts have a sweetish astringent taste, redden litmus-paper, and lose their acid upon ignition. The insoluble salts are dissolved by hydrochloric acid, with the exception of certain native compounds of alumina; the compounds of alumina which are insoluble in hydrochloric acid are decomposed and made soluble by ignition with carbonate of soda and potassa, or bisulphide of potassa. This decomposition and solution may, however, be effected also by heating them, reduced to a fine powder, with hydrochloric acid of 25 per cent., or with a mixture of 3 parts by weight of hydrated sulphuric acid, and 1 part by weight of water, in sealed glass tubes, to 392°—410° Fahrenheit, continuing the operation for two hours (A. MITSCHERLICH).

4. *Potassa* and *soda* throw down from solutions of alumina salts a bulky precipitate of HYDRATE OF ALUMINA ($Al_2O_3, 3HO$), which contains alkali and generally also an admixture of basic salt; this precipitate redissolves readily and completely in an excess of the precipitant, but from this solution it is reprecipitated by addition of chloride of ammonium, even in the cold, but more completely upon application of heat (compare § 53). The presence of salts of ammonia does not prevent the precipitation by potassa or soda.

5. *Ammonia* also produces in solutions of alumina a precipitate of HYDRATE OF ALUMINA, which contains ammonia and an admixture of basic salt; this precipitate also redissolves in a very considerable excess of the precipitant, but with difficulty only, which is the greater the larger the quantity of salts of ammonia contained in the solution. It is this deportment which accounts for the complete precipitation of hydrate of alumina from solution in potassa by an excess of chloride of ammonium.

6. *Carbonates of the alkalies* precipitate BASIC CARBONATE OF ALUMINA, which is insoluble or barely soluble in an excess of the precipitant.

7. If the solution of a salt of alumina is digested with finely pulverized *carbonate of baryta*, the greater part of the acid of the alumina salt combines with the baryta, the liberated carbonic acid escapes, and the alumina precipitates completely as HYDRATE mixed with BASIC SALT OF ALUMINA; even digestion in the cold suffices to produce this reaction.

8. If alumina or one of its compounds is ignited upon charcoal before the blowpipe, and afterwards moistened with a solution of *nitrate of protoxide of cobalt*, and then again strongly ignited, an unfused mass of a deep SKY-BLUE color is produced, which consists of a compound of the two oxides. The blue color becomes distinct only upon cooling. By candlelight it appears violet. This reaction is decisive only in the case of infusible or difficultly fusible compounds of alumina pretty free from other oxides, as solution of cobalt will often impart a blue tint to readily fusible salts, even though no alumina be present.

§ 102.

b. SESQUIOXIDE OF CHROMIUM (Cr_2O_3).

1. SESQUIOXIDE OF CHROMIUM is a green, its HYDRATE a bluish gray-green powder. The hydrate dissolves readily in acids; the non-ignited sesquioxide dissolves more difficultly, and the ignited sesquioxide is almost altogether insoluble.

2. The SALTS OF SESQUIOXIDE OF CHROMIUM have a green or violet color. Many of them are soluble in water. Most of them dissolve in hydrochloric acid. The solutions usually exhibit a fine green or a dark violet color, which latter, however, changes to green upon heating. The salts of sesquioxide of chromium with volatile acids are decomposed upon ignition, the acids being expelled. The salts of sesquioxide of chromium which are soluble in water redden litmus. Anhydrous sesquichloride of chromium is crystalline, violet-colored, insoluble in water and in acids, and volatilizes with difficulty.

3. *Potassa and soda* produce in the green as well as in the violet solutions of salts of sesquioxide of chromium a bluish-green precipitate of HYDRATE OF SESQUIOXIDE OF CHROMIUM, which dissolves readily and completely in an excess of the precipitant, imparting to the fluid an emerald-green tint. Upon *long-continued* ebullition of this solution, the whole of the hydrated sesquioxide separates again, and the supernatant fluid appears perfectly colorless. The same reprecipitation takes place if chloride of ammonium is added to the alkaline solution. Application of heat promotes the separation of the precipitate.

4. *Ammonia* produces in green solutions of salts of sesquioxide of chromium a grayish-green, in violet solutions a grayish-blue precipitate of HYDRATE OF SESQUIOXIDE OF CHROMIUM. The former precipitate dissolves in acids to a green fluid, the latter to a violet fluid. Other circumstances (concentration, way of adding the ammonia, &c.) exercise also some influence upon the composition and color of these hydrates. A small portion of the hydrates redissolves in an excess of the precipitant in the cold, imparting to the fluid a peach-blossom red tint; but if after the addition of ammonia in excess heat is applied to the mixture the precipitation is complete.

5. *Carbonates of the alkalies* precipitate BASIC CARBONATE OF SESQUIOXIDE OF CHROMIUM, which redissolves in a large excess of the precipitant.

6. *Carbonate of baryta* precipitates from solutions of sesquioxide of chromium the whole of the sesquioxide as a GREENISH HYDRATE mixed with BASIC SALT. The precipitation takes place in the cold, but is complete only after long-continued digestion.

7. If a solution of sesquioxide of chromium in solution of potassa or soda is mixed with some brown *peroxide of lead* in excess, and the mixture is boiled a short time, the sesquioxide of chromium is oxidized to chromic acid. A yellow fluid is therefore obtained on filtering, which consists of a solution of CHROMATE OF OXIDE OF LEAD in solution of potassa or soda. Upon acidifying this fluid with acetic acid, the chromate of lead separates as a yellow precipitate (CHANCEL). Very minute traces of chromic acid may be detected in this fluid with still greater certainty by acidifying with hydrochloric acid, and bringing it in contact with peroxide of hydrogen and ether. Compare chromic acid (§ 138).

8. The fusion of sesquioxide of chromium or of any of its compounds

with *nitrate of soda* and *carbonate of soda* gives rise to the formation of yellow CHROMATE OF SODA, part of the oxygen of the nitric acid separating from the nitrate of soda and converting the sesquioxide of chromium into chromic acid, which then combines with the soda. Chromate of soda dissolves in water to an intensely yellow fluid. For the reactions of chromic acid see § 138.

9. *Phosphate of soda and ammonia* dissolves sesquioxide of chromium and its salts, both in the *oxidizing* and *reducing* flame of the blowpipe, to clear beads of a faint YELLOWISH-GREEN tint, which upon cooling changes to EMERALD-GREEN. The sesquioxide of chromium and its salts show a similar reaction with *biborate of soda*. Bunsen's gas flame (§ 14) is used for the experiment, or the blowpipe flame.

§ 103.

Recapitulation and remarks.—The solubility of hydrate of alumina in solutions of potassa and soda, and its reprecipitation from the alkaline solutions by chloride of ammonium, afford a safe means of detecting alumina in the absence of sesquioxide of chromium. But if the latter substance is present, which is seen either by the color of the solution, or by the reaction with phosphate of soda and ammonia, it must be removed before alumina can be tested for. The separation of sesquioxide of chromium from alumina is effected the most completely by fusing 1 part of the mixed oxides together with 2 parts of carbonate and 2 parts of nitrate of soda, which may be done in a platinum crucible. The yellow mass obtained is boiled with water; by this process the whole of the chromium is dissolved as chromate of soda, and part of the alumina as aluminate of soda, the rest of the alumina remaining undissolved. If the solution is acidified with nitric acid, it acquires a reddish-yellow tint; if ammonia is then added to feebly alkaline reaction, the dissolved portion of the alumina separates.

The precipitation of sesquioxide of chromium effected by boiling its solution in solution of potassa or soda is also sufficiently reliable if the ebullition is continued long enough; still it is often liable to mislead in cases where only little sesquioxide of chromium is present, or where the solution contains organic matter, even though in small proportion only. I have to call attention here to the fact that the solubility of hydrated sesquioxide of chromium in an excess of cold solution of potassa or soda is considerably impaired by the presence of other oxides (protoxides of manganese, nickel, cobalt, and more particularly sesquioxide of iron). If these oxides happen to be present in preponderating proportion, these may even altogether prevent the solution of the hydrated sesquioxide of chromium in potassa or soda solution. This circumstance should never be lost sight of in the analysis of compounds containing sesquioxide of chromium. Lastly, it must be borne in mind also that alkalies produce no precipitates in the solutions of alumina if non-volatile organic substances are present, such as sugar, tartaric acid, &c. The precipitation of sesquioxide of chromium by alkalies is more especially impeded and counteracted by the presence of organic acids (oxalic acid, tartaric acid, acetic acid), owing to the formation of double salts which the alkalies fail to decompose. If organic substances are present therefore, ignite, fuse the residue with carbonate and nitrate of soda, and proceed as directed before.

Special Reactions of the rarer Oxides of the third group.

§ 104.

1. BERYLLA ($Be_2 O_3$).

Berylla is a rare earth found in the form of a silicate in phenacite, and, with other silicates, in beryl, euclase, and some other rare minerals. It is a white tasteless powder insoluble in water. The ignited earth dissolves slowly but completely in acids; it is readily soluble after fusion with bisulphate of potassa. The hydrate dissolves readily in acids. The compounds of berylla very much resemble the alumina compounds. The soluble berylla salts have a sweet astringent taste; their reaction is alkaline. The native silicates of berylla are completely decomposed by fluxing with 4 parts of carbonate of soda and potassa. *Potassa, soda, ammonia*, and *sulphide of ammonium* throw down from solution of berylla salts a white flocculent hydrate, which is insoluble in ammonia, but dissolves readily in solution of potassa or soda, from which solution it is precipitated again by chloride of ammonium; the concentrated alkaline solutions remain clear on boiling, but from more dilute alkaline solutions the whole of the berylla separates upon continued ebullition (difference between berylla and alumina). Upon continued ebullition with *chloride of ammonium*, the freshly precipitated hydrate dissolves as chloride of beryllium, with expulsion of ammonia (difference between berylla and alumina). *Carbonates of the alkalies* precipitate white carbonate of berylla, which redissolves in a great excess of the carbonates of the fixed alkalies, and in a much less considerable excess of carbonate of ammonia (most characteristic difference between berylla and alumina). Upon boiling these solutions basic carbonate of berylla separates, readily and completely from the solution in carbonate of ammonia, but only upon dilution and imperfectly from the solutions in carbonates of the fixed alkalies. *Carbonate of baryta* precipitates berylla incompletely upon cold digestion, completely upon boiling. *Oxalic acid* and *oxalates* do not precipitate berylla. Moistened with solution of nitrate of protoxide of cobalt, the berylla compounds give gray masses upon ignition.

2. THORIA (Th O).

Thoria is a very rare earth found in thorite and monacite. It is white. The ignited earth is soluble only upon heating with a mixture of 1 part of concentrated sulphuric acid and 1 part of water; but it is not soluble in other acids, not even after fusion with alkalies. The moist hydrate dissolves readily in acids, the dried hydrate only with difficulty. Thorite (silicate of thoria) is decomposed by concentrated hydrochloric acid. *Potassa, ammonia*, and *sulphide of ammonium* precipitate from solutions of thoria salts white hydrate, which is insoluble in an excess of the precipitant, even of potassa (difference between thoria and berylla). *Carbonate of potassa* and *carbonate of ammonia* precipitate basic carbonate of thoria, which readily dissolves in an excess of the precipitant in concentrated solutions, with difficulty in dilute solutions (difference between thoria and alumina). From the solution in carbonate of ammonia basic salt separates again even at 122° Fahrenheit. *Oxalic acid* produces a white precipitate (difference between thoria and berylla and alumina); the precipitate does not dissolve in oxalic acid, and barely in other acids. A concentrated solution of *sulphate of potassa* precipitates thoria solutions, slowly but completely (difference between thoria and alumina and berylla). The precipitate, which is sulphate of thoria and potassa, is insoluble in a concentrated solution of sulphate of potassa; it dissolves slowly in cold water, readily in hot water. The solution deposits basic salt upon continued boiling.

3. ZIRCONIA ($Zr_2 O_3$).

Found in zircon and some other rare minerals. A white powder insoluble in hydrochloric acid, soluble upon addition of water, after continued heating with a mixture of 2 parts of hydrated sulphuric acid and 1 part of water. The hydrate resembles hydrate of alumina. It dissolves readily in hydrochloric acid when precipitated cold, and still moist, but with difficulty only when precipitated hot, or after drying. The zirconia salts soluble in water redden litmus. The native silicates of zirconia may be decomposed by fusion with carbonate of soda. The finely elutriated silicate is fused at a high temperature, together with 4 parts of carbonate of soda. The fused mass gives to water silicate of soda, a sandy zirconate of soda being left behind, which is washed, and dissolves in hydrochloric acid. *Potassa, soda, ammonia*, and *sulphide of ammonium* precipitate from solutions of zirconia salts a flocculent hydrate, which is insoluble in an excess of the precipitant, even of soda and potassa (difference between zirconia and alumina and berylla), and is not dissolved even by boiling solution of chloride of ammonium (difference between zirconia and berylla). *Carbonates of potassa, soda*, and *ammonia*, throw down carbonate of zirconia as a flocculent precipitate, which redissolves in a large excess of carbonate of potassa, more readily in bicarbonate of potassa, and most readily in carbonate of ammonia (difference between zirconia and alumina), from which solution it precipitates again on boiling. *Oxalic acid* produces a bulky precipitate of oxalate of zirconia (difference between zirconia and alumina and berylla), which is insoluble in oxalic acid, difficultly soluble in hydrochloric acid. A concentrated solution of *sulphate of potassa* speedily produces a white precipitate of sulphate of zirconia and potassa (difference between zirconia and alumina and berylla), which—if precipitated cold—dissolves readily in a large proportion of hydrochloric acid, but is almost absolutely insoluble in water and in hydrochloric acid if precipitated hot (difference between zirconia and thoria). *Carbonate of baryta* fails to effect complete precipitation in solutions of zirconia salts, even upon boiling. *Turmeric paper* dipped into solutions of zirconia salts slightly acidified with hydrochloric or sulphuric acid, acquires a reddish-brown color after drying (difference between zirconia and thoria).

4. Yttria (Y O).

Yttria is a rare earth found in gadolinite, orthite, yttro-tantalite. The ignited earth dissolves readily in hydrochloric acid (difference between yttria and zirconia and alumina). In the pure state it is white; presence of oxides of erbium and terbium imparts a brownish-yellow tint to it. The hydrate is white; it attracts carbonic acid; if freshly precipitated, it dissolves in boiling solution of chloride of ammonium (difference between yttria and alumina and zirconia). The salts are white, with a slight shade of amethyst red. Anhydrous chloride of yttrium is not volatile (difference between yttria and alumina, berylla, thoria, and zirconia). *Potassa* precipitates white hydrate, which is insoluble in an excess of the precipitant (difference between yttria and alumina and berylla). *Ammonia* and *sulphide of ammonium* produce the same reaction. Presence of a small quantity of chloride of ammonium will not prevent the precipitation by sulphide of ammonium; but in presence of a large excess of chloride of ammonium sulphide of ammonium fails to precipitate solutions of salts of yttria. *Carbonates of the alkalies* produce a white precipitate, which dissolves with difficulty in carbonate of potassa, but more readily in bicarbonate of potassa and in carbonate of ammonia, though by no means so readily as the corresponding berylla precipitate. The solution of the pure hydrate in carbonate of ammonia deposits on boiling the whole of the yttria; if chloride of ammonium is present at the same time, this is decomposed upon continued heating, with separation of ammonia, and the precipitated yttria redissolves as chloride of yttrium. Saturated solutions of carbonate of yttria in carbonate of ammonia have a tendency to deposit carbonate of yttria and ammonia, which should be borne in mind. *Oxalic acid* produces a white precipitate (difference between yttria and alumina and berylla). The precipitate does not dissolve in oxalic acid, but it dissolves in hydrochloric acid. *Sulphate of potassa* precipitates sulphate of yttria and potassa. The precipitate, even if thrown down hot, dissolves, though slowly, in a large proportion of water (difference between yttria and zirconia); it dissolves a little more readily in solution of sulphate of potassa (difference between yttria and thoria), and still more readily in solution of ammonia salt. *Carbonate of baryta* produces no precipitate; not even on boiling. Turmeric paper is not altered by acidified solutions of salt of yttria (difference between yttria and zirconia). *Tartaric acid* does not interfere with the precipitation of yttria by alkalies (characteristic difference between yttria and alumina, berylla, thoria, and zirconia). The precipitate is tartrate of yttria. The precipitation ensues only after some time, but it is complete.

5. Oxide of Terbium (Tr O), and 6. Oxide of Erbium (E O).

These oxides are generally found associated with yttria. Upon gradual addition of ammonia to a solution containing these three bases, the oxide of erbium precipitates first, the oxide of terbium next, and the yttria last. Ignited oxide of erbium varies from yellow to orange-yellow. Oxide of terbium, which is not yet known in the pure state, appears to be white. We know as yet of no other reactions by which to separate these bases from yttria, or to distinguish between them and that earth.

7. Oxides of Cerium.

Cerium is a rare metal; it is found in the form of protoxide in cerite, orthite, &c. It forms two oxides, the protoxide (C O) and the sesquioxide ($C_2 O_3$). The hydrate of the protoxide is white, but turns yellow upon exposure to the air, by absorption of oxygen. By ignition in the air it is converted into orange-red or red sesquioxide (difference between it and the preceding earths). Hydrate of protoxide of cerium dissolves readily in acids. Ignited sesquioxide of cerium containing oxide of lanthanium and didymium dissolves readily in hydrochloric acid, with evolution of chlorine; in the pure state it dissolves very slightly in boiling hydrochloric acid, except upon addition of some alcohol (difference between oxide of cerium and thoria and zirconia); the solution contains protochloride. The salts of protoxide of cerium are colorless, occasionally with a slight shade of amethyst red; the soluble protoxide salts redden litmus. Protochloride of cerium is not volatile (difference from alumina, berylla, thoria, and zirconia). Cerite (hydrated silicate of protoxide of cerium [CO, LaO, DiO] 2, $SiO_3 + 2$ aq.) does not dissolve in aqua regia; but is decomposed by fusion with carbonate of soda, and also by concentrated sulphuric acid. *Potassa* precipitates white hydrate, which turns yellow in the air, and does not dissolve in an excess of the precipitant (difference from alumina and berylla). *Ammonia* precipitates basic salt, which is insoluble in an excess of the precipitant. *Carbonates of the alkalies* produce a white precipitate, which dissolves sparingly in an excess of carbonate of potassa, somewhat more readily in carbonate of ammonia. *Oxalic acid* produces a white precipitate; the precipitation is complete even in moderately acid solutions (difference from alumina and berylla). The precipitate is not dissolved by oxalic acid; but it dissolves in a large proportion of hydrochloric acid. A saturated solution of *sulphate of potassa* precipitates, even from somewhat acid solutions, white sulphate of potassa and protoxide of cerium (difference from alumina and berylla), which is very difficultly soluble in water and altogether insoluble in a saturated solution of sulphate of potassa (difference from yttria). The precipitate may be dissolved by boiling with a large quantity of water, to which some hydrochloric acid has been added. *Carbonate of baryta* precipitates solutions of cerium salts slowly, but completely upon long-continued action. *Tartaric acid* prevents precipitation by ammonia (difference from yttria), but not by potassa. *Borax* and *phosphate of soda and ammonia* dissolve cerium compound in the outer flame to red beads (difference from the preceding earths); the coloration gets fainter on cooling, and often disappears altogether. In the inner flame colorless beads are obtained.

8. OXIDE OF LANTHANIUM.

This oxide is generally found associated with protoxide of cerium. It is white and remains unaltered by ignition in the air (difference from protoxide of cerium). In contact with cold water it is slowly converted into a milk-white hydrate; with hot water the conversion is rapid. The oxide and its hydrate change the color of reddened litmus-paper to blue; they dissolve in boiling solution of chloride of ammonium, also in dilute acids. Oxide of lanthanium in this resembles magnesia. The salts of oxide of lanthanium are colorless; the saturated solution of sulphate of oxide of lanthanium in cold water deposits a portion of the salt already at 86° Fahrenheit (difference from protoxide of cerium). *Sulphate of potassa, oxalic acid,* and *carbonate of baryta* give the same reactions as with protoxide of cerium. *Potassa* precipitates hydrate, which is insoluble in an excess of the precipitant, and does not turn brown in the air. *Ammonia* precipitates basic salts, which pass milky through the filter on washing. The precipitate produced by carbonate of ammonia is insoluble in an excess of the precipitant (difference from protoxide of cerium). If a cold saturated solution of acetate of oxide of lanthanium is supersaturated with ammonia, the slimy precipitate repeatedly washed with cold water, and a little iodine in powder added, a blue coloration makes its appearance, which gradually pervades the entire mixture (characteristic difference between oxide of lanthanium and the other earths).

9. OXIDE OF DIDYMIUM.

This oxide, like the oxide of lanthanium and in conjunction with it, is found associated with the protoxide of cerium. After intense ignition it appears white, moistened with nitric acid and feebly ignited dark brown, after intense ignition again white. In contact with water it is slowly converted into hydrate; it rapidly attracts carbonic acid; its reaction is not alkaline; it dissolves readily in acids. The concentrated solutions have a reddish or a faint violet color. The saturated solution of the sulphate deposits salt, not at 86° Fahrenheit, but upon boiling. *Potassa* precipitates hydrate, which is insoluble in an excess of the precipitant, and does not alter in the air. *Ammonia* precipitates basic salt, which is insoluble in ammonia, but slightly soluble in chloride of ammonium. *Carbonates of the alkalies* produce a copious precipitate, which is insoluble in an excess of the precipitant, even in an excess of carbonate of ammonia (difference from protoxide of cerium), but dissolves slightly in concentrated solution of chloride of ammonium. *Oxalic acid* precipitates solutions of salt of oxide of didymium almost completely; the precipitate is difficultly soluble in cold hydrochloric acid, but dissolves in that menstruum upon application of heat. *Carbonate of baryta* precipitates oxide of didymium from its solutions slowly (more slowly than protoxide of cerium and oxide of lanthanium), and never completely. A concentrated solution of *sulphate of potassa* precipitates didymium solutions more slowly and less completely than protoxide of cerium solutions. The precipitate is insoluble in cold, difficultly soluble in hot hydrochloric acid. *Phosphate of soda and ammonia* dissolves didymium compounds in the reducing flame to amethyst-red beads, shading off to violet. With *soda* a grayish white mass is obtained in the outer flame (difference from manganese).

Perfectly satisfactory methods for effecting the separation of cerium, lanthanium, and didymium are not known. The oxide of cerium may be obtained in a state of approximate purity by treating the mixed oxides, after ignition, first with dilute, then with concentrated nitric acid, which will leave, undissolved, the greater part of the oxide of cerium. If the solution is evaporated and the residue ignited and then again treated with nitric acid, the remainder of the oxide of cerium, which had been dissolved, together with the other oxides, is now also left undissolved. The oxides of lanthanium and didymium are precipitated from the solution by an alkali; the precipitate is dissolved in sulphuric acid, water saturated with the dry saline mixture at 41° to 42·8° Fahrenheit, and the solution heated to 86° Fahrenheit. Sulphate of oxide of lanthanium separates, sulphate of oxide of didymium remains in solution.

10. Titanic Acid.

Titanium forms two oxides, sesquioxide of titanium (Ti_2O_3) and titanic acid (TiO_2). The latter is somewhat more frequently met with in analysis. It is found in the free state in rutile and anatase, in combination with bases in titanite, titaniferous iron, &c. It is found in small proportions in many iron ores, and consequently often in blast-furnace slags. The small copper-colored cubes which are occasionally found in such slags consist of a combination of cyanide of titanium with nitride of titanium. Feebly ignited titanic acid is white; it transiently acquires a lemon tint when heated; very intense ignition gives a yellowish or brownish tint to it. It is infusible, insoluble in water, and its specific gravity is 3·9 to 4·25.

a. Deportment with acids and reactions of acid solutions of titanic acid.—Ignited titanic acid is insoluble in acids, except in hydrofluoric acid and in concentrated sulphuric acid. With bisulphate of potassa it gives upon sufficiently long-continued fusion a clear mass, which dissolves in a large proportion of cold water to a clear fluid. Hydrate of titanic acid dissolves, both moist and when dried without the aid of heat, in dilute acids, especially in hydrochloric and sulphuric acids. All the solutions of titanic acid in hydrochloric or sulphuric acid, but more particularly the latter, when subjected in a highly dilute state to *long-continued* boiling, deposit titanic acid as a white powder, insoluble in dilute acids. Presence of much free acid retards the separation and diminishes the quantity of the precipitate. The precipitate which separates from the hydrochloric acid solution may, indeed, be filtered, but it will pass milky through the filter upon washing, except an acid or chloride of ammonia be added to the washing water. *Solution of potassa* throws down from solutions of titanic acid in hydrochloric or sulphuric acid hydrate of titanic acid as a bulky white precipitate, which is insoluble in an excess of the precipitant; *ammonia, sulphide of ammonium, carbonate of the alkalies,* and carbonate of baryta act in the same way. The precipitate, thrown down cold and washed with cold water, is soluble in hydrochloric acid and in dilute sulphuric acid; presence of tartaric acid prevents its formation. *Ferrocyanide of potassium* produces in acid solutions of titanic acid a dark-brown precipitate; infusion of galls a brownish precipitate, which speedily turns orange-red. *Metallic zinc* produces, in consequence of the ensuing reduction of titanic acid to sesquioxide of titanium, at first a blue coloration of the solution, afterwards a blue precipitate of hydrate of sesquioxide of titanium.

b. Reactions with alkalies.—Recently precipitated hydrate of titanic acid is almost absolutely insoluble in solution of potassa. If titanic acid is fused together with *hydrate of potassa,* and the fused mass treated with water, the solution contains a little more titanic acid. By fusion with *carbonates of the alkalies* neutral titanates of the alkalies are formed, with expulsion of carbonic acid. Water extracts from the fused mass free alkali and alkaline carbonate, leaving behind acid titanate of alkali which dissolves in hydrochloric acid. Titanic acid mixed with charcoal gives upon ignition in a *stream of chlorine* bichloride of titanium as a volatile liquid, which emits copious fumes in the air. *Phosphate of soda and ammonia* readily dissolves titanic acid in the outer flame to a clear bead of a yellowish color whilst hot, but colorless when cold. Upon long-continued exposure to a strong reducing flame this bead acquires a yellow tint, which turns to red when the bead is half cold, and to violet when quite cold.

Addition of a little tin promotes the reduction. If a small quantity of sulphate of protoxide of iron is added, the bead obtained in the reducing flame looks blood red.

11. TANTALIC ACID.

Tantalum forms two oxides, TaO_3 and TaO_5. The latter, which is called tantalic acid, is found in tantalite, yttro-tantalite, and some other rare minerals. Tantalic acid is white, and remains so upon ignition (difference between tantalic acid and titanic acid). Ignited tantalic acid has a specific gravity of from 7 to 8. Tantalic acid combines with acids and with bases.

a. Reactions with acids.—The ignited acid is insoluble in hydrochloric acid and in concentrated sulphuric acid. It fuses with sulphate of potassa to a mass which on extraction with water leaves behind tantalic acid in combination with sulphuric acid (difference between tantalic acid and titanic acid; but which affords only an imperfect means of separating the two acids from each other). Ignition in an atmosphere of ammonia converts this compound of sulphuric acid with tantalic acid into pure tantalic acid. If a solution of tantalate of an alkali is mixed with hydrochloric acid in excess, the precipitate which forms at first redissolves to an opalescent fluid. *Ammonia* and *sulphide of ammonium* throw down from it hydrated or acid tantalate of ammonia; presence of tartaric acids prevents precipitation. *Sulphuric acid* throws down from the opalescent solution sulphate of tantalic acid. If tantalates of the alkalies be strongly acidified with hydrochloric acid, and then brought into contact with *zinc*, even addition of sulphuric acid will fail to produce a blue coloration of the fluid, or at all events the coloration will only be slight; but if solid chloride of tantalum be dissolved in concentrated sulphuric acid, and water and zinc added to the solution, the fluid will turn blue, not changing to brown by standing. *b. Reactions with alkalies.*—By continued fusion with *hydrate of potassa* tantalate of potassa is formed; the fused mass dissolves in water. By fusion with *hydrate of soda* a turbid mass is obtained; a little water poured on this mass will dissolve out the excess of soda, leaving the whole of the tantalate of soda undissolved, as this latter salt is insoluble in solution of soda; but the tantalate of soda will dissolve in water after the removal of the excess of soda. Solution of soda throws down from this solution the tantalate of soda; if the precipitant be added slowly, the form of the precipitate is crystalline. Carbonic acid throws down from solutions of tantalates of the alkalies acid salts, which are not dissolved by boiling with solution of carbonate of soda. *Sulphuric acid* throws down even from dilute solutions of tantalates of the alkalies sulphate of tantalic acid; *ferrocyanide of potassium* and *infusion of galls* produce precipitates only in acidified solutions; the precipitate produced by the former is yellow, by the latter light brown. By ignition with *charcoal* in pure dry *chlorine gas* white yellow chloride of tantalum is formed, which sublimes in crystals; if the tantalic acid contains titanic acid, this reaction is attended moreover by the formation of bichloride of titanium, which emits copious fumes in the air. *Phosphate of soda and ammonia* dissolves tantalic acid to a colorless bead, which remains colorless even in the inner flame, and does not acquire a blood-red tint by addition of sulphate of protoxide of iron (difference between tantalic acid and titanic acid).

12. HYPONIOBIC ACID.

Niobium forms two oxides, viz., hyponiobic acid (Nb_2O_3) and niobic acid (NbO_2). Hyponiobic acid is a rare substance; it is found in columbite, samarskite, &c. It is white, but turns transiently yellow when ignited (difference between hyponiobic acid and tantalic acid). Specific gravity varies from 4·6 to 6·5 at the most (difference from tantalic acid). Hyponiobic acid combines with acids and with bases. *a. Acid solutions of hyponiobic acid.*—Concentrated sulphuric acid dissolves hyponiobic acid upon heating; by addition of a large proportion of cold water a clear solution is obtained, from which the hyponiobic acid separates in combination with sulphuric acid, slowly and gradually in the cold, rapidly upon boiling. By washing the precipitate with carbonate of ammonia, then with highly dilute hydrochloric acid, the hydrate is obtained —by ignition in an atmosphere of carbonate of ammonia we obtain the acid. *Ammonia* and *sulphide of ammonium* precipitate acid hyponiobate of ammonia. Hyponiobic acid is readily dissolved by fusion with bisulphate of potassa; on treating the fused mass with hot water sulphate of hyponiobic acid is left behind undissolved. *b. Alkaline solutions.*—Hyponiobic acid fuses with *hydrate of potassa* to a clear mass, which is soluble in water; with *hydrate of soda* it fuses to a turbid mass; water dissolves the excess of soda out of this mass; after the removal of the solution of soda, the hyponiobate of soda dissolves in water. Fusion with carbonate of soda gives rise to similar reactions as fusion with hydrate of soda. Solution of soda slowly added to the aqueous solution separates from it crystallized hyponiobate of soda. The solutions of the hyponiobates of the alkalies are not rendered turbid by boiling; *sulphuric acid* precipitates from them the whole of the hyponiobic acid upon boiling; the precipitation by chloride of ammonium is less complete; *hydrochloric acid* produces a precipitate which does not re-dissolve in an excess of the precipitant (difference between hyponiobic acid and titanic and tantalic acids). If *zinc* is added after the hydrochloric acid, the precipitated hyponiobic acid acquires a blue tint, which gradually changes to brown (difference from tantalic acid). *Carbonic acid* throws down acid hyponiobate of alkali, which is soluble in boiling dilute solution of carbonate of soda (means of separating hyponiobic acid from tantalic acid). *Ferrocyanide of potassium* produces no precipitate, except in acidified solutions, which it precipitates dark brown. *Infusion of galls* produces no precipitate, except in acidified solutions, which it precipitates deep orange-red.

By ignition of hyponiobic acid, mixed with charcoal, in a stream of *chlorine gas*, white solid sesquichloride of niobium, and yellow solid somewhat more volatile bichloride of niobium are obtained. *Phosphate of soda and ammonia* dissolves hyponiobic acid copiously; the bead produced in the inner flame shows a violet, blue, or brown color, according to the mode of preparation of the hyponiobic acid, and the quantity used of it; addition of sulphate or protoxide of iron imparts a blood-red tint to the bead.

For the best methods and processes of detecting many, possibly even all the oxides of the third group in presence of each other, the reader is referred to Part II., Section III.

§ 105.

FOURTH GROUP.

More common oxides of the fourth group:—OXIDE OF ZINC, PROTOXIDE OF MANGANESE, PROTOXIDE OF NICKEL, PROTOXIDE OF COBALT, PROTOXIDE OF IRON, SESQUIOXIDE OF IRON.

Rarer oxides of the fourth group:—SESQUIOXIDE OF URANIUM, OXIDES OF VANADIUM, OXIDES OF THALLIUM.

Properties of the group.—The solutions of the oxides of the fourth group, if containing a stronger free acid, are not precipitated by hydrosulphuric acid; nor are neutral solutions, at least not completely. But alkaline solutions are completely precipitated by hydrosulphuric acid; and so are other solutions if a sulphide of an alkali metal is used as the precipitant, instead of hydrosulphuric acid. The precipitated metallic sulphides corresponding to the several oxides are insoluble in water; some of them are readily soluble in dilute acids; others (sulphide of nickel and sulphide of cobalt) dissolve only with very great difficulty in these menstrua. Some of them are insoluble in sulphides of the alkali metals, others (nickel) are sparingly soluble in them, under certain circumstances, whilst others again (vanadium) are completely soluble. The oxides of the fourth group differ accordingly from those of the first and second groups in this, that their solutions are precipitated by sulphide of ammonium, and from those of the third group inasmuch that the precipitates produced by sulphide of ammonium are sulphides, and not hydrated oxides, as is the case with alumina, sesquioxide of chromium, &c.

Special Reactions of the more common Oxides of the fourth group.

§ 106.

a. OXIDE OF ZINC (ZnO).

1. METALLIC ZINC is bluish-white and very bright; when exposed to the air, a thin coating of basic carbonate of zinc forms on its surface.

It is of medium hardness, ductile at a temperature of between 212° and 302° Fah., but otherwise more or less brittle; it fuses readily on charcoal before the blowpipe, boils afterwards, and burns with a bluish-green flame, giving off white fumes, and coating the charcoal support with oxide. Zinc dissolves in hydrochloric and sulphuric acids, with evolution of hydrogen gas; in dilute nitric acid, with evolution of nitrous oxide; in more concentrated nitric acid, with evolution of nitric oxide.

2. The OXIDE OF ZINC and its HYDRATE are white powders, which are insoluble in water, but dissolve readily in hydrochloric, nitric, and sulphuric acids. The oxide of zinc acquires a lemon-yellow tint when heated, but it reassumes its original white color upon cooling. When ignited before the blowpipe, it shines with considerable brilliancy.

3. The salts of OXIDE OF ZINC are colorless; part of them are soluble in water, the rest in acids. The neutral salts of zinc which are soluble in water redden blue litmus-paper, and are readily decomposed by heat, with the exception of sulphate of zinc, which can bear a dull red heat without undergoing decomposition. Chloride of zinc is volatile at a red heat.

4. *Hydrosulphuric acid* precipitates from neutral solutions of salts of zinc a portion of the metal as white hydrated SULPHIDE of ZINC (Zn S). In acid solutions this reagent fails altogether to produce a precipitate if the free acid present is one of the stronger acids; but from a solution of oxide of zinc in acetic acid it throws down the whole of the zinc, even if the acid is present in excess.

5. *Sulphide of ammonium* throws down from *neutral*, and *hydrosulphuric acid* from *alkaline* solutions of salts of zinc, the whole of the metal as hydrated SULPHIDE OF ZINC, in the form of a white precipitate. Chloride of ammonium greatly promotes the separation of the precipitate. From very dilute solutions the precipitate separates only after long standing. This precipitate is not redissolved by an excess of sulphide of ammonium, nor by potassa or ammonia; but it dissolves readily in hydrochloric acid, nitric acid, and dilute sulphuric acid. It is insoluble in acetic acid.

6. *Potassa and soda* throw down from solutions of salts of zinc HYDRATED OXIDE OF ZINC (Zn O, H O), in the form of a white gelatinous precipitate, which is readily and completely redissolved by an excess of the precipitant. Upon boiling these alkaline solutions they remain, if concentrated, unaltered; but from dilute solutions nearly the whole of the oxide of zinc separates as a white precipitate. Chloride of ammonium does not precipitate alkaline solutions of oxide of zinc.

7. *Ammonia* also produces in solutions of oxide of zinc, if they do not contain a large excess of free acid, a precipitate of HYDRATED OXIDE OF ZINC, which readily dissolves in an excess of the precipitant. The concentrated solution turns turbid when mixed with water. On boiling the concentrated solution part of the oxide of zinc separates immediately; on boiling the dilute solution all the oxide of zinc precipitates.

8. *Carbonate of soda* produces in solutions of salts of zinc a precipitate of BASIC CARBONATE OF ZINC ($3 [ZnO, HO] + 2 [ZnO, CO_2]$ + 4 aq.), which is insoluble in an excess of the precipitant. Presence of salts of ammonia in great excess prevents the formation of this precipitate.

9. *Carbonate of ammonia* also produces in solutions of salts of zinc the same precipitate of BASIC CARBONATE OF ZINC as carbonate of soda;

but this precipitate redissolves upon further addition of the precipitant. On boiling the dilute solution oxide of zinc precipitates.

10. *Carbonate of baryta* fails to precipitate solutions of salts of zinc in the cold, with the exception of the sulphate.

11. If a mixture of oxide of zinc or one of its salts with *carbonate of soda* is exposed to the *reducing flame* of the blowpipe, the charcoal support becomes covered with a slight coating of OXIDE OF ZINC, which presents a yellow color whilst hot, and turns white upon cooling. This coating is produced by the reduced metallic zinc volatilizing at the moment of its reduction, and being reoxidized in passing through the outer flame.

12. If oxide of zinc or one of the salts of zinc is moistened with solution of *nitrate of protoxide of cobalt*, and then heated before the blowpipe, an unfused mass is obtained of a beautiful GREEN color: this mass is a compound of oxide of zinc with protoxide of cobalt. If therefore in the experiment described in 11 the charcoal is moistened around the little cavity with solution of nitrate of protoxide of cobalt, the coating appears *green* when *cold*.

§ 107.

b. PROTOXIDE OF MANGANESE (MnO).

1. METALLIC MANGANESE is whitish-gray, brittle, extremely hard, and fuses with very great difficulty. It takes a high polish, which it rather speedily loses again upon exposure to the air. At first it blues, like steel when heated, but after a time it becomes covered with a layer of brown oxide. It oxidizes only slowly in water at the common temperature, but more rapidly in boiling water. It dissolves readily in acids, the solutions contain protoxide.

2. PROTOXIDE OF MANGANESE is grayish-green; the hydrated protoxide is white. Both the protoxide and its hydrate absorb oxygen from the air, and are converted into the brown proto-sesquioxide. They are readily soluble in hydrochloric, nitric, and sulphuric acids. All the *higher oxides of manganese* without exception dissolve to protochloride, with evolution of chlorine, when heated with hydrochloric acid; to sulphate of protoxide, with evolution of oxygen, when heated with concentrated sulphuric acids.

3. The SALTS OF PROTOXIDE OF MANGANESE are colorless or pale red; part of them are soluble in water, the rest in acids. The salts soluble in water are readily decomposed by a red heat, with the exception of the sulphate. The solutions do not alter vegetable colors.

4. *Hydrosulphuric acid* does not precipitate acid solutions of salts of protoxide of manganese; neutral solutions also it fails to precipitate, or precipitates them only very imperfectly.

5. *Sulphide of ammonium* throws down from neutral, and *hydrosulphuric acid* from alkaline solutions of salts of protoxide of manganese, the whole of the metal as hydrated SULPHIDE OF MANGANESE (MnS), in form of a light flesh-colored* precipitate, which acquires a dark-brown color in the air; this precipitate is insoluble in sulphide of ammonium and in alkalies, but readily soluble in hydrochloric, nitric, and acetic acids. The separation of the precipitate is materially promoted by

---

* If the quantity of the precipitate is only trifling, the color appears yellowish-white.

addition of chloride of ammonium. From very dilute solutions the precipitate separates only after standing some time in a warm place. Solutions containing much free ammonia must first be neutralized with hydrochloric acid.

6. *Potassa, soda,* and *ammonia* produce whitish precipitates of HYDRATE OF PROTOXIDE OF MANGANESE (MnO, HO), which upon exposure to the air speedily acquire a brownish and finally a deep blackish-brown color, owing to the conversion of the hydrated protoxide into hydrated proto-sesquioxide by the absorption of oxygen from the air. Ammonia and carbonate of ammonia do not redissolve this precipitate; but presence of chloride of ammonium prevents the precipitation by ammonia altogether, and that by potassa partly. Of *already formed* precipitates solution of chloride of ammonium redissolves only those parts which have not yet undergone peroxidation. The solution of the hydrated protoxide of manganese in chloride of ammonium is owing to the disposition of the salts of protoxide of manganese to form double salts with salts of ammonia. The ammoniacal solutions of the double salts turn brown in the air, and deposit dark-brown hydrate of protosesquioxide of manganese.

7. If a few drops of a fluid containing protoxide of manganese, and free from chlorine, are sprinkled on binoxide of lead or red-lead, and nitric acid free from chlorine is added, the mixture boiled and allowed to settle, the NITRATE OF SESQUIOXIDE OF MANGANESE formed imparts a purple-red color to the fluid.

8. *Carbonate of baryta* does not precipitate protoxide of manganese from aqueous solutions of its salts upon digestion in the cold, with the exception of sulphate of protoxide of manganese.

9. If any compound of manganese, in a state of minute division, is fused with *carbonate of soda* on a platinum wire, or on a small strip of platinum foil (heated by directing the flame upon the lower surface), in the *outer* flame of the blowpipe, MANGANATE OF SODA (NaO, MnO$_3$) is formed, which makes the fused mass appear GREEN while hot, and of a BLUISH-GREEN tint after cooling, the bead at the same time becoming turbid. This reaction enables us to detect the smallest traces of manganese.

10. *Borax* and *phosphate of soda and ammonia* dissolve manganese compounds in the *outer* gas or blowpipe flame to clear VIOLET-RED beads, which upon cooling acquire an AMETHYST-RED tint: they lose their color in the *inner* flame, owing to a reduction of the sesquioxide to protoxide. The bead which borax forms with manganese compounds appears black when containing a considerable portion of sesquioxide of manganese, but that formed by phosphate of soda and ammonia never loses its transparency. The latter loses its color in the inner flame of the blowpipe far more readily than the former.

§ 108.

c. PROTOXIDE OF NICKEL (NiO).

1. METALLIC NICKEL in the fused state is silvery white, inclining to gray; it is bright, hard, malleable, ductile, difficultly fusible; it does not oxidize in the air at the common temperature, but it oxidizes slowly upon ignition; it is attracted by the magnet and may itself become magnetic. It slowly dissolves in hydrochloric acid and dilute sulphuric

acid upon the application of heat, the solution being attended with evolution of hydrogen gas. It dissolves readily in nitric acid. The solutions contain protoxide of nickel.

2. HYDRATE OF PROTOXIDE OF NICKEL is green; and remains unaltered in the air, but is converted by ignition into grayish-green PROTOXIDE OF NICKEL. Both the protoxide and its hydrate are readily soluble in hydrochloric, nitric, and sulphuric acids. But the protoxide which crystallizes in octahedrons is insoluble in acids; it dissolves however in fusing bisulphate of potassa. SESQUIOXIDE OF NICKEL is black; it dissolves in hydrochloric acid to protochloride.

3. Most of the SALTS OF PROTOXIDE OF NICKEL are yellow in the anhydrous, green in the hydrated state; their solutions are of a light green color. The soluble neutral salts slightly redden litmus-paper, and are decomposed at a red heat.

4. *Hydrosulphuric acid* does not precipitate solutions of salts of protoxide of nickel with strong acids in presence of free acids; in the absence of free acid a small portion of the nickel gradually separates as black SULPHIDE OF NICKEL (Ni S).—Acetate of protoxide of nickel is not precipitated, or scarcely at all, in presence of free acetic acid. But in the absence of free acid the greater part of the nickel is thrown down by long-continued action of hydrosulphuric acid upon the fluid.

5. *Sulphide of ammonium* produces in neutral, and *hydrosulphuric acid* in alkaline solutions of salts of protoxide of nickel, a black precipitate of hydrated SULPHIDE OF NICKEL (Ni S), which is not altogether insoluble in sulphide of ammonium, especially if containing free ammonia; the fluid from which the precipitate has been thrown down exhibits therefore usually a brownish color. Sulphide of nickel dissolves scarcely at all in acetic acid, with great difficulty in hydrochloric acid, but readily in nitro-hydrochloric acid upon application of heat.

6. *Potassa* and *soda* produce a light green precipitate of HYDRATE OF PROTOXIDE OF NICKEL (NiO, HO), which is insoluble in an excess of the precipitants, and unalterable in the air. Carbonate of ammonia dissolves this precipitate, when filtered and washed, to a greenish-blue fluid, from which potassa or soda reprecipitates the nickel as an apple-green hydrate of protoxide of nickel.

7. *Ammonia* added in small quantity to solutions of protoxide of nickel produces in them a trifling greenish turbidity; upon further addition of the reagent this redissolves readily to a blue fluid containing a compound of PROTOXIDE OF NICKEL AND AMMONIA. Potassa and soda precipitate from this solution hydrate of protoxide of nickel. Solutions containing salts of ammonia or free acid are not rendered turbid by ammonia.

8. *Cyanide of Potassium* produces a yellowish-green precipitate of CYANIDE OF NICKEL (NiCy), which redissolves readily in an excess of the precipitant as a double cyanide of nickel and potassium (NiCy + KCy); the solution is brownish-yellow. If sulphuric acid or hydrochloric acid is added to this solution, the cyanide of potassium is decomposed, and the cyanide of nickel reprecipitated. From more highly dilute solutions the cyanide of nickel separates only after some time; it is very difficultly soluble in an excess of the precipitating acids in the cold, but more readily upon boiling.

9. *Carbonate of baryta* does not precipitate protoxide of nickel from aqueous solutions of its salts, upon digestion in the cold, with the exception of sulphate of protoxide of nickel.

10. *Nitrite of potassa*, used in conjunction with acetic acid, fails to precipitate even concentrated solutions of nickel.

11. *Borax* and *phosphate of soda and ammonia* dissolve compounds of protoxide of nickel in the *outer* flame of the blowpipe to clear beads; the bead produced with borax is violet whilst hot, reddish-brown when cold; the bead produced with the phosphate of soda and ammonia is reddish, inclining to brown whilst hot, but turns yellow or reddish-yellow upon cooling. The bead which phosphate of soda and ammonia forms with salts of protoxide of nickel remains unaltered in the *inner* flame of the blowpipe, but that formed with borax turns gray and turbid from reduced metallic nickel. Upon continued exposure to the blowpipe flame the particles of nickel unite, but without fusing to a bead, and the glass becomes colorless.

§ 109.

d. Protoxide of Cobalt (CoO).

1. METALLIC COBALT in the fused state is steel-gray, pretty hard, slightly malleable, ductile, difficultly fusible, and magnetic; susceptible of polish; it oxidizes very slowly in the air at the common temperature, more rapidly at a red heat; with acids it presents the same reactions as nickel. The solutions contain protoxide of cobalt.

2. PROTOXIDE OF COBALT is an olive-green, its hydrate a pale red powder. Both dissolve readily in hydrochloric, nitric, and sulphuric acids. SESQUIOXIDE OF COBALT (Co_2O_3) is black; it dissolves in hydrochloric acid to protochloride, with evolution of chlorine.

3. The SALTS OF PROTOXIDE OF COBALT containing water of crystallization are red, the anhydrous salts mostly blue. The moderately concentrated solutions appear of a light red color, which they retain even though considerably diluted. The soluble neutral salts redden litmus-paper slightly, and are decomposed at a red heat; sulphate of protoxide of cobalt alone can bear a moderate red heat without suffering decomposition. When a solution of chloride of cobalt is evaporated, the light red color changes towards the end of the operation to blue; addition of water restores the red color.

4. *Hydrosulphuric acid* does not precipitate solutions of salts of protoxide of cobalt with strong acids, if they contain free acid; from neutral solutions it gradually precipitates part of the cobalt as black sulphide of cobalt (CoS). Acetate of protoxide of cobalt is not precipitated, or to a very slight extent, in presence of free acetic acid. But in the absence of free acid it is completely precipitated, or almost completely.

5. *Sulphide of ammonium* precipitates from neutral, and *hydrosulphuric acid* from alkaline solutions of salts of protoxide of cobalt, the whole of the metal as black hydrated SULPHIDE OF COBALT (CoS). Chloride of ammonium promotes the precipitation most materially. Sulphide of cobalt is insoluble in alkalies and sulphide of ammonium, scarcely soluble in acetic acid, very difficultly soluble in hydrochloric acid, but readily so in nitrohydrochloric acid, upon application of heat.

6. *Potassa* and *soda* produce in solutions of cobalt BLUE precipitates of BASIC SALTS OF COBALT, which turn GREEN upon exposure to the air, owing to the absorption of oxygen; upon boiling they are converted into pale red HYDRATE OF PROTOXIDE OF COBALT, which contains alkali, and generally appears rather discolored from an admixture of sesquioxide formed in the process. These precipitates are insoluble in solutions of potassa and soda; but neutral carbonate of ammonia dissolves the washed precipitates completely to intensely violet-red fluids, in which a somewhat larger proportion of potassa or soda produces a blue precipitate, the fluid still retaining its violet color.

7. *Ammonia* produces the same precipitate as potassa, but this redissolves in an excess of the ammonia to a reddish-brown fluid, from which solution of potassa or soda throws down part of the cobalt as a blue basic salt. Ammonia fails to precipitate solutions of protoxide of cobalt containing salts of ammonia or free acid.

8. Addition of *cyanide of potassium* to a solution of cobalt gives rise to the formation of a brownish-white precipitate of PROTOCYANIDE OF COBALT (Co Cy), which dissolves readily as a double cyanide of cobalt and potassium in an excess of solution of cyanide of potassium. Acids precipitate from this solution cyanide of cobalt. But if the solution is boiled with cyanide of potassium in excess, in presence of free hydrocyanic acid (liberated by addition of one or two drops of hydrochloric acid), a double compound of sesquicyanide of cobalt and cyanide of potassium (K_3, $Co_2 Cy_6 = K_3 Ckdy$) is formed, in the solution of which acids *produce no precipitate* (essential difference between cobalt and nickel).

9. *Carbonate of baryta* acts upon solutions of salts of protoxide of cobalt the same as upon solutions of salts of protoxide of nickel.

10. If *nitrite of potassa* is added in not too small proportion to a solution of protoxide of cobalt, then acetic acid to *strongly* acid reaction, and the mixture put in a moderately warm place, all the cobalt separates, from concentrated solutions immediately or very soon, from dilute solutions after some time, as NITRITE OF SESQUIOXIDE OF COBALT AND POTASSA ($Co_2 O_3$, 3 K O, 5 N O_5, 2 H O), in the form of a crystalline precipitate of a beautiful yellow color. The mode in which this precipitate forms may be seen from the following equation: 2 (CoO, SO_3) + 6 (KO, NO_3) + $\overline{A} = KO, \overline{A} + 2 KO, SO_3 + Co_2 O_3$, 3 KO, 5 NO_3 + NO_2. The precipitate is very perceptibly soluble in pure water, but altogether insoluble in more concentrated solutions of salts of potassa and in alcohol. When boiled with water it dissolves, though not copiously, to a red fluid, which remains clear upon cooling, and from which alkalies throw down hydrate of protoxide of cobalt (*Fischer, Aug. Stromeyer*). This excellent reaction enables us to distinguish and separate nickel from cobalt.

11. *Borax* dissolves compounds of cobalt both in the *inner* and *outer* flame of the blowpipe, giving clear beads of a magnificent BLUE color, which appear violet by candlelight, and almost black if the cobalt is present in considerable proportion. This test is as delicate as it is characteristic. *Phosphate of soda and ammonia* manifests with salts of cobalt before the blowpipe an analogous but less delicate reaction.

§ 110.

e. PROTOXIDE OF IRON (Fe O).

1. METALLIC IRON in the pure state has a light whitish gray color (iron containing carbon is more or less gray); the metal is hard, lustrous, malleable, ductile, exceedingly difficult to fuse, and is attracted by the magnet. In contact with air and moisture a coating of rust (hydrate of sesquioxide of iron) forms on its surface: upon ignition in the air a coat-

ing of black protosesquioxide. Hydrochloric acid and dilute sulphuric acid dissolve iron, with evolution of hydrogen gas; if the iron contains carbon, the hydrogen is mixed with carbide of hydrogen. The solutions contain protoxide. Dilute nitric acid dissolves iron in the cold to nitrate of protoxide, with evolution of nitrous oxide; at a high temperature to nitrate of sesquioxide, with evolution of nitric oxide; if the iron contains carbon, some carbonic acid is also evolved, and there is left undissolved a brown substance resembling humus, which is soluble in alkalies; under certain circumstances graphite is also left behind.

2. PROTOXIDE OF IRON is a black powder; its hydrate is a white powder, which in the moist state absorbs oxygen and speedily acquires a grayish-green, and ultimately a brownish-red color. Both the protoxide and its hydrate are readily dissolved by hydrochloric, sulphuric, and nitric acids.

3. The SALTS OF PROTOXIDE OF IRON have in the anhydrous state a white, in the hydrated state a greenish color; their solutions only look greenish when concentrated. The latter absorb oxygen when exposed to the air, and are converted into salts of the protosesquioxide, with precipitation of basic salts of sesquioxide. Chlorine or nitric acid converts them by boiling into salts of sesquioxide. The soluble neutral salts redden litmus-paper, and are decomposed at a red heat.

4. Solutions of salts of protoxide of iron made acid by strong acids are not precipitated by *hydrosulphuric acid;* nor are neutral solutions of salts of protoxide of iron acidified with weak acids precipitated by this reagent, or at the most but very incompletely; the precipitates are in that case of a black color.

5. *Sulphide of ammonium* precipitates from neutral, and hydrosulphuric acid from alkaline solutions of salts of protoxide of iron, the whole of the metal as black hydrated PROTOSULPHIDE OF IRON (FeS), which is insoluble in alkalies and sulphides of the alkali metals, but dissolves readily in hydrochloric and nitric acids: this black precipitate turns reddish-brown in the air by oxidation. To highly dilute solutions of protoxide of iron addition of sulphide of ammonium imparts a green color, and it is only after some time that the protosulphide of iron separates as a black precipitate. Chloride of ammonium promotes the precipitation most materially.

6. *Potassa* and *ammonia* produce a precipitate of HYDRATE OF PROTOXIDE OF IRON (FeO, HO), which in the first moment looks almost white, but acquires after a very short time a dirty green, and ultimately a reddish-brown color, owing to absorption of oxygen from the air. Presence of salts of ammonia prevents the precipitation by potassa partly, and that by ammonia altogether. If alkaline solutions of protoxide of iron thus obtained by the agency of salts of ammonia are exposed to the air, hydrate of protosesquioxide of iron and hydrate of sesquioxide of iron precipitate.

7. *Ferrocyanide of potassium* produces in solutions of protoxide of iron a bluish-white precipitate of FERROCYANIDE OF POTASSIUM AND IRON (K$_2$ Fe$_2$ Cfy$_3$), which, by absorption of oxygen from the air, speedily acquires a blue color. Nitric acid or chlorine converts it immediately into Prussian blue, $3 (K_2 Fe_2 Cfy_3) + 4 Cl = 3 KCl + FeCl + 2 (Fe_4 Cfy_3)$.

8. *Ferricyanide of potassium* produces a magnificently blue precipitate of FERRICYANIDE OF IRON (Fe$_3$ Cfdy). This precipitate does not differ in color from Prussian blue. It is insoluble in hydrochloric acid, but is

readily decomposed by potassa. In highly dilute solutions of salts of protoxide of iron the reagent produces simply a deep blue-green coloration.

9. *Sulphocyanide of potassium* does not alter solutions of protoxide of iron free from sesquioxide.

10. *Carbonate of baryta* does not precipitate solutions of protoxide of iron in the cold, with the exception of the sulphate of protoxide of iron.

11. *Borax* dissolves protoxide of iron compounds in the oxidizing flame, giving beads varying in color from yellow to dark red; when cold the beads vary from colorless to dark yellow. In the inner flame the beads change to bottle-green, owing to the reduction of the newly-formed sesquioxide to protosesquioxide. *Phosphate of soda and ammonia* show a similar reaction with the salts of protoxide of iron; the beads produced with this reagent lose their color upon cooling still more completely than is the case with those produced with borax; the signs of the ensuing reduction in the reducing flame are also less marked.

§ 111.

f. SESQUIOXIDE OF IRON ($Fe_2 O_3$).

1. The native crystallized SESQUIOXIDE OF IRON is steel-gray; the native as well as the artificially prepared sesquioxide of iron gives upon trituration a brownish-red powder; the color of hydrate of sesquioxide of iron is more inclined to reddish-brown. Both the sesquioxide and its hydrate, dissolve in hydrochloric, nitric, and sulphuric acids; the hydrate dissolves readily in these acids, but the anhydrous sesquioxide dissolves with greater difficulty, and completely only after long exposure to heat. PROTOSESQUIOXIDE OF IRON ($FeO, Fe_2 O_3$) is black; it dissolves in hydrochloric acid to protochloride and sesquichloride, in aqua regia to sesquichloride.

2. The neutral anhydrous SALTS OF SESQUIOXIDE OF IRON are nearly white; the basic salts are yellow or reddish-brown. The color of the solutions is brownish-yellow, and becomes reddish-yellow upon the application of heat. The soluble neutral salts redden litmus-paper, and are decomposed by heat.

3. *Hydrosulphuric acid* produces in solutions made acid by stronger acids a milky white turbidity, proceeding from separated SULPHUR; the salt of the sesquioxide being at the same time converted into salt of the protoxide: $Fe_2 O_3, 3 SO_3 + HS = 2 (FeO SO_3) + HO, SO_3 + S$. If solution of hydrosulphuric acid is rapidly added to neutral solutions, the reaction is marked by a transient blackening of the fluid, besides the separation of sulphur. From solution of neutral acetate of sesquioxide of iron hydrosulphuric acid throws down the greater part of the iron; but in presence of a sufficient quantity of free acetic acid sulphur alone separates.

4. *Sulphide of ammonium* precipitates from neutral, and hydrosulphuric acid from alkaline solutions of salts of sesquioxide of iron, the whole of the metal as black hydrated PROTOSULPHIDE OF IRON (FeS): $Fe_2 Cl_3 + 3 NH_4 S = 3 NH_4 Cl + 2 FeS + S$. In very dilute solutions the reagent produces only a blackish-green coloration. The minutely divided protosulphide of iron subsides in such cases only after long standing. Chloride of ammonium most materially promotes the precipitation. Protosulphide of iron, as already stated (§ 110, 5), is inso-

luble in alkalies and alkaline sulphides, but dissolves readily in hydrochloric and nitric acids.

5. *Potassa* and *ammonia* produce bulky reddish-brown precipitates of HYDRATE OF SESQUIOXIDE OF IRON (Fe_2O_3, HO), which are insoluble in an excess of the precipitant as well as in salts of ammonia.

6. *Ferrocyanide of potassium* produces even in highly dilute solutions a magnificently blue precipitate of FERROCYANIDE OF IRON, or Prussian blue ($Fe_4 Cfy_3$): $2 (Fe_2 Cl_3) + 3 (Cfy, 2 K) = 6 KCl + Fe_4 Cfy_3$. This precipitate is insoluble in hydrochloric acid, but is decomposed by potassa, with separation of hydrate of sesquioxide of iron.

7. *Ferricyanide of potassium* deepens the color of solutions of salts of sesquioxide of iron to reddish-brown; but it fails to produce a precipitate.

8. *Sulphocyanide of potassium* imparts to acid solutions of salts of sesquioxide of iron a most intense blood-red color, arising from the formation of a soluble SULPHOCYANIDE OF IRON. Addition of acetate of soda destroys this color, hydrochloric acid restores it again. This test is the most delicate of all; it will indicate the presence of sesquioxide of iron even in fluids which are so highly dilute that every other reagent fails to produce the slightest visible alteration. The red coloration may in such cases be detected most distinctly by resting the test-tube upon a sheet of white paper, and looking through it from the top.

9. *Carbonate of baryta* precipitates even in the cold all the iron as HYDRATE OF SESQUIOXIDE MIXED WITH A BASIC SALT.

10. The reactions before the *blowpipe* are the same as with the protoxide.

§ 112.

Recapitulation and remarks.—On observing the reactions of the several oxides of the fourth group with solution of potassa, it would appear that the separation of the oxide of zinc, which is soluble in an excess of this reagent, might be readily effected by its means; but in the actual experiment we find that rather notable quantities of oxide of zinc are thrown down with the sesquioxide of iron, protoxide of cobalt, &c. To such an extent indeed that it is often impossible to demonstrate the presence of oxide of zinc in the alkaline filtrate.

Again, the reactions of the different oxides with chloride of ammonium and an excess of ammonia would lead to the conclusion that the separation of sesquioxide of iron from the protoxides of cobalt, nickel, and manganese, and from oxide of zinc, might be readily effected by these agents. But this method also if applied to the mixed oxides is inaccurate, since greater or smaller portions of the other oxides will always precipitate along with the sesquioxide of iron; and it may therefore happen that small quantities of cobalt, manganese, &c., altogether escape detection in this process.

It is far safer therefore to separate the other oxides of the fourth group from sesquioxide of iron by carbonate of baryta, as in that case the iron is precipitated free from oxide of zinc and protoxide of manganese, and, if chloride of ammonium is added previously to the addition of the carbonate of baryta, almost entirely free also from protoxide of nickel and protoxide of cobalt.

Protoxide of manganese may conveniently be separated from the protoxides of cobalt and nickel, as well as from oxide of zinc, by treating the washed precipitated sulphides with moderately dilute acetic acid, which dissolves the sulphide of manganese, leaving the other sulphides undissolved. If the acetic acid solution is now mixed with solution of potassa, the least trace of a precipitate will be sufficient to recognise the manganese before the blowpipe with carbonate of soda.

If the sulphides left undissolved by acetic acid are now treated, after washing, with very dilute hydrochloric acid, the sulphide of zinc dissolves, leaving almost the whole of the sulphides of cobalt and nickel behind. If the fluid is then boiled, and strongly concentrated to expel the hydrosulphuric acid, and afterwards treated with solution of potassa or soda in excess, the zinc is sure to be detected in the filtrate by hydrosulphuric acid.

Cobalt may mostly be readily and safely detected in presence of nickel by the reaction with borax in the inner flame of the blowpipe; to which end the filter with the mixed sulphides of nickel and cobalt upon it is burnt in a small porcelain dish, and a portion of the residue tested with borax in the inner flame. The detection of nickel in presence of cobalt is a less easy task. The best way of effecting it is to mix the concentrated solution containing the two metals with a sufficient quantity of nitrite of potassa, then add acetic acid to strongly acid reaction, and let the mixture stand at least twelve hours in a moderately warm place; when the cobalt will separate as nitrite of sesquioxide of cobalt and potassa; the nickel may then be readily precipitated from the filtrate by soda, and tested before the blowpipe, to make quite sure of its nature.

In practical analysis we generally separate the whole of the oxides of the fourth group as sulphides by precipitation with sulphide of ammonium. It is therefore in most cases still more convenient to separate nickel and cobalt, or at least the far larger portion of these two metals at the outset. To this end the moist precipitate of the sulphides is treated with water and some hydrochloric acid, with active stirring, but without application of heat. Nearly the whole of the sulphide of nickel and sulphide of cobalt is left behind undissolved, whilst all the other sulphides are dissolved. The undissolved residue of sulphide of cobalt and sulphide of nickel is filtered and washed, and treated as directed above. By boiling the filtrate with nitric acid the iron is converted from the state of protoxide, as it existed in the solution of the sulphide, into that of sesquioxide. After the free acid has been nearly neutralized by carbonate of soda, the iron may be thrown down as basic salt either by carbonate of baryta in the cold, or by acetate of soda and boiling. Manganese and zinc alone remain in the filtrate; these metals are then also precipitated with sulphide of ammonium and some chloride of ammonium, the precipitate is filtered and washed, and the two metals are finally separated from each other by acetic acid as directed above, or, after removal of the baryta by sulphuric acid and great concentration, by solution of potassa or soda. The trifling quantities of cobalt and nickel, dissolved on the first treatment of the sulphide precipitate with dilute hydrochloric acid, remain with the sulphide of zinc in the separation of the latter from the sulphide of manganese by acetic acid—or with the protoxide of manganese if the separation of the oxide is effected by solution of potassa or soda. The sulphide of zinc may be extracted from the blackish precipitate by dilute hydrochloric acid, and the detection of the manganese in presence of the cobalt and nickel may be readily effected by means of soda in the outer flame.

Protoxide and sesquioxide of iron may be detected in presence of each other by testing for the former with ferricyanide of potassium, for the latter with ferrocyanide of potassium or, better still, with sulphocyanide of potassium.

In conclusion it is necessary to mention that alkalies fail to precipitate the oxides of the fourth group in presence of non-volatile organic substances (such as sugar, tartaric acid, &c.). We have already seen that the same remark applies to alumina. As regards sesquioxide of iron, even carbonate of baryta fails to precipitate this in presence of non-volatile organic substances.

Special Reactions of the rarer Oxides of the fourth group.

§ 113.

a. OXIDES OF URANIUM.

This metal is found in a few minerals, as pitchblende, uran-ochre, &c. The sesquioxide of the metal is used to stain glass yellowish-green. Uranium forms two oxides, viz., the protoxide (UO), and the sesquioxide (U_2O_3). The protoxide is brown; it dissolves in nitric acid to nitrate of sesquioxide. The hydrate of the sesquioxide is yellow; at about 572° Fahrenheit it loses its water and turns red; it is converted by ignition into the dark blackish-green protosesquioxide. The solutions of sesquioxide of uranium in acids are yellow. *Hydrosulphuric acid* does not alter them; *sulphide of ammonium* throws down from them, after neutralization of the free acid, a slowly subsiding precipitate, varying in color from dirty yellow to reddish-brown, and nearly blood-red, according to the presence and quantity of chloride of ammonium, ammonia, and sulphide of ammonium. Chloride of ammonium materially promotes the precipitation. The precipitate is readily soluble in acids, even in acetic acid, but insoluble in sulphide of ammonium. It does not consist of pure sulphide of uranium, but contains, besides uranium and sulphur, also ammonium, oxygen, and water. *Ammonia, potassa,* and *soda* produce yellow precipitates of sesquioxide of uranium and alkali, which are insoluble in an excess of the precipitants. *Carbonate of ammonia* and *bicarbonate of potassa* or *soda* produce yellow precipitates of carbonate of sesquioxide of uranium and alkali, which *readily redissolve in an excess of the precipitants*. Potassa and soda throw down from such solutions the whole of the sesquioxide of uranium. *Carbonate of baryta* completely precipitates solutions of sesquioxide of uranium, even in the cold. *Ferrocyanide of potassium* produces a reddish-brown precipitate (a most delicate test for uranium). *Borax* and *phosphate of soda and ammonia* give with uranium compounds in the inner flame of the blowpipe green beads, in the outer flame yellow beads, which acquire a yellowish-green tint on cooling.

b. OXIDES OF VANADIUM.

Vanadium is a rare metal, found in vanadate of lead, and occasionally in small quantity in iron and copper ores, and in the slags left by them. Vanadium forms three oxides, viz., Protoxide (VO), binoxide (VO_2), and vanadic acid (VO_3). The lower oxides are converted into the acid by heating with nitric acid or aqua regia, or by fusion with nitrate of potassa. Vanadic acid is yellowish-red; it melts at an incipient red heat, and solidifies in crystals on cooling; it is not volatile. It dissolves very sparingly in water, but the solution acts powerfully upon moist litmus-paper, imparting a decided red tint to it. Vanadic acid combines with acids and with bases. *a. Acid solutions.*—The stronger acids dissolve vanadic acid to yellow or red fluids. The solutions are often decolorized by boiling. Sulphurous acid, many metals, organic substances, &c., reduce vanadic acid and color the fluid blue by the formation of binoxide of vanadium. *Hydrosulphuric acid* does not precipitate acidified solutions, but imparts a blue tint to them, sulphur separating at the same time. *Sulphide of ammonium* imparts a brown tint to solutions of vanadic acid; on acidifying the solution with hydrochloric acid, or, better still, with sulphuric acid, brown tersulphide of vanadium separates, which dissolves in an excess of sulphide of ammonium to a reddish-brown fluid. *Ferrocyanide of potassium* produces a green precipitate, *infusion of galls,* after some time, a bluish-black precipitate. *b. Vanadates.*—Most of the neutral salts are yellow, those of the alkalies and some others are by heating with water converted into a colorless modification. The acid salts are yellowish-red. The salts can bear a red heat; most of them are soluble in water, all of them in nitric acid. The vanadate of the alkalies dissolve the more sparingly in water the more free alkali or alkaline salt is present. If the solution of a vanadate of an alkali is saturated with *chloride of ammonium*, the whole of the vanadic acid separates as white vanadate of ammonia, insoluble in solution of chloride of ammonium (most characteristic reaction). The precipitate gives by ignition vanadate of binoxide of vanadium. If an acidified solution of vanadate of alkali is shaken together with *peroxide of hydrogen*, the fluid acquires a red tint; if ether is then added, and the mixture shaken, the solution retains its color, the ether remaining colorless (most delicate reaction) (WERTHER). *Borax* dissolves vanadic acid in the inner and outer flame to a clear bead; the bead produced in the outer flame is colorless, with large quantities of vanadic acid, yellow; the bead produced in the inner flame has a beautiful green color; with larger quantities of vanadic acid it looks brownish whilst hot, and only turns green on cooling.

c. Oxides of Thallium.

Thallium is the last discovered of the known elements. It is found chiefly in iron pyrites, also in copper pyrites, and some other sulphuretted ores, and in native sulphur. It is found accumulated in the dust of the flues leading to the sulphuric acid chambers where the furnaces are fed with such ores. Thallium is a lead-like metal. It is soft, melts readily, and is volatile at a white heat. When bent or twisted it emits the same peculiar crackling sound as tin. It does not decompose pure water, but decomposes it after addition of acid. It forms a basic oxide, a sesquioxide, and a peroxide. *Protoxide of thallium* dissolves in water forming an alkaline fluid. The solutions of its salts are not precipitated by alkalies nor by alkaline carbonates. The sulphate, nitrate, and carbonate of protoxide of thallium are white, soluble in water, and readily crystallizable. The phosphate is a crystalline precipitate almost insoluble in water and alkaline solutions, but readily soluble in mineral acids. *Chromate of potassa* produces a pale yellow, and bichromate of potassa a deep orange precipitate almost insoluble in water and acids. *Hydrochloric acid*, or *soluble chlorides*, precipitate white chloride of thallium, which is nearly insoluble in water, hydrochloric acid or ammonia; it is soluble in boiling water and crystallizes out on cooling; hot nitric acid dissolves it permanently. Iodide of potassium precipitates it yellow, and bromide of potassium white. *Hydrosulphuric acid* does not precipitate acid solutions, and only very imperfectly when neutral. *Sulphide of ammonium* throws down brownish black sulphide of thallium which readily collects in lumps; it is insoluble in ammonia, in alkaline sulphides and in cyanide of potassium: it dissolves sparingly in hydrochloric acid, readily in sulphuric acid and nitric acid; when moist it oxidizes rapidly in the air, being converted into sulphate. Sulphide of thallium is more fusible than the metal. *Zinc* precipitates from thallium solutions the metal in small crystalline leaflets or scales. *Colorless flames* are colored intensely green by thallium compounds. The *spectrum* of thallium consists of a single most characteristic line of a magnificent emerald-green color, which nearly coincides with Ba, δ (See Table I.) With minute quantities of thallium the spectral reaction is but of very short duration. Spectrum analysis affords by far the best means for the detection of thallium. Thalliferous pyrites give the green line mostly at once. In native sulphur thallium is detected the most readily by first removing the principal part of the sulphur by means of sulphide of carbon, and then examining the residue.

§ 114.

FIFTH GROUP.

More common oxides of the fifth group:—OXIDE OF SILVER, SUBOXIDE OF MERCURY, OXIDE OF MERCURY, OXIDE OF LEAD, TEROXIDE OF BISMUTH, OXIDE OF COPPER, OXIDE OF CADMIUM.

Rarer oxides of the fifth group;—OXIDES OF PALLADIUM, RHODIUM, OSMIUM, RUTHENIUM.

Properties of the group.—The sulphides corresponding to the oxides of this group are insoluble both in dilute acids and in alkaline sulphides.[*] The solutions of these oxides are therefore completely precipitated by

---

[*] Consult however the paragraphs on oxide of copper and suboxide and oxide of mercury, as the latter remark applies only partially to them.

hydrosulphuric acid, no matter whether they be neutral, or contain free acid or free alkali. The fact that the solutions of the oxides of the fifth group are precipitated by hydrosulphuric acid in presence of a free stronger acid, distinguishes them from the oxides of the fourth group and from the oxides of all the preceding groups.

For the sake of greater clearness and simplicity, we divide the oxides of this group into two classes, and distinguish,

1. OXIDES PRECIPITABLE BY HYDROCHLORIC ACID, viz., oxide of silver, suboxide of mercury, oxide of lead.

2. OXIDES NOT PRECIPITABLE BY HYDROCHLORIC ACID, viz., oxide of mercury, oxide of copper, teroxide of bismuth, oxide of cadmium.

Lead must be considered in both classes, since the sparing solubility of its chloride might lead to confounding its oxide with suboxide of mercury and oxide of silver, without affording us on the other hand any means of effecting its perfect separation from the oxides of the second division.

Special Reactions of the more common Oxides of the fifth group.

FIRST DIVISION OF THE FIFTH GROUP; OXIDES WHICH ARE PRECIPITATED BY HYDROCHLORIC ACID.

§ 115.

a. OXIDE OF SILVER (Ag O.)

1. METALLIC SILVER is white, very lustrous, moderately hard, highly malleable, ductile, rather difficultly fusible. It is scarcely oxidized by fusion in the air. Nitric acid dissolves silver readily; the metal is insoluble in dilute sulphuric acid and in hydrochloric acid.

2. OXIDE OF SILVER is a grayish-brown powder; it is not altogether insoluble in water, and dissolves readily in dilute nitric acid. It forms no hydrate. It is decomposed by heat into metallic silver and oxygen gas. The black suboxide of silver (Ag_2O) and the binoxide (AgO_2) are likewise decomposed by heat into metallic silver and oxygen.

3. The SALTS OF OXIDE OF SILVER are non-volatile and colorless; many of them acquire a black tint upon exposure to light. The soluble neutral salts do not alter vegetable colors, and are decomposed at a red heat.

4. *Hydrosulphuric acid* and *sulphide of ammonium* precipitate from solutions of salts of silver black SULPHIDE OF SILVER (AgS) which is insoluble in dilute acids, alkalies, alkaline sulphides, and cyanide of potassium. Boiling nitric acid decomposes and dissolves this precipitate readily, with separation of sulphur.

5. *Potassa* and *soda* precipitate from solutions of salts of silver OXIDE OF SILVER in the form of a grayish-brown powder, which is insoluble in an excess of the precipitants, but dissolves readily in ammonia.

6. *Ammonia,* if added in very small quantity to neutral solutions of oxide of silver, throws down the *oxide* as a brown precipitate, which readily redissolves in an excess of ammonia. Acid solutions of silver are not precipitated.

7. *Hydrochloric acid* and *soluble metallic chlorides* produce in solutions of salts of silver a white curdy precipitate of CHLORIDE OF SILVER (AgCl). In very dilute solutions these reagents impart at first simply a bluish-white opalescent appearance to the fluid; but after long

standing in a warm place the chloride of silver collects at the bottom of the vessel. By the action of light the white chloride of silver first acquires a violet tint, and ultimately turns black; it is insoluble in nitric acid, but dissolves readily in ammonia as ammonio-chloride of silver, from which double compound the chloride of silver is again separated by acids. Concentrated hydrochloric acid and concentrated solutions of chlorides of the alkali metals dissolve chloride of silver to a very perceptible amount, more particularly upon application of heat; but the dissolved chloride separates again upon dilution. Upon exposure to heat chloride of silver fuses without decomposition, giving upon cooling a transparent horny mass.

8. If compounds of silver mixed with *carbonate of soda* are exposed on a charcoal support to the *inner* flame of the blowpipe, white brilliant ductile metallic globules are obtained, with or without a slight dark red incrustation of the charcoal.

§ 116.

b. Suboxide of Mercury (Hg_2O).

1. Metallic mercury is grayish-white, lustrous, fluid at the common temperature; it solidifies at $-40°$, and boils at $680°$ Fah. It is insoluble in hydrochloric acid; in dilute cold nitric acid it dissolves to nitrate of suboxide, in more concentrated hot nitric acid to nitrate of oxide of mercury.

2. Suboxide of mercury is a black powder, readily soluble in nitric acid. It is decomposed by the action of heat, the mercury volatilizing in the metallic state. It forms no hydrate.

3. The salts of suboxide of mercury volatilize upon ignition; most of them suffer decomposition in this process. Subchloride and subbromide of mercury volatilize unaltered. Most of the salts of suboxide of mercury are colorless. The soluble salts in the neutral state redden litmus-paper. Nitrate of suboxide of mercury is decomposed by addition of much water into a light yellow insoluble basic and a soluble acid salt.

4. *Hydrosulphuric acid* and *sulphide of ammonium* produce black precipitates of subsulphide of mercury (Hg_2S), which are insoluble in dilute acids, sulphide of ammonium, and cyanide of potassium. Monosulphide of sodium, in presence of some caustic soda, dissolves this subsulphide, with separation of metallic mercury. Bisulphide of sodium dissolves the subsulphide without separation of metallic mercury. The solutions contain sulphide of mercury (HgS). Subsulphide of mercury is readily decomposed and dissolved by nitrohydrochloric acid, but not by boiling concentrated nitric acid.

5. *Potassa, soda,* and *ammonia* produce in solutions of salts of suboxide of mercury black precipitates, which are insoluble in an excess of the precipitants. The precipitates produced by the fixed alkalies consist of suboxide of mercury; whilst those produced by ammonia consist of basic compounds of mercury with ammonia or amidogen.

6. *Hydrochloric acid* and *soluble metallic chlorides* precipitate from solutions of salts of suboxide of mercury subchloride of mercury (Hg_2Cl) as a fine powder of dazzling whiteness. Cold hydrochloric acid and cold nitric acid fail to dissolve this precipitate; it dissolves however, although very difficultly and slowly, upon long-continued boiling with these acids, being resolved by hydrochloric acid into chloride

of mercury and metallic mercury, which separates; and converted by nitric acid into chloride of mercury and nitrate of oxide of mercury. Nitrohydrochloric acid and chlorine water dissolve the subchloride of mercury readily, converting it into chloride. Ammonia and potassa decompose the subchloride of mercury, separating from it, the former a compound of protamide of mercury with protochloride of mercury ($Hg_2 NH_2$, $Hg_2 Cl$), the latter suboxide of mercury.

7. If a drop of a neutral or slightly acid solution of suboxide of mercury is put on a *clean and smooth surface of copper*, and washed off after some time, the spot will afterwards, on being gently rubbed with cloth, paper, &c., appear white and lustrous like silver. The application of a gentle heat to the copper causes the metallic mercury precipitated on its surface to volatilize, and thus removes the silvering.

8. *Protochloride of tin* produces in solutions of suboxide of mercury a gray precipitate of METALLIC MERCURY, which may be united into globules by boiling the metallic deposit, after decanting the fluid, with hydrochloric acid, to which a little protochloride of tin may also be added.

9. If an intimate mixture of an anhydrous compound of mercury with anhydrous *carbonate of soda* is introduced into a sealed glass tube, and covered with a layer of carbonate of soda, and the tube is then strongly heated, the mercurial compound invariably undergoes decomposition, and METALLIC MERCURY separates, forming a coat of gray sublimate above the heated part of the tube. The minute particles of mercury may be united into larger globules by rubbing this coating with a glass rod.

§ 117.

c. OXIDE OF LEAD (Pb O).

1. METALLIC LEAD is bluish-gray; its surface recently cut exhibits a metallic lustre; it is soft, malleable, readily fusible. It evaporates at a white heat. Fused upon charcoal before the blowpipe it forms a coating of yellow oxide on the support. Hydrochloric acid and moderately concentrated sulphuric acid act upon it but little, even with the aid of heat; but dilute nitric acid dissolves it readily, more particularly on heating.

2. OXIDE OF LEAD is a yellow or reddish-yellow powder, looking brownish-red whilst hot, and fusible at a red heat. Hydrated oxide of lead is white. Both the oxide and its hydrate dissolve readily in nitric and acetic acids. SUBOXIDE OF LEAD (Pb_2 O) is black, MINIUM (2 PbO, PbO_2) red, BINOXIDE (PbO_2) brown. They are all of them converted into the oxide by ignition in the air. The binoxide is not dissolved by heating with nitric acid, but it dissolves readily in that menstruum on addition of some spirit of wine. The solution contains nitrate of oxide of lead.

3. The SALTS OF OXIDE OF LEAD are non-volatile; most of them are colorless; the neutral soluble salts redden litmus-paper, and are decomposed at a red heat. If chloride of lead is ignited in the air, part of it volatilizes, and leaves behind a mixture of oxide of lead and chloride of lead.

4. *Hydrosulphuric acid* and *sulphide of ammonium* produce in solutions of salts of lead black precipitates of SULPHIDE OF LEAD (Pb S), which are insoluble in *cold* dilute acids, in alkalies, alkaline sulphides, and

cyanide of potassium. Sulphide of lead is decomposed by hot nitric acid. If the acid was dilute, the whole of the lead is obtained in solution as nitrate of oxide of lead, and sulphur separates—if the acid was fuming, the sulphur is also completely oxidized, and insoluble sulphate of lead alone is obtained;—if the acid was of medium concentration, both processes take place, a portion of the lead being obtained in solution as nitrate of lead, whilst the remainder separates as sulphate of lead, together with the unoxidized sulphur. In solutions of salts of lead containing a large excess of a concentrated mineral acid, hydrosulphuric acid produces a precipitate only after the addition of water or after neutralization of the free acid by an alkali. If a solution of lead is precipitated by hydrosulphuric acid in presence of a large quantity of free hydrochloric acid, a red precipitate is formed, consisting of chloride and sulphide of lead, which is however converted by an excess of hydrosulphuric acid into black sulphide of lead.

5. *Potassa, soda,* and *ammonia* throw down BASIC SALTS OF LEAD in the form of white precipitates, which are insoluble in ammonia and difficultly soluble in potassa and soda. In solutions of acetate of lead ammonia (free from carbonic acid) does not immediately produce a precipitate, owing to the formation of a soluble triacetate of lead.

6. *Carbonate of soda* throws down from solutions of salts of lead a white precipitate of BASIC CARBONATE OF LEAD [*e.g.*, $6(PbO, CO_2) + PbO, HO$], which is insoluble in an excess of the precipitant and also in cyanide of potassium.

7. *Hydrochloric acid* and *soluble chlorides* produce in concentrated solutions of salts of lead heavy white precipitates of CHLORIDE OF LEAD $(PbCl)$, which are soluble in a large amount of water, especially upon application of heat. This chloride of lead is converted by ammonia into basic chloride of lead $(PbCl, 3PbO + HO)$, which is also a white powder, but almost absolutely insoluble in water. In dilute nitric and hydrochloric acids chloride of lead is more difficultly soluble than in water.

8. *Sulphuric acid* and *sulphates* produce in solutions of salts of lead white precipitates of SULPHATE OF LEAD (PbO, SO_3), which are nearly insoluble in water and dilute acids. From dilute solutions, especially from such as contain much free acid, the sulphate of lead precipitates only after some time, frequently only after a long time. It is advisable to add a considerable *excess* of dilute sulphuric acid, as this tends to increase the delicacy of the reaction, sulphate of lead being more insoluble in dilute sulphuric acid than in water. The separation of small quantities of sulphate of lead is best effected by evaporating, after the addition of the sulphuric acid, as far as practicable on the water-bath, and then treating the residue with water. Sulphate of lead is slightly soluble in concentrated nitric acid; it dissolves with difficulty in boiling concentrated hydrochloric acid, but more readily in solution of potassa. It dissolves also pretty readily in the solutions of some of the salts of ammonia, particularly in solution of acetate of ammonia; dilute sulphuric acid precipitates it again from these solutions.

9. *Chromate of potassa* produces in solutions of salt of lead a yellow precipitate of CHROMATE OF LEAD (PbO, CrO_3), which is readily soluble in potassa, but difficultly so in dilute nitric acid.

10. If a mixture of a compound of lead with *carbonate of soda* is exposed on a charcoal support to the *reducing flame of the blowpipe,* soft

malleable METALLIC GLOBULES OF LEAD are readily produced, the charcoal becoming covered at the same time with a slight yellow incrustation of OXIDE OF LEAD.

§ 118.

Recapitulation and remarks.—The metallic oxides of the first division of the fifth group are most distinctly characterized in their corresponding chlorides; since the different reactions of these chlorides with water and ammonia afford us a simple means both of detecting them and of effecting their separation from one another. For if the precipitate containing the three metallic chlorides is boiled with a somewhat large quantity of water, or boiling water is repeatedly poured over it on the filter, the chloride of lead dissolves, whilst the chloride of silver and the subchloride of mercury remain undissolved. If these two chlorides are then treated with ammonia, the subchloride of mercury is converted into the black basic salt, insoluble in an excess of the ammonia, described in § 116, 5, whilst the chloride of silver dissolves readily in the ammonia, and precipitates from this solution again upon addition of nitric acid. When operating upon small quantities it is advisable first to expel the greater part of the ammonia by heat. In the aqueous solution of chloride of lead the metal may be readily detected by sulphuric acid.

SECOND DIVISION OF THE MORE COMMON OXIDES OF THE FIFTH GROUP: OXIDES WHICH ARE NOT PRECIPITATED BY HYDROCHLORIC ACID.

Special Reactions.

§ 119.

a. OXIDE OF MERCURY (HgO).

1. OXIDE OF MERCURY is generally crystalline, and has a bright red color, which upon reduction to powder changes to a pale yellowish-red; the oxide precipitated from solutions of the nitrate or from solutions of the chloride forms a yellow powder. Upon exposure to heat it transiently acquires a deeper tint; at a dull red heat it is resolved into metallic mercury and oxygen. Both the crystalline and non-crystalline oxide dissolve readily in hydrochloric acid and in nitric acid.

2. The SALTS OF OXIDE OF MERCURY volatilize upon ignition; they suffer decomposition in this process; chloride, bromide, and iodide of mercury volatilize unaltered. Most of the salts of oxide of mercury are colorless. The soluble neutral salts redden litmus-paper. The nitrate and sulphate of oxide of mercury are decomposed by a large quantity of water into soluble acid and insoluble basic salts.

3. Addition of a very small quantity of *hydrosulphuric acid* or *sulphide of ammonium* produces in solutions of oxide of mercury, after shaking, a perfectly white precipitate. Addition of a somewhat larger quantity of these reagents causes the precipitate to acquire a yellow, orange, or brownish-red color, according to the less or greater proportion added; an excess of the precipitant produces a black precipitate of SULPHIDE OF MERCURY (HgS). This progressive variation of color from white to black, which depends on the proportion of the hydrosulphuric acid or sulphide of ammonium added, distinguishes the oxide of mercury from all other bodies. The white precipitate which forms at first con-

sists of a double compound of sulphide of mercury with the still undecomposed portion of the salt of oxide of mercury (in a solution of chloride of mercury, for instance, $HgCl + 2\ HgS$); the gradually increasing admixture of black sulphide causes the precipitate to pass through the several gradations of color above mentioned. Sulphide of mercury is not dissolved by sulphide of ammonium, nor by potassa or cyanide of potassium; it is altogether insoluble in hydrochloric acid and in nitric acid, even upon boiling. It dissolves completely in sulphide of potassium and sulphide of sodium, in presence of some caustic soda or potassa; it is readily decomposed and dissolved by nitrohydrochloric acid. In solutions of oxide of mercury containing a large excess of concentrated mineral acid, hydrosulphuric acid produces a precipitate only after addition of water.

4. *Potassa* added in small quantity produces in neutral or slightly acid solutions of oxide of mercury a reddish-brown precipitate, which acquires a yellow tint if the reagent is added in excess. The *reddish-brown* precipitate is a BASIC SALT; the *yellow* precipitate consists of OXIDE OF MERCURY. An excess of the precipitant does not redissolve these precipitates. In very acid solutions this reaction does not take place at all, or at least the precipitation is very incomplete. In presence of salts of ammonia potassa produces in solutions of salts of oxide of mercury, instead of reddish-brown or yellow, *white* precipitates. The precipitate thrown down by potassa from a solution of chloride of mercury containing an excess of chloride of ammonium is of analogous composition to the precipitate produced by ammonia (see 5).

5. *Ammonia* produces in solutions of salts of oxide of mercury white precipitates quite analogous to those produced by potassa in presence of chloride of ammonium; thus, for instance, ammonia precipitates from solutions of chloride of mercury a DOUBLE COMPOUND OF CHLORIDE OF MERCURY AND AMIDE OF MERCURY ($HgCl + Hg\ NH_2$).

6. *Protochloride of tin* added in small quantity to solution of chloride of mercury, or to solutions of salts of oxide of mercury in presence of hydrochloric acid, throws down SUBCHLORIDE OF MERCURY ($2\ HgCl + Sn\ Cl = Hg_2Cl + SnCl_2$). By addition of a larger quantity of the reagent the pure precipitated subchloride is reduced to METAL ($Hg_2Cl + SnCl = Hg_2 + SnCl_2$). The precipitate, which was white at first, acquires therefore now a gray tint, and may, after it has subsided, be readily united into globules of metallic mercury by boiling with hydrochloric acid.

7. The salts of oxide of mercury show the same reaction as the salts of the suboxide with metallic *copper* and when heated together with *carbonate of soda* in a glass tube.

§ 120.

b. OXIDE OF COPPER (CuO).

1. METALLIC COPPER has a peculiar red color, and a strong lustre; it is moderately hard, malleable, ductile, rather difficultly fusible; in contact with water and air it becomes covered with a green crust of basic carbonate of oxide of copper; upon ignition in the air it becomes coated over with black oxide. In hydrochloric acid and dilute sulphuric acid it is insoluble or nearly so, even upon boiling. Nitric acid dissolves the metal readily. Concentrated sulphuric acid converts it into sulphate of oxide of copper, with evolution of sulphurous acid.

2. SUBOXIDE OF COPPER is red, its hydrate yellow; both change to oxide upon ignition in the air. On treating the suboxide with dilute sulphuric acid metallic copper separates, whilst sulphate of oxide of copper dissolves; on treating suboxide of copper with hydrochloric acid white subchloride of copper is formed, which dissolves in an excess of the acid, but is reprecipitated from this solution by water.

3. OXIDE OF COPPER is a black fixed powder; its hydrate (CuO, HO) is of a light blue color. Both the oxide of copper and its hydrate dissolve readily in hydrochloric, sulphuric, and nitric acids.

4. Most of the neutral SALTS OF OXIDE OF COPPER are soluble in water; the soluble salts redden litmus, and suffer decomposition when heated to gentle redness, with the exception of the sulphate, which can bear a somewhat higher temperature. They are usually white in the anhydrous state; the hydrated salts are usually of a blue or green color, which their solutions continue to exhibit even when much diluted.

5. *Hydrosulphuric acid* and *sulphide of ammonium* produce in alkaline, neutral, and acid solutions of salts of oxide of copper, brownish-black precipitates of SULPHIDE OF COPPER (CuS). This sulphide is insoluble in dilute acids and caustic alkalies. Hot solutions of sulphide of potassium and sulphide of sodium fail also to dissolve it or dissolve it only to a very trifling extent; but it is a little more soluble in sulphide of ammonium. The latter reagent is therefore not well adapted to effect the perfect separation of sulphide of copper from other metallic sulphides. Sulphide of copper is readily decomposed and dissolved by boiling nitric acid, but it remains altogether unaffected by boiling dilute sulphuric acid. It dissolves completely in solution of cyanide of potassium. In solutions of salts of copper which contain an excess of a concentrated mineral acid hydrosulphuric acid produces a precipitate only after the addition of water.

6. *Potassa* or *soda* produces in solutions of salts of oxide of copper a light blue bulky precipitate of HYDRATE OF OXIDE OF COPPER (CuO, HO). If the solution is highly concentrated, and the precipitant is added in excess, the precipitate turns black after the lapse of some time, and loses its bulkiness, even in the cold; but the change takes place immediately if the precipitate is boiled with the fluid in which it is suspended (and which must, if necessary, be diluted for the purpose). In this process the (CuO, HO) hydrated oxide is converted into the $(3 CuO, HO)$ hydrated oxide.

7. *Carbonate of soda* produces in solutions of salts of oxide of copper a greenish-blue precipitate of HYDRATED BASIC CARBONATE OF COPPER $(CuO, CO_2 + CuO, HO)$, which upon boiling changes to brownish-black hydrate of oxide of copper, and dissolves in ammonia to an azure-blue, and in cyanide of potassium to a brownish fluid.

8. *Ammonia* added in small quantity to solutions of neutral salts of oxide of copper produces a greenish-blue precipitate, consisting of a BASIC SALT OF COPPER. This precipitate redissolves readily upon further addition of ammonia to a perfectly clear fluid of a magnificent azure-blue, which owes its color to the formation of a BASIC DOUBLE SALT OF AMMONIO-OXIDE OF COPPER. Thus, for instance, in a solution of sulphate of oxide of copper ammonia produces a precipitate of $NH_3, CuO + NH_4O, SO_3$. In solutions containing a certain amount of free acid ammonia produces no precipitate, but this azure-blue coloration makes its appearance at once the instant the ammonia predominates. The blue color

ceases to be perceptible only in very dilute solutions. Potassa produces in such blue solutions in the cold, after the lapse of some time, a precipitate of blue hydrate of oxide of copper; but upon boiling the fluid this reagent precipitates the whole of the copper as black hydrated oxide. Carbonate of ammonia shows the same reactions with salts of copper as pure ammonia.

9. *Ferrocyanide of potassium* produces in moderately dilute solutions a reddish-brown precipitate of FERROCYANIDE OF COPPER (Cu_2, Cfy), which is insoluble in dilute acids, but suffers decomposition when acted upon by potassa. In very highly dilute solutions the reagent produces only a reddish coloration of the fluid.

10. *Metallic iron* when brought into contact with concentrated solutions of salts of copper is almost immediately covered with a coppery-red coating of METALLIC COPPER; very dilute solutions produce this coating only after some time. Presence of a little free acid accelerates the reaction.

If a fluid containing copper and a little free hydrochloric acid is poured into a small *platinum dish* (the lid of a platinum crucible), and a small piece of *zinc* is introduced, the bright platinum surface speedily becomes covered with a COATING OF COPPER; even with *very* dilute solutions this coating is clearly discernible.

11. If a mixture of a compound of copper with *carbonate of soda* is exposed on a charcoal support to the *inner flame of the blowpipe*, METALLIC COPPER is obtained, without incrustation of the charcoal. The best method of freeing this copper from the particles of charcoal is to triturate the fused mass in a small mortar with water, and to wash off the charcoal powder, when the coppery-red metallic particles will be left behind.

12. If copper, or some alloy containing copper, or a trace of a salt of copper, or even simply the loop of a platinum wire dipped in a highly dilute copper solution, is introduced into the fusion zone of the *gas flame*, or exposed to the inner *blowpipe flame*, the upper or outer portion of the flame shows a magnificent emerald-green tint. Addition of hydrochloric acid to the sample considerably heightens the beauty and delicacy of this reaction.

13. *Borax* and *phosphate of soda and ammonia* readily dissolve oxide of copper in the outer gas- or blowpipe-flame. The beads are green while hot, blue when cold. In the inner flame the bead produced with borax appears colorless, that produced with phosphate of soda and ammonia turns dark green; both acquire a brownish-red tint upon cooling.

§ 121.

c. TEROXIDE OF BISMUTH (BiO_3).

1. BISMUTH has a reddish tin-white color and moderate metallic lustre; it is of medium hardness, brittle, readily fusible; fused upon a charcoal support it forms a coating of yellow teroxide on the surface of the charcoal. It dissolves readily in nitric acid, but is nearly insoluble in hydrochloric acid and altogether so in dilute sulphuric acid. Concentrated sulphuric acid converts it into sulphate of teroxide of bismuth, with evolution of sulphurous acid.

2. The TEROXIDE OF BISMUTH is a yellow powder, which transiently acquires a deeper tint when heated. It fuses at a red heat. Hydrate

of teroxide of bismuth is white. Both the teroxide and its hydrate dissolve readily in hydrochloric, sulphuric, and nitric acids. The grayish-black binoxide of bismuth (BiO_2) and the red bismuthic acid (BiO_3) are converted into teroxide by ignition in the air. By heating with nitric acid they are converted into nitrate of teroxide of bismuth.

3. The SALTS OF TEROXIDE OF BISMUTH are non-volatile; most of them are decomposed at a red heat. Terchloride of bismuth is volatile. The salts of teroxide of bismuth are colorless or white; some of them are soluble in water, others insoluble. The soluble salts in the neutral state redden litmus paper; they are decomposed by a large quantity of water into insoluble basic salts, which separate, whilst the greater portion of the acid remains in solution together with some teroxide of bismuth.

4. *Hydrosulphuric acid* and *sulphide of ammonium* produce in neutral and acid solutions black precipitates of TERSULPHIDE OF BISMUTH (BiS_3), which are insoluble in dilute acids, alkalies, alkaline sulphides, and cyanide of potassium, but are readily decomposed and dissolved by boiling nitric acid. In solutions of salts of bismuth which contain a considerable excess of hydrochloric or nitric acid hydrosulphuric acid produces a precipitate only after the addition of water.

5. *Potassa* and *ammonia* throw down from solutions of salts of bismuth HYDRATE OF TEROXIDE OF BISMUTH as a white precipitate, which is insoluble in an excess of the precipitant.

6. *Carbonate of soda* throws down from solutions of salts of bismuth BASIC CARBONATE OF TEROXIDE OF BISMUTH (BiO_3, CO_2), as a white bulky precipitate, which is insoluble in an excess of the precipitant, and equally so in cyanide of potassium.

7. *Chromate of potassa* precipitates from solutions of salts of bismuth CHROMATE OF TEROXIDE OF BISMUTH (BiO_3, $2 CrO_3$) as a yellow powder. This substance differs from chromate of lead in being readily soluble in dilute nitric acid and insoluble in potassa.

8. *Dilute sulphuric acid* fails to precipitate even only moderately dilute solutions of nitrate of teroxide of bismuth. On evaporating with an excess of sulphuric acid on the water-bath to dryness, a white saline mass is left, which always dissolves readily and to a clear fluid in water acidified with sulphuric acid (characteristic difference between teroxide of bismuth and oxide of lead). After long standing (several days occasionally) basic sulphate of teroxide of bismuth (BiO_3, SO_3, $+ 2$ aq.) separates from this solution in white microscopic needle-shaped crystals, which dissolve in nitric acid.

9. The reaction which characterizes the teroxide of bismuth more particularly is the decomposition of its neutral salts by *water*, which is attended with separation of insoluble basic salts. The addition of a large amount of water to solutions of salts of bismuth causes the immediate formation of a dazzling white precipitate, provided there be not too much free acid present. This reaction is the most sensitive with terchloride of bismuth, as the BASIC CHLORIDE OF BISMUTH ($BiCl_3$, $2 BiO_3$) is almost absolutely insoluble in water. Where water fails to precipitate nitric acid solutions of bismuth, owing to the presence of too much free acid, a precipitate will almost invariably make its appearance immediately upon addition of solution of chloride of sodium. Presence of tartaric acid does not interfere with the precipitation of bismuth solutions by water.

10. If a mixture of a compound of bismuth with *carbonate of soda*

is exposed on a charcoal support to the *reducing flame*, brittle GLOBULES OF BISMUTH are obtained, which fly into pieces under the stroke of a hammer. The charcoal becomes covered at the same time with a slight incrustation of TEROXIDE OF BISMUTH, which is orange-colored whilst hot, yellow when cold.

§ 122.

d. OXIDE OF CADMIUM (Cd O.)

1. METALLIC CADMIUM has a tin-white color; it is lustrous, not very hard, malleable, ductile; it fuses at a temperature below red heat, and volatilizes at a temperature somewhat above the boiling point of mercury, and may accordingly easily be sublimed in a glass tube. Heated on charcoal before the blowpipe it takes fire and burns, emitting brown fumes of oxide of cadmium, which form a coating on the charcoal. Hydrochloric acid and dilute sulphuric acid dissolve it, with evolution of hydrogen; but nitric acid dissolves it most readily.

2. OXIDE OF CADMIUM is a yellowish-brown, fixed powder; its hydrate is white. Both the oxide and its hydrate dissolve readily in hydrochloric, nitric, and sulphuric acids.

3. The SALTS OF OXIDE OF CADMIUM are colorless or white; some of them are soluble in water. The soluble salts in the neutral state redden litmus-paper, and are decomposed at a red heat.

4. *Hydrosulphuric acid* and *sulphide of ammonium* produce in alkaline, neutral, and acid solutions of salts of cadmium, bright yellow precipitates of SULPHIDE OF CADMIUM (Cd S), which are insoluble in dilute acids, alkalies, alkaline sulphides, and cyanide of potassium (difference from copper). They are readily decomposed and dissolved by boiling nitric acid, as well as by boiling hydrochloric acid and by boiling dilute sulphuric acid (difference between cadmium and copper). In solutions of salts of cadmium containing a large excess of acid, hydrosulphuric acid produces a precipitate only after dilution with water.

5. *Potassa* and *soda* produce in solutions of salts of cadmium white precipitates of HYDRATE OF OXIDE OF CADMIUM (CdO, HO), which are insoluble in an excess of the precipitants.

6. *Ammonia* likewise precipitates from solutions of salts of cadmium white HYDRATE OF OXIDE OF CADMIUM, which however redissolves readily and completely to a colorless fluid in an excess of the precipitant.

7. *Carbonate of soda* and *carbonate of ammonia* produce white precipitates of CARBONATE OF CADMIUM (CdO, CO$_2$), which are insoluble in an excess of the precipitants. The presence of salts of ammonia does not prevent the formation of these precipitates. The precipitated carbonate of cadmium dissolves readily in solution of cyanide of potassium. From dilute solutions the precipitate separates only after some time.

8. If a mixture of a compound of cadmium with *carbonate of soda* is exposed on a charcoal support to the *reducing flame*, the charcoal becomes covered with a reddish-brown coating of OXIDE OF CADMIUM, owing to the instant volatilization of the reduced metal and its subsequent reoxidation in passing through the oxidizing flame. The coating is seen most distinctly after cooling.

§ 123.

Recapitulation and remarks.—The perfect separation of the metallic oxides of the second division of the fifth group from suboxide of mercury and oxide of silver may, as already stated, be effected by means of hydrochloric acid; but this agent fails to separate them completely from oxide of lead. Traces of salt of oxide of mercury, which are at first retained by the precipitated chloride of silver by surface attraction, are dissolved out completely by washing (G. J. MULDER). The oxide of mercury is distinguished from the other oxides of this division by the insolubility of the corresponding sulphide in boiling nitric acid. This property affords a convenient means for its separation. Only care must always be taken to free the sulphides completely by washing from all traces of hydrochloric acid or a chloride that may happen to be present, before proceeding to boil them with nitric acid. Moreover, the reactions with protochloride of tin or with metallic copper, as well as those in the dry way, will, after the previous removal of the suboxide, always readily indicate the presence of oxide of mercury. When the moist way is chosen, the sulphide of mercury is dissolved most conveniently by heating it with hydrochloric acid and a few crystals of chlorate of potassa.

From the still remaining oxides the oxide of lead is separated by addition of sulphuric acid. The separation is the most complete if the fluid, after addition of dilute sulphuric acid in excess, is evaporated on the water-bath, the residue diluted with water, slightly acidified with sulphuric acid, and the undissolved sulphate of lead filtered off immediately. The sulphate of lead may be further examined in the dry way by the reaction described in § 117, 10, or also as follows:—Pour over a small portion of the sulphate of lead a little of a solution of chromate of potassa, and apply heat, which will convert the white precipitate into yellow chromate of lead. Wash this, add a little solution of potassa or soda, and heat; the precipitate will now dissolve to a clear fluid; by acidifying this fluid with acetic acid, a yellow precipitate of chromate of lead will again be produced. After the removal of the oxides of mercury and lead, the teroxide of bismuth may be separated from oxide of copper and oxide of cadmium by addition of ammonia in excess, as the latter two oxides are soluble in an excess of this agent. If the filtered precipitate is dissolved in one or two drops of hydrochloric acid on a watchglass, and water added, the appearance of a milky turbidity is a confirmation of the presence of teroxide of bismuth.—The presence of a notable quantity of oxide of copper is revealed by the blue color of the ammoniacal solution; smaller quantities are detected by evaporating the ammoniacal solution nearly to dryness, adding a little acetic acid, and then ferrocyanide of potassium. The separation of oxide of copper from oxide of cadmium may be effected by acting on the sulphides with cyanide of potassium or with boiling dilute sulphuric acid (5 parts of water to 1 part of concentrated sulphuric acid). The solution obtained of the two sulphides is then precipitated by hydrosulphuric acid, and the precipitate separated from the fluid by decantation or filtration. On treating the precipitate now with some water and a small lump of cyanide of potassium, the sulphide of copper will dissolve, leaving the yellow sulphide of cadmium undissolved in the residue. By boiling the precipitate of the mixed sulphides, on the other hand, with dilute sulphuric acid, the sulphide of copper remains undissolved, whilst the sulphide of cadmium is obtained in solution. Hydrosulphuric acid will therefore now throw down from the filtrate yellow sulphide of cadmium (A. W. HOFMANN).

Special Reactions of the rarer Oxides of the fifth group.

§ 124.

a. PROTOXIDE OF PALLADIUM (Pd O).

PALLADIUM is a rare metal. It is found in the metallic state, occasionally alloyed with gold and silver, but more particularly in platinum ores. It greatly resembles platinum, but is somewhat darker in color. It fuses with great difficulty. Heated in the air to dull redness it becomes covered with a blue film: but it recovers its light color and metallic lustre upon more intense ignition. It is sparingly soluble in pure nitric acid, but dissolves somewhat more readily in nitric acid containing nitrous acid; it dissolves very sparingly in boiling concentrated sulphuric acid, but readily in nitrohydrochloric acid. It combines with 1 and 2 eq. of oxygen to protoxide and binoxide. PROTOXIDE OF PALLADIUM is black, its hydrate dark-brown; both are by intense ignition resolved into oxygen and metallic palladium. BINOXIDE OF PALLADIUM (Pd O_2) is black; by heating with dilute hydrochloric acid it is dissolved to protochloride, with evolution of chlorine. The SALTS OF PROTOXIDE OF PALLADIUM are mostly soluble in water; they are brown or reddish-brown; their concentrated solutions are reddish-brown, their dilute solutions yellow. Water precipitates from a solution of nitrate of protoxide of palladium containing a slight excess of acid a brown-colored basic salt. The oxygen salts, as well as the protochloride, are decomposed by ignition, leaving metallic palladium behind. *Hydrosulphuric acid* and *sulphide of ammonium* throw down from acid or neutral solutions of salts of protoxide of palladium black protosulphide of palladium, which dissolves neither in sulphide of ammonium nor in boiling hydrochloric acid, and with difficulty in boiling nitric acid, but readily in nitrohydrochloric acid. From the solution of the protochloride *potassa* precipitates a brown basic salt, soluble in an excess of the precipitant; *ammonia* flesh-colored *ammonio-protochloride of palladium* (Pd Cl, N H_3); *cyanide of mercury* yellowish-white gelatinous *protocyanide of palladium*, soluble in hydrochloric acid and in ammonia (this reaction is particularly characteristic). *Protochloride of tin* produces, in absence of free hydrochloric acid, a brownish-black precipitate; in presence of free hydrochloric acid, a red-colored solution, which speedily turns brown and ultimately green, and upon addition of water brownish-red. *Sulphate of protoxide of iron* produces a deposit of palladium on the sides of the glass. *Iodide of potassium* precipitates black protiodide of palladium (this reaction also is very characteristic). *Chloride of potassium* precipitates from highly concentrated solutions of protoxide of palladium potassio-protochloride of palladium (K Cl, Pd Cl), in the form of golden-yellow needles, which dissolve readily in water to a dark-red fluid, but are insoluble in absolute alcohol.

b. SESQUIOXIDE OF RHODIUM ($R_2 O_3$).

RHODIUM is found in small quantity in platinum ores. It is a steel-gray hard and brittle metal. It occurs also as a gray powder. In this latter state it is converted by ignition in the air into protoxide (R O), then into protosesquioxide; but upon more intense ignition it again loses the absorbed oxygen. None of the acids dissolve rhodium; even in aqua regia this metal is soluble only when alloyed with platinum, copper, &c., but not when alloyed with gold or silver. Fusing hydrate of phosphoric acid and fusing bisulphate of potassa dissolve it to salt of sesquioxide. SESQUIOXIDE OF RHODIUM is black, its hydrate greenish-gray or brown; it is insoluble in acids, but dissolves in fusing hydrate of phosphoric acid and in fusing bisulphate of potassa. The solutions are rose-red. *Hydrosulphuric acid* and *sulphide of ammonium* precipitate on long-continued action, more particularly with the aid of heat, brown sulphide of rhodium, which is insoluble in sulphide of ammonium, but dissolves in boiling hydrochloric and nitric acids. *Hydrate of potassa* precipitates brown hydrate only on boiling. If some alcohol is added to the solution made alkaline by potassa, rhodium shortly precipitates as a black powder; in presence of a larger excess of potassa the rhodium takes a longer time to separate. *Ammonia* produces after some time a yellow precipitate, soluble in hydrochloric acid. *Zinc* precipitates black metallic rhodium. All solid rhodium compounds give by ignition in *hydrogen* rhodium in the metallic state, which is well characterized by its insolubility in aqua regia, its solubility in fusing bisulphate of potassa, and the reaction of this solution with potassa and alcohol.

c. OXIDES OF OSMIUM.

OSMIUM is a rare metal; it is found in platinum ores as a native alloy of osmium and iridium. It is generally obtained as a black powder, or gray and with metallic lustre; it is infusible. The metal, the PROTOXIDE (Os O), and the BINOXIDE (Os O_2) burn readily when heated to redness in the air, and give OSMIC ACID (OsO_4), which volatilizes and makes its presence speedily known by its peculiar exceedingly irritating and offensive smell, resembling that of chlorine and iodine (highly characteristic). If a little osmium on a strip of platinum plate is held in the outer mantle of a *gas or alcohol flame*, at half height, the flame becomes most strikingly luminous. Even minute traces of osmium may by this reaction be detected in alloys of iridium and osmium; but the reaction is in that case only momentary; it may however be reproduced by holding the sample first in the reducing flame, then again in the outer mantle. *Nitric acid*, more particularly red fuming nitric acid, and aqua regia dissolve osmium to osmic acid. Application of heat promotes the solution, which is however attended in that case with volatilization of osmic acid. Very intensely ignited osmium is insoluble in acids. It is fused with nitrate of potassa, and the fused mass distilled with nitric acid; the osmic acid is found in the distillate. By heating osmium in *chlorine gas* volatile green protochloride of osmium and still more volatile red bichloride of osmium are formed. BICHLORIDE OF OSMIUM in solution is rapidly decomposed, if no alkaline chloride is present, into hydrochloric acid, osmic acid, and osmium metal. All osmium compounds give osmium metal by ignition in a current of *hydrogen gas*. Anhydrous OSMIC ACID is white, crystalline, fusible by gentle heat; it boils at about 212° Fahrenheit; the fumes have a most irritating action upon the nose and eyes. Heated with water osmic acid fuses, and dissolves only slowly. The solution has scarcely acid reaction; it has a strong irritating and offensive smell. *Alkalies* color the solution yellow and remove the smell, which however is immediately restored by heating with nitric acid or hydrochloric acid; by heating in a distilling apparatus the osmic acid is obtained in the distillate (most characteristic). In the evaporation of a solution of osmate of alkali, OSMOUS ACID (OsO_2) is formed, more particularly in presence of an excess of alkali. Addition of alcohol promotes the reduction. Hydrosulphuric acid precipitates brown tetrasulphide of osmium, which separates only in presence of a stronger free acid; the precipitate does not dissolve in sulphide of ammonium. *Nitrite of soda* imparts to osmium solutions a deep blue-violet tint, and the fluid gradually deposits black osmium. *Sulphate of protoxide of iron* throws down black osmium; *formic acid* produces the same precipitate; *zinc* also and many other metals in presence of a stronger free acid. POTASSIO-BICHLORIDE OF OSMIUM dissolves very sparingly in cold, a little more readily in hot water; it is insoluble in spirit of wine; the solution in water acquires a deep blue tint by heating with *tannic acid* (characteristic); *formate of soda* precipitates black osmium upon application of heat.

d. OXIDES OF RUTHENIUM.

RUTHENIUM is found in small quantity in platinum ores. It is a grayish-white brittle and very difficultly fusible metal. It is barely acted upon by aqua regia; fusing bisulphate of potassa fails altogether to affect it. By ignition in the air it is converted into bluish-black sesquioxide of ruthenium (Ru_2O_3), insoluble in acids; by ignition with chloride of potassium in a current of chlorine gas into potassio-sesquichloride of ruthenium; by fusion with nitrate of potassa, with hydrate of potassa, or with chlorate of potassa, into rhutenate of potassa (KO, RuO_3). The fused mass obtained in the latter case is greenish-black, and dissolves to an orange-colored fluid, which tinges the skin black, from causing reduction and separation of black oxide. Acids throw down from the solution black OXIDE, which dissolves in hydrochloric acid to an orange-yellow fluid. This solution is resolved by heat into hydrochloric acid and brownish-black oxide. In a concentrated state it gives with chloride of potassium and chloride of ammonium crystalline glossy-violet precipitates, which on boiling with water deposit black oxyprotochloride. *Potassa* precipitates black hydrate of sesquioxide of ruthenium, which is insoluble in alkalies, but dissolves in acids. *Hydrosulphuric acid gas* causes at first no alteration; but after some time the fluid acquires an azure-blue tint, and deposits brown sulphide of ruthenium (very characteristic). *Sulphide of ammonium* produces brownish-black precipitates, barely soluble in an excess of the precipitant. *Sulphocyanide* of potassium produces—in the absence of other metals of the platinum ores—after some time a red coloration, which gradually changes to purple-red, and upon heating to a fine violet tint (very characteristic). *Zinc* produces at first an azure-blue coloration, which subsequently disappears, ruthenium being deposited at the same time in the metallic state.

§ 125.

SIXTH GROUP.

More common oxides of the sixth group:—TEROXIDE OF GOLD, BINOXIDE OF PLATINUM, PROTOXIDE OF TIN, BINOXIDE OF TIN, TEROXIDE OF ANTIMONY, ARSENIOUS ACID AND ARSENIC ACID.

Rarer oxides of the sixth group:—Oxides of IRIDIUM, MOLYBDENUM, TUNGSTEN, TELLURIUM, SELENIUM.

The higher oxides of the elements belonging to the sixth group are all of them more or less strongly pronounced acids. But we class them here with the bases, as they cannot well be separated from the lower degrees of oxidation of the same elements, to which they are very closely allied in their reactions with hydrosulphuric acid.

Properties of the group.—The sulphides corresponding to the oxides of the sixth group are insoluble in dilute acids. These combine with alkaline sulphides to soluble sulphur salts, in which they perform the part of the acid. Hydrosulphuric acid precipitates these oxides therefore, like those of the fifth group, completely from acidified solutions. The precipitated sulphides differ however from those of the fifth group in this, that they dissolve in sulphide of ammonium, sulphide of potassium, &c., and are reprecipitated from these solutions by addition of acids.

We divide the more common oxides of this group into two classes, and distinguish,

1. OXIDES WHOSE CORRESPONDING SULPHIDES ARE INSOLUBLE IN HYDROCHLORIC ACID AND IN NITRIC ACID, and are reduced to the metallic state upon fusion in conjunction with nitrate and carbonate of soda: viz., TEROXIDE OF GOLD and BINOXIDE OF PLATINUM.

2. OXIDES WHOSE CORRESPONDING SULPHIDES ARE SOLUBLE IN BOILING HYDROCHLORIC ACID OR NITRIC ACID, and are upon fusion with nitrate and carbonate of soda converted into oxides or acids, which then combine with the soda: viz., TEROXIDE OF ANTIMONY, PROTOXIDE and BINOXIDE OF TIN, ARSENIOUS and ARSENIC ACIDS.

FIRST DIVISION.

Special Reactions.

§ 126.

a. TEROXIDE OF GOLD (AuO_3).

1. METALLIC GOLD has a reddish-yellow color and a high metallic lustre: it is rather soft, exceedingly malleable and ductile, difficultly fusible; it does not oxidize upon ignition in the air, and is insoluble in hydrochloric, nitric, and sulphuric acids; but it dissolves in fluids containing or evolving chlorine, *e.g.*, in nitrohydrochloric acid. The solution contains terchloride of gold.

2. TEROXIDE OF GOLD is a blackish-brown, its HYDRATE a chestnut-brown powder. Both are reduced by light and heat, and dissolve readily in hydrochloric acid, but not in dilute oxygen acids. Concentrated nitric and sulphuric acids dissolve a little TEROXIDE OF GOLD; water reprecipitates it from these solutions. PROTOXIDE OF GOLD (AuO) is violet-black; it is decomposed by heat into gold and oxygen.

3. SALTS OF GOLD with oxygen acids are nearly unknown. The

HALOID SALTS of gold are yellow, and their solutions continue to exhibit this color up to a high degree of dilution. The whole of them are readily decomposed by ignition. Neutral solution of terchloride of gold reddens litmus-paper.

4. *Hydrosulphuric acid* precipitates from neutral or acid solutions of gold the whole of the metal, from cold solutions as black TERSULPHIDE OF GOLD (AuS_3), from boiling solutions as PROTOSULPHIDE OF GOLD (AuS). The precipitates are insoluble in hydrochloric acid and in nitric acid, but soluble in nitrohydrochloric acid. They are insoluble in colorless sulphide of ammonium, but soluble in yellow sulphide of ammonium, and more readily still in yellow sulphide of sodium or sulphide of potassium.

5. *Sulphide of ammonium* precipitates brownish-black TERSULPHIDE OF GOLD (AuS_3), which redissolves in an excess of the precipitant only if the latter contains an excess of sulphur.

6. *Ammonia* produces, though only in concentrated solutions of gold, reddish-yellow precipitates of AURATE OF AMMONIA (fulminating gold). The more acid the solution and the greater the excess of ammonia added the more gold remains in solution.

7. *Protochloride of tin* containing an admixture of *bichloride* (which may be easily prepared by mixing solution of protochloride of tin with a little chlorine-water), produces even in extremely dilute solutions of gold, a purple-red precipitate (or coloration at least), which sometimes inclines rather to violet or to brownish-red. This precipitate, which has received the name of PURPLE OF CASSIUS, is insoluble in hydrochloric acid. It is assumed to be a hydrated compound of binoxide of tin and protoxide of gold with protoxide and binoxide of tin ($AuO, SnO_2 + SnO, SnO_2 + 4 HO$).

8. *Salts of protoxide of iron* reduce the teroxide of gold in its solutions, and precipitate METALLIC GOLD in form of a most minutely divided brown powder. The fluid in which the precipitate is suspended appears of a blackish-blue color by transmitted light. The dried precipitate shows metallic lustre when pressed with the blade of a knife.

9. *Potassa or soda* added in excess to a solution of terchloride of gold leaves the fluid clear; upon addition of *tannic acid* a deep black precipitate of protoxide of gold (AuO) separates, which subsides completely after some time.

§ 127.

b. BINOXIDE OF PLATINUM (PtO_2).

1. METALLIC PLATINUM has a light steel-gray color; it is very lustrous, moderately hard, very difficultly fusible; it does not oxidize upon ignition in the air, and is insoluble in hydrochloric, nitric, and sulphuric acids. It dissolves in nitrohydrochloric acid, especially upon heating. The solution contains bichloride of platinum ($PtCl_2$).

2. BINOXIDE OF PLATINUM is a blackish-brown, its hydrate a reddish-brown powder. Both are reduced by heat; they are both readily soluble in hydrochloric acid, and difficultly soluble in oxygen acids. The HYDRATE OF PROTOXIDE OF PLATINUM (PtO) is black; it is by ignition reduced to the metallic state.

3. The SALTS OF BINOXIDE OF PLATINUM are decomposed at a red heat. They are yellow. BICHLORIDE OF PLATINUM is reddish-brown, its solution reddish-yellow, which tint it retains up to a high degree of dilu-

tion. The solution reddens litmus-paper. Exposure to a very low red heat converts bichloride of platinum to protochloride (PtCl); application of a stronger red heat reduces it to the metallic state. Solution of bichloride of platinum containing protochloride has a deep dark brown color.

4. *Hydrosulphuric acid* throws down from acid and neutral solutions, but always only after the lapse of some time, a blackish-brown precipitate of BISULPHIDE OF PLATINUM (PtS_2). If the solution is heated after the addition of the hydrosulphuric acid the precipitate forms immediately. It dissolves in a great excess of alkaline sulphides, more particularly of the higher degrees of sulphuration. Bisulphide of platinum is insoluble in hydrochloric acid and in nitric acid; but it dissolves in nitrohydrochloric acid.

5. *Sulphide of ammonium* produces the same precipitate; this redissolves completely, though slowly and with difficulty, in a large excess of the precipitant if the latter contains an excess of sulphur. Acids reprecipitate the bisulphide of platinum unaltered from the reddish-brown solution.

6. *Chloride of potassium* and *chloride of ammonium* (and accordingly of course also potassa and ammonia in presence of hydrochloric acid) produce in not too highly dilute solutions of bichloride of platinum yellow crystalline precipitates of POTASSIO and AMMONIO-BICHLORIDE OF PLATINUM, which are as insoluble in acids as in water, but are dissolved by heating with solution of potassa. From dilute solutions these precipitates are obtained by evaporating the fluid mixed with the precipitants on the water-bath to dryness, and treating the residue with a little water or with dilute spirit of wine. Upon ignition ammonio-bichloride of platinum leaves spongy platinum behind. Potassio-bichloride leaves platinum and chloride of potassium. The decomposition of the latter compound is complete only if the ignition is effected in a current of hydrogen gas or with addition of some oxalic acid.

7. *Protochloride of tin* imparts to solutions of bichloride of platinum containing much free hydrochloric acid an intensely dark brownish-red color, owing to a reduction of the bichloride of platinum to simple chloride. But the reagent produces no precipitate in such solutions.

8. *Sulphate of protoxide of iron* does not precipitate solution of bichloride of platinum, except upon very long-continued boiling, in which case the platinum ultimately suffers reduction.

§ 128.

Recapitulation and remarks.—The reactions of gold and platinum enable us, at least partially, to detect these two metals in the presence of many other oxides, and more particularly where platinum and gold are present in the same solution. In the latter case the solution is most conveniently evaporated to dryness at a gentle heat with chloride of ammonium, and the residue treated with spirit of wine, in order to obtain the gold in solution and the platinum in the residue. The precipitate will thus give platinum by ignition, and the gold may be precipitated from the solution by sulphate of protoxide of iron, after removing the spirit of wine by evaporation.

SECOND DIVISION OF THE SIXTH GROUP.

Special Reactions.

§ 129.

a. Protoxide of Tin (Sn O).

1. TIN has a light grayish-white color and a high metallic lustre; it is soft and malleable; when bent it produces a crackling sound. Heated in the air it absorbs oxygen and is converted into grayish-white binoxide; heated on charcoal before the blowpipe it forms a white coating on the support. Concentrated hydrochloric acid dissolves tin to protochloride, with evolution of hydrogen gas; nitrohydrochloric acid dissolves it, according to circumstances, to bichloride or to a mixture of proto- and bichloride. Tin dissolves with difficulty in dilute sulphuric acid; concentrated sulphuric acid converts it, with the aid of heat, into sulphate of binoxide; moderately concentrated nitric acid oxidizes it readily, particularly with the aid of heat; the white binoxide formed (hydrate of metastannic acid, $Sn O_2 2HO$) does not redissolve in an excess of the acid.

2. PROTOXIDE OF TIN is a black or grayish-black powder; its hydrate is white. Protoxide of tin is reduced by fusion with cyanide of potassium. It is readily soluble in hydrochloric acid. Nitric acid converts it into hydrate of metastannic acid which is insoluble in an excess of the acid.

3. The SALTS OF PROTOXIDE OF TIN are colorless; they are decomposed by heat. The soluble salts, in the neutral state, redden litmus-paper. The salts of protoxide of tin rapidly absorb oxygen from the air, and are partially or entirely converted into salts of binoxide. Protochloride of tin, no matter whether in crystals or in solution, also absorbs oxygen from the air, which leads to the formation of insoluble oxy-protochloride of tin and bichloride of tin. Hence a solution of protochloride of tin becomes speedily turbid if the bottle is often opened and there is only little free acid present; it is therefore only quite recently prepared protochloride of tin which will completely dissolve in water free from air, whilst crystals of protochloride of tin that have been kept for any time will dissolve to a clear fluid only in water containing hydrochloric acid.

4. *Hydrosulphuric acid* throws down from neutral and acid solutions of salts of protoxide of tin a dark brown precipitate of hydrated PROTOSULPHIDE OF TIN (Sn S), which is insoluble, or nearly so, in monosulphide of ammonium, but dissolves readily in the higher yellow sulphide. Acids precipitate from this solution yellow bisulphide of tin, mixed with sulphur. Protosulphide of tin dissolves also in solution of soda or potassa. Acids precipitate from these solutions brown protosulphide. Boiling hydrochloric acid dissolves it, with evolution of hydrogen; boiling nitric acid converts it into insoluble hydrate of metastannic acid. Alkaline solutions of protosalts of tin are not, or at least only imperfectly, precipitated by hydrosulphuric acid. Presence of a very large quantity of free hydrochloric acid may also prevent precipitation of solutions of salts of protoxide of tin by hydrosulphuric acid.

5. *Sulphide of ammonium* produces the same precipitate of hydrated PROTOSULPHIDE OF TIN.

6. *Potassa, soda, ammonia,* and *carbonates of the alkalies,* produce in solutions of salts of protoxide of tin a white bulky precipitate of HYDRATE OF PROTOXIDE OF TIN (Sn O, H O), which redissolves readily in an excess of potassa or soda, but is insoluble in an excess of the other precipitants. If the solution of hydrate of protoxide of tin in potassa is briskly evaporated a compound of binoxide of tin and potassa is formed, which remains in solution, whilst metallic tin precipitates; but upon evaporating slowly crystalline anhydrous protoxide of tin separates.

7. *Terchloride of gold* produces in solutions of protochloride of tin and in solutions of salts of protoxide of tin mixed with hydrochloric acid, upon addition of some nitric acid (without application of heat), a precipitate or coloration of PURPLE OF CASSIUS. (Compare § 126, 7.)

8. Solution of *chloride of mercury,* added in excess, produces in solutions of protochloride or protoxide of tin a white precipitate of SUBCHLORIDE OF MERCURY, owing to the protosalt of tin withdrawing from the chloride of mercury half of its chlorine.

9. If a fluid containing protoxide or protochloride of tin is added to a mixture of *ferricyanide of potassium* and *sesquichloride of iron* a precipitate of *Prussian blue* separates immediately, owing to the reduction of the ferricyanide ($Fe_2 Cfdy$) to ferrocyanide ($Fe_4 Cfy_3$). ($Fe_4 Cfy_4$*) $+ 2 H Cl + 2 Sn Cl = Fe_4 Cfy_3 + H_2 Cfy + 2 Sn Cl_2$). This reaction is extremely delicate, but it can be held to be decisive only in cases where no other reducing agent is present.

10. If compounds of protoxide of tin, mixed with *carbonate of soda* and some *borax,* or, better still, with a mixture of equal parts of *carbonate of soda* and *cyanide of potassium,* are exposed on a charcoal support to the *inner blowpipe flame* malleable grains of METALLIC TIN are obtained. The best way of making quite sure of the real nature of these grains is to triturate them and the surrounding parts of charcoal with water in a small mortar, pressing heavily upon the mass; then to wash the charcoal off from the metallic particles. Upon strongly heating the grains of metallic tin on a charcoal support the latter becomes covered with a coating of white binoxide.

§ 130.

b. BINOXIDE OF TIN ($Sn O_2$).

1. BINOXIDE OF TIN is a powder varying in color from white to straw-yellow, and which upon heating transiently assumes a brown tint. It forms two different series of compounds with acids, bases, and water. The hydrate precipitated by alkalies from solution of bichloride of tin dissolves readily in hydrochloric acid; whilst that formed by the action of nitric acid upon tin—hydrate of metastannic acid—remains undissolved. But if it is boiled for some time with hydrochloric acid it takes up acid; if the excess of the acid is then poured off, and water added, a clear solution is obtained. The aqueous solution of the common bichloride of tin is not precipitated by concentrated hydrochloric acid, whilst that acid produces in the aqueous solution of the metastannic chloride a white precipitate of the latter compound. The solution of the common bichloride of tin is not colored yellow by addition of protochloride of tin, as is the case in a remarkable degree if the solution contains meta-

---

* $2 (Fe_2 Cfdy) = Fe_4 Cfy_4$; for $Cfdy = C_{12} N_6 Fe_2 = 2 Cfy$.

stannic chloride (LÖWENTHAL). The dilute solutions of both chlorides of tin give upon boiling precipitates of the hydrates corresponding to the chlorides.

2. The SALTS OF BINOXIDE OF TIN are colorless. The soluble salts are decomposed at a red heat; in the neutral state they redden litmus paper. Bichloride of tin is a volatile liquid, strongly fuming in the air.

3. *Hydrosulphuric acid* throws down from all acid and neutral solutions of salts of binoxide of tin, particularly upon heating, a *white* flocculent precipitate if the solution of the binoxide is in excess; a *faintly yellow* precipitate if the hydrosulphuric acid is in excess. The former (the white precipitate) may safely be assumed, in the case of a solution of bichloride of tin, to consist of a mixture of bichloride and bisulphide of tin (it has not however as yet been analysed); the latter (the yellow precipitate) consists of hydrated BISULPHIDE OF TIN (SnS_2). Alkaline solutions are not precipitated by hydrosulphuric acid; presence of a very large quantity of hydrochloric acid can also prevent precipitation. The bisulphide of tin dissolves readily in potassa or soda, alkaline sulphides, and concentrated boiling hydrochloric acid, as also in aqua regia. It dissolves with some difficulty in pure ammonia, and is nearly insoluble in carbonate of ammonia. Concentrated nitric acid converts it into insoluble hydrate of metastannic acid. Upon deflagrating bisulphide of tin with nitrate and carbonate of soda sulphate of soda and binoxide of tin are obtained. If a solution of bisulphide of tin in potassa is boiled with teroxide of bismuth tersulphide of bismuth and binoxide of tin are formed, which latter substance remains dissolved in the potassa solution.

4. *Sulphide of ammonium* produces the same precipitate of hydrated BISULPHIDE OF TIN; the precipitate redissolves readily in an excess of the precipitant. From this solution acids reprecipitate the bisulphide of tin unaltered.

5. *Potassa, soda,* and *ammonia, carbonate of soda* and *carbonate of ammonia* produce in solutions of salts of binoxide of tin white precipitates which, according to the nature of the solutions, consist of hydrate of binoxide of tin, or of hydrate of metastannic acid. Both dissolve readily in an excess of solution of potassa or soda.

6. *Sulphate of soda* or *nitrate of ammonia* (in fact, most neutral salts of the alkalies), when added in excess, throw down from solutions of both modifications of binoxide of tin, provided they are not too acid, the whole of the tin as HYDRATED BINOXIDE or HYDRATED METASTANNIC ACID. Heating promotes the precipitation: $SnCl_2 + 4NaO, SO_3 + 4HO = SnO_2, 2HO + 2NaCl + 2(NaO, HO, 2SO_3)$.

7. *Metallic zinc* precipitates from solutions of bichloride of tin, in the absence of free acid, first some metallic tin, then oxychloride; but in presence of a sufficient quantity of free hydrochloric acid METALLIC TIN in the shape of small gray scales, or as a spongy mass. If the operation is conducted in a platinum dish, no blackening of the latter is observed (difference between tin and antimony).

8. The compounds of the binoxide of tin show the same reactions before the blowpipe as those of the protoxide. Binoxide of tin is also readily reduced when fused with cyanide of potassium in a glass tube or in a crucible.

§ 131.

c. TEROXIDE OF ANTIMONY (Sb O_3).

1. METALLIC ANTIMONY has a bluish tin-white color and is very lustrous; it is hard, brittle, readily fusible. When heated on charcoal before the blowpipe it emits thick white fumes of teroxide of antimony, which form a coating on the charcoal; this combustion continues for some time even after the removal of the metal from the flame; it is the most distinctly visible if a current of air is directed with the blowpipe directly upon the sample on the charcoal. But if the sample on the support is kept steady, that the fumes may ascend straight, the metallic grain becomes surrounded with a net of brilliant acicular crystals of teroxide of antimony. Nitric acid oxidizes antimony readily: the dilute acid converting it almost entirely into teroxide, boiling concentrated acid into antimonic acid; neither of the two is altogether insoluble in nitric acid; traces of antimony are therefore always found in the acid fluid filtered from the precipitate. Hydrochloric acid, even boiling, does not attack antimony. In nitrohydrochloric acid the metal dissolves readily. The solution contains terchloride of antimony (SbCl_3), or pentachloride of antimony (SbCl_5), according to the degree of concentration of the acid and the duration of the action.

2. According to the different modes of its preparation, TEROXIDE OF ANTIMONY occurs either in the form of white and brilliant crystalline needles, or as a grayish-white powder. It fuses at a moderate red heat; when exposed to a higher temperature it volatilizes without decomposition. It is almost insoluble in nitric acid, but dissolves readily in hydrochloric and tartaric acids. No separation of iodine takes place on boiling it with hydrochloric acid (free from chlorine) and iodide of potassium (free from iodic acid) BUNSEN. Teroxide of antimony is easily reduced to the metallic state by fusion with cyanide of potassium.

3. ANTIMONIC ACID (Sb O_5) is pale yellow; its hydrates are white. Both the acid and its hydrates redden moist litmus-paper; they are only very sparingly soluble in water, and insoluble in nitric acid, but dissolve pretty readily in hot concentrated hydrochloric acid: the solution contains pentachloride of antimony (SbCl_5), and turns turbid upon addition of water. On boiling antimonic acid with hydrochloric acid and iodide of potassium iodine separates, which dissolves in the hydriodic acid present to a brown fluid (BUNSEN). Upon ignition antimonic acid loses oxygen, and is converted into antimonate of teroxide of antimony (SbO_3, SbO_5). Of the antimonates the potassa and ammonia salts are almost the only ones soluble in water: acids precipitate hydrate of antimonic acid from the solutions, chloride of sodium throws down from them antimonate of soda (§ 90, 2).

4. The greater part of the SALTS OF TEROXIDE OF ANTIMONY are decomposed upon ignition; the haloid salts volatilize readily and unaltered. The soluble neutral salts of antimony redden litmus-paper. With a large quantity of water they give insoluble basic salts and acid solutions containing teroxide of antimony. Thus, for instance, *water* throws down from solutions of terchloride of antimony in hydrochloric acid a white bulky precipitate of BASIC TERCHLORIDE OF ANTIMONY (powder of Algaroth), SbCl_3, 5SbO_3, which after some time becomes heavy and crystalline. Tartaric acid dissolves this precipitate readily, and therefore

prevents its formation if mixed with the solution previously to the addition of the water. It is by this property that the basic terchloride of antimony is distinguished from the basic salts of bismuth formed under similar circumstances.

5. *Hydrosulphuric acid* precipitates from acid solutions of teroxide of antimony (if the quantity of free mineral acid present is not too large), the whole of the metal as orange-red amorphoses TERSULPHIDE OF ANTIMONY ($Sb S_3$). In alkaline solutions this reagent fails to produce a precipitate or, at least, it precipitates them only imperfectly; neutral solutions also are only imperfectly thrown down by it. The tersulphide of antimony produced is readily dissolved by potassa and by alkaline sulphides, especially if the latter contain an excess of sulphur; it is but sparingly soluble in ammonia, and, if free from pentasulphide of antimony, almost insoluble in bicarbonate of ammonia. It is insoluble in dilute acids, as also in acid sulphite of potassa. Concentrated boiling hydrochloric acid dissolves it, with evolution of hydrosulphuric acid gas. By heating in the air it is converted into a mixture of antimonate of teroxide of antimony with tersulphide of antimony. By deflagration with nitrate of soda it gives sulphate and antimonate of soda. If a potassa solution of tersulphide of antimony is boiled with teroxide of bismuth tersulphide of bismuth precipitates, and teroxide of antimony dissolved in potassa remains in the solution. On fusing tersulphide of antimony with cyanide of potassium metallic antimony and sulphocyanide of potassium are produced. If the operation is conducted in a small tube expanded into a bulb at the lower end, or in a stream of carbonic acid gas (see § 132, 12), no sublimate of antimony is produced. But if a mixture of tersulphide of antimony with carbonate of soda or with cyanide of potassium and carbonate of soda is heated in a glass tube in a stream of hydrogen gas (compare § 132, 4), a mirror of antimony is deposited on the inner surface of the tube, immediately behind the spot occupied by the mixture.

From a solution of antimonic acid in hydrochloric acid sulphuretted hydrogen throws down pentasulphide of antimony (SbS_5), which dissolves readily when heated with solution of soda or ammonia, and equally so in concentrated boiling hydrochloric acid, with evolution of hydrosulphuric acid gas and separation of sulphur, but dissolves only very sparingly in cold bicarbonate of ammonia.

6. *Sulphide of ammonium* produces in solutions of teroxide of antimony an orange-red precipitate of TERSULPHIDE OF ANTIMONY, which readily redissolves in an excess of the precipitant if the latter contains an excess of sulphur. Acids throw down from this solution pentasulphide of antimony (SbS_5). However, the orange color appears in that case usually of a lighter tint, owing to an admixture of free sulphur.

7. *Potassa, soda, ammonia, carbonate of soda* and *carbonate of ammonia* throw down from solutions of terchloride of antimony, and also of simple salts of teroxide of antimony—but far less completely, and mostly only after some time, from solutions of tartar emetic or analogous compounds—a white bulky precipitate of TEROXIDE OF ANTIMONY, which redissolves pretty readily in an excess of potassa or soda, but requires the application of heat for its re-solution in carbonate of potassa, and is altogether insoluble in ammonia.

8. *Metallic zinc* precipitates from all solutions of teroxide of antimony, if they contain no free nitric acid, METALLIC ANTIMONY as a black

powder. If a few drops of a solution of antimony, containing some free hydrochloric acid, are put into a platinum dish (the lid of a platinum crucible), and a fragment of zinc is introduced, hydrogen is evolved and antimony separates, staining the part of the platinum covered by the liquid brown or black, even in the case of very dilute solutions: this new reaction I can therefore recommend as being equally delicate and characteristic. Cold hydrochloric acid fails to remove the stain, heating with nitric acid removes it immediately.

9. If a solution of teroxide of antimony in solution of potassa or soda is mixed with solution of *nitrate of silver*, a deep black precipitate of SUBOXIDE OF SILVER forms along with the grayish-brown precipitate of oxide of silver. Upon now adding ammonia in excess the oxide is redissolved, whilst the suboxide is left undissolved (*H. Rose*). The formation of the suboxide of silver in this process is explained as follows: $KO, SbO_3 + 4 AgO = KO, SbO_5 + 2 Ag_2O$. This exceedingly delicate reaction affords more especially also an excellent means of detecting teroxide of antimony in presence of antimonic acid.

10. If a solution of teroxide of antimony is introduced into a flask in which hydrogen gas is being evolved from pure *zinc* and dilute *sulphuric acid* the zinc oxidizes not only at the expense of the oxygen of the water, but also at the expense of that of the teroxide of antimony, and antimony separates accordingly in the metallic state; but a portion of the metal combines in the moment of its separation with the liberated hydrogen of the water, forming ANTIMONETTED HYDROGEN GAS (SbH_3). If this operation is conducted in a gas-evolution flask, connected by means of a perforated cork with the limb of a bent tube of which the other limb ends in a finely drawn-out point, pinched off at the top,* and the hydrogen passing through the fine aperture of the tube is ignited after the atmospheric air is completely expelled, the flame appears of a bluish-green tint, which is imparted to it by the antimony separating in a state of intense ignition upon the decomposition of the antimonetted hydrogen; white fumes of teroxide of antimony rise from the flame, which condense readily upon cold substances, and are not dissolved by water. But if a cold body, such as a porcelain dish (which answers the purpose best), is now depressed upon the flame, METALLIC ANTIMONY is deposited upon the surface in a state of the most minute division, forming a deep black and almost lustreless spot. If the middle part of the tube through which the gas is passing is heated to redness the bluish-green tint of the flame decreases in intensity, and a metallic mirror of antimony of silvery lustre is formed within the tube on both sides of the heated part.

As the acids of arsenic give under the same circumstances similar stains of metallic arsenic, it is always necessary to carefully examine the spots produced, in order to ascertain whether they really consist of antimony or contain any of that metal. With stains deposited on a porcelain dish the object in view is most readily attained by treating them with a solution of chloride of soda (a compound of hypochlorite of soda with chloride of sodium, prepared by mixing a solution of chloride of lime with carbonate of soda in excess, and filtering); which will immediately dissolve arsenical stains, leaving those proceeding from antimony untouched, or, at least, removing them only after a very pro-

---

* In accurate experiments it is advisable to use *Marsh's* apparatus (§ 132, 10).

tracted action. A mirror within the glass tube, on the other hand, may be tested by heating it whilst the current of hydrogen gas still continues to pass through the tube: if the mirror volatilizes only at a higher temperature, and the hydrogen gas then issuing from the tubes does not smell of garlic; if it is only with a strong current that the ignited gas deposits spots on porcelain, and the mirror before volatilizing fuses to small lustrous globules distinctly discernible through a magnifying glass, —the presence of antimony may be considered certain. Or the metals may be identified by conducting through the tube a *very slow* stream of dry hydrosulphuric acid gas, and heating the mirror, by means of a spirit-lamp, proceeding from the outer to the inner border, and accordingly in an opposite direction to that of the gaseous current. The antimonial mirror is by this means converted into tersulphide of antimony, which appears of a more or less reddish-yellow color, and almost black when in thick layers. If a feeble stream of dry hydrochloric acid gas is now transmitted through the glass tube, the tersulphide of antimony, if present in thin layers only, disappears immediately; if the incrustation is somewhat thicker it takes a short time to dissipate it. The reason for this is, that the tersulphide of antimony decomposes readily with hydrochloric acid, and the terchloride of antimony formed is exceedingly volatile in a stream of hydrochloric acid gas. If the gaseous current is now conducted into some water the presence of antimony in the latter fluid may readily be proved by means of hydrosulphuric acid. By this combination of reactions antimony may be distinguished with positive certainty from all other metals. The reaction which hydrogen gas containing antimonetted hydrogen shows with solution of nitrate of silver will be found in § 134, 6.

11. If a mixture of a compound of antimony with *carbonate of soda* and *cyanide of potassium* is exposed on a charcoal support to the *reducing flame of the blowpipe*, brittle globules of METALLIC ANTIMONY are produced, which may be readily recognised by the peculiar reactions that mark their oxidation (compare § 131, 1).

§ 132.

d. ARSENIOUS ACID (AsO_3).

1. METALLIC ARSENIC has a blackish-gray color and high metallic lustre, which it retains in dry air, but loses in moist air, becoming covered with suboxide; the metallic arsenic of commerce looks therefore rather dull, with a dim bronze lustre on the planes of crystallization. Arsenic is not very hard, but very brittle: at a dull red heat it volatilizes without fusion. The fumes have a most characteristic odor of garlic, which proceeds from the suboxide of arsenic formed. Heated with free access of air arsenic burns—at an intense heat with a bluish flame—emitting white fumes of arsenious acid, which condense on cold bodies. If arsenic is heated in a glass tube sealed at the lower end the greater part of it volatilizes unoxidized, and recondenses above the heated spot as a lustrous black sublimate (arsenical mirror); a very thin coating of the sublimate appears of a brownish-black color. In contact with air and water arsenic oxidizes slowly to arsenious acid. Weak nitric acid converts it, with the aid of heat, into arsenious acid, which dissolves only sparingly in an excess of the acid; strong nitric acid converts it partially into arsenic acid. It is insoluble in hydrochloric acid and dilute

sulphuric acid; concentrated boiling sulphuric acid oxidizes it to arsenious acid, with evolution of sulphurous acid.

2. ARSENIOUS ACID generally presents the appearance either of a transparent vitreous or of a white porcelain-like mass. By trituration it gives a heavy, white, gritty powder. When heated it volatilizes in white inodorous fumes. If the operation is conducted in a glass tube a sublimate is obtained consisting of small brilliant octahedrons and tetrahedrons. Arsenious acid is only difficultly moistened by water; it comports itself in this respect like a fatty substance. It is sparingly soluble in cold, but more readily in hot water. It is copiously dissolved by hydrochloric acid, as well as by solution of soda and potassa. Upon boiling with nitrohydrochloric acid it dissolves to arsenic acid. It is highly poisonous.

3. The ARSENITES are mostly decomposed upon ignition either into arsenates and metallic arsenic, which volatilizes, or into arsenious acid and the base with which it was combined. Of the arsenites those only with alkaline bases are soluble in water. The insoluble arsenites are dissolved, or at least decomposed, by hydrochloric acid. Anhydrous terchloride of arsenic ($AsCl_3$) is a colorless volatile liquid, fuming in the air, which will bear the addition of a little water, but is decomposed by a larger amount into arsenious acid, which partly separates, and hydrochloric acid, which retains the rest of the arsenious acid in solution. If a solution of arsenious acid in hydrochloric acid is evaporated by heat, chloride of arsenic escapes along with the hydrochloric acid.

4. *Hydrosulphuric acid* colors aqueous solutions of arsenious acid yellow, but produces no precipitate in them; it fails equally to precipitate aqueous solutions of neutral arsenites of the alkalies; but upon addition of a stronger acid a bright yellow precipitate of TERSULPHIDE OF ARSENIC (AsS_3) forms at once. The same precipitate forms in like manner in the hydrochloric acid solution of arsenites insoluble in water. Even a large excess of hydrochloric acid does not prevent complete precipitation. Alkaline solutions are not precipitated. The precipitate is readily and completely dissolved by pure alkalies, alkaline carbonates and bicarbonates, and also by alkaline sulphides; but it is nearly insoluble in hydrochloric acid, even though concentrated and boiling. Boiling nitric acid decomposes and dissolves the precipitate readily.

If recently precipitated tersulphide of arsenic is digested with sulphurous acid and acid sulphite of potassa the precipitate is dissolved; upon heating the solution to boiling the fluid turns turbid, owing to the separation of sulphur, which upon continued boiling is for the greater part redissolved. The fluid contains, after expulsion of the sulphurous acid, arsenite and hyposulphite of potassa: $2 AsS_3 + 8 (KO, SO_2) = 2 (KO, AsO_3) + 6 (KO, S_2O_2) + S_2 + 7 SO_2$ (BUNSEN).

The deflagration of tersulphide of arsenic with carbonate of soda and nitrate of soda gives rise to the formation of arsenate and sulphate of soda. If a solution of tersulphide of arsenic in potassa is boiled with hydrated carbonate or basic nitrate of teroxide of bismuth tersulphide of bismuth and arsenite of potassa are produced.

If a mixture of tersulphide of arsenic with from 3 to 4 parts of carbonate of soda, made into a paste with some water, is spread over small glass splinters, and these, after being well dried, are rapidly heated to redness in a glass tube through which dry hydrogen gas is transmitted, a large portion of the arsenic present is reduced to the metallic state and expelled if the temperature applied is sufficiently high. Part of the reduced arsenic forms a metallic mirror on the inner surface of the tube, the remainder is carried away suspended in the hydrogen gas; the minute particles of arsenic impart a bluish tint to the flame when the gas is kindled, and form stains of arsenic upon the surface of a porcelain dish depressed upon the flame. The fusion of the mixture of tersulphide of arsenic with carbonate of soda first gives rise to the formation of a double tersulphide of arsenic and sulphide of sodium, and of arsenite of soda ($2 AsS_3 + 4 NaO, CO_2 = 3 NaS, AsS_3 + NaO, AsO_3 + 4 CO_2$). Upon heating these products the arsenite of soda is resolved into arsenic and arsenate of soda ($5 AsO_3 = 2 As + 3 AsO_5$), and the tersulphide of arsenic and sulphide of sodium into arsenic and pentasulphide of arsenic and sulphide of sodium ($5 AsS_3 = 2 As + 3 AsS_5$); and by the action of the hydrogen the arsenate of soda is also converted into hydrate of soda, arsenic, and water. The whole of the arsenic is accordingly expelled, except that portion of the metal which constitutes a component part of the double pentasulphide of arsenic and sulphide of sodium formed in the process, a sulphur salt which is not decomposed by hydrogen (H. ROSE).

This method of reduction gives indeed very accurate results, but it does not enable us to distinguish arsenic from antimony with a sufficient degree of certainty, nor to detect the one in presence of the other. (Compare § 131, 5.)

The operation is conducted in the apparatus illustrated by

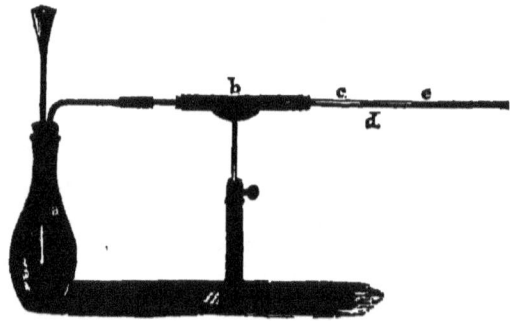

Fig. 26.

a is the evolution flask, *b* a tube containing chloride of calcium, *c* the tube in which, at the point *d*, the glass splinter with the mixture of tersulphide of arsenic and carbonate of soda is placed; this tube is made of difficultly fusible glass free from lead. When the apparatus is completely filled with pure hydrogen gas *d* is exposed to a very gentle heat at first, in order to expel all the moisture which may still be present, and then suddenly to a very intense heat,* to prevent the sublimation of undecomposed tersulphide of arsenic. The metallic mirror is deposited near the point *e*.—Another method of effecting the reduction of tersulphide of arsenic to the metallic state, which combines with the very highest degree of delicacy the advantage of precluding the possibility of

* The flame of the gas-lamp with chimney, or of the blowpipe, answers the purpose best.

confounding arsenic with antimony, will be found described in number 12 of this §.

5. *Sulphide of ammonium* also causes the formation of TERSULPHIDE OF ARSENIC. In neutral and alkaline solutions, however, the tersulphide formed does not precipitate, but remains dissolved as a double sulphide of arsenic and ammonium (tersulphide of arsenic and sulphide of ammonium). From this solution it precipitates immediately upon the addition of a free acid.

6. *Nitrate of silver* leaves aqueous solutions of arsenious acid perfectly clear, or at least produces only a trifling yellowish-white turbidity in them; but if a little ammonia is added a yellow precipitate of ARSENITE OF SILVER (3 AgO, AsO$_3$) separates. The same precipitate forms of course immediately upon the addition of nitrate of silver to the solution of a neutral arsenite. The precipitate dissolves readily in nitric acid as well as in ammonia, and is not insoluble in nitrate of ammonia; if therefore a small quantity of the precipitate is dissolved in a large amount of nitric acid, and the latter is afterwards neutralized with ammonia, the precipitate does not make its appearance again, as it remains dissolved in the nitrate of ammonia formed.

7. *Sulphate of copper* produces under the same circumstances as the nitrate of silver a yellowish-green precipitate of ARSENITE OF COPPER.

8. If to a solution of arsenious acid in an excess of concentrated solution of soda or potassa, or to a solution of an alkaline arsenite mixed with *caustic potassa* or *soda*, a few drops of a dilute solution of sulphate of copper are added, a clear blue fluid is obtained, which upon boiling deposits a red precipitate of SUBOXIDE OF COPPER, leaving arsenate of potassa in solution. This reaction is exceedingly delicate, provided not too much of the solution of sulphate of copper be used. Even should the red precipitate of suboxide of copper be so exceedingly minute as to escape detection by transmitted light, yet it will always be discernible with great distinctness upon looking in at the top of the test-tube. Of course this reaction, although really of great importance in certain instances as a confirmatory proof of the presence of arsenious acid, and more particularly also as a means of distinguishing that acid from arsenic acid, is yet entirely inapplicable for the *direct detection* of arsenic, since grape sugar and other organic substances also produce suboxide of copper from salts of oxide of copper in the same manner.

9. If a solution of arsenious acid mixed with hydrochloric acid is heated with a perfectly clean slip of *copper* or *copper wire*, an IRON-GRAY film of metallic arsenic is deposited on the copper, even in *highly dilute* solutions; when this film increases in thickness it peels off in black scales. If the coated copper, after washing off the free acid, is heated with solution of ammonia the film peels off from the copper, and separates in form of minute spangles (REINSCH). Let it be borne in mind that these are not pure arsenic, but consist of an ARSENIDE OF COPPER (Cu$_2$As). If the substance, either simply dried or oxidized by ignition in a current of air (which is attended with escape of some arsenious acid), is heated, there escapes relatively but little arsenic, alloys richer in copper being left behind (FRESENIUS, LIPPERT). It is only after the presence of arsenic in the alloys has been fully demonstrated that this reaction can be considered a decisive proof of the presence of that metal, as antimony and other metals will under the same circumstances also precipitate in a similar manner upon copper.

138 ARSENIOUS ACID.

10. If an acid or neutral solution of arsenious acid or any of its compounds is mixed with *zinc, water*, and *dilute sulphuric acid* ARSENETTED HYDROGEN (AsH_3) is formed, in the same manner as compounds of antimony give under analogous circumstances antimonetted hydrogen. (Compare § 131, 10.) This reaction affords us a most delicate test for the detection of even the most minute quantities of arsenic.

The operation is conducted in the apparatus illustrated by Fig. 27, or in one of similar construction.* *a* is the evolution flask, *b* a bulb intended to receive the water carried along with the gaseous current, *c* a tube filled with cotton and small lumps of chloride of calcium for drying the gas. This tube is connected with *b* and *d* by india-rubber tubes which have been boiled in solution of soda; *d* should have an inner diameter of 7 millimetres (see Fig. 28), and must be made of difficultly fusible glass

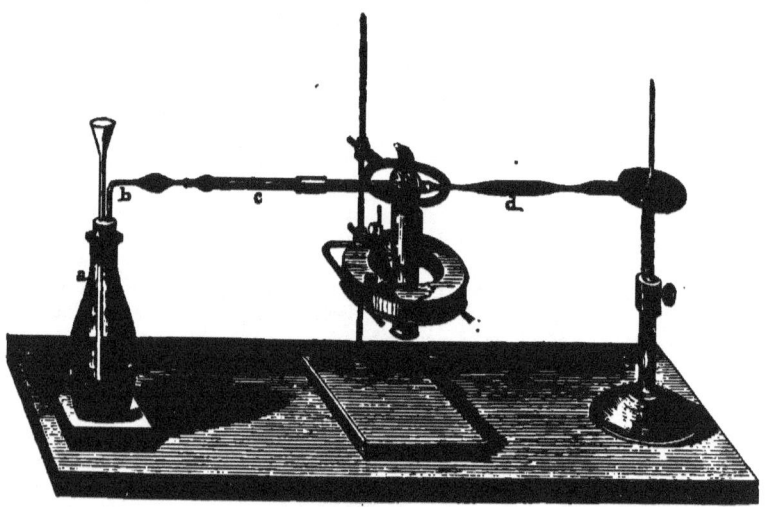

Fig. 27.

free from lead. In experiments requiring great accuracy the tube should be drawn out as shown in Fig. 27. The operation is now commenced by evolving in *a* a moderate and uniform current of hydrogen gas, from pure granulated zinc and pure sulphuric acid diluted with 3 parts of water. Addition of a few drops of bichloride of platinum will be found useful. When the evolution of hydrogen has proceeded for some time, so that it may safely be concluded the air has been completely expelled from the apparatus, the gas is kindled at the open end of the tube *d*. It is advisable to wrap a piece of cloth round the flask before kindling the gas, to guard against accidents in case of an explosion. It is now absolutely necessary first to ascertain whether the zinc and the sulphuric acid are quite free from any admixture of arsenic. This is

Fig. 28.

* I use the very convenient form of *Marsh's* apparatus recommended by Otto in his excellent *Manual of Chemistry*.

done by depressing a porcelain dish horizontally upon the flame to make it spread over the surface: if the hydrogen contains arsenetted hydrogen brownish or brownish-black stains of arsenic will appear on the porcelain; the non-appearance of such stains may be considered as a proof of the freedom of the zinc and sulphuric acid from arsenic. In very accurate experiments, however, additional evidence is required to ensure the positive certainty of the purity of the reagents employed; for this purpose the part of the tube d shown in Fig. 27 over the flame is heated to redness with a *Berzelius* or gas-lamp, and kept some time in a state of ignition: if no arsenical coating makes its appearance in the narrowed part of the tube the agents employed may be pronounced free from arsenic, and the operation proceeded with, by pouring the fluid to be tested for arsenic through the funnel tube into the flask, and afterwards some water to rinse the tube. Only a very little of the fluid ought to be poured in at first, as in cases where the quantity of arsenic present is considerable, and a somewhat large supply of the fluid is poured into the flask, the evolution of gas often proceeds with such violence as to stop the further progress of the experiment.

Now if the fluid contains an oxygen compound of arsenic or arsenic in combination with a salt radical, there is immediately evolved, along with the hydrogen, arsenetted hydrogen, which at once imparts a bluish tint to the flame of the kindled gas, owing to the combustion of the particles of arsenic separating from the arsenetted hydrogen in passing through the flame. At the same time white fumes of arsenious acid arise, which condense upon cold objects. If a porcelain plate is now depressed upon the flame the separated and not yet reoxidized arsenic condenses upon the plate in black stains, in a very similar manner to antimony. (See § 131, 10.) The stains formed by arsenic incline however more to a blackish-brown tint, and show a bright metallic lustre; whilst the antimonial stains are of a deep black color and but feebly lustrous. The arsenical stains may be distinguished moreover from the antimonial stains by solution of chloride of soda—hypochlorite of soda with chloride of sodium—(compare § 131, 10), which will at once dissolve arsenical stains, leaving antimonial stains unaffected, or removing them only after a considerable time.

If the heat of a *Berzelius* or gas-lamp is now applied to the part of the tube d shown in Fig. 27 over the flame, a brilliant arsenical mirror makes its appearance in the narrowed portion of the tube behind the heated part; this mirror is of a darker and less silvery-white hue than that produced by antimony under similar circumstances; from which it is moreover distinguished by the facility with which it may be dissipated in a current of hydrogen gas without previous fusion, and by the characteristic odor of garlic emitted by the escaping (unkindled) gas. If the gas is kindled whilst the mirror in the tube is being heated the flame will, even with a very slight current of gas, deposit arsenical stains on a porcelain plate.

The reactions and properties just described are amply sufficient to enable us to distinguish between arsenical and antimonial stains and mirrors; but they will often fail to detect arsenic with positive certainty in presence of antimony. In cases of this kind the following process will serve to set at rest all possible doubt as to the presence or absence of arsenic:—

Heat the long tube through which the arsenetted hydrogen passes to redness in several parts, to produce distinct metallic mirrors; then transmit through the tube a very weak stream of dry hydrosulphuric acid gas, and heat the metallic mirrors with a common spirit-lamp, proceeding from the outer towards the inner border. If arsenic alone is present yellow tersulphide of arsenic is formed inside the tube; if antimony alone is present an orange-red or black tersulphide of antimony is produced; and if the mirror consisted of both metals the two sulphides appear side by side, the sulphide of arsenic as the more volatile lying invariably before the sulphide of antimony. If you now transmit through the tube containing the sulphide of arsenic or the sulphide of antimony, or both sulphides together, dry hydrochloric gas, without applying heat, no alteration will take place if sulphide of arsenic alone is present, even though the gas be transmitted through the tube for a considerable time. If sulphide of antimony alone is present this will entirely disappear, as already stated, § 131, 10, and if both sulphides are present the sulphide of antimony will immediately volatilize, whilst the yellow sulphide of arsenic will remain. If a small quantity of ammonia is now drawn into the tube the sulphide of arsenic is dissolved, and may thus be readily distinguished from sulphur which may have separated. My personal experience has convinced me of the infallibility of these combined tests for the detection of arsenic.

The reaction of hydrogen containing arsenetted hydrogen with solution of nitrate of silver will be found in § 134, 6.

Marsh was the first who suggested the method of detecting arsenic by the production of arsenetted hydrogen.

11. If a small lump of arsenious acid (*a*) be introduced into the pointed end of a drawn-out glass tube (Fig. 29), a fragment of quite recently burnt charcoal (*b*) pushed down the tube to within a short distance of the arsenious acid, and first the charcoal then the arsenious acid heated to redness by the flame of a spirit-lamp, a MIRROR OF METALLIC ARSENIC will form at *c*, owing to the reduction of the arsenious acid vapor by the red-hot charcoal. If the tube be now cut between *b* and *c* and then heated in an inclined position, with the cut end *c* turned upwards, the metallic mirror will volatilize, emitting the characteristic odor of garlic. This is both the simplest and safest way of detecting pure arsenious acid.

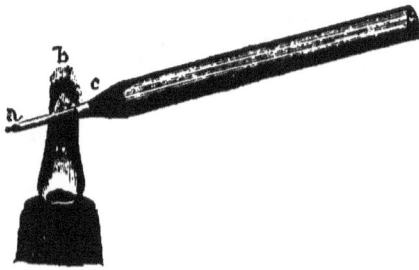

Fig. 29.

12. If arsenites, or arsenious acid, or tersulphide of arsenic are fused together with a mixture of equal parts of dry *carbonate of soda* and *cyanide of potassium* the whole of the arsenic is reduced to the metallic

state, and so is the base also, if easily reducible; the eliminated oxygen converts part of the cyanide of potassium into cyanate of potassa. In the reduction of tersulphide of arsenic sulphocyanide of potassium is formed. The operation is conducted as follows:—introduce the perfectly dry arsenical compound into the bulb of a small bulb-tube (Fig. 30), and cover it with six times the quantity of a perfectly dry mixture of equal parts of carbonate of soda and of cyanide of potassium. The whole quantity must not much more than half-fill the bulb, otherwise the fusing cyanide of potassium is likely to ascend into the tube. Heat the bulb now gently with a gas or spirit-lamp; should some water still escape upon gently heating the mixture wipe the inside of the tube *perfectly dry* with a twisted slip of paper. It is of the highest importance for the success of the experiment to bestow great care upon expelling the water, drying the mixture, and wiping the tube clean and dry. Apply now a strong heat to the bulb, to effect the reduction of the arsenical compound, and continue this for some time, as the arsenic often requires

Fig. 30.

some time for its complete sublimation. The mirror which is deposited at b is of exceeding purity. It is obtained from all arsenites whose bases remain either altogether unaffected, or are reduced to such metallic arsenides as lose their arsenic partly or totally upon the simple application of heat. This method deserves to be particularly recommended on account of its simplicity and neatness, as well as for the accuracy of the results attainable by it, even in cases where only very minute quantities of arsenic are present. It is more especially adapted for the direct production of arsenic from tersulphide of arsenic, and is in this respect superior in simplicity and accuracy to all other methods hitherto suggested. The delicacy of the reaction may be very much heightened by heating the mixture in a stream of dry carbonic acid gas. A series of experiments made by *Dr. V. Babo* and myself has shown that the most accurate and satisfactory results are obtained in the following manner:—

Figs. 31 and 32 show the apparatus in which the process is conducted.

A is a capacious flask intended for the evolution of carbonic acid; it is half-filled with water and lumps of solid limestone or marble (not chalk, as this would not give a constant stream of gas). B is a smaller flask containing concentrated sulphuric acid. The flask A is closed with a double-perforated cork, into one aperture of which is inserted a funnel-tube (a), which reaches nearly to the bottom of the flask; into the other perforation is fitted a tube (b), which serves to conduct the evolved gas into the sulphuric acid in B, where it is thoroughly freed from moisture. The tube c conducts the dried gas into the reduction-tube C, of which Fig. 32 gives a representation on the scale of one-third of the actual length. The tubes which I employ for the purpose in my own experiments have an inner diameter of eight millimetres.

Fig. 31.

When the apparatus is fully prepared for use triturate the perfectly dry sulphide of arsenic or arsenite in a slightly heated mortar with about twelve parts of a well-dried mixture consisting of three parts of carbonate of soda and one part of cyanide of potassium. Put the powder upon a narrow slip of card-paper bent into the shape of a gutter, and push this into the reduction-tube down to e; turn the tube now half-way round its axis, which will cause the mixture to drop into the tube between e and d, every other part remaining perfectly clean. Connect the tube now with the gas-evolution apparatus, and evolve a moderate stream of carbonic acid, by pouring some hydrochloric acid into the flask A. Heat the tube C in its whole length very gently with a spirit-lamp until the mixture in it is quite dry; when every trace of water is expelled, and the gas stream has become so slow that the single bubbles pass through the sulphuric acid in B at intervals of one second, heat the reduction-tube C, to redness at c (Fig. 32), by means of a spirit

Fig. 32.

or gas lamp; when c is red-hot, apply the flame of a gas or a larger spirit lamp to the mixture, proceeding from d to e, until the whole of the arsenic is expelled. The far greater portion of the volatilized arsenic recondenses at h, whilst a small portion only escapes through i, imparting to the surrounding air the peculiar odor of garlic. Advance the flame of the second lamp slowly and gradually up to c, by which means the whole of the arsenic which may have condensed in the wide part of the tube is driven to h. When you have effected this, close the tube at the point i by fusion, and apply heat, proceeding from i towards h, by which means the extent of the mirror is narrowed, whilst its beauty and lustre are correspondingly increased. In this manner perfectly distinct mirrors of arsenic may be produced from as little as the $\frac{1}{300}$th part of a grain of tersulphide of arsenic. No mirrors are obtained by this process from tersulphide of antimony, nor from any other compound of antimony.

13. If arsenious acid or one of its compounds is exposed on a charcoal support to the *reducing flame of the blowpipe* a highly characteristic garlic odor is emitted, more especially if some carbonate of soda is added to the examined sample. This odor has its origin in the reduction and re-oxidation of the arsenic, and enables us to detect very minute quantities. This test, however, like all others that are based upon the mere indications of the sense of smell, cannot be implicitly relied on.

§ 133.

e. ARSENIC ACID (AsO_5).

1. ARSENIC ACID is a transparent or white mass, which gradually deliquesces in the air, and dissolves slowly but copiously in water. It fuses at a gentle red heat without suffering decomposition; but at a higher temperature it is resolved into oxygen and arsenious acid, which volatilizes. It is highly poisonous.

2. Most of the ARSENATES are insoluble in water. Of the so-called neutral arsenates those with alkaline bases alone are soluble in water. Most of the neutral and basic arsenates can bear a strong red heat without suffering decomposition. The acid arsenates lose their excess of acid upon ignition, the free acid being resolved into arsenious acid and oxygen.

3. *Hydrosulphuric acid* fails to precipitate alkaline and neutral solutions of arsenates; but in acidified solutions it causes first reduction of the arsenic acid to arsenious acid, with separation of sulphur, then precipitation of tersulphide of arsenic. This process continues until the whole of the arsenic is thrown down as AsS_3, mixed with 2 S (WACKENRODER, LUDWIG, H. ROSE). The action never takes place immediately, and in dilute solutions frequently only after the lapse of a considerable time (twelve to twenty-four hours, for instance). Heating (to about 158° F.) greatly accelerates the action. If a solution of arsenic acid, or of an arsenate, is mixed with sulphurous acid, or with sulphite of soda and some hydrochloric acid, the sulphurous acid is converted into sulphuric acid, and the arsenic acid reduced to arsenious acid; application of heat promotes the change. If hydrosulphuric acid is now added the whole of the arsenic is immediately thrown down as tersulphide.

4. *Sulphide of ammonium* converts the arsenic acid in neutral and alkaline solutions of arsenates into pentasulphide of arsenic, which remains in solution as ammonio-pentasulphide of arsenic (pentasulphide of arsenic and sulphide of ammonium). Upon the addition of an acid to the solution this double sulphide is decomposed, and pentasulphide of arsenic precipitates. The separation of this precipitate proceeds more rapidly than is the case when acid solutions of arsenates are precipitated with hydrosulphuric acid. It is promoted by heat. The precipitate formed is AsS_5, instead of consisting of a mixture of AsS_3 with S_2, as in precipitation with hydrosulphuric acid.

5. *Nitrate of silver* produces under the circumstances stated § 132, 6,

a highly characteristic reddish-brown precipitate of ARSENATE OF SILVER (3 AgO, AsO$_5$), which is readily soluble in dilute nitric acid and in ammonia, and dissolves also slightly in nitrate of ammonia. Accordingly, if a little of the precipitate is dissolved in a large proportion of nitric acid, neutralization with ammonia often fails to reproduce the precipitate.

6. *Sulphate of copper* produces under the circumstances stated § 132, 7, a greenish-blue precipitate of ARSENATE OF COPPER (2 CuO, HO, AsO$_5$).

7. If a solution of arsenic acid mixed with some hydrochloric acid is heated with a clean slip of *copper* the metal remains perfectly clean (WERTHER, REINSCH); but if to one volume of the solution two volumes of concentrated hydrochloric acid are added, a gray film is deposited on the copper, the same as with arsenious acid. The reaction is under these circumstances equally delicate with arsenic acid as with arsenious acid (REINSCH).

8. With *zinc* in presence of sulphuric acid, with *cyanide of potassium*, and before the *blowpipe*, the compounds of arsenic acid comport themselves in the same way as those of arsenious acid. If the reduction of arsenic acid by zinc is effected in a platinum dish the platinum does not turn black, as is the case in the reduction of antimony by zinc (§ 131, 8).

9. If a solution of arsenic acid, or of an arsenate soluble in water, is added to a clear mixture of *sulphate of magnesia, chloride of ammonium*, and a sufficient quantity of *ammonia*, a crystalline precipitate of ARSENATE OF AMMONIA AND MAGNESIA (2 MgO, NH$_4$, O, AsO$_5$ + 12 aq) separates; from concentrated solutions immediately, from dilute solutions after some time. If a small portion of the precipitate is dissolved on a watch-glass in a drop of nitric acid, a little nitrate of silver added, and the solution touched with a glass rod dipped in ammonia, brownish-red arsenate of silver is formed (difference between arsenate and phosphate of magnesia and ammonia).

§ 134.

Recapitulation and remarks.—I will here describe first the different ways best adapted to effect the detection or separation of tin, antimony, and arsenic, when present together in the same compound or mixture, and afterwards the most reliable means of distinguishing between the several oxides of each of the three metals.

1. If you have a mixture of sulphide of tin, sulphide of antimony, and sulphide of arsenic, triturate 1 part of it, together with 1 part of dry carbonate of soda and 1 part of nitrate of soda, and transfer the mixed powder gradually to a small porcelain crucible containing 2 parts of nitrate of soda kept in a state of fusion at a not over-strong heat; oxidation of the sulphides ensues, attended with slight deflagration. The fused mass contains binoxide of tin, arsenate and antimonate of soda, with sulphate, carbonate, nitrate, and nitrite of soda. You must take care not to raise the heat to such a degree, nor continue the fusion so long, as to lead to a reduction of the nitrite of soda to the caustic state, that there may not be formed stannate of soda soluble in water. Upon treating the mass with a little cold water binoxide of tin and antimonate of soda remain undissolved, whilst arsenate of soda and the other salts are dissolved. If the filtrate is acidified with nitric acid, and

heat is applied to remove carbonic acid and nitrous acid, the arsenic acid may be detected and separated, either with nitrate of silver, according to § 133, 5, or with a mixture of sulphate of magnesia, chloride of ammonium, and ammonia, according to § 133, 9.

If the undissolved residue, consisting of binoxide of tin and antimonate of soda, is, after being washed once with cold water and three times with dilute spirits of wine, treated with some hydrochloric acid in the lid of a platinum crucible, and a gentle heat applied, the mass is either completely dissolved or, if the tin is present in a large proportion, a white residue is left undissolved. If, regardless of the presence of this latter, a fragment of zinc is added, the compounds are reduced to the metallic state, when the antimony will at once reveal its presence by blackening the platinum. If, after the evolution of hydrogen has nearly stopped, the remainder of the zinc is taken away, and the contents of the lid are heated with some hydrochloric acid, the tin dissolves the protochloride, whilst the antimony is left undissolved in the form of black flakes. The tin may then be more distinctly tested in the solution, with chloride of mercury, or with a mixture of sesquichloride of iron and ferricyanide of potassium, and the antimony, after solution in a little aqua regia, with hydrosulphuric acid. As this method of detecting arsenic, tin, and antimony in presence of each other, forms one of the processes in the systematic course of analysis, I have here simply explained the principle upon which it is based, and refer for the details of the process to the first section of Part II.

2. If the mixed sulphides, after being freed from the greater part of the adhering water, by laying the filter containing them on blotting paper, are treated with fuming hydrochloric acid, with application of a gentle heat, the sulphides of antimony and tin dissolve, whilst the sulphide of arsenic is left almost completely undissolved. By treating this with ammonia, and evaporating the solution obtained, with addition of a small quantity of carbonate of soda, an arsenical mirror may easily be produced from the residue, by means of cyanide of potassium and carbonate of soda in a stream of carbonic acid gas (§ 132, 12). The solution, which contains the tin and the antimony, may be treated as stated in 1.

If a great excess of antimony is present the solution may also be mixed with sesquicarbonate of ammonia in excess, and the mixture boiled; when a large proportion of the antimony will dissolve, leaving binoxide of tin behind, mixed with but little teroxide of antimony; in which undissolved residue the tin may now be the more readily detected by the method described in 1 (BLOXAM).

3. If the mixed sulphides are digested at a gentle heat with some common solid carbonate of ammonia and water sulphide of arsenic dissolves, whilst the sulphides of antimony and tin remain undissolved. But even this separation is not quite absolute, as traces of sulphide of antimony are apt to pass into the solution, whilst some sulphide of arsenic remains in the residue. The sulphide of arsenic precipitating from the alkaline solution upon acidifying this latter with hydrochloric acid must therefore, especially if consisting only of a few flakes, after washing, be treated with ammonia, the solution evaporated, with addition of a small quantity of carbonate of soda, and the residue fused with cyanide of potassium in a stream of carbonic acid, to make quite sure by the pro-

duction of an arsenical mirror. The residue, insoluble in carbonate of ammonia, should be treated as directed in 2.

4. If sulphide of antimony, sulphide of tin, and sulphide of arsenic are dissolved in sulphide of potassium, a *large* excess of a concentrated solution of sulphurous acid added, the mixture digested for some time on the water-bath, boiled until all sulphurous acid is expelled, then filtered, the filtrate contains all the arsenic as arsenious acid (which may be precipitated from it by hydrosulphuric acid), whilst tersulphide of antimony and bisulphide of tin are left behind undissolved (BUNSEN). These latter may then be treated as directed in 2.

5. In the analysis of alloys, binoxide of tin, teroxide of antimony, and arsenic acid are often obtained together as a residue insoluble in nitric acid. The best way is to fuse this residue with hydrate of soda in a silver crucible, to treat the mass with water, and add one-third (by volume) of spirit of wine ; then to filter the fluid off from the antimonate of soda, which remains undissolved, and wash the latter with spirit of wine mixed with a few drops of solution of carbonate of soda. The filtrate is acidified with hydrochloric acid, and the tin and arsenic are then precipitated as sulphides, with the aid of heat. On heating the precipitated sulphides in a stream of hydrosulphuric acid gas the whole of the tin is left as sulphide, whilst the sulphide of arsenic volatilizes, and may be received in solution of ammonia (H. ROSE).

6. For the most accurate way of separating antimony and arsenic, and distinguishing between the two metals, viz., by treating with hydrosulphuric acid the mirror produced by MARSH's method, and separating the resulting sulphides by means of hydrochloric acid gas, I refer to § 132, 10. Antimony and arsenic may, however, when mixed together in form of hydrogen compounds, be separated also in the following way : Conduct the gases mixed with an excess of hydrogen, first through a tube containing glass splinters moistened with solution of acetate of lead, to retain the hydrochloric and hydrosulphuric gas, then in a slow stream into a solution of nitrate of silver. All the antimony in the gas falls down as black antimonide of silver (Ag_3Sb), whilst the arsenic passes into the solution as arsenious acid, with reduction of the silver, and may be detected in the fluid as arsenite of silver, by cautious addition of ammonia, or—after precipitating the excess of silver by hydrochloric acid—by means of hydrosulphuric acid. In the precipitated antimonide of silver, which is often mixed with much silver, the antimony may be most readily detected, by heating the precipitate—thoroughly freed from arsenious acid by boiling with water—with tartaric acid and water to boiling. This will dissolve the antimony alone, which may then be readily detected by means of hydrosulphuric acid in the solution acidified with hydrochloric acid (A. W. HOFMANN).

7. *Protoxide* and *binoxide of tin* may be detected and identified in presence of each other, by testing one portion of the solution containing both oxides, for the protoxide with chloride of mercury, terchloride of gold or a mixture of ferricyanide of potassium and sesquichloride of iron, and another portion for the binoxide, by pouring it into a concentrated hot solution of sulphate of soda.

8. *Teroxide of antimony* in presence of *antimonic acid* may be identified by the reaction described in § 131, 9. *Antimonic acid* in presence of *teroxide of antimony*, by heating the teroxide suspected to contain an admixture of the acid, but without any other ad-

mixture, with hydrochloric acid and iodide of potassium (§ 131, 2 and 3).

9. *Arsenious acid* and *arsenic acid* in the same solution may be distinguished by means of nitrate of silver. If the precipitate contains little arsenate and much arsenite of silver it is necessary, in order to identify the former, to add cautiously and drop by drop most highly dilute nitric acid, which dissolves the yellow arsenite of silver first.

A still safer way to detect small quantities of arsenic acid in presence of arsenious acid is to precipitate the solution which contains the two acids, with a mixture of sulphate of magnesia, chloride of ammonium, and ammonia (§ 133, 9), by which means an actual separation of the two acids is also effected. The *immediate* precipitation of arsenic acid from an acidified solution by hydrosulphuric acid unaided by the application of heat affords also a ready means of distinguishing between the two acids, as this reaction differs considerably in the case of arsenious acid. Arsenious acid in presence of arsenic acid may also be identified by the reduction of oxide of copper effected by its agency in alkaline solutions. To ascertain the degree of sulphuration of a sulphide of arsenic in a sulphur salt, boil the alkaline solution of the salt under examination with hydrate of teroxide of bismuth, filter off from the tersulphide of bismuth formed, and test the filtrate for arsenious and arsenic acids. To distinguish between the ter- and pentasulphide of arsenic, extract first the sulphur which may be present by means of sulphide of carbon, then dissolve the residue in ammonia, add nitrate of silver in excess, filter off the sulphide of silver, and observe whether arsenite or arsenate of silver is formed upon addition of ammonia.

Special Reactions of the rarer Oxides of the Sixth Group.

§ 135.

a. OXIDE OF IRIDIUM (IrO_2).

IRIDIUM is found in combination with platinum and other metals in platinum ores, also, and more especially, as a native alloy of osmium and iridium. Alloyed with platinum, it has of late been employed for crucibles, &c. Iridium resembles platinum, but is brittle; it fuses with extreme difficulty. In the compact state, or reduced at a red heat by hydrogen, it dissolves in no acid, not even in aqua regia (difference between iridium and gold and platinum); reduced in the moist way, say by formic acid, or largely alloyed with platinum, it dissolves in aqua regia to bichloride ($IrCl_2$). *Acid sulphate of potassa* in a state of fusion will oxidize, but not dissolve it (difference between iridium and rhodium). It oxidizes by fusion with *hydrate of soda*, with access of air, or by fusion with nitrate of soda. The compound of sesquioxide of iridium (Ir_2O_3) with soda, which is formed in this process, dissolves partially in water; by heating with aqua regia it gives a deep-black solution of bichloride of iridium and chloride of sodium.

If iridium in powder is mixed with chloride of sodium, the mixture heated to insipient redness, and treated with *chlorine gas*, sodio-bichloride of iridium is formed, which dissolves in water to a deep reddish-brown fluid. *Potassa*, added in excess, decolorizes the solution, a little brownish-black potassio-bichloride of iridium precipitating at the same time. If the solution is heated, and exposed some time to the air, it acquires at first a reddish tint, which changes afterwards to azure blue (characteristic difference between iridium and platinum); if the solution is now evaporated to dryness, and the residue treated with water, a colorless fluid is obtained, with a blue deposit left undissolved. *Hydrosulphuric acid* in the first place decolorizes solutions of bichloride of iridium, protochloride is formed, with separation of sulphur, and finally brown sulphide of iridium precipitates. *Sulphide of ammonium* produces the same precipitate, which redissolves readily in an excess of the precipitant. *Chloride of ammonium* precipitates from more concentrated solutions ammonio-bichloride of iridium in form of a blackish-red powder consisting of microscopic octahedrons. *Protochloride of tin* produces a light-brown precipitate. *Sulphate of protoxide of iron* decolorizes, but fails to precipitate solutions of bichloride of iridium ; *zinc* precipitates black iridium.

b. OXIDES OF MOLYBDENUM.

MOLYBDENUM is not largely disseminated in nature, and is found only in moderate quantities, more especially as sulphide of molybdenum and as molybdate of lead (yellow lead ore). Since the use of molybdate of ammonia as a means of detecting and determining phosphoric acid, molybdenum has acquired greater importance for practical chemistry. MOLYBDENUM is silvery white ; it fuses with very great difficulty. The PROTOXIDE of the metal (Mo O) is black, the BINOXIDE (Mo O_2) dark brown. By heating in the air, or treating with nitric acid, the metal and the two oxides are converted into MOLYBDIC ACID (Mo O_3). Molybdic acid is a white porous mass, which in water separates as fine scales ; it fuses at a red-heat ; in close vessels it volatilizes only at a very high temperature, in the air easily at a red-heat, subliming to transparent laminæ and needles. The non-ignited acid dissolves in acids. The solutions are colorless ; in contact with *zinc* or *tin* they first turn blue, then green, and ultimately black, with separation of protoxide of molybdenum ; by digestion with *copper* the sulphuric acid solution acquires a blue, the hydrochloric acid solution a brown tint. The reaction often takes place only after some time. *Ferrocyanide of potassium* produces a reddish-brown precipitate, *infusion of galls* a green precipitate. *Hydrosulphuric acid*, added in small proportion, imparts a blue tint to solutions of molybdic acid ; added in larger proportion it produces a brownish-black precipitate ; the fluid over the latter at first appears green. But after being allowed to stand for some time, and heated, additional quantities of hydrosulphuric acid being repeatedly conducted into it, the whole of the molybdenum present will ultimately though slowly separate as black tersulphide of molybdenum (Mo S_3). The precipitated tersulphide of molybdenum dissolves in sulphides of the alkali metals ; acids reprecipitate from the sulphur salts the sulphur acid (Mo S_3). Application of heat promotes the separation. By heating to redness in the air, or by heating with nitric acid, sulphide of molybdenum is converted into molybdic acid.

Molybdic acid dissolves readily in solutions of *pure alkalies* and *carbonates of the alkalies;* from concentrated solutions *nitric acid* or *hydrochloric acid* throws down molybdic acid, which redissolves upon further addition of the precipitant. The solutions of molybdates of the alkalies are colored yellow by *hydrosulphuric acid*, and give afterwards, upon addition of acids, a brownish-black precipitate. For the deportment of molybdic acid with phosphoric acid and ammonia, see § 142, 10.

If a fragment of *zinc* is put into a hydrochloric acid solution of molybdic acid, and a few drops of solution of *sulphocyanide of potassium* are added, and as much hydrochloric or sulphuric acid as will suffice to bring on a slight evolution of hydrogen, the fluid acquires a carmine-red tint, which, however, is not very persistent (C. D. BRAUN).

Molybdic acid volatilizes when heated on charcoal in the *oxidizing flame*, coating the charcoal with a yellow, often crystalline, powder, which turns white on cooling. In the *reducing flame* the acid suffers reduction to the metallic state, the molybdenum is obtained as a gray powder by washing the charcoal support. Sulphide of molybdenum gives in the oxidizing flame sulphurous acid and an incrustation of molybdic acid on the charcoal support.

c. OXIDES OF WOLFRAMIUM OR TUNGSTEN.

WOLFRAMIUM or tungsten metal is not widely disseminated in nature, and is found only in very moderate quantity. It occurs most frequently as tungstate of lime and in the mineral *wolfram*, which is a double tungstate of protoxide of iron and protoxide of manganese. TUNGSTEN, or WOLFRAMIUM, obtained by the reduction of tungstic acid in a stream of hydrogen gas at an intense red heat, is an iron-gray powder, which fuses with extreme difficulty, and is converted by ignition in the air into tungstic acid (W O_3), by ignition in a stream of chlorine gas into a sublimate of red bichloride of tungsten (W Cl_2). The latter in contact with water gives first hydrochloric acid and binoxide of tungsten (W O_2), which, however, oxidizes subsequently in the air to blue oxide (W O_2, W O_3). Tungsten is insoluble in acids, even in aqua regia, and also in solutions of potassa ; it dissolves, however, in solution of potassa mixed with hypochlorite of alkali. BINOXIDE OF TUNGSTEN is black ; by intense ignition with free access of air it is converted into tungstic acid. TUNGSTIC ACID is lemon-yellow, fixed, insoluble in water and in acids. By fusing tungstic acid with acid sulphate of potassa, and treating the fused mass with water, an acid solution is obtained, which contains no tungstic acid. After the removal of this solution, the residue, consisting of tungstate of potassa, dissolves. Tungstates of the alkalies, soluble in water, are formed readily by fusion, with carbonated alkalies, with difficulty by boiling with solutions of alkalies. *Hydrochloric acid, nitric acid,* and *sulphuric acid* produce in the solution of these tungstates white precipitates, which are insoluble in an excess of the acids (difference from molybdic acid), but soluble in ammonia. Upon evaporating with an excess of hydrochloric acid to dryness, and treating the residue with water, the tungstic acid is left undissolved. *Chloride of barium, chloride of calcium, nitrate of silver, nitrate of suboxide of mercury* produce white precipitates. *Ferrocyanide of potassium,* with addition of some acid, colors the fluid deep brownish-red, and after some time produces a precipitate of the same color. *Tincture of galls,* with a little acid added, produces a brown precipitate. *Hydrosulphuric acid* barely precipitates acid solutions. *Sulphide of ammonium* fails to precipitate solutions of tungstates of alkalies; upon acidifying the mixture light-brown tersulphide of tungsten (WS_3) precipitates, which is slightly soluble in pure water, but insoluble in water containing salts. *Protochloride of tin* produces a yellow precipitate; on acidifying with hydrochloric acid, and applying heat, this precipitate acquires a beautiful blue color (highly delicate and characteristic reaction). If solutions of tungstates of alkalies are mixed with hydrochloric acid, or, better still, with an excess of phosphoric acid, and *zinc* is added, the fluid acquires a beautiful blue color. *Phosphate of soda and ammonia* dissolves tungstic acid. The bead, exposed to the oxidizing flame, appears clear, varying from colorless to yellowish; in the reducing flame it acquires a pure blue color, and upon addition of sulphate of protoxide of iron a blood-red color. By mixing with a little *carbonate of soda,* and exposing in the cavity of the charcoal support to the reducing flame, tungsten in powder is obtained, which may be washed off the charcoal. The tungstates which are insoluble in water may, most of them, be decomposed by digestion with acids. The mineral Wolfram, which strongly resists the action of acid, is fused with carbonated alkali, when water will dissolve out of the fused mass the tungstate of alkali formed.

d. Oxides of Tellurium.

TELLURIUM is not widely disseminated, and is found in small quantities only in the native state, or alloyed with other metals, or as tellurous acid. It is a white brittle, but readily fusible metal, which may be sublimed in a glass tube. Heated in the air it burns with a greenish-blue flame, emitting thick white fumes of tellurous acid. Tellurium is insoluble in hydrochloric acid, but dissolves readily in nitric acid to tellurous acid (TeO_2). Tellurium in powder dissolves in cold concentrated sulphuric acid to a purple-colored fluid, from which it separates again upon addition of water. TELLUROUS ACID is white; at a gentle red heat it fuses to a yellow fluid; it is volatilized by strong ignition in the air, forming no crystalline sublimate. The anhydrous acid dissolves readily in hydrochloric acid, sparingly in nitric acid, freely in solution of potassa, slowly in ammonia, barely in water. The hydrate of tellurous acid is white; it is perceptibly soluble in cold water, and dissolves in hydrochloric acid and in nitric acid. By addition of water white hydrate is thrown down from the solution, and from the nitric acid solution nearly the whole of the tellurous acid separates after some time as a crystalline precipitate, even without addition of water. *Pure alkalies* and *carbonates of the alkalies* throw down from the hydrochloric acid solution white hydrate, which is soluble in an excess of the precipitant. *Hydrosulphuric acid* produces in acid solutions a brown precipitate (TeS_2) (in color like protosulphide of tin), which dissolves very freely in sulphide of ammonium. *Sulphite of soda, protochloride of tin,* and *zinc* precipitate black metallic tellurium. TELLURIC ACID (TeO_3) is formed by fusing tellurium or tellurites with nitrates and carbonates of the alkalies. The fused mass is soluble in water. The solution remains clear upon acidifying with hydrochloric acid in the cold; but upon boiling chlorine is disengaged, and tellurous acid formed, and the solution is therefore now precipitated by water if the excess of acid is not too great. If tellurium, its sulphide, or an oxygen compound of the metal is fused with *cyanide of potassium* in a stream of hydrogen, a double cyanide of tellurium and potassium is formed. The fused mass dissolves in water, but a current of air throws down from the solution the whole of the tellurium (difference and means of separating tellurium from selenium). By fusion of tellurous or telluric acid with *carbonate of potassa* and *charcoal,* telluride of potassium is formed, which with acids evolves stinking telluretted hydrogen gas. Upon exposing tellurium compounds mixed with carbonate of soda to the *inner blowpipe flame,* reduction, volatilization, and reoxidation take place, and the charcoal support becomes accordingly covered with white tellurous acid.

e. Oxides of Selenium.

SELENIUM is a rare substance; it occurs in nature in the form of selenides of metals. It is found occasionally in the dust of roasting-furnaces, and also in the Nordhausen oil of vitriol. It resembles sulphur in some respects, tellurium in others, and stands thus on the border between the metals and the non-metallic elements. Fused selenium is grayish-black; it volatilizes at a higher temperature, and may be sublimed. Heated in the air it burns to selenious acid (SeO_2), exhaling a characteristic smell of decaying horse-radish. Concentrated sulphuric acid dissolves selenium without oxidizing it; upon diluting the solution the selenium falls down in red flakes. Nitric acid and aqua regia dissolve selenium to SELENIOUS ACID. Sublimed anhydrous selenious acid appears in form of white four-sided needles, its hydrate in form of crystals resembling those of nitrate of potassa. Both the acid and its hydrate dissolve readily in water to a strongly acid fluid. Of the neutral salts only those with the alkalies are soluble in water; the solutions have alkaline reactions. All selenites dissolve readily in nitric acid, with the exception of the selenites of lead and silver, which dissolve with difficulty. *Hydrosulphuric acid* produces in solutions of selenious acid or of selenites (in presence of free hydrochloric acid) a yellow precipitate of SULPHIDE OF SELENIUM (?) which, upon heating, turns reddish-yellow, and is soluble in sulphide of ammonium. *Chloride of barium* produces (after neutralization of the free acid, should any be present) a white precipitate of *selenite of baryta*, which is soluble in hydrochloric acid and in nitric acid. *Protochloride of tin* or *sulphurous acid*, with addition of hydrochloric acid, produces a red precipitate of SELENIUM, which turns gray at a high temperature. SELENIC ACID is formed by heating selenium or its compounds with carbonates and nitrates of the alkalies. The fused mass dissolves in water; the solution remains clear upon acidifying with hydrochloric acid; when concentrated by boiling, it evolves chlorine, whilst the selenic acid is reduced to selenious acid. By fusing selenium or its compounds with *cyanide of potassium* in a stream of hydrogen gas, a double cyanide of selenium and potassium is obtained, from which the selenium is not eliminated by the action of the air (as is the case with tellurium); it separates, however, upon long-continued boiling, after addition of hydrochloric acid. When exposed on a charcoal support to the *reducing flame*, the selenites evolve selenium, exhaling at the same time a most characteristic odor of decaying horse-radish, which unmistakeably betrays their presence.

B.—REACTIONS OR DEPORTMENT OF THE ACIDS AND THEIR RADICALS WITH REAGENTS.

§ 136.

The reagents which serve for the detection of the acids are divided, like those used for the detection of the bases, into GENERAL REAGENTS, *i.e.*, such as indicate the GROUP to which the acid under examination belongs; and SPECIAL REAGENTS, *i.e.*, such as serve to effect the detection and identification of the INDIVIDUAL ACIDS. The groups into which we classify the various acids can scarcely be defined and limited with the same degree of precision as those into which the bases are divided.

The two principal groups into which acids are divided are those of INORGANIC and ORGANIC ACIDS. We base this division upon those characteristics by which, irrespectively of theoretical considerations, the ends of analysis are most easily attained. We select therefore here, as the characteristic mark to guide us in the classification into organic and inorganic acids, the deportment which the various acids manifest at a high temperature, and call *organic* those acids of which the salts—(particularly those which have an alkali or an alkaline earth for base)—are decomposed upon ignition, the decomposition being attended with separation of carbon.

By selecting this deportment at a high temperature as the distinctive characteristic of organic acids, we are enabled to determine at once by a most simple preliminary experiment the class to which an acid belongs. The salts of organic acids with alkalies or alkaline earths are converted into carbonates when heated to redness.

Before proceeding to the special study of the several acids considered in this work, I give here, the same as I have done with the bases, a general view of the whole of them classified in groups.

CLASSIFICATION OF ACIDS IN GROUPS.

I. INORGANIC ACIDS.

FIRST GROUP:

Division *a. Chromic acid* (sulphurous and hyposulphurous acids, iodic acid).
Division *b. Sulphuric acid* (hydrofluosilicic acid).
Division *c. Phosphoric acid, boracic acid, oxalic acid, hydrofluoric acid* (phosphorous acid).
Division *d. Carbonic acid, silicic acid.*

SECOND GROUP:

Chlorine and *hydrochloric acid; bromine* and *hydrobromic acid; iodine* and *hydriodic acid; cyanogen* and *hydrocyanic acid,* together with *hydroferro-* and *hydroferricyanic acids; sulphur* and *hydrosulphuric acid* (nitrous acid, hypochlorous acid, chlorous acid, hypophosphorous acid).

THIRD GROUP:

Nitric acid, chloric acid (perchloric acid).

II. ORGANIC ACIDS.

FIRST GROUP:

Oxalic acid, tartaric acid, citric acid, malic acid (racemic acid).

SECOND GROUP:

Succinic acid, benzoic acid.

THIRD GROUP:

Acetic acid, formic acid (propionic acid, butyric acid, lactic acid).

The acids printed in italics are more frequently met with in the examination of minerals, waters, ashes of plants, industrial products, medicines, &c.; the others are more rarely met with.

I. INORGANIC ACIDS.

§ 137.

First Group.

ACIDS WHICH ARE PRECIPITATED FROM NEUTRAL SOLUTIONS BY CHLORIDE OF BARIUM.

This group is again subdivided into four divisions, viz.:
1. Acids which are decomposed in acid solution by hydrosulphuric acid, and to which attention has therefore been directed already in the testing for bases, viz., CHROMIC ACID (sulphurous acid and hyposulphurous acid, the latter because it is decomposed and detected by the mere addition of hydrochloric acid to the solution of one of its salts; and also iodic acid).*

---

* To this first division of the first group of inorganic acids belong properly also all the oxygen compounds of a distinctly pronounced acid character, which have been discussed already with the Sixth Group of the metallic oxides (acids of arsenic, antimony, selenium, &c.). But as the reaction of these compounds with hydrosulphuric acid tends to lead to confounding them rather with other metallic oxides than with other acids, it appeared the safer course to class these compounds, which may be said to stand between the bases and the acids, with the metallic oxides.

2. Acids which are not decomposed in acid solution by hydrosulphuric acid, and the baryta compounds of which are insoluble in hydrochloric acid: SULPHURIC ACID (hydrofluosilicic acid).

3. Acids which are not decomposed in acid solution by hydrosulphuric acid, and the baryta compounds of which dissolve in hydrochloric acid, apparently WITHOUT DECOMPOSITION, inasmuch as the acids cannot be completely separated from the hydrochloric acid solution by heating or evaporation; these are PHOSPHORIC ACID, BORACIC ACID, OXALIC ACID, HYDROFLUORIC ACID (phosphorous acid). (Oxalic acid belongs more properly to the organic group. We consider it, however, here with the acids of the inorganic class, as the property of its salts to be decomposed upon ignition without actual carbonization may lead to its being overlooked as an organic acid.)

4. Acids which are not decomposed in acid solution by hydrosulphuric acid, and the baryta salts of which are soluble in hydrochloric acid WITH DECOMPOSITION (separation of the acid): CARBONIC ACID, SILICIC ACID.

First Division of the First Group of the Inorganic Acids.

§ 138.

CHROMIC ACID (CrO_3).

1. CHROMIC ACID appears as a scarlet-red crystalline mass, or in the form of distinct acicular crystals. Upon ignition it is resolved into sesquioxide of chromium and oxygen. It deliquesces rapidly upon exposure to the air. It dissolves in water, imparting to the fluid a deep reddish-brown tint, which remains still visible in very dilute solutions.

2. The CHROMATES are all red or yellow, and for the most part insoluble in water. Part of them are decomposed upon ignition; those with alkaline bases are fixed, and are soluble in water; the solutions of the neutral alkaline chromates are yellow, those of the alkaline bichromates are reddish-brown. These tints are still visible in highly dilute solutions. The yellow color of the solution of a neutral salt changes to reddish-brown on the addition of a mineral acid, owing to the formation of an acid chromate.

3. *Hydrosulphuric acid* acting upon the acidified solution of a chromate produces first a brownish coloration of the fluid, then a green coloration, arising from the salt of sesquioxide of chromium formed; this change of color is attended with separation of sulphur, which imparts a milky appearance to the fluid ($KO, 2 CrO_3 + 4 SO_3 + 3 HS = KO, SO_3 + Cr_2O_3, 3 SO_3 + 3 HO + 3 S$). Heat promotes this reaction, part of the sulphur being in that case converted into sulphuric acid.

4. Chromic acid may also be reduced to sesquioxide of chromium by means of many other substances, and more particularly by *sulphurous acid*, or by heating with *hydrochloric acid*, especially upon the addition of alcohol (in which case chloride of ethyle and aldehyde are evolved); also by *metallic zinc*, or by heating with *tartaric acid, oxalic acid*, &c. All these reactions are clearly characterized by the change of the red or yellow color of the solution to the green tint of the salt of sesquioxide of chromium.

5. *Chloride of barium* produces in aqueous solutions of chromates a yellowish-white precipitate of CHROMATE OF BARYTA (BaO, CrO_3), which is soluble in dilute hydrochloric acid and nitric acid.

6. *Nitrate of silver* produces in aqueous solutions of chromates a dark purple-red precipitate of CHROMATE OF SILVER (AgO, CrO_3), which is soluble in nitric acid and in ammonia; in slightly acid solutions it produces a precipitate of BICHROMATE OF SILVER (AgO, 2 CrO_3).

7. *Acetate of lead* produces in an aqueous or acetic acid solution of a chromate a yellow precipitate of CHROMATE OF LEAD (PbO, CrO_3), which is soluble in potassa, but only sparingly soluble in dilute nitric acid. Upon heating with alkalies the yellow neutral salt is converted into basic red chromate of lead (2 PbO, $Cr.O_3$).

8. If a very dilute acid solution of *peroxide of hydrogen** (about 6 or 8 cubic centimetres) is covered with a layer of ether (about half a centimetre thick), and a fluid containing chromic acid is added, the solution of peroxide of hydrogen acquires a fine blue color. By inverting the test-tube, closed with the thumb, repeatedly, without much shaking, the solution becomes colorless, whilst the ether acquires a blue color. The latter reaction is particularly characteristic. One part of chromate of potassa in 40,000 parts of water suffices to produce it distinctly (STORER); presence of vanadic acid materially impairs the delicacy of the test (WERTHER)—compare § 113, *b*. The blue coloration is in all probability caused, not by perchromic acid (the existence of which is altogether doubtful), but by a combination of chromic acid with peroxide of hydrogen. After some time reduction of the chromic acid to sesquioxide of chromium takes place, and at the same time decoloration of the ether.

9. If insoluble chromates are fused together with *carbonate of soda* and *nitrate of soda*, and the fused mass is treated with water, the fluid obtained appears YELLOW from the alkaline chromate which it holds in solution; upon the addition of an acid the yellow colour changes to reddish-brown. The oxides are left either in the pure state or as carbonates, unless they are soluble in the caustic soda formed from the nitrate.

10. The compounds of chromic acid show the same reactions with *phosphate of soda and ammonia* and with *borax* in the blowpipe flame, as the compounds of sesquioxide of chromium.

11. Very minute quantities of chromic acid may be detected by mixing with the fluid, slightly acidified with sulphuric acid, a little *tincture of guaiacum* (1 part of the resin to 100 parts of alcohol of 60 per cent.), when an intense blue coloration of the fluid will at once make its appearance, speedily vanishing again, however, where mere traces of chromic acid are present (H. SCHIFF).

Chromic acid being reduced by hydrosulphuric acid to sesquioxide of chromium, this acid is in the course of analysis always found already in the examination for bases. The intense color of the solutions containing chromic acid, the excellent reaction with peroxide of hydrogen, and the characteristic precipitates produced by solutions of salts of lead and salts of silver, afford moreover ready means for its detection.

* Solution of peroxide of hydrogen may be easily prepared by triturating a fragment of peroxide of barium (about the size of a pea) with some water, and adding it with stirring to a mixture of about 30 cubic centimetres of hydrochloric acid, and 120 cubic centimetres of water. The solution keeps a long time without suffering decomposition. In default of peroxide of barium, impure peroxide of sodium may be used instead, which is obtained by heating a fragment of sodium in a small porcelain dish until it takes fire, and letting it burn.

Rarer Acids of the First Division of the First Group.

§ 139.

a. SULPHUROUS ACID (SO_2)

SULPHUROUS ACID is a colorless, uninflammable gas, which exhales the stifling odor of burning sulphur. It dissolves copiously in water. The solution has the odor of the gas, reddens litmus-paper, and bleaches Brazil-wood paper. It absorbs oxygen from the air, and is thereby converted into sulphuric acid. The salts of sulphurous acid are colorless. Of the neutral sulphites those with alkaline base only are readily soluble in water; many of the sulphites insoluble or sparingly soluble in water dissolve in an aqueous solution of sulphurous acid, but fall down again upon boiling. All the sulphides evolve sulphurous acid when treated with *sulphuric acid* or *hydrochloric acid*. *Chlorine water* dissolves most sulphites to sulphates. *Chloride of barium* precipitates neutral sulphites, but not free sulphurous acid. The precipitate dissolves in hydrochloric acid. *Hydrosulphuric acid* decomposes free sulphurous acid, water and pentathionic acid being formed and free sulphur eliminated, which latter separates from the fluid. If a trace of sulphurous acid or of a sulphite is introduced into a flask in which hydrogen is being evolved from *zinc* and *hydrochloric acid*, hydrosulphuric acid is immediately evolved along with the hydrogen, and the gas now produces a black coloration or a black precipitate in a solution of acetate of lead to which has been added a sufficient quantity of solution of soda to redissolve the precipitate which forms at first. Sulphurous acid is a powerful reducing agent; it reduces chromic acid, permanganic acid, chloride of mercury (to subchloride), decolorizes iodide of starch, produces a blue precipitate in a mixture of ferricyanide of potassium and sesquichloride of iron, &c. With a hydrochloric acid solution of *protochloride of tin* a brown precipitate of PROTOSULPHIDE OF TIN is formed after some time. If an aqueous solution of an alkaline sulphite is mixed with acetic acid just to give it an incipient acid reaction, and is then added to a relatively large amount of solution of sulphate of zinc, mixed with a very small quantity of nitroprusside of sodium, the fluid acquires a red color if the quantity of the sulphite present is not too inconsiderable, but when the quantity of the sulphite is very minute the coloration makes its appearance only after addition of some solution of ferrocyanide of potassium. If the quantities are not altogether too minute, a purple-red precipitate will form at once upon the addition of the ferrocyanide of potassium (BÖDEKER). Hyposulphites of the alkalies do not show this reaction.

b. HYPOSULPHUROUS ACID (S_2O_2).

This acid does not exist in the free state. Most of its salts are soluble in water. The solutions of most hyposulphites may be boiled, without suffering decomposition; hyposulphite of lime is resolved upon boiling into sulphite of lime and sulphur. If *hydrochloric acid* or *sulphuric acid* is added to the solution of a hyposulphite, the fluid remains at first clear and inodorous, but after a short time—the shorter the more concentrated the solution—it becomes more and more turbid, owing to the separation of sulphur, and exhales the odor of sulphurous acid. Application of heat promotes this decomposition. *Nitrate of silver* produces a white precipitate of HYPOSULPHITE OF SILVER, which is soluble in an excess of the hyposulphite; after a little while (upon heating almost immediately) this precipitate turns black, being decomposed into sulphide of silver and sulphuric acid. Hyposulphite of soda dissolves chloride of silver; upon the addition of an acid the solution remains clear at first, but after some time, and immediately upon boiling, sulphide of silver separates. *Chloride of barium* produces a white precipitate, which is soluble in much water, more especially hot water, and is decomposed by hydrochloric acid.

Where it is required to find sulphites and hyposulphites of the alkalies in presence of alkaline sulphides, as is often the case, solution of sulphate of zinc is first added to the fluid until the sulphide is decomposed; the sulphide of zinc is then filtered off, and one part of the filtrate is tested for hyposulphurous acid by addition of acid, another portion for sulphurous acid with nitroprusside of potassium, &c.

c. IODIC ACID (IO_5).

IODIC ACID crystallizes in white, six-sided tables; at a moderate heat it is resolved into iodine vapor and oxygen; it is readily soluble in water. The salts are decomposed upon ignition, being resolved either into oxygen and a metallic iodide, or into iodine, oxygen, and metallic oxide; the iodates with an alkaline base alone dissolve readily in water. *Chloride of barium* throws down from solution of iodates of the alkalies a white precipitate of IODATE OF BARYTA, which is soluble in nitric acid; *nitrate of silver* a white granular-crystalline precipitate of IODATE OF SILVER, which dissolves readily in ammonia, but only sparingly in nitric acid. *Hydrosulphuric acid* throws down from solutions of iodic acid IODINE, which then dissolves in hydriodic acid; the precipitation is attended with separation of sulphur. If an excess of hydrosulphuric acid is added, the fluid loses its color, and a further separation of sulphur takes place, the iodine being converted into hydriodic acid. Iodic acid combined with bases is also decomposed by hydrosulphuric acid. *Sulphurous acid* throws down IODINE, which upon addition of an excess of the acid is converted into hydriodic acid.

Second Division of the First Group of the Inorganic Acids.

§ 140.

Sulphuric Acid (SO_3).

1. Anhydrous sulphuric acid is a white feathery-crystalline mass, which emits strong fumes upon exposure to the air; hydrated sulphuric acid forms an oily liquid, colorless and transparent like water. Both the anhydrous and hydrated acid char organic substances, and combine with water in all proportions, the process of combination being attended with considerable elevation of temperature, and in the case of the anhydrous acid with a hissing noise.

2. The neutral sulphates are readily soluble in water with the exception of the sulphates of baryta, strontia, lime, and lead. The basic sulphates of the oxides of the heavy metals which are insoluble in water dissolve in hydrochloric acid or in nitric acid. Most of the sulphates are colorless or white. The sulphates of the alkalies are not decomposed by ignition. The other sulphates are acted upon in different ways by a red heat, some of them being readily decomposed, others with difficulty, and some resisting decomposition altogether.

3. *Chloride of barium* produces even in exceedingly dilute solutions of sulphuric acid and of the sulphates a finely-pulverulent, heavy, white precipitate of SULPHATE OF BARYTA (BaO, SO_3), which is insoluble in dilute hydrochloric acid and nitric acid. From very dilute solutions the precipitate separates only after standing some time. Concentrated acids and concentrated solutions of many salts impair the delicacy of the reaction.

4. *Acetate of lead* produces a heavy white precipitate of SULPHATE OF LEAD (PbO, SO_3) which is sparingly soluble in dilute nitric acid, but dissolves completely in hot concentrated hydrochloric acid.

5. The salts of sulphuric acid with the alkaline earths which are insoluble in water and acids are converted into CARBONATES, by fusion with *alkaline carbonates*. But the sulphate of lead is reduced to the state of PURE OXIDE when treated in this manner. Both the conversion of the former into carbonates and the reduction of the latter to the state of oxide are attended with the formation of an alkaline sulphate. The sulphates of the alkaline earths and sulphate of lead are also resolved into insoluble carbonates and soluble alkaline sulphate by digestion or boiling with concentrated solutions of carbonates of the alkalies (comp. §§ 95, 96, 97).

6. Upon fusing sulphates with *carbonate of soda* on charcoal in the inner flame of the blowpipe the sulphuric acid is reduced, and sulphide of sodium formed, which may be readily recognised by the odor of hydro-

sulphuric acid emitted upon moistening the sample and the part of the charcoal into which the fused mass has penetrated, and adding some acid. If the fused mass is transferred to a clean silver plate, or a polished silver coin, and then moistened with water and some acid, a black stain of sulphide of silver is immediately formed.

Remarks.—The characteristic and exceedingly delicate reaction of sulphuric acid with salts of baryta renders the detection of this acid an easier task than that of almost any other. It is simply necessary to take care not to confound with sulphate of baryta precipitates of chloride of barium, and particularly of nitrate of baryta, which are formed upon mixing aqueous solutions of these salts with fluids containing a large proportion of free hydrochloric acid or free nitric acid. It is very easy to distinguish these precipitates from sulphate of baryta, since they redissolve immediately upon diluting the acid fluid with water. It is a rule that should never be departed from, in testing for sulphuric acid with chloride of barium, to dilute the fluid largely; a little hydrochloric acid should also be added, which counteracts the adverse influence of many salts, as, for instance, citrates of the alkalies. Where very minute quantities of sulphuric acid are to be detected the fluid should be allowed to stand several hours at a gentle heat, the trace of sulphate of baryta formed will in that case be found deposited at the bottom of the vessel. When the least uncertainty exists about the nature of the precipitate produced by chloride of barium in presence of hydrochloric acid, the reaction sub 6 will at once set all doubt at rest. To detect free sulphuric acid in presence of a sulphate the fluid under examination is mixed with a very little cane-sugar, and the mixture evaporated to dryness in a porcelain dish at 212° Fah. If free sulphuric acid was present a black residue remains, or, in the case of most minute quantities, a blackish-green residue. Other free acids do not decompose cane-sugar in this way.

§ 141.

HYDROFLUOSILICIC ACID (H F, Si F$_2$).

Hydrofluosilicic acid is a very acid fluid; upon evaporation on platinum it volatilizes completely as fluoride of silicon and hydrofluoric acid. When evaporated on glass it etches the latter. With bases it forms water and silico-fluorides of the metals, which are most of them soluble in water, redden litmus-paper, and are resolved upon ignition into metallic fluorides and fluoride of silicon.

Chloride of barium forms a crystalline precipitate with hydrofluosilicic acid (§ 95, 6). Chloride of strontium and acetate of lead form no precipitates with this acid; *salts of potassa* precipitate transparent GELATINOUS SILICO-FLUORIDE OF POTASSIUM; *ammonia* in excess precipitates HYDRATED SILICIC ACID, with formation of fluoride of ammonium. By heating metallic silico-fluorides with concentrated *sulphuric acid* dense fumes are emitted in the air, arising from the evolution of hydrofluoric and silicofluoric gas. If the experiment is conducted in a platinum vessel covered with glass the fumes ETCH the glass (§ 146, 5); the residue contains the sulphates formed.

Third Division of the First Group of the Inorganic Acids.

§ 142.

a. PHOSPHORIC ACID (P O$_5$).

1. Phosphorus is a colorless, transparent, solid body, of 2·089 specific gravity; it has a waxy appearance. Taken internally it acts as a virulent poison. It fuses at 113°, and boils at 554° Fah. By the influence of light phosphorus kept under water turns first yellow, then red, and is

finally covered with a white crust. If phosphorus is exposed to the air at the common temperature, it exhales a highly characteristic and most disagreeable odor, and is gradually entirely oxidized to phosphorous acid; this process is attended with the formation of dense fumes of nitrite of ammonia, and in the dark with strong phosphorescence. Phosphorus very readily takes fire spontaneously, and burns with a luminous flame, being converted into phosphoric acid, which is dissipated for the most part in white fumes through the surrounding air. Nitric acid and nitrohydrochloric acid dissolve phosphorus pretty readily upon heating. The solutions contain at first, besides phosphoric acid, also phosphorous acid. Hydrochloric acid does not dissolve phosphorus. If phosphorus is boiled with solution of soda or potassa, or with milk of lime, hypophosphites and phosphates are formed, whilst spontaneously inflammable phosphuretted hydrogen gas escapes. If a substance containing unoxidized phosphorus is placed at the bottom of a flask, and a slip of paper moistened with solution of nitrate of silver is by means of a cork loosely inserted into the mouth, suspended inside the flask, and a gentle heat applied (from 86° to 104° Fah.), the paper slip will turn black in consequence of the reducing action of the phosphorus fumes, even though only a most minute quantity of phosphorus should be present. If after the termination of the reaction the blackened part of the paper is boiled with water, the undecomposed portion of the silver salt precipitated with hydrochloric acid, the fluid filtered, and the filtrate evaporated as far as practicable on the water-bath, the presence of phosphoric acid in the residue may be shown by means of the reactions described in 2, &c. (J. Scherer). It must be borne in mind that the silver salt is blackened also by hydrosulphuric acid, formic acid, volatile products of putrefaction, &c.; and also that the detection of phosphoric acid in the slip of paper can be of value only where the latter and the filtering paper were perfectly free from phosphorus. As regards the deportment of phosphorus upon boiling with dilute sulphuric acid, and in a hydrogen gas evolution apparatus supplied with zinc and dilute sulphuric acid, see Part II. Section II., V., III.

2. Anhydrous PHOSPHORIC ACID is a white, snowlike mass, which rapidly deliquesces in the air, and dissolves in water with a hissing noise. It forms with water and bases three different series of compounds: viz., with three equivalents of water or base hydrate of tribasic phosphoric acid or common phosphates; with two equivalents of water or base hydrate of pyrophosphoric acid or pyrophosphates; with one equivalent of water or base hydrate of metaphosphoric acid or metaphosphates.

The phosphates which we generally meet with in nature and in analytical investigations belong, as a rule, to the tribasic series; we therefore make them alone the object of a fuller study in this place, devoting a supplemental paragraph to a briefer consideration of monobasic and bibasic phosphoric acids and their salts.

3. The HYDRATE of TRIBASIC PHOSPHORIC ACID ($3HO, PO_5$) forms colorless and pellucid crystals, which deliquesce rapidly in the air to a syrupy non-caustic liquid. The action of heat changes it into hydrated pyro- or metaphosphoric acid, according as either one or two equivalents of water are expelled. Heated in an open platinum dish the hydrate of common phosphoric acid, if pure, volatilizes completely, though with difficulty, in white fumes.

4. The action of heat fails to decompose the TRIBASIC PHOSPHATES with fixed bases, but converts them into pyrophosphates if they contain one equivalent of basic water or ammonia, and into metaphosphates if they contain two equivalents. Of the tribasic phosphates those with alkaline base alone are soluble in water, in the neutral state. The solutions manifest alkaline reaction. If pyro- or metaphosphates are fused with carbonate of soda the fused mass contains the phosphoric acid invariably in the tribasic state.

5. *Chloride of barium* produces in aqueous solutions of the neutral or basic phosphates of the alkalies, but not in solutions of the hydrate, a white precipitate of PHOSPHATE OF BARYTA [$2\,BaO, HO, PO_5$; or $3\,BaO, PO_5$],* which is soluble in hydrochloric acid and in nitric acid, but sparingly soluble in chloride of ammonium.

6. *Solution of sulphate of lime* produces in neutral or alkaline solutions of phosphates, but not in solutions of the hydrate, a white precipitate of PHOSPHATE OF LIME ($2\,CaO, HO, PO_5$ or $3\,CaO, PO_5$), which dissolves readily in acids, even in acetic acid, and is soluble also in chloride of ammonium.

7. *Sulphate of magnesia* produces in concentrated neutral solutions of phosphates of the alkalies a white precipitate of PHOSPHATE OF MAGNESIA ($2\,MgO, HO, PO_5 + 14$ aq.), which often separates only after some time; upon boiling, a precipitate of basic salt ($3\,MgO, PO_5 + 5$ aq.) is thrown down immediately. The latter precipitate forms also upon addition of sulphate of magnesia to the solution of a basic alkaline phosphate. But if *sulphate of magnesia*, mixed with a sufficient quantity of chloride of ammonium to leave the solution clear upon addition of ammonia, is added to a solution of free phosphoric acid or of an alkaline phosphate, and ammonia in excess is then added, a white, crystalline, and quickly subsiding precipitate of BASIC PHOSPHATE OF MAGNESIA AND AMMONIA ($2\,MgO, NH_4O, PO_5 + 12$aq.) is formed, even in highly dilute solutions. This precipitate is insoluble in ammonia and most sparingly soluble in chloride of ammonium, but dissolves readily in acids, even in acetic acid. It makes its appearance often only after the lapse of some time; stirring promotes its separation (§ 98, 7). The reaction can be considered decisive only if no arsenic is present (§ 133, 9).

8. *Nitrate of silver* throws down from solutions of neutral and basic alkaline phosphates a light-yellow precipitate of PHOSPHATE OF SILVER ($3\,AgO, PO_5$), which is readily soluble in nitric acid and in ammonia. If the solution contained a basic phosphate the fluid in which the precipitate is suspended manifests a neutral reaction; whilst the reaction is acid if the solution contained a neutral phosphate. The acid reaction in the latter case arises from the circumstance that the nitric acid receives, for the 3 equivalents of oxide of silver which it yields to the phosphoric acid, only 2 eq. of alkali and 1 eq. of water; and as the latter does not neutralize the acid properties of the nitric acid, the solution becomes acid.

9. If to a solution containing phosphoric acid and the *least possible excess* of hydrochloric or nitric acid a tolerably large amount of acetate of soda is added, and then a drop of *sesquichloride of iron*, a yellowish-

---

* Precipitates of the former composition are produced in solutions containing an alkaline phosphate with two equivalents of a fixed base or ammonia; whilst precipitates of the latter composition are formed in solutions which contain an alkaline phosphate with three equivalents of a fixed base or ammonia.

white, flocculent-gelatinous precipitate of PHOSPHATE OF SESQUIOXIDE OF IRON (Fe_2O_3, PO_4 + 4 aq.) is formed. An excess of sesquichloride of iron must be avoided, as acetate of sesquioxide of iron (of red color) would thereby be formed, in which the precipitate is not insoluble. This reaction is of importance, as it enables us to detect the phosphoric acid in phosphates of the alkaline earths; but it can be held to be decisive only if no arsenic is present, as this shows the same reaction. To effect the complete separation of the phosphoric acid from the alkaline earths, a sufficient quantity of sesquichloride of iron is added to impart a reddish color to the solution, which is then boiled (whereby the whole of the sesquioxide of iron is thrown down, partly as phosphate, partly as basic acetate), and filtered hot. The filtrate contains the alkaline earths as chlorides. If you wish to detect, by means of this reaction, phosphoric acid in presence of a large proportion of sesquioxide of iron, boil the hydrochloric acid solution with sulphite of soda until the sesquichloride is reduced to protochloride, which reduction is indicated by the decoloration of the solution; add carbonate of soda until the fluid is nearly neutral, then acetate of soda, and finally one drop of sesquichloride of iron. The reason for this proceeding is, that acetate of protoxide of iron does not dissolve phosphate of sesquioxide of iron.

10. If a few cubic centimetres of the solution of *molybdate of ammonia* in nitric acid (§ 52) are poured into a test-tube, and a little of a fluid is added containing phosphoric acid in neutral or acid solution, a light-yellow finely-pulverulent precipitate forms at once or after a very short time, even in the cold, if the quantity of phosphoric acid is not too inconsiderable; this precipitate speedily subsides to the bottom of the tube, or is deposited on the sides. With exceedingly minute quantities of phosphorus, as *e.g.* 0·00002 grm. = about 0·0003 grain, a few hours must be allowed for the manifestation of the reaction, which should be aided also by applying a gentle heat, but not higher than 104° F. If no other coloring substances are present, the fluid above the precipitate appears colorless. The fluid to be tested for phosphoric acid should not be added in larger proportion than an equal volume to that of the molybdate of ammonia solution; a mere yellow coloration of the fluid should never be considered to prove the presence of phosphoric acid.

The yellow precipitate contains MOLYBDIC ACID, AMMONIA, WATER, and a little PHOSPHORIC ACID (about 3 per cent.). As it is insoluble in dilute acids only in presence of an excess of molybdic acid, addition of phosphoric acid in excess will necessarily altogether prevent its formation, which should be borne in mind. Presence of certain organic substances, *e.g.* tartaric acid, will also prevent the precipitation. The precipitate, after subsiding, may be readily recognised even in dark-colored fluids. By washing it with the solution of molybdate of ammonia with which the precipitation has been effected, dissolving in ammonia, and adding a mixture of sulphate of magnesia, chloride of ammonium, and ammonia, phosphate of magnesia and ammonia is produced. By conducting the operation in the manner above stated, phosphoric acid cannot well be confounded with any other acid; since arsenic acid gives in the cold no precipitate with solution of molybdate of ammonia in nitric acid, though it gives one upon application of heat, and more especially upon boiling (the fluid above this precipitate appears yellow); and silicic acid shows no reaction with it in the cold, and gives only a yellow coloration on heating, but no precipitate.

11. If a finely-powdered substance containing phosphoric acid (or a metallic phosphide) is intimately mixed with 5 parts of a flux consisting of 3 parts of carbonate of soda, 1 part of nitrate of potassa, and 1 part of silicic acid, the mixture fused in a platinum spoon or crucible, the fused mass boiled with water, the solution obtained decanted, carbonate of ammonia added to it, the fluid boiled again, and the silicic acid which is thereby precipitated filtered off, the filtrate now holds in solution alkaline phosphate, and may accordingly be tested for phosphoric acid as directed in 7, 8, 9, or 10.

12. *White of egg* is not precipitated by solution of hydrate of tribasic phosphoric acid, nor by solutions of tribasic phosphates mixed with acetic acid.

§ 143.

α. *Bibasic phosphoric acid.* The solution of the hydrate $2HO, PO_5$ is converted by boiling into solution of the hydrate $3HO, PO_5$. The solutions of the salts bear heating without suffering decomposition; but upon boiling with a strong acid the phosphoric acid is converted into the tribasic state. If the salts are fused with carbonate of soda in excess tribasic phosphates are produced. Of the neutral pyrophosphates only those with alkaline bases are soluble in water; the acid salts (*e.g.*, NaO, HO, PO_5) are by ignition converted into metaphosphates (NaO, PO_5). *Chloride of barium* fails to precipitate the free acid; from solutions of the salts it precipitates PYROPHOSPHATE OF BARYTA ($2BaO, PO_5$). *Nitrate of silver* throws down from a solution of the hydrate, especially upon addition of an alkali, a white, earthy-looking precipitate of PYROPHOSPHATE OF SILVER ($2AgO, PO_5$), which is soluble in nitric acid and in ammonia. *Sulphate of magnesia* precipitates PYROPHOSPHATE OF MAGNESIA ($2MgO, PO_5$). The precipitate dissolves in an excess of the pyrophosphate, as well as in an excess of the sulphate of magnesin. Ammonia fails to precipitate it from these solutions. Upon boiling the solution it separates again. *White of egg* is not precipitated by solution of the hydrate nor by solutions of the salts mixed with acetic acid. *Molybdate of ammonia*, with addition of hydrochloric acid, fails to produce a precipitate.

β. *Monobasic phosphoric acid.* Five sorts of monobasic phosphates are known, and the hydrates also of most of these have been produced. The several reactions by which to distinguish between these I will not enter upon here, and confine myself to the simple observation that the monobasic phosphoric acids differ from the bibasic and tribasic phosphoric acids in this, that the solutions of the hydrates of the monobasic acids precipitate *white of egg* at once, and the solutions of their salts after addition of acetic acid. Those hydrates and salts which are precipitated by *nitrate of silver* produce with that reagent a white precipitate. A mixture of *sulphate of magnesia*, chloride of ammonium, and ammonia fails to precipitate the monobasic phosphoric acids and their salts, or produces precipitates soluble in chloride of ammonium. All monobasic phosphates yield upon fusion with carbonate of soda tribasic phosphate of soda.

§ 144.

b. BORACIC ACID (BO_3).

1. Boracic acid, in the anhydrous state, is a colorless, fixed glass, fusible at a red heat; hydrate of boracic acid ($3HO, BO_3$) is a porous, white mass; in the crystalline state ($HO, BO_3 + 2$ aq.), it presents small scaly laminæ. It is soluble in water and in spirit of wine; upon evaporating the solutions a large proportion of boracic acid volatilizes along with the aqueous and alcoholic vapors. The solutions redden litmus-paper, and impart to turmeric-paper a faint brown-red tint, which acquires intensity upon drying. The borates are not decomposed upon ignition; those with alkaline bases alone are readily soluble in water. The solutions of borates of the alkalies are colorless, and all of them, even those of the acid salts, manifest alkaline reaction.

2. *Chloride of barium* produces in solutions of borates, if not too

highly dilute, a white precipitate of BORATE OF BARYTA, which is soluble in acids and ammoniacal salts. The formula of this precipitate, when thrown down from solutions of neutral borates, is $BaO, BO_3 + aq.$; when thrown down from solutions of acid borates, $3 BaO, 5 BO_3 + 6 aq.$ (H. ROSE).

3. *Nitrate of silver* produces in concentrated solutions of neutral borates of the alkalies a white precipitate, inclining slightly to yellow from admixture of free oxide of silver ($AgO, BO_3 + HO$); in concentrated solutions of acid borates a white precipitate of $3 AgO, 4 BO_3$. Dilute solutions of borates give with nitrate of silver a grayish-brown precipitate of oxide of silver (H. ROSE). All these precipitates dissolve in nitric acid and in ammonia.

4. If dilute *sulphuric acid* or *hydrochloric acid* is added to highly concentrated, hot prepared solutions of alkaline borates, the BORACIC ACID separates upon cooling, in the form of shining crystalline scales.

5. If *alcohol* is poured over free boracic acid or a borate—with addition, in the latter case, of *a sufficient quantity of concentrated sulphuric acid* to liberate the boracic acid—and the alcohol is kindled, the flame appears of a very distinct YELLOWISH-GREEN color, especially upon stirring the mixture; this tint is imparted to the flame by the ignited boracic acid which volatilizes with the alcohol. The delicacy of this reaction may be considerably heightened by heating the dish which contains the alcoholic mixture, kindling the alcohol, allowing it to burn for a short time, then extinguishing the flame, and afterwards rekindling it. At the first flickering of the flame its borders will now appear green, even though the quantity of the boracic acid be so minute that it fails to produce a perceptible coloring of the flame when treated in the usual manner. As salts of copper also impart a green tint to the flame of alcohol, the copper which might be present must first be removed by means of hydrosulphuric acid. Presence of metallic chlorides also may lead to mistakes, as the chloride of ethyle formed in that case colors the borders of the flame greenish.

6. If a solution of boracic acid, or of a borate with an alkali or an alkaline earth for base, is mixed with hydrochloric acid to slight, but distinct, acid reaction, and a slip of *turmeric paper* is half dipped into it, and then dried on a watch-glass at 212° Fah., the dipped half shows a peculiar RED tint (H. ROSE).

This reaction is very delicate; care must be taken not to confound the characteristic red coloration with the blackish-brown color which turmeric-paper acquires when moistened with rather concentrated hydrochloric acid, and then dried; nor with the brownish-red coloration which sesquichloride of iron, or a hydrochloric acid solution of molybdate of ammonia, gives to turmeric paper, more particularly upon drying. By moistening turmeric-paper reddened by boracic acid with a solution of an alkali or an alkaline carbonate, the color is changed to bluish-black or greenish-black; but a little hydrochloric acid will at once restore the brownish-red color (A. VOGEL, H. LUDWIG).

7. If a substance containing boracic acid is reduced to a fine powder, this, with addition of a drop of water, mixed with 3 parts of a flux composed of $4\frac{1}{2}$ parts of bisulphate of potassa and 1 part of finely pulverized fluoride of calcium, free from boracic acid, and the paste exposed on the loop of a platinum wire in the outer mantle of *Bunsen's* gas flame, or at the apex of the inner flame of the blowpipe, fluoride of boron escapes,

which imparts to the flame—though only for a few instants—a green tint. With readily decomposed compounds the reaction may be obtained by simply moistening the sample with hydrofluosilicic acid, and holding it in the flame.

8. Boracic acid or borates, fused together with carbonate of soda, give, when placed in the flame of the *spectrum apparatus*, a spectrum of four strong green and blue lines of equal width, and placed at equal distances. $BO\alpha$ is yellowish-green, and coincides with $Ba \gamma$; $BO\beta$ is light green, and coincides with $Ba\beta$; $BO\gamma$ is faint bluish-green, and almost coincides with the blue barium line. $BO\delta$ is very faint blue, and does not quite reach $Sr\delta$. Presence of alkali and alkaline earths does not prevent the reaction.

§ 145.

c. Oxalic Acid $(C_4 O_6 = \overline{O})$.

1. The HYDRATE OF OXALIC ACID $(2 HO, C_4 O_6)$ is a white powder; the crystallized acid $(2 HO, C_4 O_6 + 2 \text{ aq.})$ forms colorless rhombic prisms. Both dissolve readily in water and in spirit of wine. By heating rapidly in open vessels part of the hydrated acid undergoes decomposition, whilst another portion volatilizes unaltered. The fumes of the volatilizing acid are very irritating and provoke coughing. If the hydrate is heated in a test-tube the greater part of it sublimes unaltered.

2. The whole of the OXALATES undergo decomposition at a red heat, the oxalic acid being converted into carbonic acid and carbonic oxide. Those with an alkali or an alkaline earth for base are in this process converted into carbonates (if pure, almost without separation of charcoal). Oxalate of magnesia is converted into pure magnesia even by a very gentle red heat. The oxalates with metallic bases leave either the pure metal or the oxide behind, according to the greater or less degree of reducibility of the metallic oxide. The alkaline oxalates, and also some of the oxalates with metallic bases, are soluble in water.

3. *Chloride of barium* produces in neutral solutions of oxalates a white precipitate of OXALATE OF BARYTA $(2 BaO, C_4 O_6 + 2 \text{ aq.})$, which dissolves very sparingly in water, more readily in water containing chloride of ammonium, acetic acid, or oxalic acid, freely in nitric acid and in hydrochloric acid; ammonia reprecipitates it from the latter solutions unaltered.

4. *Nitrate of Silver* produces in neutral solutions of oxalic acid and of alkaline oxalates a white precipitate of OXALATE OF SILVER $(2 AgO, C_4 O_6)$, which is readily soluble in concentrated hot nitric acid and also in ammonia, but dissolves with difficulty in dilute nitric acid, and is most sparingly soluble in water.

5. *Lime-water* and all the *soluble salts of lime*, and consequently also *solution of sulphate of lime*, produce in even highly dilute solutions of free oxalic acid or of oxalates of the alkalies, white finely pulverulent precipitates of OXALATE OF LIME $(2 CaO, C_4 O_6 + 2 \text{ aq.}$, and occasionally $2 CaO, C_4 O_6 + 6 \text{ aq.})$, which dissolve readily in hydrochloric acid and in nitric acid, but are nearly insoluble in oxalic acid and in acetic acid, and almost absolutely insoluble in water. The presence of salts of ammonia does not interfere with the formation of these precipitates. Addition of ammonia considerably promotes the precipitation of the free oxalic acid by salts of lime. In *highly dilute* solutions the precipitate is only formed after some time.

6. If hydrated oxalic acid (or an oxalate), in the dry state, is heated with an excess of *concentrated sulphuric acid*, the latter withdraws from the oxalic acid its constitutional water, and thus causes its decomposition into CARBONIC ACID and CARBONIC OXIDE ($C_4O_6 = 2\,CO + 2\,CO_2$), the two gases escaping with effervescence. If the quantity operated upon is not too minute the escaping carbonic oxide gas may be kindled; it burns with a blue flame. Should the sulphuric acid acquire a dark color in this reaction, this is a proof that the oxalic acid contained some organic substance in admixture.

7. If oxalic acid or an oxalate is mixed with some finely pulverized *binoxide of manganese* (which must be free from carbonates), a little water added and a few drops of sulphuric acid, a lively effervescence ensues, caused by the escaping CARBONIC ACID [$2\,MnO_2 + C_4O_6 + 2\,SO_3 = 2\,(MnO, SO_3) + 4\,CO_2$].

8. If oxalates of alkaline earths are boiled with a concentrated solution of *carbonate of soda*, and the fluid filtered, the oxalic acid is obtained in the filtrate in combination with soda, whilst the precipitate contains the base as carbonate. With oxalates containing for their base oxides of heavy metals, this operation is not always sure to attain the desired object, as many of these oxalates, *e.g.* oxalate of protoxide of nickel, will partially dissolve in the alkaline fluid, with formation of a double salt. Metals of this kind should therefore be separated as sulphides.

§ 146.

d. HYDROFLUORIC ACID (HF).

1. Anhydrous HYDROFLUORIC ACID is a colorless corrosive gas, which fumes in the air, and is freely absorbed by water. Liquid hydrofluoric acid is distinguished from all other acids by the exclusive property it possesses of dissolving crystallized silicic acid, and also the silicates which are insoluble in hydrochloric acid. Fluoride of silicon and water are formed in the process of solution ($SiO_2 + 2\,HF = SiF_2 + 2\,HO$). Hydrofluoric acid decomposes with metallic oxides in the same manner, metallic fluorides and water being formed.

2. The FLUORIDES of the alkali metals are soluble in water; the solutions have an alkaline reaction. The fluorides of the metals of the alkaline earths are either altogether insoluble in water, or they dissolve in that menstruum only with very great difficulty. Fluoride of aluminium is readily soluble. Most of the fluorides corresponding to the oxides of the heavy metals are very sparingly soluble in water, as, for instance, the fluorides of copper, lead, and zinc; many other of the fluorides of the heavy metals dissolve in water without difficulty, as, for instance, the sesquifluoride of iron, protofluoride of tin, fluoride of mercury, &c. Many of the fluorides insoluble or difficultly soluble in water dissolve in free hydrofluoric acid; others do not. Most of the fluorides bear ignition in a crucible without suffering decomposition.

3. *Chloride of barium* precipitates aqueous solutions of free hydrofluoric acid, but much more completely solutions of fluoride of the alkalies. The bulky white precipitate of FLUORIDE OF BARIUM (BaF) is almost absolutely insoluble in water, but dissolves in large quantities of hydrochloric acid or nitric acid, from which solutions ammonia fails to precipitate it, or throws it down only very incompletely, owing to the dissolving action of the neutral ammonia salts.

4. *Chloride of calcium* produces in aqueous solutions of hydrofluoric acid or of fluorides a gelatinous precipitate of FLUORIDE OF CALCIUM (Ca F), which is so transparent as at first to induce the belief that the fluid has remained perfectly clear. Addition of ammonia promotes the complete separation of the precipitate. The precipitated fluoride of calcium is almost absolutely insoluble in water, and only very slightly soluble in hydrochloric acid and nitric acid in the cold; it dissolves somewhat more largely upon boiling with hydrochloric acid. Ammonia produces no precipitate in the solution, or only a very trifling one, as the salt of ammonia formed retains it in solution. Fluoride of calcium is scarcely more soluble in free hydrofluoric acid than in water. It is insoluble in alkaline fluids.

5. If a finely pulverized fluoride, no matter whether soluble or insoluble, is treated in a platinum crucible with just enough *concentrated sulphuric acid* to make it into a thin paste, the crucible covered with the convex face of a watch-glass of hard glass coated with bees-wax, which has been removed again in some places by tracing lines in it with a pointed piece of wood, the hollow of the glass filled with water, and the crucible gently heated for the space of half an hour or an hour, the exposed lines will, upon the removal of the wax, be found more or less deeply ETCHED into the glass.* If the quantity of hydrofluoric acid disengaged by the sulphuric acid was very minute, the etching is often invisible upon the removal of the wax; it will, however, in such cases reappear when the plate is breathed upon. This reappearance of the etched lines is owing to the unequal capacity of condensing water which the etched and the untouched parts of the plate respectively possess. The impressions which thus appear upon breathing on the glass may, however, owe their origin to other causes; therefore, though their non-appearance may be held as a proof of the absence of fluorine, their appearance is not a positive proof of the presence of that element. At all events, they ought only to be considered of value where they can be developed again after the glass has been properly washed with water, dried, and wiped.†

This reaction (5) fails if there is too much silicic acid present, or if the body under examination is not decomposed by sulphuric acid. In such cases the one or the other of the two following methods is resorted to, according to circumstances.

6. If we have to deal with a fluoride *decomposable by sulphuric acid*, but mixed with a large proportion of silicic acid, the fluorine in it may be detected by heating the mixture in a test-tube with *concentrated sulphuric acid*, as FLUOSILICIC GAS is evolved in this process, which forms dense white fumes in moist air. If the gas is conducted into water through a bent tube moistened inside, the latter has its transparency more or less impaired, owing to the separation of silicic acid. If the

---

* The coating with wax may be readily effected by heating the glass cautiously, putting a small piece of wax upon the convex face, and spreading the fused mass equally over it. The removal of the wax coating is effected by heating the glass gently, and wiping the wax off with a cloth.

† J. NICKLÈS states that etchings on glass may be obtained with all kinds of sulphuric acid, and, in fact, with all acids suited to effect evolution of hydrofluoric acid. I have tried watch-glasses of Bohemian glass with sulphuric and other acids, but could get no etchings in confirmation of this statement. Still, proper caution demands that before using the sulphuric acid, it should first be positively ascertained that its fumes will not etch glass.

quantity operated upon is rather considerable, hydrate of silicic acid separates in the water, and the fluid is rendered acid by hydrofluosilicic acid.

The following process answers best for the detection of smaller quantities of fluorine. Heat the substance with concentrated sulphuric acid in a flask closed with a cork with double perforation, bearing two tubes, of which one reaches down to the bottom of the flask, whilst the other terminates immediately under the cork. Conduct through the longer tube a slow stream of dry air into the flask, and conduct this, upon its re-issuing through the other tube, into a U-shaped tube containing a little ammonia, and connected at the other end with an aspirator. The silicofluoric gas, which escapes along with the air, decomposes with the ammonia, more particularly upon the application of a gentle heat towards the end of the process, fluoride of ammonium and hydrated silicic acid being formed. Filter, evaporate in a platinum crucible to dryness, and examine the residue by the method described in 5.—For more difficultly decomposable substances bisulphate of potassa is used instead of sulphuric acid, and the mixture, to which some marble is added (to ensure a continuous slight evolution of gas), heated to fusion, and kept in that state for some time.

7. Compounds not decomposable by sulphuric acid must first be fused with four parts of carbonate of soda and potassa. The fused mass is treated with water, the solution filtered, the filtrate concentrated by evaporation, allowed to cool, transferred to a platinum or silver vessel, hydrochloric acid added to feebly acid reaction, and the fluid allowed to stand until the carbonic acid has escaped. It is then supersaturated with ammonia, heated, filtered into a bottle, chloride of calcium added to the still hot fluid, the bottle closed, and allowed to stand at rest. If a precipitate separates after some time it is collected on a filter, dried, and examined by the method described in 5 (H. ROSE).

8. Minute quantities of metallic fluorides in minerals, slags, &c., may also be readily detected by means of the *blowpipe*. To this end bend a piece of platinum foil in gutter-shape, insert it in a glass tube as shown in Fig. 33, introduce the finely triturated substance mixed with powdered phosphate of soda and ammonia fused on charcoal, and let the blowpipe flame play upon it in a manner to make the products of combustion pass into the tube. If fluorides of metals are present hydrofluoric acid gas is evolved, which betrays its presence by its pungent odor, the dimming of the glass tube (which becomes perceptible only after cleaning and drying), and the yellow tint which the acid air issuing from the tube imparts to a moist slip of Brazil-wood paper* (BERZELIUS, SMITHSON). When silicates containing metallic fluorides are treated in this manner gaseous fluoride of silicon is formed, which also colors yellow a moist slip of Brazil-wood paper inserted in the tube, and leads to silicic acid being deposited within the tube. After washing and drying the tube, it appears here and there dimmed. In the case of minerals containing water the presence of even a small proportion of metallic fluorides will usually suffice upon heating, even without addition of phosphate of soda and ammonia, to color yellow a moistened slip of Brazil-wood paper inserted in the tube (BERZELIUS).

Fig. 33.

---

* Prepared by moistening slips of fine printing-paper with decoction of Brazil-wood.

§ 147.

Recapitulation and remarks.—The baryta compounds of the acids of the third division are dissolved by hydrochloric acid, apparently without undergoing decomposition; alkalies therefore reprecipitate them unaltered, by neutralizing the hydrochloric acid. The baryta compounds of the acids of the first division of the first group show, however, the same deportment; these acids must, therefore, if present, be removed before any conclusion regarding the presence of phosphoric acid, boracic acid, oxalic acid, or hydrofluoric acid, can be drawn from the reprecipitation of a salt of baryta by alkalies. But even leaving this point altogether out of the question no great value is to be placed on this reaction, not even so far as the simple detection of these acids is concerned, and far less still as regards their separation from other acids, since ammonia fails to reprecipitate from hydrochloric acid solutions the salts of baryta in question, and more particularly the borate of baryta and the fluoride of barium, if the solution contains any considerable proportion of free acid or of an ammoniacal salt. *Boracic acid* is well characterized by the coloration which it imparts to the flame of alcohol, and also by its action on turmeric-paper. The latter reaction is more particularly suited for the detection of very minute traces of boracic acid. Oxides of the heavy metals, if present, are most conveniently removed first by hydrosulphuric acid or sulphide of ammonium. Before proceeding to concentrate dilute solutions of boracic acid the acid must be combined with an alkali, otherwise a large portion of it will volatilize along with the aqueous vapors.

The detection of *phosphoric acid* in compounds soluble in water is not difficult; the reaction with sulphate of magnesia is the best adapted for the purpose. The detection of phosphoric acid in insoluble compounds cannot be effected by means of magnesia solution. Sesquichloride of iron (§ 142, 9) is well suited for the detection of phosphoric acid in its salts with the alkaline earths, and more particularly for the separation of the acid from the alkaline earths; the nitric acid solution of molybdate of ammonia is more especially adapted to effect the detection of phosphoric acid in presence of alumina and sesquioxide of iron. I must repeat again that both these reactions demand the *strictest* attention to the directions given in § 142, 9 and 10. If present in combination with metallic oxides of the fourth, fifth, or sixth group, it may be isolated, either by the method given in § 142, 11, or simply by removing the bases by precipitating them with hydrosulphuric acid or sulphide of ammonium.

Oxalic acid may always be easily detected in aqueous solutions of oxalates of the alkalies, by solution of sulphate of lime. The formation of a finely pulverulent precipitate, insoluble in acetic acid, leaves hardly a doubt on the point, as racemic acid alone, which occurs so very rarely, gives the same reaction. In case of doubt the oxalate of lime may be readily distinguished from the paratartrate, or racemate, by simple ignition, with exclusion of air, as the decomposed paratartrate leaves a considerable proportion of charcoal behind; the paratartrate dissolves moreover in cold solution of potassa or soda, in which oxalate of lime is insoluble. The deportment of the oxalates with sulphuric acid, or with binoxide of manganese and sulphuric acid, affords also sufficient means to confirm the results of other tests. In insoluble salts the oxalic acid is detected most safely by decomposing the insoluble compound by boiling with solution of carbonate of soda, or in oxalates with oxides of the heavy metals, by hydrosulphuric acid or sulphide of ammonium (§ 145, 8). I must finally also call attention here to the fact that there are certain soluble oxalates which are not precipitated by salts of lime ; these are more particularly oxalate of sesquioxide of chromium, and oxalate of sesquioxide of iron. Their non-precipitation is owing to the circumstance that these salts form soluble double salts with oxalate of lime. In salts decomposable by sulphuric acid, the *hydrofluoric acid* is readily detected ; only it must be borne in mind that an over large proportion of sulphuric acid impedes the free evolution of hydrofluoric gas, and thus impairs the delicacy of the reaction ; also that the glass cannot be distinctly etched if, instead of hydrofluoric gas, fluosilicic gas alone is evolved ; and therefore, in the case of compounds abounding in silica, the safer way is to try, besides the reaction given § 146, 5, also the one given in § 146, 6.—In silicates which are not decomposed by sulphuric acid the presence of fluorine is often overlooked, because the analyst omits to examine the compound carefully by the method given § 146, 7.

§ 148.

PHOSPHOROUS ACID (PO_3).

Anhydrous phosphorous acid is a white powder, which admits of sublimation, and burns when heated in the air. It forms with a small proportion of water a thickish fluid, which crystallizes by long standing. Heat decomposes it into hydrated phosphoric acid and phosphuretted hydrogen gas, which does not spontaneously take fire. It freely dissolves in water. Of the salts those with alkaline base are readily soluble in water. All the others sparingly soluble ; the latter dissolve in dilute acids. All the salts are decomposed by ignition into phosphates, which are left behind, and hydrogen gas, or a mixture of hydrogen and phosphuretted hydrogen, which escapes. With *nitrate of silver* separation of metallic silver takes place, more especially upon addition of ammonia and application of heat ; with *nitrate of suboxide of mercury*, under the same circumstances, separation of metallic mercury. From *chloride of mercury* in excess phosphorous acid throws down subchloride of mercury after some time, more apidly upon heating. *Chloride of barium* and *chloride of calcium* produce in not over-dilute solutions of phosphorous acid, upon addition of ammonia, white precipitates, which are soluble in acetic acid. A mixture of sulphate of magnesia, chloride of ammonium, and ammonia will precipitate only somewhat more concentrated solutions. *Acetate of lead* throws down white phosphite of lead, insoluble in acetic acid. By heating to boiling with *sulphurous acid* in excess phosphoric acid is formed, attended by separation of sulphur.

In contact with *zinc* and *dilute sulphuric* acid phosphorous acid gives a mixture of hydrogen gas with phosphuretted hydrogen, which accordingly fumes in the air, burns with an emerald-green color, and precipitates phosphide of silver from solution of nitrate of silver.

Fourth Division of the First Group of the Inorganic Acids.

§ 149.

a. CARBONIC ACID (CO_2).

1. CARBON is a solid tasteless and inodorous body. The very highest degrees of heat alone can effect its fusion and volatilization (DESPRETZ). All carbon is combustible, and yields carbonic acid when burnt with a sufficient supply of oxygen or atmospheric air. In the diamond the carbon is crystallized, transparent, pellucid, exceedingly hard, difficultly combustible ; in the form of graphite it is opaque, blackish-gray, soft, greasy to the touch, difficultly combustible, and stains the fingers ; as charcoal

produced by the decomposition (destructive distillation) of organic matters it is black, opaque, noncrystalline—occasionally dense, shining, and difficultly combustible, and occasionally porous, dull, and readily combustible.

2. CARBONIC ACID, at the common temperature and common atmospheric pressure, is a colorless gas of far higher specific gravity than atmospheric air, so that it may be poured from one vessel into another. It is almost inodorous, has a sourish taste, and reddens moist litmus-paper; but the red tint disappears again upon drying. Carbonic acid is readily absorbed by solution of potassa; it dissolves pretty copiously in water.

3. The AQUEOUS SOLUTION OF CARBONIC ACID has a feebly acid and pungent taste; it transiently imparts a red tint to litmus-paper, and colors solution of litmus wine-red; it loses its carbonic acid when shaken with air in a half-filled bottle, and more completely still upon application of heat. Some of the CARBONATES lose their carbonic acid by ignition; those with colorless oxides are white or colorless. Of the neutral carbonates only those with alkaline bases are soluble in water. The solutions manifest a very strong alkaline reaction. Besides the carbonates with alkaline bases, those also with an alkaline earth for base, and some of those with a metallic base, dissolve as acid or bicarbonates.

4. The carbonates are decomposed by all *free acids* soluble in water, with the exception of hydrocyanic acid and hydrosulphuric acid. The decomposition of the carbonates by acids is attended with EFFERVESCENCE, the carbonic acid being disengaged as a colorless and almost inodorous gas, which transiently imparts a reddish tint to litmus-paper. It is necessary to apply the decomposing acid in excess, especially when operating upon carbonates with alkaline bases, since the formation of bicarbonates will frequently prevent effervescence if too little of the decomposing acid be added.—Substances which it is intended to test for carbonic acid by this method should first be heated with a little water, to prevent any mistake which might arise from the escape of air-bubbles upon treating the dry substances with the acid. Where there is reason to apprehend the escape of carbonic acid upon boiling with water, lime water should be used instead of pure water. If it is wished to determine by a direct experiment whether the disengaged gas is really carbonic acid or not, this may be readily accomplished by dipping the end of a glass rod in baryta water, and inserting the rod into the test-tube, bringing the moistened end near the surface of the fluid in the tube, when ensuing turbidity of the baryta water on the glass rod will prove that the evolved gas is really carbonic acid, since

5. *Lime-water* and *baryta-water*, brought into contact with carbonic acid or with soluble carbonates, produce white precipitates of neutral CARBONATE OF LIME (CaO, CO_2), or neutral CARBONATE OF BARYTA (BaO, CO_2). In testing for free carbonic acid the reagents ought always to be added in excess, as the *acid* carbonates of the alkaline earths are soluble in water. The precipitated carbonates of lime and baryta dissolve in acids, with effervescence, and are not reprecipitated from such solutions by ammonia, after the complete expulsion of the carbonic acid by ebullition. As lime-water dissolves very minute quantities of carbonate of lime, the detection of exceedingly minute traces of carbonic acid requires the use of a lime-water saturated with carbonate of lime by long digestion with the latter salt (WELTER, BERTHOLLET).

6. *Chloride of calcium* and *chloride of barium* immediately produce in

solutions of neutral alkaline carbonates, precipitates of CARBONATE OF LIME or of CARBONATE OF BARYTA; in dilute solutions of bicarbonates these precipitates are formed only upon ebullition; with free carbonic acid these reagents give no precipitate.

§ 150.

b. SILICIC ACID (SiO_3).

1. SILICIC ACID is colorless or white, even in the hottest blowpipe flame unalterable and infusible. It fuses in the flame of the oxyhydrogen blowpipe. It is met with in two modifications (more correctly speaking, in the crystalline and in the amorphous state). It is insoluble in water and acids, with the exception of hydrofluoric acid; whilst its hydrate is soluble in acids, but only at the moment of its separation. The amorphous silicic acid and the hydrate dissolve in hot aqueous solutions of caustic alkalies and of fixed alkaline carbonates; but the crystallized acid is insoluble or nearly so in these fluids. If either of the two is fused with pure alkalies or alkaline carbonates a basic silicate of the alkali is obtained, which is soluble in water, and from which acids again separate hydrated silicic acid. The SILICATES with alkaline bases alone are soluble in water.

2. The solutions of the alkaline silicates are decomposed by all *acids*. If a large proportion of hydrochloric acid is added at once to even concentrated solutions of alkaline silicates the separated silicic acid remains in solution; but if the hydrochloric acid is added gradually drop by drop, whilst stirring the fluid, the greater part of the silicic acid separates as gelatinous hydrate. The more dilute the fluid, the more silicic acid remains in solution, and in *highly* dilute solutions no precipitate is formed. But if the solution of an alkaline silicate, mixed with hydrochloric or nitric acid in excess, is evaporated to dryness silicic acid separates in proportion as the acid escapes; upon treating the residue with hydrochloric acid and water the silicic acid remains in the free state (or, if the temperature in the process of drying was restricted to 212°, as hydrate, $HO, 4 SiO_3$), as an insoluble white powder. Chloride of ammonium produces in not over-dilute solutions of alkaline silicates precipitates of hydrate of silicic acid (containing alkali). Heating promotes the separation.

3. Some of the silicates insoluble in water are decomposed by hydrochloric acid or nitric acid, others are not affected by these acids, not even upon boiling. In the decomposition of the former the greater portion of the silicic acid separates usually as gelatinous, more rarely as pulverulent hydrate. To effect the complete separation of the silicic acid, the hydrochloric acid solution, with the precipitated hydrate of silicic acid suspended in it, is evaporated to dryness, the residue heated with stirring, at a uniform temperature above the boiling point of water until no more acid fumes escape, then moistened with hydrochloric acid, heated with water, and the fluid containing the bases filtered from the residuary insoluble silicic acid.

4. Of the silicates not decomposed by hydrochloric acid many, *e.g.*, kaolin, are completely decomposed by heating with a mixture of 8 parts of hydrated sulphuric acid and 3 parts of water, the decomposition being attended with separation of silicic acid in the pulverulent form; many others are acted upon to some extent by this reagent.

5. If a silicate, reduced to a fine powder, is fused with 4 parts of *carbonate of potassa and soda* until the evolution of carbonic acid has ceased, and the fused mass is then boiled with water, the greater part of the silicic acid dissolves as alkaline silicate, whilst the alkaline earths, the earths proper (with the exception of alumina and baryta, which pass more or less completely into the solution), and the heavy metallic oxides are left undissolved. If the fused mass is treated with water, then, without previous filtration, hydrochloric or nitric acid added to strongly acid reaction, and the fluid treated as directed in 3, the silicic acid is left undissolved, whilst the bases are dissolved. If the powdered silicate is fused with 4 parts of hydrate of baryta, the fused mass digested with water, with addition of hydrochloric or nitric acid, and the acid solution treated as directed in 3, the silicic acid separates, and the bases, especially also the alkalies, are found in the filtrate.

6. If *hydrofluoric acid*, in concentrated aqueous solution or in the gaseous state, is made to act upon silicic acid, fluosilic gas escapes ($SiO_3 + 2HF = SiF_2 + 2HO$); dilute acid dissolves silica to hydrofluosilicic acid ($SiO_3 + 3HF = SiF_2, HF + 2HO$). Hydrofluoric acid acting upon silicates gives rise to the formation of silicofluorides ($CaO, SiO_3 + 3HF = SiF_2, CaF + 3HO$), which by heating with hydrated sulphuric acid are changed to sulphates, with evolution of hydrofluoric and fluosilicic gas. If the powdered silicate is mixed with 5 parts of fluoride of calcium in powder, the mixture made into a paste with hydrated sulphuric acid, and heat applied (best in the open air) until no more fumes escape, the whole of the silicic acid present volatilizes as fluosilicic gas. The bases present are found in the residue as sulphates, mixed with sulphate of lime.

7. If silicic acid or a silicate is fused with *carbonate of soda* on the loop of a platinum wire FROTHING is observed in the fusing bead, owing to the disengagement of carbonic acid. If the proper proportion of carbonate of soda is not exceeded the bead of silicate of soda formed in the process will remain transparent on cooling.

8. *Phosphate of soda and ammonia*, in a state of fusion, fails nearly altogether to dissolve silicic acid. If therefore silicic acid or a silicate is fused, in small fragments, with phosphate of soda and ammonia on a platinum wire the bases are dissolved, whilst the silicic acid separates and floats about in the clear bead as a more or less transparent mass, exhibiting the shape of the fragment used in the experiment.

§ 151.

Recapitulation and remarks.—Free carbonic acid is readily known by its reaction with lime-water; the carbonates are easily detected by the evolution of a nearly inodorous gas, which takes place when they are treated with acids. When operating upon compounds which evolve other gases besides carbonic acid, the disengaged gas is to be tested with lime-water or baryta-water. Silicic acid, both in the free state and in silicates, may usually be readily detected by the reaction with phosphate of soda and ammonia. It differs moreover from all other bodies in the form in which it is always obtained in analyses, by its insolubility in acids (except hydrofluoric acid), and its solubility in boiling solutions of pure alkalies and alkaline carbonates, and from many bodies by completely volatilizing upon repeated evaporation in a platinum dish, with hydrofluoric acid or fluoride of ammonium and sulphuric acid.

SECOND GROUP OF THE INORGANIC ACIDS.

ACIDS WHICH ARE PRECIPITATED BY NITRATE OF SILVER, BUT NOT BY CHLORIDE OF BARIUM: *Hydrochloric Acid, Hydrobromic Acid, Hydriodic Acid, Hydrocyanic Acid, Hydroferro-* and *Hydroferricyanic Acid, Hydrosulphuric Acid* (Nitrous Acid, Hypochlorous Acid, Chlorous Acid, Hypophosphorous Acid).

The silver compounds corresponding to the acids of this group are insoluble in dilute nitric acid. The acids of this group decompose with metallic oxides, the metals combining with the chlorine, bromine, cyanogen, iodine, or sulphur, whilst the oxygen of the metallic oxide forms water with the hydrogen of the hydracid.

§ 152.

a. HYDROCHLORIC ACID (H Cl).

1. CHLORINE is a heavy yellowish-green gas of a disagreeable and suffocating odor, which has a most injurious action upon the respiratory organs; it destroys vegetable colors (litmus, indigo-blue, &c.); it is not inflammable, and supports the combustion of few bodies only. Minutely-divided antimony, tin, &c., spontaneously ignite in it, and are converted into chlorides. It dissolves pretty freely in water; the chlorine water formed has a faint yellowish-green color, smells strongly of the gas, bleaches vegetable colors, is decomposed by the action of light (§ 27), and loses its smell when shaken with mercury; in this process the latter is converted into a mixture of subchloride and metallic mercury. Small quantities of free chlorine may be readily detected in a fluid, by adding the latter to a solution of pure protoxide of iron mixed with sulphocyanide of potassium; the solution is at once colored red by the action of the free chlorine;—or, also—where no nitrous acid is present—by adding the fluid to a dilute solution of iodide of potassium mixed with starch-paste. (See § 154, 9.)

2. HYDROCHLORIC ACID, at the common temperature and common atmospheric pressure, is a colorless gas, which forms dense fumes in the air, is suffocating and very irritant, and dissolves in water with exceeding facility. The more concentrated solution (fuming hydrochloric acid) loses a large portion of its gas upon heating.

3. The neutral METALLIC CHLORIDES are readily soluble in water, with the exception of chloride of lead, chloride of silver, and subchloride of mercury; most of the chlorides are white or colorless. Many of them volatilize at a high temperature, without suffering decomposition; others are decomposed upon ignition, and many of them are fixed at a moderate red heat.

4. *Nitrate of silver* produces in even highly dilute solutions of free hydrochloric acid or of metallic chlorides white precipitates of CHLORIDE OF SILVER (Ag Cl), which upon exposure to light change first to violet, then to black; they are insoluble in dilute nitric acid, but dissolve readily in ammonia as well as in cyanide of potassium, and fuse without decomposition when heated. (Compare § 115, 7.)

5. *Nitrate of suboxide of mercury* and *acetate of lead* produce in solutions containing free hydrochloric acid or metallic chlorides precipitates of SUBCHLORIDE OF MERCURY (Hg$_2$ Cl) and CHLORIDE OF LEAD (Pb Cl). For the properties of these precipitates see § 116, 6, and § 117, 7.

6. If hydrochloric acid is heated with *binoxide of manganese*, or a

chloride with *binoxide of manganese* and *sulphuric acid*, chlorine gas is evolved, which may be readily recognised by its odor, its YELLOWISH-GREEN color, and its bleaching action upon vegetable colors. The best way of testing the latter is to expose to the gas a slip of moist litmus-paper, or a slip of moist paper colored with solution of indigo.

7. If a metallic chloride is triturated together with *chromate of potassa*, the mixture treated with *concentrated sulphuric acid* in a tubulated retort, and a gentle heat applied, a deep brownish-red gas is copiously evolved (CHLOROCHROMIC ACID, $Cr O_2 Cl$), which condenses into a fluid of the same color, and passes over into the receiver. If this distillate is mixed with ammonia in excess, a yellow-colored liquid is produced; the yellow tint is imparted to the fluid by the chromate of ammonia which forms in this process ($Cr O_2 Cl + 2 NH_3 O = NH_4 Cl + NH_4 O$, $Cr O_3$). Upon addition of an acid the color of the solution changes to a reddish-yellow, owing to the formation of acid chromate of ammonia.

8. In the metallic chlorides insoluble in water and nitric acid the chlorine is detected by fusing them with carbonate of soda and potassa, and treating the fused mass with water, which will dissolve, besides the excess of the alkaline carbonate, the chloride of the alkali metal formed in the process of fusion.

9. If in a bead of *phosphate of soda and ammonia* on a platinum wire *oxide of copper* be dissolved in the outer blowpipe flame in sufficient quantity to make the mass nearly opaque, a trace of a substance containing chlorine added to it while still in fusion, and the bead then exposed to the reducing flame, a fine BLUE-COLORED flame, inclining to PURPLE, will be seen encircling it so long as chlorine is present (BERZELIUS).

§ 153.

b. HYDROBROMIC ACID (H Br).

1. BROMINE is a heavy reddish-brown fluid of a very disagreeable chlorine-like odor; it boils at 145·4° F., and volatilizes rapidly even at the common temperature. The vapor of bromine is brownish-red. Bromine bleaches vegetable colors like chlorine; it is pretty soluble in water, but dissolves more readily in alcohol, and very freely in ether. The solutions are yellowish-red.

2. HYDROBROMIC ACID GAS, its AQUEOUS SOLUTION, and the METALLIC BROMIDES offer in their general deportment a great analogy to the corresponding chlorides.

3. *Nitrate of silver* produces in aqueous solutions of hydrobromic acid or of bromides a yellowish-white precipitate of BROMIDE OF SILVER (Ag Br), which changes to gray upon exposure to light; this precipitate is insoluble in dilute nitric acid, and somewhat sparingly soluble in ammonia, but dissolves with facility in cyanide of potassium.

4. *Nitrate of protoxide of palladium*, but not protochloride of palladium, produces in neutral solutions of metallic bromides a reddish-brown precipitate of PROTOBROMIDE OF PALLADIUM (Pd Br). In concentrated solutions this precipitate is formed immediately; in dilute solutions it makes its appearance only after standing some time.

5. *Nitric acid* decomposes hydrobromic acid and the bromides, with the exception of bromide of silver and bromide of mercury, upon the application of heat, and liberates the bromine, by oxidizing the hydrogen

or the metal. The liberated bromine colors the solution yellow or yellowish-red. When operating upon bromides in the solid state or in concentrated solution, brownish-red (if diluted, brownish-yellow) vapors of bromine gas escape at the same time, which, if evolved in sufficient quantity, condense in the cold part of the test-tube to small drops. In the cold, nitric acid, even the red fuming, fails to liberate the bromine contained in rather dilute solutions of metallic bromides, nor is it liberated by solution of hyponitric acid in hydrated sulphuric acid, or by hydrochloric acid and nitrite of potassa.

6. *Chlorine*, in the *gaseous* state or in *solution*, immediately liberates bromine in the solutions of its compounds; the fluid assuming a yellowish-red tint if the quantity of the bromine present is not too minute. A large excess of chlorine must be avoided, since this will cause formation of chloride of bromine, which will destroy the color wholly or nearly so. This reaction is made much more delicate by addition of a fluid which dissolves bromine and does not mix with water, as sulphide of carbon or chloroform. Mix the neutral or feebly acid solution in a test tube with a little of one of these fluids, sufficient to form a large drop at the bottom, then add dilute chlorine-water drop by drop, and shake the tube. With appreciable quantities of bromine, *e.g.* 1 part in 1000 parts of water, the drop at the bottom acquires a reddish-yellow tint; with very minute quantities (1 part of bromine in 30,000 parts of water), a pale yellow tint, which, however, is still distinctly discernible. Ether was formerly used for this reaction; this agent is *by no means* so well suited for it. A large excess of chlorine-water must be avoided in this experiment also, and it must always be ascertained first whether the chlorine-water, mixed with a large quantity of water and some sulphide of carbon or chloroform, and shaken, will leave these reagents quite uncolored. If not, the chlorine-water is not suited for the intended purpose. If the solution of bromine in sulphide of carbon or chloroform (or ether) is mixed with some solution of potassa, the mixture shaken, and heat applied, the yellow color disappears, and the solution now contains bromide of potassium and bromate of potassa. By evaporation and ignition the bromate of potassa is converted into bromide of potassium, and the ignited mass may then be further tested as directed in 7.

7. If bromides are heated with *binoxide of manganese* and *hydrated sulphuric acid*, BROWNISH-RED VAPORS OF BROMINE are evolved. Presence of chlorides in large proportion is not favorable to the reaction and requires addition of some water, and the sulphuric acid to be added gradually in *very small* quantities. If the bromine is present only in very minute quantity, the color of these vapors is not visible. But if the mixture is heated in a small retort, and the evolved vapors are transmitted through a long glass condenser, the color of the bromine vapors may generally be seen by looking lengthways through the tube, and the first drops of the distillate are also colored yellow. The vapors and the first drops of the distillate passing over should be received in a test-tube containing some starch moistened with water; since

8. If moistened *starch* is brought into contact with free bromine, more especially in form of vapor, YELLOW BROMIDE OF STARCH is formed. The coloration is not always instantaneous. The reaction is rendered most delicate by sealing the test-tube which contains the moistened starch and the fluid under examination, and then cautiously inverting it, so as to cause the moist starch to occupy the upper part of the tube

whilst the fluid occupies the bottom. The presence of even the slightest trace of bromine will now, in the course of from twelve to twenty-four hours, impart a yellow tint to the starch, which, however, after some time, will again disappear. The reaction may be called forth in a simple manner, almost with the same degree of delicacy, by gently heating the fluid containing some free bromine, or also the original mixture of bromide, binoxide of manganese, and sulphuric acid, in a very small beaker, covered with a watch-glass with a slip of paper attached to the lower side, moistened with starch paste, and strewed over with starch powder.

9. If sulphuric acid is poured over a mixture of a bromide with *chromate of potassa*, and heat is then applied, a brownish-red gas is evolved, exactly as in the case of chlorides. But this gas consists of pure BROMINE, and therefore the fluid passing over does not turn yellow, but becomes colorless upon supersaturation with ammonia.

10. In the metallic bromides which are insoluble in water and nitric acid, the bromine is detected in the same way as the chlorine in the corresponding chlorides.

11. If a *phosphate of soda and ammonia bead saturated with oxide of copper* is mixed with a substance containing bromine, and then ignited in the inner blowpipe flame, the flame is colored BLUE, inclining to GREEN, more particularly at the edges (BERZELIUS).

§ 154.

c. HYDRIODIC ACID (HI).

1. IODINE is a solid soft body of a peculiarly disagreeable odor. It is generally seen in the form of black, shining, crystalline scales. It fuses at a gentle heat; at a somewhat higher temperature it is converted into iodine vapor, which has a beautiful violet-blue color, and condenses upon cooling to a black sublimate. It is very sparingly soluble in water, but readily in alcohol and ether as well as in solution of iodide of potassium in water. The aqueous solution is of a light-brown, the alcoholic, ethereal, and iodide of potassium solutions are a deep red-brown color. Iodine destroys vegetable colors only slowly and imperfectly; it stains the skin brown; with starch it forms a compound of an intensely deep blue color. This compound is formed invariably where iodine vapor or a solution containing free iodine comes in contact with starch, best with starch-paste. It is decomposed by alkalies, as well as by chlorine and bromine.

2. HYDRIODIC ACID GAS resembles hydrochloric and hydrobromic acid gas; it dissolves copiously in water. The colorless hydrated hydriodic acid turns speedily to a reddish-brown in contact with the air, water being formed, and a solution of iodine in hydriodic acid.

3. The IODIDES also correspond in many respects with the chlorides. Of the iodides of the heavy metals, however, many more are insoluble in water than is the case with the corresponding chlorides. Many iodides have characteristic colors, *e.g.* iodide of lead, subiodide and iodide of mercury.

4. *Nitrate of silver* produces in aqueous solutions of hydriodic acid and of iodides yellowish-white precipitates of IODIDE OF SILVER (AgI), which blacken on exposure to light; these precipitates are insoluble in dilute nitric acid, and *very sparingly soluble in ammonia*, but dissolve readily in cyanide of potassium.

5. *Protochloride of palladium* and *nitrate of protoxide of palladium* produce even in very dilute solutions of hydriodic acid or metallic iodides a brownish-black precipitate of PROTIODIDE OF PALLADIUM (PdI), which dissolves to a trifling extent in saline solutions (solution of chloride of sodium, chloride of magnesium, &c.), but is insoluble or nearly so in dilute cold hydrochloric and nitric acids.

6. A solution of 1 part of *sulphate of oxide of copper* and $2\frac{1}{2}$ parts of *sulphate of protoxide of iron* throws down from neutral aqueous solutions of the iodides SUBIODIDE OF COPPER (Cu_2I), in the form of a dirty-white precipitate. The addition of ammonia promotes the complete precipitation of the iodine. Chlorides and bromides are not precipitated by this reagent.

7. Pure *nitric acid*, free from nitrous acid, decomposes hydriodic acid or iodides only when acting upon them in its concentrated form, particularly when aided by the application of heat. But *nitrous acid* and *hyponitric acid* decompose hydriodic acid and iodides with the greatest facility even in the most dilute solutions. Colorless solutions of iodides therefore acquire immediately a brownish-red color upon addition of some red fuming nitric-acid, or of a mixture of this with concentrated sulphuric acid, or, better still, upon addition of a solution of hyponitric acid in hydrated sulphuric acid, or of nitrite of potassa and some sulphuric or hydrochloric acid. From more concentrated solutions the iodine separates under these circumstances in the form of small black plates or scales, whilst nitric oxide gas and iodine vapors escape.

8. As the blue coloration of iodide of starch remains still visible in much more highly dilute solutions than the yellow color of solutions of iodine in water, the delicacy of the reaction just now described (7) is considerably heightened by mixing the fluid to be tested for iodine first with some thin clear *starch-paste*, then adding one or two drops of dilute sulphuric acid, to make the fluid acid, and adding finally one or the other of the reagents given in 7. Of the solution of hyponitric acid in sulphuric acid a single drop on a glass rod suffices to produce the reaction most distinctly. I can therefore strongly recommend this reagent, which was first proposed by *Otto*. Red fuming nitric acid must be added in somewhat larger quantity, to call forth the reaction in its highest intensity; this reagent therefore is not well adapted to detect *very minute* quantities of iodine. The reaction with nitrite of potassa also is very delicate. The fluid to be tested is mixed with dilute sulphuric acid or with hydrochloric acid to distinctly acid reaction, and a drop or two of a concentrated solution of nitrite of potassa is then added. In cases where the quantity of iodine present is very minute the fluid turns reddish, instead of blue. An excess of the fluid containing nitrous acid or hyponitric acid does not materially impair the delicacy of the reaction. As iodide of starch dissolves in hot water to a colorless liquid, the fluids must of necessity be cold; the colder they are the more delicate the reaction. To attain the highest degree of delicacy, cool the fluid with ice, let the starch deposit, and place the test-tube upon white paper to observe the reaction (compare also § 157, recapitulation and remarks).

9. *Chlorine gas* and *chlorine-water* decompose compounds of iodine also, setting the iodine free; but if the chlorine is applied in excess the liberated iodine combines with it to colorless chloride of iodine. A dilute solution of the metallic iodide, mixed with starch-paste, acquires

therefore upon addition of a little chlorine-water at once a blue tint, but becomes colorless again upon addition of more chlorine-water. As it is therefore difficult not to exceed the proper limit, especially where the quantity of iodine present is only small, chlorine-water is not well-adapted for the detection of minute quantities of iodine.

10. If a solution containing hydriodic acid or an iodide is mixed with *chloroform* or *sulphide of carbon*, so as to leave a few drops undissolved, and one of the agents by which iodine is liberated (a drop of a solution of hyponitric acid in hydrated sulphuric acid—hydrochloric acid and nitrite of potassa—chlorine water, &c.) is added, the mixture vigorously shaken, and then allowed to stand at rest, the chloroform or the sulphide of carbon, colored a deeper or lighter violet-red by the iodine dissolved in it, separates and subsides to the bottom. This reaction also is exceedingly delicate. If a solution containing free iodine is mixed with *ether*, and the mixture shaken, the ether acquires a reddish-brown or yellow color. The color imparted to the ether by iodine is much more intense than that imparted to that fluid by an equal quantity of bromine.

11. If metallic iodides are heated with *concentrated sulphuric acid*, or with *sulphuric acid* and *binoxide of manganese*, or with *sulphuric acid* and *chromate of potassa*, iodine separates, which may be known by the color of its vapor and, in the case of very minute quantities, also by its action upon a slip of paper coated with starch-paste.

12. The iodides which are insoluble in water and nitric acid comport themselves upon fusion with *carbonate of soda and potassa* in the same manner as the corresponding chlorides.

13. A *phosphate of soda and ammonia bead*, saturated with oxide of copper, when mixed with a substance containing iodine, and ignited in the inner blowpipe flame, imparts an intense GREEN color to the flame.

§ 155.

d. HYDROCYANIC ACID (HCy).

1. CYANOGEN is a colorless gas of a peculiar penetrating odor; it burns with a crimson flame, and is pretty soluble in water.

2. HYDROCYANIC ACID is a colorless, volatile, inflammable liquid, the odor of which resembles that of bitter almonds; it is miscible with water in all proportions; in the pure state it speedily suffers decomposition. It is extremely poisonous.

3. The cyanides with alkalies and alkaline earths are soluble in water; the solutions smell of hydrocyanic acid. They are readily decomposed by acids, even by carbonic acid; but ignition fails to decompose them if access of air is excluded. When fused with oxides of lead, copper, antimony, tin, &c., the cyanides reduce these oxides, and are converted into cyanates. Only a few of the cyanides with heavy metals are soluble in water; all of them are decomposed by ignition, the cyanides of the noble metals being converted into cyanogen gas and metal, the cyanides of the other heavy metals into nitrogen gas and metallic carbides. Many of the cyanides with heavy metals are not decomposed by dilute oxygen acids, and only with difficulty by concentrated nitric acid. By heating and evaporation with concentrated sulphuric acid all cyanides are decomposed; hydrochloric acid decomposes a few of them; hydrosulphuric acid decomposes many cyanides.

4. The CYANIDES have a great tendency to combine with each other; hence most of the cyanides of the heavy metals dissolve in cyanide of potassium. The resulting compounds are either:

a. True double salts, compounds of the second order, *e.g.*, $KCy + NiCy$. From solutions of such double salts acids, by decomposing the cyanide of potassium, precipitate the metallic cyanide which was combined with it. —Or they are:

b. Simple haloid salts, compounds of the first order, in which a metal, *e.g.*, potassium, is combined with a compound radical consisting of cyanogen and another metal (iron, cobalt, manganese, chromium). Compounds of this kind are the ferro- and the ferricyanide of potassium, K_2Cy_2Fe or K_2Cfy, and $K_3Cy_3Fe_2$ or K_3Cfdy, cobalticyanide of potassium, $K_3Cy_3Co_2$, &c. From solutions of compounds of this nature dilute acids do not separate metallic cyanides in the cold. If the potassium is replaced by hydrogen, peculiar hydracids are formed, which must not be confounded with hydrocyanic acid.

We will now first consider the reactions of hydrocyanic acid and the simple cyanides, then, in an appendix to this paragraph, those of hydroferro- and hydroferricyanic acid.

5. *Nitrate of silver* produces in solutions of free hydrocyanic acid and of cyanides of the alkali metals white precipitates of CYANIDE OF SILVER ($AgCy$), which are readily soluble in cyanide of potassium, dissolve with some difficulty in ammonia, and are insoluble in dilute nitric acid; these precipitates are decomposed by ignition, leaving metallic silver with some paracyanide of silver.

6. If a solution of *sulphate of protoxide of iron* which has been for some time in contact with the air, is added to a solution of free hydrocyanic acid no alteration takes place; but if *solution of potassa* or *soda* is now added a bluish-green precipitate forms, which consists of a mixture of Prussian-blue (Fe_4Cfy_3), and hydrate of protosesquioxide of iron. Upon now adding hydrochloric acid (best after previous application of heat) the hydrate of protosesquioxide of iron dissolves, whilst the PRUSSIAN-BLUE remains undissolved. If only a very minute quantity of hydrocyanic acid is present the fluid simply appears green after the addition of the hydrochloric acid, and it is only after long standing that a trifling blue precipitate separates from it. The same reactions are observed when sulphate of protoxide of iron is mixed with the solution of an alkaline cyanide, and hydrochloric acid is then added.

7. If a liquid containing a little hydrocyanic acid or cyanide of potassium is mixed with sufficient yellow sulphide of ammonium to impart a yellowish tint to the fluid, then with a little ammonia, and the mixture is warmed in a porcelain dish, with renewal of the water if necessary, until it has become colorless, the excess of sulphide of ammonia is decomposed or volatilized, the fluid contains now sulphocyanide of ammonium, and after being acidified with hydrochloric acid (which must not be attended with disengagement of hydrosulphuric gas), acquires a blood-red tint upon addition of sesquichloride of iron (LIEBIG). This reaction is exceedingly delicate. The following formula expresses the transformation of hydrocyanic acid into sulphocyanide of ammonium: $NH_4S_2 + 2(NH_4O) + 2HCy = 2(NH_4, CyS_2) + NH_4S + 2HO$. If an acetate is present the reaction takes place only upon addition of more hydrochloric acid.

8. Neither of the above methods will serve to effect the detection of cyanogen in cyanide of mercury. To detect cyanogen in that compound,

the solution is mixed with hydrosulphuric acid : sulphide of mercury precipitates, and the solution contains free hydrocyanic acid. In solid cyanide of mercury the cyanogen is most readily detected by heating in a glass tube. (Compare 3.)

Appendix.

a. Hydroferrocyanic acid ($H_2 Cfy$). Hydroferrocyanic acid is soluble in water. Some of the ferrocyanides, as those containing metals of the alkalies and alkaline earths, are soluble in water ; but the greater part of them are insoluble in that menstruum. All the ferrocyanides are decomposed by ignition ; where they are not quite anhydrous, hydrocyanic acid, carbonic acid, and ammonia escape, otherwise nitrogen and occasionally cyanogen. In solutions of hydroferrocyanic acid or of soluble ferrocyanides *sesquichloride of iron* produces a blue precipitate of FERROCYANIDE OF IRON ($Fe_4 Cfy_3$) ; *sulphate of oxide of copper* a brownish-red precipitate of FERROCYANIDE OF COPPER ($Cu_2 Cfy$) ; *nitrate of silver* a white precipitate of FERROCYANIDE OF SILVER ($Ag_2 Cfy$), which is insoluble in nitric acid and in ammonia, but dissolves in cyanide of potassium. Insoluble ferrocyanides of metals are decomposed by boiling with solution of soda, ferrocyanide of sodium being formed and the oxides thrown down, unless they are soluble in solution of soda. When heated with 3 parts of sulphate and 1 part of nitrate of ammonia, they yield sulphates of the metals contained in them, the whole of the cyanogen volatilizing in form of cyanide of ammonium and the products of its decomposition (BOLLEY).

b. Hydroferricyanic acid ($H_3 Cfdy$). Hydroferricyanic acid and many of the ferricyanides are soluble in water ; all ferricyanides are decomposed by ignition in a similar manner as the ferrocyanides. In the aqueous solutions of hydroferricyanic acid and its salts *sesquichloride of iron* produces no blue precipitate ; but *sulphate of protoxide of iron* produces a blue precipitate of PROTOFERRICYANIDE OF IRON (3 Fe, Cfdy) ; *sulphate of copper* a yellowish-green precipitate of FERRICYANIDE OF COPPER (3 Cu, Cfdy), which is insoluble in hydrochloric acid ; *nitrate of silver* an orange-colored precipitate of FERRICYANIDE OF SILVER (3 Ag, Cfdy), which is insoluble in nitric acid, but dissolves readily in ammonia and in cyanide of potassium. The insoluble ferricyanides of metals are decomposed by boiling with solution of soda, the metallic oxides being thrown down ; in the fluid filtered off from them either ferrocyanide of sodium alone is found, or a mixture of ferro- with ferricyanide of sodium. By heating with sulphate and nitrate of ammonia the ferricyanides are decomposed the same as the ferrocyanides.

§ 156.

e. HYDROSULPHURIC ACID (H S).

Sulphuretted Hydrogen.

1. SULPHUR is a solid, brittle, friable, tasteless body, insoluble in water. It occurs occasionally in the form of yellow or brownish crystals, or crystalline masses of a yellow or brownish color, and occasionally in that of a yellow or yellowish-white or grayish-white powder. It melts at a moderate heat ; upon the application of a stronger heat it is converted into brownish-yellow vapors, which in cold air condense to a yellow powder, and on the sides of the vessel to drops. Heated in the air it burns with bluish flame to sulphurous acid, which betrays its presence

in the air at once by its suffocating odor. Concentrated nitric acid, nitrohydrochloric acid, and a mixture of chlorate of potassa and hydrochloric acid dissolve sulphur gradually, with the aid of a moderate heat, and convert it into sulphuric acid; in boiling solution of soda sulphur dissolves to a yellow fluid, which contains sulphide of sodium and hyposulphite of soda; in ammonia sulphur is almost absolutely insoluble.

2. HYDROSULPHURIC ACID, at the common temperature and under common atmospheric pressure, is a colorless inflammable gas, soluble in water, which may be readily recognised by its characteristic smell of rotten eggs; it transiently imparts a red tint to litmus-paper.

3. Of the SULPHIDES only those with alkalies and alkaline earths are soluble in water. These, as well as the sulphides of iron, manganese, and zinc, are decomposed by dilute mineral acids, with evolution of hydrosulphuric acid gas, which may be readily detected by its peculiar smell, and by its action upon solution of lead (see 4). The decomposition of higher sulphides is attended also with separation of sulphur in a finely-divided state; the white precipitate may be readily distinguished from similar precipitates by its deportment on heating. Part of the sulphides of the metals of the fifth and sixth groups are decomposed by concentrated and boiling hydrochloric acid, with evolution of hydrosulphuric acid gas, whilst others are not dissolved by hydrochloric acid, but by concentrated and boiling nitric acid. The compounds of sulphur with mercury, gold, and platinum, resist the action of both acids, but dissolve readily in nitrohydrochloric acid. Upon the solution of sulphides in nitric acid, and in nitrohydrochloric acid, sulphuric acid is formed, and the process of solution is moreover attended, in most cases, with separation of sulphur. Many metallic sulphides, more especially those of a higher degree of sulphuration, give a sublimate of sulphur when heated in a test-tube.

4. If hydrosulphuric acid, in the gaseous state or in solution, is brought into contact with *nitrate of silver* or *acetate of lead*, black precipitates of SULPHIDE OF SILVER or SULPHIDE OF LEAD are formed. In cases therefore where the odor of the gas fails to afford sufficient proof of the presence of hydrosulphuric acid, these reagents will remove all doubt. If the hydrosulphuric acid is present in the gaseous form the air suspected to contain it is tested by placing in it a small slip of paper moistened with solution of neutral acetate of lead and a little ammonia; if the gas is present the slip becomes covered with a brownish-black shining film of sulphide of lead. To detect a trace of an alkaline sulphide in presence of a free alkali or an alkaline carbonate, the best way is to mix the fluid with a solution of oxide of lead in solution of soda, which is prepared by mixing solution of acetate of lead with solution of soda until the precipitate which forms at first is redissolved.

5. If a fluid containing hydrosulphuric acid or an alkaline sulphide is mixed with solution of soda, then with *nitroprusside of sodium*,* it acquires a fine reddish-violet tint. The reaction is very delicate; but that with solution of oxide of lead in solution of soda is still more sensitive.

6. If metallic sulphides are exposed to the *oxidizing flame of the blowpipe*, the sulphur burns with a blue flame, emitting at the same time the well-known odor of sulphurous acid. If a metallic sulphide is heated in a glass tube open at both ends, in the upper part of which a slip of

* Nitroprusside of sodium being a reagent which can very well be dispensed with, I have omitted giving it a place among the reagents.

blue litmus-paper is inserted, and the tube is held in a slanting position during the operation, the escaping sulphurous acid reddens the litmus-paper.

7. If a finely-pulverized metallic sulphide is boiled in a porcelain dish with solution of potassa, and the mixture heated to incipient fusion of the hydrate of potassa, or if the test specimen is fused in a platinum spoon with hydrate of potassa, and the mass is, in either case, dissolved in a little water, a piece of bright silver (a polished coin) put into the solution, and the fluid warmed, a brownish-black film of sulphide of silver forms on the metal. This film may be removed afterwards by rubbing the metal with leather and quicklime (v. KOBELL).

8. If the powder of a sulphide which hydrochloric acid will not, or only with difficulty, decompose is mixed in a small cylinder, or in a wide-necked flask, with an equal volume of finely divided iron free from sulphur (ferrum alcoholisatum), and some moderately dilute hydrochloric acid (1 volume of concentrated acid to 1 volume of water) is poured over the mixture, in a layer a few lines thick, hydrosulphuric acid escapes along with the hydrogen. This may be easily detected by placing a slip of paper moistened with solution of acetate of lead, and dried again, under the cork, so that the bottom is covered by it, the ends of the slip projecting on both sides, and then loosely inserting the cork into the mouth of the flask. Realgar, orpiment, and molybdenite do not show this reaction (v. KOBELL).

§ 157.

Recapitulation and remarks.—Most of the acids of the first group are also precipitated by nitrate of silver, but the precipitates cannot well be confounded with the silver compounds of the acids of the second group, since the former are *soluble* in dilute nitric acid, whilst the latter are *insoluble* in that menstruum. The presence of hydrosulphuric acid interferes more or less with the testing for the other acids of the second group; this acid must therefore, if present, be removed first before the testing for the other acids can be proceeded with. The removal of the hydrosulphuric acid, when present in the free state, may be effected by simple ebullition; and when present in the form of an alkaline sulphide, by the addition of a metallic salt, such as will not precipitate any of the other acids, or at least will not precipitate them from acid solutions. Hydriodic and hydrocyanic acids may be detected, even in presence of hydrochloric or hydrobromic acid, by the equally characteristic and delicate reactions with starch (with addition of a fluid containing nitrous acid), and with solution of protosesquioxide of iron. But the detection of chlorine and bromine is more or less difficult in presence of iodine and cyanogen. These latter must therefore, if present, be removed first before the proper tests for chlorine and bromine can be applied. The separation of the cyanogen may be readily effected by converting the whole of the radicals present into salts of silver, and igniting the mixed compound produced: the cyanide of silver is decomposed in this process, whilst the chloride, bromide, and iodide of silver remain unaltered. Upon fusing the ignited residue with carbonate of soda and potassa, and boiling the fused mass with water, chloride, bromide, and iodide of the alkali metals are obtained in solution. The fused silver compounds may also be readily decomposed with zinc; all that is required for this purpose is to pour water over the mixed compound, to add a little sulphuric acid and a fragment of zinc, to let the mixture stand some time, and finally to filter the solution of chloride, bromide, or iodide of zinc off the separated metallic silver.

The iodine may be separated from the chlorine and bromine, by treating the mixed silver compound with ammonia, but more accurately by precipitating the iodine as protiodide of copper. From bromine the iodine is separated most accurately by protochloride of palladium, which only precipitates the iodine ; from chlorine it is separated by nitrate of protoxide of palladium.

Bromine in presence of iodine and chlorine may be identified by the following simple operation : Mix the fluid with a few drops of dilute sulphuric acid, then with some starch-paste, and add a little red fuming nitric acid or, better still, a solution of hyponitric acid in sulphuric acid, whereupon the iodine reaction will show itself immediately. Add now chlorine-water drop by drop until that reaction has disappeared ; then add some more chlorine-water to set the bromine also free, which may then be separated and identified by means of chloroform or bisulphide of carbon. Or the liberated iodine may also be taken up with chloroform or bisulphide of carbon, and chlorine-water cautiously added, when the violet-red coloration imparted by the iodine gradually fades away, and after its disappearance the brownish-yellow color given by the bromine is distinctly visible.

Metallic chlorides are detected in presence of metallic bromides by the reaction with chromate of potassa and sulphuric acid.

In conclusion I have to state that, besides the above-mentioned, many other agents by which to effect the liberation of iodine have been proposed, and may be employed ; thus, for instance, iodic acid or alkaline iodate and hydrochloric acid (LIEBIG); sesquichloride of iron and sulphuric acid, or bichloride of platinum, with addition of some hydrochloric acid (HEMPEL) ; permanganate of potassa in slightly acidified solution (HENRY), &c. With respect to these agents I have to observe that iodic acid must be used with the greatest caution, as a, in presence of reducing substances iodine is set free from the reagent, and, b, an excess of iodic acid will at once put an end to the reaction.—Sesquichloride of iron, with addition of sulphuric acid, will not act immediately upon very dilute solutions ; but after a time the reaction will make its appearance, revealing the presence of even the minutest trace of iodine ; the delicacy of the reaction is not materially impaired by an excess of the reagents. Permanganate of potassa acts immediately, even in the most dilute solutions. However, as a fluid colored by minute traces of iodide of starch is also apt to look reddish, the coloration imparted by the permanganic acid alone may lead to mistakes. From six to twelve hours should therefore always be allowed to elapse before judging of the actual nature of the coloration. The *modus operandi* may of course be modified in various ways to increase the delicacy of the starch reaction ; interesting particulars upon this point may be found in MORIN's paper on the subject in the Journal f. prakt. Chem., 78, 1, and in HEMPEL's paper, in the Annal. der Chem. und Pharm., 107, 102.

§ 158.
Rarer Acids of the Second Group.
1. NITROUS ACID (NO_3).

Nitrous acid, in the free state, at the common temperature, is a brownish-red gas. In contact with water it is converted into nitric acid, which dissolves, and nitric oxide gas, which remains undissolved, and, where the quantity of water is not very large, escapes in part ($3 NO_3 = NO_5 + 2 NO_2$). The nitrites are decomposed by ignition; many of them are soluble in water. When nitrites or concentrated solutions of nitrites are treated with dilute sulphuric acid it is not nitrous acid gas which is evolved, but nitric oxide gas, and the evolution is attended with formation of nitric acid. In solutions of nitrites of the alkalies *nitrate of silver* produces a white precipitate, which dissolves in a very large proportion of water, especially upon application of heat; *sulphate of protoxide of iron*, upon addition of a small quantity of acid, produces a dark blackish-brown coloration, which is due to the nitric oxide gas dissolving in the solution of the sulphate of protoxide of iron. *Hydrosulphuric acid* produces in solutions containing nitrous acid, after neutralization by an acid of the free alkali, should any be present, a copious precipitate of sulphur, the reaction being attended also with formation of nitrate of ammonia. *Pyrogallic acid* imparts a brown color to even very dilute solutions of nitrites acidified with sulphuric acid (SCHÖNBEIN). But the most delicate reagent for nitrous acid is solution of iodide of potassium mixed with starch paste, especially upon addition of sulphuric acid (PRICE, SCHÖNBEIN). Water containing the one hundred thousandth part of nitrite of potassa, together with free sulphuric acid, is colored distinctly blue by iodide of starch in a few seconds, and a few minutes suffice to produce the same effect in water containing as little as one part of nitrite of potassa in one million parts of water. Let it be borne in mind that this reaction is reliable only where no other substance is present that might exercise a decomposing action upon iodide of starch, such, for instance, as iodic acid, sesquioxide of iron, &c.

2. HYPOCHLOROUS ACID (ClO).

Hypochlorous acid, at the common temperature, is a deep yellowish-green gas of a disagreeable irritating odor, similar to that of chlorine. It dissolves in water; the dilute aqueous solution bears distillation. The hypochlorites are usually found in combination with metallic chlorides, as is the case, for instance, in chloride of lime, *eau de Javelle*, &c. The solutions of hypochlorites undergo alteration by boiling, the hypochlorite being resolved into chloride of the metal and chlorate of the oxide, attended, in the case of concentrated, but not in that of dilute solutions, with evolution of oxygen. If a solution of chloride of lime is mixed with hydrochloric acid or sulphuric acid, chlorine is disengaged, whilst addition of a little nitric acid leads to the liberation of hypochlorous acid. *Nitrate of silver* throws down from solution of chloride of lime chloride of silver (the hypochlorite of silver, which forms at first, is speedily resolved into chloride of silver and chlorate of silver: $3(AgO, ClO) = AgO, ClO_5 + 2AgCl$); *nitrate of lead* produces a precipitate which from its original white changes gradually to orange-red, and ultimately, owing to formation of binoxide, to brown; *salts of protoxide of manganese* give brown-black precipitates of hydrate of binoxide of manganese. Solution of permanganate of potassa is not decolorized. Solutions of *litmus* and *indigo* are decolorized even by the alkaline solutions of hypochlorites, but still more rapidly and completely upon addition of an acid. If a solution of arsenious acid in hydrochloric acid is colored blue with indigo tincture, and a solution of chloride of lime is added, with active stirring, the decoloration will take place only after the whole of the arsenious acid has been converted into arsenic acid.

3. CHLOROUS ACID (ClO_3).

Chlorous acid is a yellowish-green gas of a peculiar and very disagreeable odor; it is soluble in water. The solution has an intensely yellow color, even when highly dilute. Most of the chlorites are soluble in water; the solutions readily suffer decomposition, the chlorites being resolved into chlorides and chlorates. *Nitrate of silver* precipitates white chlorite of silver, which is soluble in much water. A solution of *permanganate of potassa* is immediately decomposed, and hydrate of sesquioxide of manganese separates after some time. *Tincture of litmus* and *tincture of indigo* are instantly decolorized, even if mixed with arsenious acid in excess. If a slightly acidified dilute solution of a *salt of protoxide of iron* is mixed with a dilute solution of chlorous acid, the fluid transiently acquires an amethyst tint, and assumes only after the lapse of a few seconds the yellowish coloration of salts of sesquioxide of iron (LENSSEN).

4. HYPOPHOSPHOROUS ACID (PO).

The concentrated solution of hypophosphorous acid is of syrupy consistence, and resembles that of phosphorous acid (see § 148), with which it also has this in common, that it is resolved by heating, with exclusion of air, into hydrate of phosphoric acid and not spontaneously inflammable phosphuretted hydrogen gas. All hyposulphites are soluble in water; by ignition all of them are resolved into phosphate and phosphuretted hydrogen gas, which in most cases is spontaneously inflammable. *Chloride of barium, chloride of calcium,* and *acetate of lead* fail to precipitate solutions of hypophosphites (difference from phosphorous acid). *Nitrate of silver* gives with hypophosphites at first a white precipitate of hypophosphite of silver, which turns black even at the common temperature, but more rapidly on heating, the change of color being attended with separation of metallic silver. From *chloride of mercury in excess* hypophosphorous acid precipitates, slowly in the cold, more rapidly on heating, subchloride of mercury. Brought together with *zinc* and *dilute sulphuric acid* hypophosphorous acid gives hydrogen gas mixed with phosphuretted hydrogen. (Compare § 148, phosphorous acid.)

THIRD GROUP OF THE INORGANIC ACIDS.

ACIDS WHICH ARE NOT PRECIPITATED BY SALTS OF BARYTA NOR BY SALTS OF SILVER: *Nitric Acid, Chloric Acid* (Perchloric Acid).

§ 159.

a. NITRIC ACID (NO_5).

1. ANHYDROUS NITRIC ACID crystallizes in six-sided prisms. It fuses at 85·2° F., and boils at about 113° F. (DEVILLE). The pure HYDRATE is a colorless exceedingly corrosive fluid, which emits fumes in the air, exercises a rapidly destructive action upon organic substances, and colors nitrogenous matter intensely yellow. Hydrate of nitric acid containing nitrous acid has a red colour.

2. All the NEUTRAL SALTS of nitric acid are soluble in water; only some of the basic nitrates are insoluble in this menstruum. All nitrates without exception undergo decomposition at an intense red heat. Those with alkaline bases yield at first oxygen, and change to nitrites, which are then further resolved into oxygen and nitrogen; the others yield oxygen and nitrous or hyponitric acid.

3. If a nitrate is thrown upon *red-hot charcoal*, or if charcoal or some organic substance, paper for instance, is brought into contact with a nitrate in fusion, DEFLAGRATION takes place, *i.e.*, the charcoal burns at the expense of the oxygen of the nitric acid, the combustion being attended with vivid scintillation.

4. If a mixture of a nitrate with *cyanide of potassium* in powder is heated on platinum foil, a vivid DEFLAGRATION ensues, attended with distinct ignition and detonation. Even very minute quantities of nitrates may be detected by this reaction.

5. If a nitrate is mixed with *copper filings*, and the mixture heated in a test-tube with concentrated sulphuric acid, the air in the tube acquires a yellowish-red tint, owing to the nitric oxide gas which is liberated upon the oxidation of the copper by the nitric acid combining with the oxygen of the air to nitrous acid. The coloration may be observed most distinctly by looking lengthways through the tube.

6. If the solution of a nitrate is mixed with an equal volume of concentrated *sulphuric acid*, free from nitric and hyponitric acid, the mixture allowed to cool, and a concentrated solution of *sulphate of protoxide of iron* then cautiously added to it so that the fluids do not mix, the stratum where the two fluids are in immediate contact shows at first a purple, afterwards a brown color, or, in cases where only a very minute quantity of nitric acid is present, a reddish color. In this process the nitric acid is decomposed by the protoxide of iron, three-fifths of its oxygen combining with the protoxide and converting a portion of it

into sesquioxide, whilst the remaining nitric oxide combines with the remaining portion of the protoxide of iron, and forms with it a peculiar compound, which dissolves in water, imparting a brownish-black color to the fluid. A similar reaction is observed in presence of selenious acid; but on mixing the fluid, and letting it stand, red selenium separates (WITTSTOCK).

7. If some hydrochloric acid is boiled in a test-tube, one or two drops of very dilute *solution of indigo and sulphuric acid* added, and the mixture boiled again, the fluid remains blue (provided the hydrochloric acid was free from chlorine). If a nitrate, solid or in solution, is now added to the faint light-blue fluid, and the mixture heated again to boiling, the color disappears owing to the decomposition of the indigo blue. This is a most delicate reaction. It must be borne in mind, however, that several other substances also cause decoloration of solution of indigo—free chlorine more particularly produces this effect.

8. If a little *brucia* is dissolved in concentrated sulphuric acid, and a small quantity of a fluid containing nitric acid added to the solution, the latter immediately acquires a magnificent red color. This reaction is exceedingly delicate.

9. Very minute quantities of nitric acid may be detected also by reducing the nitric acid first to nitrous acid, which may be effected both in the moist and in the dry way; in the former by heating the solution of the nitric acid or of the nitrate for some time with finely-divided zinc, best with zinc amalgam, and then filtering (SCHÖNBEIN); in the dry way by fusing the substance under examination with carbonate of soda and potassa at a moderate heat, extracting the mass, after cooling, with water, and filtering. Upon adding either of the filtrates to a solution of iodide of potassium mixed with starch-paste and dilute sulphuric acid, the fluid acquires a blue color from iodide of starch (comp. § 158, 1).

§ 160.

b. CHLORIC ACID (ClO_5).

1. CHLORIC ACID, in its most highly concentrated solution, is a colorless or slightly yellowish oily fluid; its odor resembles that of nitric acid. It first reddens litmus, then bleaches it. Dilute chloric acid is colorless and inodorous.

2. All CHLORATES are soluble in water. When chlorates are heated to redness, the whole of their oxygen escapes and metallic chlorides remain.

3. Heated with *charcoal* or some organic substance the chlorates DEFLAGRATE, and this with far greater violence than the nitrates.

4. If a mixture of a chlorate with *cyanide of potassium* is heated on platinum foil, DEFLAGRATION takes place, attended with strong detonation and ignition, even though the chlorate be present only in very small quantity. This experiment should be made with very minute quantities only.

5. If the solution of a chlorate is colored light-blue with *solution of indigo* in sulphuric acid, a little dilute sulphuric acid added, and a solution of sulphite of soda dropped cautiously into the blue fluid, the color of the indigo disappears immediately. The cause of this equally characteristic and delicate reaction is, that the sulphurous acid deprives the chloric acid of its oxygen, thus setting free chlorine or a lower oxide of it, which then decolorizes the indigo.

6. If chlorates are treated with moderate dilute *hydrochloric acid* the constituents of the two acids transpose, forming water, chlorine, and bichlorochloric acid ($2\,ClO_3$, ClO_2). Application of heat promotes the reaction. The test-tube in which the experiment is made becomes filled in this process with a greenish-yellow gas of a very disagreeable odor resembling that of chlorine; the hydrochloric acid acquires a greenish-yellow color. If the hydrochloric acid is colored blue with indigo solution, the presence of very minute quantities of chlorates will suffice to destroy the indigo color at once.

7. If a little chlorate is added to a few drops of *concentrated sulphuric acid* in a watch-glass, two-thirds of the metallic oxide are converted into a sulphate, and the remaining one-third into perchlorate; this conversion is attended moreover with liberation of chlorochloric acid, which imparts an intensely yellow tint to the sulphuric acid, and betrays its presence also by its odor and the greenish color of the evolved gas $[3\,(KO, ClO_5) + 4\,SO_3 = 2\,(KO, 2\,SO_3) + KO, ClO_7 + (ClO_3, ClO_2)]$. The application of heat must be avoided in this experiment, and the quantities operated upon should be very small, since otherwise the decomposition might take place with such violence as to cause an explosion.

§ 161.

Recapitulation and remarks.—Of the reactions which have been suggested to effect the detection of nitric acid those with sulphate of protoxide of iron and sulphuric acid, with copper filings and sulphuric acid, with brucia, and also those based upon the reduction of the nitrates to nitrites, give the most positive results; with regard to deflagration with charcoal, detonation with cyanide of potassium, and decoloration of solution of indigo, we have seen that these reactions belong equally to chlorates as to nitrates, and are consequently decisive only where no chloric acid is present. The presence of free nitric acid in a fluid may be detected by evaporating in a porcelain dish on the water-bath to dryness, having first thrown in a few quill-cuttings: yellow coloration of these indicates the presence of nitric acid (RUNGE). The best way to ascertain whether chloric acid is present or not is to ignite the sample under examination, with addition of carbonate of soda, dissolve the mass, and test the solution with nitrate of silver. If a chlorate is present, this is converted into a chloride upon ignition, and nitrate of silver will now precipitate chloride of silver from the solution. However, the process is thus simple only if no chloride is present along with the chlorate. But in presence of a chloride the latter must be removed first by adding nitrate of silver to the solution as long as a precipitate continues to form, and filtering the fluid from the precipitate; the filtrate is then, after addition of pure carbonate of soda, evaporated to dryness, and the residue ignited. It is, however, generally unnecessary to pursue this circuitous way, since the reactions with concentrated sulphuric acid, and with indigo and sulphurous acid, are sufficiently marked and characteristic to afford positive proof of the presence of chloric acid, even in presence of nitrates.—The best way of detecting nitric acid in presence of a large proportion of chloric acid is to mix the compound under examination with carbonate of soda in excess, evaporate ignite the residue gently, but sufficiently long to convert the chlorate into chloride, and then test the residue for nitric acid, or, as the case may be, for nitrous acid.

§ 162.

PERCHLORIC ACID (ClO_7).

Pure anhydrous perchloric acid is a colorless mobile fluid, which forms dense white fumes in the air, and explodes with great violence when dropped on woodcharcoal (ROSCOE). The hydrate crystallizes in needles; the concentrated aqueous solution is oily and heavy. The dilute solution gives by distillation first water, then dilute acid, and finally concentrated acid. All perchlorates are soluble in water, most of them freely. They are all decomposed by ignition, those with alkaline bases leaving chlorides behind, with disengagement of oxygen. *Salts of potassa* produce in not too dilute solutions a white crystalline precipitate of perchlorate of potassa (KO, ClO_7), which is sparingly soluble in water, insoluble in spirit of wine. *Baryta salts* and *silver salts* are not precipitated. Concentrated sulphuric acid fails to decompose perchloric acid in the cold, and decomposes it with difficulty on heating (difference from chloric acid). Hydrochloric acid, nitric acid, and sulphurous acid fail to decompose aqueous solutions of perchloric acid or perchlorates; indigo tincture, therefore, previously added to it, is not decolorized (difference from all other acids of chlorine).

II. ORGANIC ACIDS.

First Group.

ACIDS WHICH ARE INVARIABLY PRECIPITATED BY CHLORIDE OF CALCIUM: *Oxalic Acid, Tartaric Acid* (Paratartaric or Racemic Acid), *Citric Acid, Malic Acid.*

§ 163.

a. OXALIC ACID.

For the reactions of oxalic acid I refer to § 145.

b. TARTARIC ACID ($2HO, C_8H_4O_{10}$).

1. The HYDRATE of TARTARIC ACID forms colorless crystals of an agreeable acid taste, which are persistent in the air, and soluble in water and in spirit of wine. Heated to 212° F., tartaric acid loses no water; heated to 338° F., it fuses; at a higher temperature it becomes carbonized, emitting during the process a very peculiar and highly characteristic odor, which resembles that of burnt sugar. Aqueous solution of tartaric acid, as also of almost all tartrates, turns the plane of polarization of light towards the right.

2. The TARTRATES with alkaline base are soluble in water, and so are those with the metallic oxides of the third and fourth groups. Evaporated on the water-bath to syrupy consistence, the solution of tartrate of sesquioxide of iron deposits a pulverulent basic salt. Those of the tartrates which are insoluble in water dissolve in hydrochloric or nitric acid. The tartrates suffer decomposition when heated to redness; charcoal separates, and the same peculiar odor is emitted as attends the carbonization of free tartaric acid.

3. If to a solution of tartaric acid, or to that of an alkaline tartrate, solution of *sesquioxide of iron, protoxide of manganese,* or *alumina* is added in not too large proportion, and then ammonia or potassa, no precipitation of sesquioxide of iron, protoxide of manganese or alumina will ensue, since the double tartrates formed are not decomposed by alkalies. Tartaric acid prevents also the precipitation of several other oxides by alkalies.

4. Free tartaric acid produces with *salts of potassa,* and more particularly with the acetate, a sparingly soluble precipitate of BITARTRATE OF POTASSA. A similar precipitate is formed when acetate of potassa and free acetic acid are added to the solution of a neutral tartrate. The acid tartrate of potassa dissolves readily in alkalies and mineral acids;

tartaric acid and acetic acid do not increase its solubility in water. The separation of the bitartrate of potassa precipitate is greatly promoted by shaking, or by rubbing the sides of the vessel with a glass rod. The delicacy of the reaction may be heightened by concentrating the solution of the tartaric acid. By adding a drop of a concentrated solution of acetate of potassa to one or two drops of a highly concentrated solution of tartaric acid on a watch-glass, and stirring the mixture with a glass rod, crystals will at once be deposited on the rubbed part. Addition of an equal volume of alcohol heightens the delicacy of the reaction.

5. *Chloride of calcium* throws down from solutions of neutral tartrates a white precipitate of TARTRATE OF LIME ($2 CaO, C_8 H_4 O_{10} + 8$ aq.). Presence of ammoniacal salts retards the formation of this precipitate for a more or less considerable space of time. Agitation of the fluid or friction on the sides of the vessel promotes the separation of the precipitate. The precipitate is crystalline, or invariably assumes a crystalline form after some time; it dissolves in a cold not over dilute solution of potassa or soda, pretty free from carbonic acid, to a clear fluid. But upon boiling this solution, the dissolved tartrate of lime separates again in the form of a gelatinous precipitate, which redissolves upon cooling.

6. *Lime-water* produces in solutions of neutral tartrates—and also in a solution of free tartaric acid, if added to alkaline reaction—white precipitates which, flocculent at first, assume afterwards a crystalline form; so long as they remain flocculent they are readily dissolved by tartaric acid as well as by solution of chloride of ammonium. From these solutions the tartrate of lime separates again, after the lapse of several hours, in the form of small crystals deposited upon the sides of the vessel.

7. *Solution of sulphate of lime* fails to produce a precipitate in a solution of tartaric acid; in solutions of neutral tartrates of the alkalies, it produces a trifling precipitate after the lapse of some time.

8. If solution of ammonia is poured upon even a very minute quantity of tartrate of lime, a small fragment of crystallized *nitrate of silver* added, and the mixture slowly and gradually heated, the sides of the test-tube are covered with a bright coating of metallic silver. If, instead of a crystal, solution of nitrate of silver be used, or heat be applied more rapidly, the reduced silver will separate in a pulverulent form (ARTHUR CASSELMANN).

9. *Acetate of lead* produces white precipitates in solutions of tartaric acid and its salts. The precipitate ($2 PbO, C_8 H_4 O_{10}$) dissolves readily in nitric acid and in ammonia.

10. *Nitrate of silver* does not precipitate free tartaric acid; but in solutions of neutral tartrates it produces a white precipitate of TARTRATE OF SILVER ($2 AgO, C_8 H_4 O_{10}$), which dissolves readily in nitric acid and in ammonia; upon boiling it turns black, owing to ensuing reduction of the silver to the metallic state.

11. Upon heating hydrated tartaric acid or a tartrate with *hydrate of sulphuric acid*, the latter acquires a brown color almost simultaneously with the evolution of gas.

§ 164.

c. CITRIC ACID ($3 HO, C_{12} H_8 O_{11}$).

1. CRYSTALLIZED CITRIC ACID, obtained by the cooling of its solution, has the formula $3 HO, C_{12} H_8 O_{11} + 2$ aq. It crystallizes in pellucid, colorless and inodorous crystals of an agreeable strongly acid taste, which

dissolve readily in water and in spirit of wine, and effloresce slowly in the air. Heated to 212° F. the crystallized acid loses its water of crystallization; when subjected to the action of a stronger heat, it fuses at first, and afterwards carbonizes, with evolution of pungent acid fumes, the odor of which may be readily distinguished from that emitted by tartaric acid upon carbonization.

2. The CITRATES with alkaline base are readily soluble in water, as well in the neutral as in the acid state; solution of citric acid therefore is not precipitated by acetate of potassa. The compounds of citric acid with such of the metallic oxides as are weak bases, sesquioxide of iron, for instance, are also freely soluble in water. Evaporated on the water-bath to syrupy consistence the solution of citrate of sesquioxide of iron deposits no solid salt. Citrates, like tartrates, and for the same reason, prevent the precipitation of sesquioxide of iron, protoxide of manganese, alumina, &c., by alkalies.

3. *Chloride of calcium* fails to produce a precipitate in solution of free citric acid, even upon boiling; but a precipitate of NEUTRAL CITRATE OF LIME ($3 CaO, C_{12}H_5O_{11} + 4$ aq.) forms immediately upon saturating with potassa or soda the concentrated solution of citric acid, mixed with chloride of calcium in excess. The precipitate is insoluble in potassa, but dissolves freely in solution of chloride of ammonium; upon boiling this chloride of ammonium solution, neutral citrate of lime of the same composition separates again in the form of a white crystalline precipitate, which however is now no longer soluble in chloride of ammonium. If a solution of citric acid mixed with chloride of calcium is saturated with ammonia, a precipitate will form in the cold only after many hours' standing; but upon boiling the clear fluid, neutral citrate of lime of the properties just stated will suddenly precipitate. By heating citrate of lime with ammonia and nitrate of silver the latter salt is not reduced, or only to a trifling extent.

4. *Lime-water* produces no precipitate in cold solutions of citric acid or of citrates. But upon heating the solution to boiling, with a tolerable excess of hot prepared lime-water, a white precipitate of CITRATE OF LIME is formed, of which the greater portion redissolves upon cooling.

5. *Acetate of lead*, when added in excess to a solution of citric acid, produces a white precipitate of CITRATE OF LEAD ($3 PbO, C_{12}H_5O_{11}$), which, after washing, dissolves readily in ammonia.

6. *Nitrate of silver* produces in solutions of neutral citrates of the alkalies a white flocculent precipitate of CITRATE OF SILVER ($3 AgO, C_{12}H_5O_{11}$), which does not turn black on boiling.

7. Upon heating citric acid or citrates with concentrated sulphuric acid, carbonic oxide and carbonic acid escape at first, the sulphuric acid retaining its natural color; upon continued ebullition, however, the solution acquires a dark color, and sulphurous acid is evolved.

§ 165.

d. MALIC ACID ($2 HO, C_8H_4O_8$).

1. HYDRATE OF MALIC ACID crystallizes with great difficulty, forming crystalline crusts, which deliquesce in the air, and dissolve readily in water and in alcohol. Exposed to a temperature of 302° F., hydrated malic acid is slowly converted into hydrated fumaric acid, with loss of two equivalents of water ($2 HO, C_8H_2O_6$); heated to 356° F., malic

acid is resolved into water, MALEIC ACID ($2\,HO, C_8H_2O_6$), which volatilizes, and FUMARIC ACID ($2\,HO, C_8H_2O_6$), which remains. By raising the temperature to above 392° F. the fumaric acid is finally also volatilized. This deportment of malic acid is highly characteristic. If the experiment is made in a small spoon, pungent acid vapors are evolved with frothing effervescence; if the experiment is made in a small tube, the maleic acid first, and afterwards the fumaric acid also will condense to crystals in the colder part of the tube.

2. Malic acid forms with most bases salts soluble in water. The acid malate of potassa is not very difficultly soluble in water; acetate of potassa fails therefore to precipitate solutions of malic acid. Malic acid prevents, like tartaric acid, the precipitation of sesquioxide of iron, &c., by alkalies.

3. *Chloride of calcium* fails to produce a precipitate in solutions of free malic acid. Even after saturation with ammonia or soda no precipitate is formed. But upon boiling, a precipitate of MALATE OF LIME ($2\,CaO, C_8H_4O_8 + 6$ aq.) separates from concentrated solutions. If the precipitate is dissolved in a very little hydrochloric acid, ammonia added to the solution, and the fluid boiled, the malate of lime separates again; but if it is dissolved in a somewhat larger quantity of hydrochloric acid, it will not reprecipitate, after addition of ammonia in excess, even upon continued boiling. Alcohol precipitates it immediately from a solution of the kind. Malate of lime, when heated with ammonia and nitrate of silver, fails absolutely or nearly so to effect the reduction of the latter to the metallic state.

4. *Lime-water* produces no precipitate in solutions of free malic acid, nor in solutions of malates. The fluid remains perfectly clear even upon boiling, provided the lime-water is prepared with boiling water.

5. *Acetate of lead* throws down from solutions of malic acid and of malates a white precipitate of MALATE OF LEAD ($2\,PbO, C_8H_4O_8 + 6$ aq.). The precipitation is the most complete if the fluid is neutralized by ammonia, as the precipitate is slightly soluble in free malic acid and acetic acid, and also in ammonia. If the fluid in which the precipitate is suspended is heated to boiling, a portion of the precipitate dissolves, and the remainder fuses to a mass resembling resin melted under water. This reaction is distinctly marked only if the malate of lead is tolerably pure; if mixed with other salts of lead—if, for instance, ammonia is added to alkaline reaction—it is only imperfect or fails altogether to make its appearance.

6. *Nitrate of silver* throws down from solutions of neutral malates of the alkalies a white precipitate of MALATE OF SILVER, which upon boiling turns a little gray.

7. Upon heating malic acid with concentrated sulphuric acid, carbonic acid and carbonic oxide gas are evolved at first; the fluid then turns brown and ultimately black, with evolution of sulphurous acid.

§ 166.

Recapitulation and remarks.—Of the organic acids of this group *oxalic acid* is characterized by the instant precipitation of its lime-salt from its solution in hydrochloric acid by ammonia, and also by acetate of soda, as well as by the immediate precipitation of the free acid by solution of sulphate of lime. *Tartaric acid* is characterized by the sparing solubility of the acid potassa salt, the solubility of the lime-salt in cold solution of soda and of potassa, the reaction of the lime-salt with ammonia and nitrate of silver, and the peculiar odor which the acid and its salts emit upon heating. It is most safely detected in presence of the other acids by means of acetate of potassa (§ 163, 4). *Citric acid* is distinctly characterized by its reaction with lime-water, or with chloride of calcium and ammonia in presence of chloride of ammonium. But in this the absence or the previous removal of oxalic and tartaric acids is always presupposed. *Malic acid* would be sufficiently characterized by the deportment of malate of lead when heated under water, were this reaction more sensitive, and not so easily prevented by the presence of other acids. The safest means of identifying malic acid is to convert it into maleic acid and fumaric acid by heating in a glass tube; but this conversion can be effected successfully only with pure hydrate of malic acid. Malate of lead is sparingly soluble in ammonia, whilst the citrate and tartrate of lead dissolve freely in that agent; this different deportment of the lead salts of the acids affords also a means of distinguishing between them. If only one of the four acids is present in a solution, lime-water will suffice to indicate which of the four is present; since malic acid is not precipitated by this reagent, citric acid only upon boiling, tartaric acid and oxalic acid already in the cold; and the tartrate of lime redissolves upon addition of chloride of ammonium, whilst the oxalate does not. If the four acids together are present in a solution, the oxalic acid and tartaric acid are precipitated first by chloride of calcium and ammonia, in presence of chloride of ammonium (the tartrate of lime separates under these circumstances completely only after some time; it is separated from the oxalate by treating with solution of soda); the citrate of lime is then thrown down by boiling, and the malate finally by means of spirit of wine. The precipitate produced by spirit of wine must never be taken positively for malate of lime, without further proof, since the sulphate and other salts of lime are also precipitated by that agent under the same circumstances. Positive conviction can only be attained by the production of hydrate of malic acid from the lime-salt. To effect this the precipitate is dissolved in acetic acid, spirit of wine added, and the fluid filtered if necessary. The filtrate is precipitated with acetate of lead, the fluid neutralized with ammonia, the precipitate washed, stirred in water, decomposed by hydrosulphuric acid, and the filtrate evaporated to dryness. Where a small quantity of citric acid or malic acid is to be detected in presence of a large proportion of tartaric acid, the best way is to remove the latter first by acetate of potassa, with addition of an equal volume of strong alcohol. The other acids may then be completely precipitated in the filtrate by chloride of calcium and ammonia if the quantity of the alcohol is a little increased.

§ 167.

RACEMIC ACID, OR PARATARTARIC ACID ($2 H O, C_8 H_4 O_{10}$).

The formula of crystallized racemic acid is $2 H O, C_8 H_4 O_{10} + 2$ aq. The crystallization water escapes slowly in the air, but rapidly at 212° F. (difference between racemic acid and tartaric acid). To solvents the racemic acid comports itself like the tartaric acid. The racemates also show very similar deportment to that of the tartrates. However, many of them differ in the amount of water they contain, and in form and solubility from the corresponding tartrates. The aqueous solution of racemic acid and the racemates exercises no diverting action upon polarized light. *Chloride of calcium* precipitates from the solutions of free racemic acid and of racemates RACEMATE OF LIME ($2 Ca O, C_8 H_4 O_{10} + 8$ aq.), as a white crystalline powder. Ammonia throws down the precipitate from its solution in hydrochloric acid, either immediately or at least very speedily (difference between racemic acid and tartaric acid). It dissolves in solution of soda and potassa, but is reprecipitated from this solution by boiling (difference between racemic acid and oxalic acid). *Lime-water* added in excess produces immediately a white precipitate insoluble in chloride of ammonium (difference between racemic and tartaric acid). *Solution of sulphate of lime* does not immediately produce a precipitate in a solution of racemic acid (difference between racemic and oxalic acid); however, after ten or fifteen minutes, racemate of lime separates (difference between racemic acid and tartaric acid); in solutions of neutral racemates the precipitate forms immediately. With *salts of potassa* racemic acid comports itself like tartaric acid. By letting racemate of soda and potassa or soda and ammonia crystallize, two kinds of crystals are obtained, which resemble each other as the image reflected by the mirror resembles the object reflected. The one kind of crystals contain common tartaric acid (which turns the plane of polarization of light towards the right); the other kind contains antitartaric acid, *i.e.* an acid which is the same in every respect as tartaric acid, with this exception only, that it turns the polarized light towards the left. If the two kinds of crystals are redissolved, the solution shows again the reactions of racemic acid.

SECOND GROUP OF THE ORGANIC ACIDS.

ACIDS WHICH CHLORIDE OF CALCIUM FAILS TO PRECIPITATE UNDER ANY CIRCUMSTANCES, BUT WHICH ARE PRECIPITATED FROM NEUTRAL SOLUTIONS BY SESQUICHLORIDE OF IRON: *Succinic Acid, Benzoic Acid.*

§ 168.

a. SUCCINIC ACID ($2 HO, C_8 H_4 O_6$).

1. HYDRATE OF SUCCINIC ACID forms colorless and inodorous prisms or tables of slightly acid taste, which are readily soluble in water, alcohol, and ether, sparingly soluble in nitric acid, and volatilize when exposed to the action of heat, leaving only a little charcoal behind. The officinal acid has an empyreumatic odor, and leaves a somewhat larger carbonaceous residue upon volatilization. Succinic acid is not destroyed by heating with nitric acid, and may therefore be easily obtained in the pure state by boiling with that acid for half an hour, by which means the oil of amber, if present, will be destroyed. By sublimation crystalline needles of silky lustre are obtained. The hydrate loses water in this process, so that by repeated sublimation anhydrous acid is ultimately obtained. Heated in the air succinic acid burns with a blue flame, free from soot.

2. The SUCCINATES are decomposed at a red heat; those which have an alkali or alkaline earth for base are converted into carbonates in this process, the change being attended with separation of charcoal. Most of the succinates are soluble in water.

3. *Sesquichloride of iron* produces in solutions of neutral succinates of the alkalies a brownish pale red bulky precipitate of SUCCINATE OF SESQUIOXIDE OF IRON ($Fe_2 O_3, C_8 H_4 O_8$); one-third of the succinic acid is liberated in this reaction, and retains part of the precipitate in solution if the fluid is filtered off hot. The precipitate dissolves readily in mineral acids; ammonia decomposes it, causing the separation of a less bulky precipitate of a highly basic succinate of sesquioxide of iron, and combining with the greater portion of the acid to succinate of ammonia, which dissolves.

4. *Acetate of lead* gives with succinic acid a white precipitate of neutral SUCCINATE OF LEAD ($2 PbO, C_8 H_4 O_8$), which is very sparingly soluble in water, acetic acid, and succinic acid, but dissolves freely in

solution of acetate of lead and in nitric acid. Treated with ammonia the neutral succinate of lead is converted into a basic salt ($6\, PbO, C_8 H_4 O_8$).

5. A mixture of *alcohol, ammonia*, and *solution of chloride of barium* produces in solutions of free succinic acid and of succinates a white precipitate of SUCCINATE OF BARYTA ($2\, BaO, C_8 H_4 O_8$).

6. *Nitrate of suboxide of mercury* and *nitrate of silver* also precipitate the succinates; the precipitates, however, are not possessed of any characteristic properties.

§ 169.

b. BENZOIC ACID ($HO, C_{14} H_5 O_3$).

1. Pure HYDRATE OF BENZOIC ACID forms inodorous white scales or needles, or simply a crystalline powder. It fuses when heated, and afterwards volatilizes completely. The fumes of benzoic acid cause a peculiar irritating sensation in the throat, and provoke coughing; when cautiously cooled, they condense to brilliant needles; when kindled, they burn with a luminous sooty flame. The common official hydrate of benzoic acid has the odor of benzoin, and leaves a small carbonaceous residue upon volatilization. Hydrate of benzoic acid is very sparingly soluble in cold water, but it dissolves pretty freely in hot water and in alcohol. Addition of water imparts therefore a milky turbidity to a saturated solution of benzoic acid in alcohol.

2. Most of the BENZOATES are soluble in water; only those with weak bases, *e.g.*, sesquioxide of iron, are insoluble. The soluble benzoates have a peculiar pungent taste. The addition of a *strong acid* to concentrated aqueous solutions of benzoates displaces the benzoic acid, which separates as hydrate in the form of a dazzling white sparingly soluble powder. Benzoic acid is expelled in the same way from the insoluble benzoates by such strong acids as form soluble salts with the bases with which the benzoic acid is combined.

3. *Sesquichloride of iron* precipitates solutions of free benzoic acid incompletely; solutions of neutral benzoates of the alkalies completely. The precipitate of BENZOATE OF SESQUIOXIDE OF IRON [$2\, Fe_2 O_3, 3\, (C_{14} H_5 O_3) + 15\, aq.$], is bulky, flesh-colored, insoluble in water. It is decomposed by ammonia in the same manner as succinate of sesquioxide of iron, from which salt it differs in this, that it dissolves in a little hydrochloric acid, with separation of the greater portion of the benzoic acid.

4. *Acetate of lead* fails to precipitate free benzoic acid and benzoate of ammonia, at least immediately; but it produces white flocculent precipitates in solutions of benzoates with a fixed alkaline base.

5. A mixture of *alcohol, ammonia*, and *solution of chloride of barium* produces NO precipitate in solutions of free benzoic acid or of the alkaline benzoates.

§ 170.

Recapitulation and remarks.—Succinic and benzoic acids are distinguished from all other acids by the facility with which they may be sublimed, and by their deportment with sesquichloride of iron. They are distinguished from one another by the different color of their salts with sesquioxide of iron, and also by their different deportment with chloride of barium and alcohol; but principally by their different degrees of solubility, succinic acid being readily soluble in water, whilst benzoic acid is very difficult of solution. Succinic acid is seldom perfectly pure, and may therefore often be detected by the odor of oil of amber which it emits.

The detection of the two acids, when present in the same solution with other acids, may be effected as follows: precipitate with sesquichloride of iron, warm the washed precipitate with ammonia, filter, concentrate the solution, divide it into two parts, and mix one part with hydrochloric acid, the other with chloride of barium and alcohol.

Succinic acid and benzoic acid do not prevent the precipitation of sesquioxide of iron, alumina, &c., by alkalies.

THIRD GROUP OF THE ORGANIC ACIDS.

ACIDS WHICH ARE NOT PRECIPITATED BY CHLORIDE OF CALCIUM NOR BY SESQUICHLORIDE OF IRON : *Acetic Acid, Formic Acid* (Lactic Acid, Propionic Acid, Butyric Acid).

§ 171.

a. ACETIC ACID ($HO, C_4 H_3 O_3$).

1. The HYDRATE OF ACETIC ACID forms transparent crystalline scales, which fuse at 62·6° F. to a colorless fluid of a peculiar pungent and penetrating odor, and exceedingly acid taste. When exposed to the action of heat it volatilizes completely, forming pungent inflammable vapors, which burn with a blue flame. It is miscible with water in all proportions; it is to such mixtures of the acid with water that the name of acetic acid is commonly applied. The hydrate of acetic acid is also soluble in alcohol.

2. The ACETATES undergo decomposition at a red heat; among the products of this decomposition we generally find hydrate of acetic acid, and almost invariably acetone ($C_3 H_6 O_2$). The acetates of the alkalies and alkaline earths are converted into carbonates in this process; of the acetates with metallic bases many leave the metal behind in the pure state, others in the form of oxide. Most of the residues which the acetates leave upon ignition are carbonaceous. Nearly all acetates dissolve in water and in alcohol; most of them are readily soluble in water, a few only are difficult of solution in that menstruum. If acetates are distilled with dilute sulphuric acid, the free acetic acid is obtained in the distillate.

3. If *sesquichloride of iron* is added to acetic acid, and the acid is then nearly saturated with ammonia, or if a neutral acetate is mixed with sesquichloride of iron, the fluid acquires a deep dark red color, owing to the formation of ACETATE OF SESQUIOXIDE OF IRON. By boiling the fluid becomes colorless if it contains an excess of acetate, the whole of the sesquioxide of iron precipitating as a basic acetate, in the form of brown-yellow flakes. Ammonia precipitates from it the whole of the sesquioxide of iron as hydrate. By addition of hydrochloric acid a fluid which appears red from the presence of acetate of sesquioxide of iron turns yellow (difference from sulphocyanide of iron).

4. Neutral acetates (but not free acetic acid of a certain degree of dilution) give with *nitrate of silver* white crystalline precipitates of ACETATE OF SILVER ($AgO, C_4 H_3 O_3$), which are very sparingly soluble

in cold water. They dissolve more easily in hot water, but separate again upon cooling, in the form of very fine crystals. Ammonia dissolves them readily; free acetic acid does not increase their solubility in water.

5. *Nitrate of suboxide of mercury* produces in solutions of acetic acid, and more readily still in solutions of acetates, white scaly crystalline precipitates of ACETATE OF SUBOXIDE OF MERCURY ($Hg_2 O, C_4 H_3 O_3$), which are sparingly soluble in water and acetic acid in the cold, but dissolve without difficulty in an excess of the precipitant. The precipitates dissolve in water upon heating, but separate again upon cooling, in the form of small crystals; in this process the salt undergoes partial decomposition: a portion of the mercury separates in the metallic state, and imparts a gray color to the precipitate. If the acetate of suboxide of mercury is boiled with dilute acetic acid, instead of water, the quantity of the metallic mercury which separates is exceedingly minute.

6. *Chloride of mercury* produces no precipitate of subchloride of mercury with acetic acid or acetates upon heating.

7. By heating acetates with *concentrated sulphuric acid* HYDRATE OF ACETIC ACID is evolved, which may be known by its pungent odor. But if the acetates are heated with a mixture of about equal volumes of *concentrated sulphuric acid* and *alcohol*, ACETIC ETHER ($C_4 H_5 O, C_4 H_3 O_3$) is formed. The odor of this ether is highly characteristic and agreeable; it is most distinct upon shaking the mixture when somewhat cooled, and is much less liable to lead to mistakes than the pungent odor of the free acetic acid.

8. If acetates are distilled with dilute sulphuric acid, and the distillate is digested with an excess of *oxide of lead*, part of the latter dissolves as basic acetate of lead, which may be readily recognised by its alkaline reaction.

§ 172.

b. FORMIC ACID ($HO, C_2 HO_3$).

1. The HYDRATE OF FORMIC ACID is a transparent and colorless slightly fuming liquid of a characteristic and exceedingly penetrating odor. When cooled to below 32° F., it crystallizes in colorless plates. It is miscible in all proportions with water and with alcohol. When exposed to the action of heat, it volatilizes completely; the vapors are inflammable and burn with a blue flame.

2. The FORMATES, like the corresponding acetates, leave upon ignition either carbonates, oxides, or metals behind, the process being attended with separation of charcoal, and escape of carbide of hydrogen, carbonic acid, and water. All the compounds of formic acid with bases are soluble in water; alcohol also dissolves many of them, but not all.

3. Formic acid shows the same reaction with *sesquichloride of iron* as acetic acid.

4. *Nitrate of silver* fails to precipitate free formic acid, and decomposes the alkaline formates only in concentrated solutions. The white sparingly soluble, crystalline precipitate of FORMATE OF SILVER ($AgO, C_2 HO_3$) acquires very rapidly a darker tint, owing to the separation of metallic silver. Complete reduction of the oxide of silver to the metallic state takes place, even in the cold, after the lapse of some time; but immediately upon applying heat to the fluid containing the precipitated formate of silver. The same reduction of the oxide of silver to

the metallic state takes place in a solution of free formic acid, and also in solutions of formates so dilute that the addition of the nitrate of silver fails to produce a precipitate in them. But it does not take place in presence of an excess of ammonia. The rationale of this reduction is as follows: the formic acid, which may be looked upon as a compound of carbonic oxide with water, deprives the oxide of silver of its oxygen, thus causing the formation of carbonic acid, which escapes, and of water, whilst the reduced silver separates in the metallic state.

5. *Nitrate of suboxide of mercury* gives no precipitate with free formic acid; but in concentrated solutions of alkaline formates this reagent produces a white sparingly soluble precipitate of FORMATE OF SUBOXIDE OF MERCURY (Hg_2O, C_2HO_3), which rapidly becomes gray, owing to the separation of metallic mercury. Complete reduction ensues, even in the cold, after the lapse of some time, but is immediate upon application of heat. This reduction is also attended with the formation of carbonic acid and water, and takes place, the same as with the oxide of silver, both in solutions of free formic acid and in fluids so highly dilute that the formate of suboxide of mercury is retained in solution.

6. If formic acid or an alkaline formate is heated with *chloride of mercury* to from 140° to 158° F. SUBCHLORIDE OF MERCURY precipitates. Presence of free hydrochloric acid or of somewhat considerable quantities of alkaline chlorides will prevent the reaction.

7. If formic acid or a formate is heated with *concentrated sulphuric acid* the formic acid is resolved into water and carbonic oxide gas, which latter escapes with effervescence and, if kindled, burns with a blue flame. The fluid does not turn black in this process. The rationale of the decomposition of the formic acid is this: the sulphuric acid withdraws from the formic acid the water or the oxide necessary for its existence, and thus occasions a transposition of its elements ($C_2HO_3 = 2 CO + HO$). Upon heating formates with dilute sulphuric acid in a distilling apparatus free formic acid is obtained in the distillate, and may mostly be readily detected by its odor. Upon heating a formate with a mixture of sulphuric acid and alcohol formic ether is evolved, which is characterized by its peculiar arrack-like smell.

8. If dilute formic acid is heated with *oxide of lead*, the latter dissolves. On cooling the solution, which, if necessary, is concentrated by evaporation, the FORMATE OF LEAD (PbO, C_2HO_3) separates in brilliant prisms or needles.

§ 173.

Recapitulation and remarks.—Acetic acid and formic acid may be distilled over with water, and form with sesquioxide of iron soluble neutral salts which dissolve in water, imparting to the fluid a blood-red color, and are decomposed upon boiling. These reactions distinguish the two acids of the third group from the other organic acids. From each other the two acids are distinguished by the odor of their hydrates and ethyle compounds, and by their different reactions with salts of silver and salts of mercury, oxide of lead, and concentrated sulphuric acid. The separation of acetic acid from formic acid is effected by heating the mixture of the two acids with an excess of oxide of mercury or oxide of silver. Formic acid reduces the oxides, and suffers decomposition, being resolved into carbonic oxide and water; whilst the acetic acid combines with the oxides, forming acetates, which remain in solution.

§ 174.

Rarer Acids of the third group of Organic Acids.

1. LACTIC ACID ($2HO, C_{12}H_{10}O_{10}$).

Lactic acid is developed in animal fluids, vegetable matters that have turned sour, &c. Pure hydrate of lactic acid is an inodorous syrupy liquid; it has a sour and biting taste. When it is slowly heated in a retort to 266° F., water containing a little hydrated lactic acid distils over, leaving a residue of anhydrous lactic acid ($C_{12}H_{10}O_{10}$), which at a temperature between 482° F. and 572° F. is decomposed into carbonic oxide, carbonic acid, lactide, and other products.—Hydrate of lactic acid dissolves freely in water, alcohol, and ether. Upon boiling the aqueous solution a little lactic acid volatilizes along with the aqueous vapour. All the lactates are soluble in water, the greater part of them, however, only sparingly; it is the same with regard to spirit of wine; but they are all insoluble in ether. The production of some of these salts and the inspection and examination of their form under the microscope supply the means for the detection of lactic acid; lactate of lime and lactate of zinc are the best suited for this purpose. Lactate of lime may be conveniently prepared from animal or vegetable liquids by the following method devised by SCHERER:—Dilute the liquid, if necessary, with water, mix with baryta-water, and filter. Distil the filtrate with some sulphuric acid (to remove volatile acids), digest the residue several days with strong alcohol, distil the acid solution with a little milk of lime, filter warm from the excess of lime and the sulphate of lime, conduct carbonic acid into the filtrate, heat once more to boiling, filter from the precipitated carbonate of lime, evaporate the filtrate, warm the residue with strong alcohol, filter, and let the neutral filtrate stand several days to give the lactate of lime time to crystallize. Should the quantity of lactic acid present be insufficient to allow the formation of crystals, evaporate the fluid to syrupy consistence, mix with strong alcohol, let the mixture stand some time, decant or filter the alcoholic solution into a vessel that can be closed, and add gradually a small quantity of ether. The process will cause even minute traces of lactate of lime to separate from the fluid. Lactate of lime shows under the microscope the form of minute crystalline needles aggregated in tufts with short stalks, pairs of them always being joined at the stalked ends, so as to look like brushes united together.—Lactate of zinc deposited quickly from its solution shows under the microscope the form of spherical groups of needles. The slow evaporation of solution of lactate of zinc gives first crystals resembling clubs truncated at both ends; these crystals gradually increase in size; the two ends apparently diminish, whilst the middle parts increase in size (FUNKE).

2. PROPIONIC ACID ($HO, C_6H_5O_3$), and BUTYRIC ACID ($HO, C_8H_7O_3$).

PROPIONIC ACID is formed under a great variety of circumstances; it is chiefly found in fermented liquids. The pure hydrate of the acid crystallizes in minute plates; it boils at from 284° F. to 287·6° F.; it dissolves readily in water. Propionic acid floats as an oily stratum on aqueous solution of phosphoric acid and on solution of chloride of calcium. It has a peculiar smell, which reminds both of butyric and acetic acid. Upon distilling the aqueous solution, the propionic acid passes over into the distillate. BUTYRIC ACID is frequently found in animal and vegetable matter, more particularly also in fermented liquids of various kinds. The pure hydrate of the acid is a colorless, mobile, corrosive, intensely sour fluid, of a disagreeable odor, a combination of the smell of rancid butter and acetic acid; it boils at 320° F. It is miscible in all proportions with water and alcohol. It is separated from the concentrated aqueous solution by chloride of calcium, concentrated acids, &c., in the form of a fluid oil. The smell of butyric acid is particularly strong in the aqueous solution of the acid. Upon distilling the aqueous solution, the acid passes over along with the aqueous vapors.

Propionic acid and butyric acid are often found associated with formic acid and acetic acid in fermented liquids, in guano, and in many mineral waters. The detection of the several acids may in such cases be effected as follows: dilute the substance sufficiently with water, acidify with sulphuric acid, and distil; saturate the distillate with baryta-water, evaporate to dryness, and treat the residue repeatedly with boiling alcohol of 85 per cent. This will leave the formate of baryta and part of the acetate, the remainder of the acetate, together with the propionate and butyrate, dissolving in the alcohol. Evaporate the alcoholic solution, dissolve the residue in water, decompose cautiously with sulphate of silver, boil, filter, and let the fluid (which ought rather to contain a little undecomposed baryta salt than any sulphate of silver) evaporate under the desiccator. Take out separately the crystals which form first, those which form after, and those which form last, and examine them to ascertain their nature. Acetate of silver emits upon solution in concentrated sulphuric acid the odor of acetic acid, and gives no oily drops; propionate and butyrate of silver emit the peculiar odor of the acids, and give oily drops, which, however, with minute quantities are visible only under the microscope. To distinguish positively between propionic and butyric acid, it is indispensable to determine the amount of silver in the separated silver salts, and to fix by this the atomic weight of the acids. If much acetate of baryta has passed into the solution, with a small quantity only of butyrate and propionate, the baryta is first completely precipitated with sulphuric acid from the aqueous solution of the baryta salts soluble in alcohol, half of the acid fluid neutralized with soda, the other half added, the fluid then distilled, the distillate, which now contains principally propionic and butyric acids, saturated with baryta, then decomposed with sulphate of silver, and the remaining part of the process conducted as above.

PART II.

SYSTEMATIC COURSE

OF

QUALITATIVE CHEMICAL ANALYSIS.

PART II.

PRELIMINARY REMARKS

ON THE

COURSE OF QUALITATIVE ANALYSIS IN GENERAL, AND ON THE PLAN OF THIS PART OF THE PRESENT WORK IN PARTICULAR.

THE knowledge of reagents and of the deportment and reaction of other bodies with them enables us to ascertain at once whether a simple compound of which the physical properties permit an inference as to its nature, is in reality what we suspect it to be. Thus, for instance, a few simple reactions suffice to show whether a body which appears to be calcareous spar is really carbonate of lime, and that another which we hold to be gypsum *is* actually sulphate of lime. This knowledge usually suffices also to ascertain whether a certain body is present or not in a compound; for instance, whether or not a white powder contains subchloride of mercury. But if our design is to ascertain the chemical nature of a substance entirely unknown to us—if we wish to discover *all* the constituents of a mixture or chemical compound—if we intend to prove that, besides certain bodies which we have detected in a mixture or compound, no other substance *can* possibly be present—if consequently a *complete qualitative analysis* is our object, the mere knowledge of the reagents, and of the reactions of other bodies with them, will not suffice for the attainment of this end; this requires the additional knowledge of a systematic and progressive course of analysis, in other words, the knowledge of the *order* and *succession* in which solvents, and general and special reagents, should be applied, both to effect the speedy and certain detection of every component element of a compound or mixture, and to prove with certainty the absence of all other substances. If we do not possess the knowledge of this systematic course, or if, in the hope of attaining our object more rapidly, we adhere to no method whatever in our investigations and experiments, analyzing becomes (at least in the hands of a novice) mere guess-work, and the results obtained are no longer the fruits of scientific calculation, but mere matters of accident, which sometimes may prove lucky hits, and at others total failures.

Every analytical investigation must therefore be based upon a definite method. But it is not by any means necessary that this method should be the same in all cases. Practice, reflection, and a due attention to circumstances will, on the contrary, generally lead to the adoption of different methods for different cases. However, all analytical methods agree in this, that the substances present or supposed to be present in a compound or mixture are in the first place classed into certain groups, which are then again subdivided, until the individual detection of the various substances present is finally accomplished. The diversity of analytical methods depends partly on the order and succession in which reagents are applied, and partly on their selection.

Before we can venture upon inventing methods of our own for individual cases, we must first make ourselves thoroughly conversant with a certain definite course or system of chemical analysis in general. This system must have passed through the ordeal of experience, and must be adapted to every imaginable case, so that afterwards, when we have acquired some practice in analysis, we may be able to determine which modification of the general method will in certain given cases most readily and rapidly lead to the attainment of the object in view.

The exposition of such a systematic course, adapted to all cases, tested by experience, and combining simplicity with the greatest possible security, is the object of the *First Section* of the second part of this work.

The elements and compounds comprised in it are the same which we have studied in Part I., with the exception of those discussed more briefly, and marked by the use of smaller type.

The *First Section* of the Second Part consists of PRACTICAL INSTRUCTIONS IN ANALYSIS, wherein I have laid down a systematic course which, with due care and attention, will, by progressive steps, lead speedily and safely to the attainment of the end in view.

The subdivisions of this practical course are, 1, Preliminary examinations; 2, Solution; 3, Actual examination.

The third subdivision (the *actual examination*) is again subdivided into, (1) Examination of compounds in which but *one* base and *one* acid are assumed to be present; and, (2) Examination of mixtures or compounds in which *all* the substances treated of in the present work are assumed to be present. With respect to the latter section I have to remark that where the preliminary examination has not clearly demonstrated the absence of certain groups of substances, the student cannot safely disregard any of the paragraphs to which reference is made in consequence of the reactions observed. In cases where the intention is simply to test a compound or mixture for certain substances, and not to ascertain all its constituents, it will be easy to select the particular numbers which ought to be attended to.

As the construction of a universally applicable systematic course of analysis requires due regard to, and provision for, every contingency that may possibly arise, it is self-evident that, though in the system here laid down the various bodies comprised in it have been assumed to be mixed up together in every conceivable way, it was absolutely indispensable to proceed throughout upon the supposition that no foreign organic matters whatever were present, since the presence of such matters would of course tend to prevent or obscure many reactions, and variously modify others.

Although the general analytical course laid down here is devised and arranged in a manner to suit all possible contingencies, with a very few exceptions, still there are special cases in which it may be advisable to modify it. A preliminary treatment of the substance is also sometimes necessary, before the actual analysis can be proceeded with; the presence of coloring or slimy organic matters more especially requires certain preliminary operations.

The *Second Section* of this Part will be found to contain a detailed description of the special methods employed to effect the analysis of a few important compounds and mixtures which chemists are frequently called upon to examine. Some of these methods show how the analytical processes become simplified as the number of substances decreases to which regard must be had in the analysis.

In conclusion, as an intelligent and successful pursuit of analysis is possible only with an accurate knowledge of the principles whereon the detection and separation of bodies depend, since this knowledge alone can furnish the student with a guide to the selection of the proper reagents, and the order in which they ought to be applied, I have given in the *Third Section* of the Second Part an explanation and elucidation of the general analytical process, with numerous additions to the practical operations. As this third section may properly be regarded as the key to the first and second sections, I strongly recommend students to make themselves early and thoroughly acquainted with it. I have devoted a special section to this theoretical explanation of the process, as I think it will be understood better in a connected form than it would have been by explanatory additions to the several paragraphs, which, moreover, might have materially interfered with the plainness and perspicuity of the plan of the practical process.

I have also in this third section taken occasion to point out in what residues, solutions, precipitates, &c., which are obtained in the systematic course of analysis, the more rarely occurring elements may be expected to be met with; and also to give instructions how to proceed with a view to ensure the detection of these bodies also systematically.

SECTION I.

PRACTICAL PROCESS FOR THE ANALYSIS OF COMPOUNDS AND MIXTURES IN GENERAL.

I. Preliminary Examination.*

§ 175.

1. EXAMINE, in the first place, the physical properties, such as the color, shape, hardness, gravity, odor, &c., of the substance intended for analysis, since these will often enable you in some measure to infer its nature. Before proceeding to the application of any chemical process, you must always consider how much of the substance to be analyzed you have at command, since it is necessary, at this early period of the examination, to calculate the quantity which may safely be used in the preliminary investigation. A reasonable economy is in all cases advisable, even though you may possess the substance in large quantities;

1†

---

* Consult also the observations and additions in the Third Section of Part II.
† These marginal numbers are simply intended to facilitate reference.

but, under all circumstances, let it be a fixed rule never to use at once the whole of what you possess of a substance, but always to keep a portion of it for unforeseen contingencies, and for confirmatory experiments.

A. THE BODY UNDER EXAMINATION IS SOLID.

I. IT IS NEITHER A PURE METAL NOR AN ALLOY.

§ 176.

1. The substance is fit for examination if in powder or in minute crystals; but in the case of larger crystals or solid pieces, a portion must, if practicable, be first reduced to *fine powder*. Bodies of the softer kind may be triturated in a porcelain mortar; those of a harder nature must first be broken into small pieces in a steel mortar, or upon a steel anvil, and the pieces then be triturated in an agate mortar.

2. PUT SOME OF THE POWDER INTO A GLASS TUBE, SEALED AT ONE END, ABOUT SIX CENTIMETRES LONG AND FIVE MILLIMETRES WIDE, AND HEAT first gently over the spirit or gas-lamp, then intensely in the blowpipe flame. The reactions resulting may lead to many positive or probable conclusions regarding the nature of the substance. The following are the most important of these reactions, to which particular attention ought to be paid; it often occurs that several of them are observed in the case of one and the same substance.

a. THE SUBSTANCE REMAINS UNALTERED: absence of organic matters, salts containing water of crystallization, readily fusible matters, and volatile bodies.

b. THE SUBSTANCE DOES NOT FUSE AT A MODERATE HEAT, BUT SIMPLY CHANGES COLOR. From white to yellow, turning white again on cooling, indicates OXIDE OF ZINC; from white to yellowish brown, turning to a dirty light yellow on cooling, indicates BINOXIDE OF TIN; if the color changes from white to brownish-red, turning to yellow on cooling, and the body is fusible at a red heat, this indicates the presence of OXIDE OF LEAD; if the color changes from white, or pale yellow, to orange yellow, or to a deeper and more reddish tint up to reddish-brown, turning pale yellow on cooling, and the body fuses at an intense red heat, this indicates the presence of TEROXIDE OF BISMUTH; if the color changes from red to black, turning reddish-brown again on cooling, this indicates the presence of SESQUIOXIDE OF IRON; if the color changes from yellow to dark orange, and the body fuses at an intense heat, this indicates NEUTRAL CHROMATE OF POTASSA, &c.

c. THE SUBSTANCE FUSES WITHOUT EXPULSION OF AQUEOUS VAPORS. If by intense heating, gas (oxygen) is evolved, and a small fragment of charcoal thrown in is energetically consumed, NITRATES or CHLORATES may be assumed to be present.

d. AQUEOUS VAPORS ARE EXPELLED, WHICH CONDENSE IN THE COLDER PART OF THE TUBE: this indicates the presence either (α) of SUBSTANCES CONTAINING WATER OF CRYSTALLIZATION, in which case they will generally readily fuse, and re-solidify after expulsion of the water; many of these swell considerably whilst yielding up their water, *e. g.* (borax, alum); or (β) of decomposable

HYDRATES, in which case the bodies often will not fuse; or (γ) of anhydrous salts, holding water MECHANICALLY ENCLOSED between their lamellæ—in which case the bodies will decrepitate; or (δ) of bodies with moisture externally adhering to them.

Test the reaction of the condensed fluid in the tube; if it is alkaline, ammonia may be assumed to be present; if acid, a volatile acid (sulphuric acid, sulphurous acid, hydrofluoric acid, hydrochloric, hydrobromic, hydriodic acids, nitric acid, &c.).

e. GASES OR FUMES ESCAPE. Observe whether they have a color, a smell, an acid or alkaline reaction, whether they are inflammable, &c.

8

aa. OXYGEN. The disengagement of this gas indicates the presence of peroxides, chlorates, nitrates, &c. A glimmering slip of wood is relighted in the gaseous current.

bb. SULPHUROUS ACID. This is often produced by the decomposition of sulphates; it may be known by its peculiar odor and by its acid reaction.

cc. HYPONITRIC ACID, resulting from the decomposition of nitrates, especially with oxides of the heavy metals; it may be known by the brownish-red color of the fumes.

dd. CARBONIC ACID. The evolution of carbonic acid indicates the presence of carbonates decomposable by heat. The gas evolved is colorless and tasteless, non-inflammable; a drop of limewater on a watch-glass becomes turbid on exposure to the gaseous current.

ee. CARBONIC OXIDE GAS. The escape of this gas indicates the presence of oxalates and also of formates. The gas burns with a blue flame. In the case of oxalates the carbonic oxide evolved is generally mixed with carbonic acid, and is therefore more difficult to kindle: in the case of formates the evolution of the gas is attended with marked carbonization. Oxalates evolve carbonic acid when brought into contact with binoxide of manganese, a little water, and some concentrated sulphuric acid, on a watch-glass; formates evolve no carbonic acid under similar circumstances.

ff. CYANOGEN. The evolution of cyanogen gas denotes the presence of cyanides decomposable by heat. The gas may be known by its odor, and by the crimson flame with which it burns.

gg. HYDROSULPHURIC ACID GAS. The escape of hydrosulphuric acid gas indicates the presence of sulphides containing water; the gas may be readily known by its odor.

hh. AMMONIA, resulting from the decomposition of ammoniacal salts, or also of cyanides or nitrogenous organic matters, in which latter cases browning or carbonization of the substance takes place, and either cyanogen or offensive empyreumatic oils escape with the ammonia.

f. A SUBLIMATE FORMS. This indicates the presence of volatile bodies: the following are those more frequently met with :—

9

aa. SULPHUR. Eliminated from mixtures or from many of the metallic sulphides. Sublimes in reddish-brown drops, which solidify on cooling, and turn yellow or yellowish-brown.

bb. AMMONIA SALTS give white sublimates; heated with soda and a drop of water on platinum-foil they evolve ammonia.

cc. MERCURY and compounds of mercury. METALLIC MERCURY forms globules; SULPHIDE OF MERCURY is black, but acquires a red tint when rubbed; CHLORIDE OF MERCURY fuses before volatilizing; SUBCHLORIDE OF MERCURY sublimes without previous fusion; the sublimate, which is yellow whilst hot, turns white on cooling. The red IODIDE OF MERCURY gives a yellow sublimate.

dd. ARSENIC and compounds of that metal. METALLIC ARSENIC forms the well-known arsenical mirror; ARSENIOUS ACID forms small shining crystals; the SULPHIDES OF ARSENIC give sublimates which are reddish-yellow whilst hot, and turn yellow on cooling.

ee. TEROXIDE OF ANTIMONY fuses to a yellow liquid before subliming. The sublimate consists of brilliant needles.

ff. BENZOIC ACID and SUCCINIC ACID. These may be known by the odor of their fumes.

gg. HYDRATED OXALIC ACID. White crystalline sublimate, thick fumes in the tube. Heating a small sample on platinum-foil with a drop of concentrated sulphuric acid gives rise to a copious evolution of gas.

g. CARBONIZATION TAKES PLACE: organic substances. This **10** is always attended with evolution of gases (acetates evolve acetone) and water, which latter has an alkaline or acid reaction. If the residue effervesces with acids, whilst the original substance did not show this reaction, organic acids may be assumed to be present in combination with alkalies or alkaline earths.

3. PLACE A SMALL PORTION OF THE SUBSTANCE ON A CHARCOAL **11** SUPPORT (in the cavity scooped out for the purpose), AND EXPOSE TO THE INNER BLOWPIPE FLAME.

As most of the reactions described under 2 (3—10) are also produced by this process, I will here enumerate only those which result more particularly and exclusively from its application. Evolution of sulphurous acid, when the flame plays upon the sample or the coal, generally indicates the presence of a sulphide. The following are the reactions which will permit pretty accurate conclusions.

a. THE BODY FUSES, AND IS ABSORBED BY THE CHARCOAL **12** OR FORMS A BEAD IN THE CAVITY, not attended by incrustation: this denotes more particularly the presence of salts of the alkalies.

b. AN INFUSIBLE WHITE RESIDUE REMAINS on the charcoal, **13** either at once or after previous melting in the water of crystallization. This indicates more particularly the presence of baryta, strontia, lime, magnesia, alumina, oxide of zinc (which appears yellow whilst hot), and silicic acid. Among these substances STRONTIA, LIME, MAGNESIA, and OXIDE OF ZINC are distinguished by strong luminosity in the blowpipe flame. Moisten the white residue with a drop of solution of nitrate of protoxide of cobalt, and expose again to a strong heat. If the mass assumes a fine blue tint this indicates the presence of ALUMINA; if a reddish tint, of MAGNESIA; if a green color, of OXIDE OF ZINC. If SILICIC ACID is present the mass also assumes a faint bluish tint, which must not be confounded with that proceeding from the presence of alumina.

In the case *a* or *b* the preliminary examination for alkalies and alkaline earths may be completed by inspecting the colors which the substances impart to flame. For this purpose a little of the substance is attached to the loop of a fine platinum wire, moistened repeatedly with sulphuric acid, dried cautiously near the border of the flame, and then held in the fusing zone of BUNSEN's gas-flame. The colorations caused by the alkalies will make their appearance first, followed—after volatilization of the alkalies, by that of baryta, and finally—after moistening with hydrochloric acid—by those of strontia and lime. For details see § 92 and § 99.

14 *c.* THE SUBSTANCE LEAVES AN INFUSIBLE RESIDUE OF ANOTHER COLOR, OR REDUCTION TO THE METALLIC STATE TAKES PLACE, OR AN INCRUSTATION FORMS ON THE CHARCOAL. Mix a portion of the powder with carbonate of soda, and heat on charcoal in the reducing flame; observe the residue in the cavity, as well as the incrustation on the charcoal.

15 *a.* The sustained application of a strong flame produces a metallic globule, without incrustation of the charcoal; this indicates the presence of GOLD or COPPER. The oxides of platinum, iron, cobalt, and nickel are indeed also reduced, but they yield no metallic globules.

16 *β.* The charcoal support is coated with an incrustation, either with or without simultaneous formation of a metallic globule.

aa. The incrustation is *white*, at some distance from the test specimen, and is very readily dissipated by heat, emitting a garlic-like odor: ARSENIC.

bb. The incrustation is *white*, nearer the test specimen than in *aa*, and may be driven from one part of the support to another: ANTIMONY. Metallic globules are generally observed at the same time, which continue to evolve white fumes long after the blowpipe jet is discontinued, and upon cooling become surrounded with crystals of teroxide of antimony; the globules are brittle.

cc. The incrustation is *yellow* whilst hot, but turns white on cooling; it is pretty near the test specimen, and is with difficulty volatilized: ZINC.

dd. The incrustation has a *faint yellow* tint whilst hot, but turns white on cooling; it surrounds the test specimen, and both the inner and outer flame fail to volatilize it: TIN. The metallic globules formed at the same time, but only in a strong reducing flame, are bright, readily fusible, and ductile.

ee. The incrustation has a *lemon-yellow* color, turning on cooling to sulphur-yellow; heated in the reducing flame it leaves its place with a blue gleam: LEAD. Readily fusible, ductile globules are formed at the same time with the incrustation.

ff. The incrustation is of a *dark orange-yellow* color whilst hot, which changes to lemon-yellow on cooling; heated in the reducing flame it leaves its place without a blue gleam: BISMUTH. The metallic globules formed at the same time as the incrustation are readily fusible and brittle.

gg. The incrustation is reddish-brown, in thin layers, orange-yellow; it volatilizes without a colored gleam: CADMIUM.

hh. The incrustation is dark-red: SILVER. Where lead and antimony are present at the same time, the incrustation is crimson. In cases where a reduction to the metallic state has taken place, moisten the sample with water, scoop it out, triturate in a small agate dish, and wash off the charcoal particles with water—when the gold will be obtained in yellow, the copper in coppery-red, the silver in nearly white, the tin in grayish-white, the lead in whitish-gray minute plates or strips, the bismuth as a reddish-gray, the zinc as a bluish-white, the antimony as a gray powder. When copper and tin, or copper and zinc, are present at the same time, yellow alloys are occasionally formed.

4. FUSE A SMALL PORTION TOGETHER WITH A BEAD OF MICROCOSMIC SALT, AND EXPOSE FOR SOME TIME TO THE OUTER FLAME OF THE BLOWPIPE. 17

    a. THE SUBSTANCE DISSOLVES READILY AND RATHER COPIOUSLY TO A CLEAR BEAD (WHILST HOT).

        a. The hot bead is colored: 18

BLUE, by candlelight inclining to violet—COBALT;

GREEN, upon cooling blue; in the reducing flame, after cooling, red—COPPER;

GREEN, particularly fine on cooling, unaltered in the reducing flame—CHROMIUM;

BROWNISH-RED, on cooling light yellow or colorless; in the reducing flame red whilst hot, yellow whilst cooling, then greenish—IRON;

DARK-YELLOW to REDDISH, turning lighter or altogether colorless on cooling; in the reducing flame unaltered—NICKEL;

YELLOWISH-BROWN, on cooling changing to light yellow or losing its color altogether; in the reducing flame almost colorless (especially after contact with tin), blackish-gray on cooling—BISMUTH;

BRIGHT YELLOWISH to OPAL, when cold rather dull; in the reducing flame whitish-gray—SILVER;

AMETHYST-RED, especially on cooling; colorless in the reducing flame, not quite clear—MANGANESE.

        β. *The hot bead is colorless:* 19

IT REMAINS CLEAR ON COOLING: ANTIMONY, ALUMINA, ZINC, CADMIUM, LEAD, LIME, MAGNESIA; the latter five metals, when added in somewhat large proportion to the microcosmic salt, give enamel-white beads; the bead of oxide of lead saturated is yellowish;

IT BECOMES ENAMEL-WHITE ON COOLING, even where only a smaller portion of the powder has been added to the microcosmic salt: BARYTA, STRONTIA.

    b. THE SUBSTANCE DISSOLVES SLOWLY AND ONLY IN SMALL QUANTITY: 20

        a. The bead is colorless, and remains so even after cooling; the undissolved portion looks semi-transparent; upon addition of a little sesquioxide of iron it acquires the characteristic color of an iron bead : SILICIC ACID.

β. The bead is colorless, and remains so after addition of a little sesquioxide of iron : TIN.

c. THE SUBSTANCE DOES NOT DISSOLVE, BUT FLOATS (IN THE METALLIC STATE) IN THE BEAD : GOLD, PLATINUM. 21

5. MINERALS ARE EXAMINED FOR FLUORINE AS DIRECTED' § 146, 8.

As the body under examination may consist of a mixture of the most dissimilar matters, it is impossible to give well defined cases that shall offer at the same time the advantage of general applicability. If, therefore, reactions are observed in an experiment which proceed from a combination of two or several cases, the conclusions drawn from these reactions must of course be modified accordingly.

After the termination of the preliminary examination, proceed to the solution of the substance, as directed § 180 (32).

§ 177.

II. THE SUBSTANCE IS A METAL OR AN ALLOY.

1. HEAT A SMALL PORTION OF THE SUBSTANCE WITH WATER ACIDULATED WITH ACETIC ACID. 22

If HYDROGEN GAS is evolved this indicates the presence of a light metal (possibly also of manganese in the metallic state).

2. HEAT A SAMPLE OF THE SUBSTANCE ON CHARCOAL IN THE REDUCING FLAME OF THE BLOWPIPE, and watch the reactions; for instance, whether the substance fuses, whether an incrustation is formed, or an odor emitted, &c. 23

By this operation the following metals may be detected with greater or less certainty : ARSENIC by the smell of garlic; MERCURY by its fluidity ; ANTIMONY, ZINC, LEAD, BISMUTH, CADMIUM, TIN, SILVER, by fusing, with incrustation of the charcoal (comp. 16); COPPER by the green coloration of the outer flame. Further conclusions can be formed only in cases where the substance contains only one metal, and this in a state of purity or approaching purity ; thus, for instance, GOLD fuses without incrustation ; PLATINUM, IRON, MANGANESE, NICKEL, and COBALT, if pure, do not fuse in the blowpipe flame.

3. HEAT A SAMPLE OF THE SUBSTANCE BEFORE THE BLOWPIPE IN A GLASS TUBE SEALED AT ONE END. 24

a. NO SUBLIMATE IS FORMED IN THE COLDER PART OF THE TUBE : absence of mercury.

b. A SUBLIMATE IS FORMED : presence of MERCURY, CADMIUM, or ARSENIC. The sublimate of mercury, which consists of small globules, cannot possibly be confounded with that of cadmium or arsenic.

After the termination of the preliminary examination, proceed to the solution of the substance as directed § 181 (42).

§ 178.

B. THE SUBSTANCE UNDER EXAMINATION IS A FLUID.

1. EVAPORATE A SMALL PORTION OF THE FLUID in a platinum dish, or in a small porcelain crucible, to ascertain whether it actually 25

contains any matter in solution; if a residue remains, examine this as directed § 176.

2. TEST WITH LITMUS-PAPER (blue and red). 26

   a. THE FLUID REDDENS BLUE LITMUS-PAPER. This reaction may be caused by a free acid or an acid salt, as well as by a metallic salt soluble in water. To distinguish between these two cases, pour a small quantity of the fluid into a watch-glass, and dip into it a small glass rod, after moistening the extreme point of the latter with dilute solution of carbonate of soda; if the fluid remains clear, or if the precipitate which may form at first, redissolves upon stirring the liquid, this proves the presence of a free acid or of an acid salt; but if the fluid becomes turbid and remains so, this generally denotes the presence of a soluble metallic salt.

   b. REDDENED LITMUS-PAPER TURNS BLUE: this indicates the 27 presence of a free alkali or an alkaline carbonate, free alkaline earths, alkaline sulphides, and of a number of other salts containing an alkali or, may be, an alkaline earth, in combination with a weak acid.

3. SMELL THE FLUID, or, should this fail to give satisfactory 28 results, DISTIL, to ascertain whether the simple solvent present is water, alcohol, ether, &c. If you find it is not water, evaporate the solution to dryness, and treat the residue as directed § 176.

4. If the solution is aqueous, and manifests an acid reaction, 29 DILUTE A PORTION OF IT LARGELY WITH WATER. Should this impart a milky and turbid appearance to it, the presence of ANTIMONY, BISMUTH (possibly also of tin) may be inferred. Comp. § 121, 9, and § 131, 4.

After the termination of the preliminary examination, proceed to 30 the actual examination. If the solution is aqueous, with neutral reaction, it can only contain substances soluble in water; but if it has an acid reaction, arising from the presence of free acid, the actual examination must be conducted with due regard to the possible presence also of bodies soluble in acids, though insoluble in water. Proceed accordingly with neutral aqueous solutions as directed § 182, with acid solutions as directed § 185, if you are quite sure that there is only one acid and one base present; but where there is reason to suppose the presence of several bases and acids, proceed as directed § 189. With fluids of alkaline reaction, proceed as directed § 182, unless there be reason to suppose the presence of more than one acid and more than one base, when the instructions given in § 189 must be followed.

II. SOLUTION OF BODIES, OR CLASSIFICATION OF SUBSTANCES, ACCORDING TO THEIR DEPORTMENT WITH CERTAIN SOLVENTS.*

§ 179.

Water and acids (hydrochloric acid, nitric acid, aqua regia) are 31 the solvents used to classify simple or compound substances, and to isolate the component parts of mixtures. We divide the various substances into three classes, according to their respective deportment with these solvents.

First class.—SUBSTANCES SOLUBLE IN WATER.

* Consult the remarks in the third section.

Second class.—SUBSTANCES INSOLUBLE OR SPARINGLY SOLUBLE IN WATER, BUT SOLUBLE IN HYDROCHLORIC ACID, NITRIC ACID, OR AQUA REGIA.

Third class.—SUBSTANCES INSOLUBLE OR SPARINGLY SOLUBLE IN WATER AS WELL AS IN HYDROCHLORIC ACID, NITRIC ACID, AND AQUA REGIA.

The solution of alloys being more appropriately effected in a different manner from that pursued with other bodies, I shall give a special method for these substances (see § 181).

The process of solution is conducted in the following manner.

A. THE SUBSTANCE UNDER EXAMINATION IS NEITHER A METAL NOR AN ALLOY.

§ 180.

1. Put about a gramme (15·45 grains) of the finely pulverized **32** substance under examination into a small flask or a test-tube, add from ten to twelve times the amount of distilled water, and heat to boiling over a spirit or gas-lamp.

a. THE SUBSTANCE DISSOLVES COMPLETELY. In that case it **33** belongs to the first class; regard must be had to what has been stated in the chapter on the preliminary examination (30) with respect to reactions. Treat the solution either as directed § 182, or as directed § 189, according as either one or several acids and bases are supposed to be present.

b. AN INSOLUBLE RESIDUE REMAINS, EVEN AFTER PROTRACTED **34** BOILING. Let the residue subside, and filter the fluid off, if practicable in such a manner as to retain the residue in the test-tube; evaporate a few drops of the clear filtrate on platinum foil; if nothing remains, the substance is completely insoluble in water; in which case proceed as directed (35). But if a residue remains, the substance is at least partly soluble; in which case boil again with water, filter, add the filtrate to the original solution, and treat the fluid, according to circumstances, either as directed § 182, or as directed § 189. Wash the residue with water, and proceed as directed (35).

2. Treat a small portion of the residue which has been boiled **35** with water (34) with dilute hydrochloric acid. If it does not dissolve, heat to boiling, and if this fails to effect complete solution, decant the fluid into another test-tube, boil the residue with concentrated hydrochloric acid, and, if it dissolves, add the solution to the fluid in the other test-tube.

The reactions which may manifest themselves in this operation, and which ought to be carefully observed, are, (a) Effervescence, which indicates the presence of carbonic acid or hydrosulphuric acid; (β) Evolution of chlorine, which indicates the presence of peroxides, chromates, &c.; (γ) Emission of the odor of hydrocyanic acid, which indicates the presence of insoluble cyanides. The analysis of the latter bodies being effected in a somewhat different manner, a special paragraph will be devoted to them (see § 204).

a. THE RESIDUE IS COMPLETELY DISSOLVED BY THE HYDRO- **36** CHLORIC ACID (except perhaps that sulphur separates, which may be known by its color and light specific gravity, and may, after boiling some time longer, be removed by filtration; or that gelatinous hydrate of silicic acid separates). Proceed, according as there is reason to suppose the presence of one or of several bases and acids, either as directed § 185, or as directed § 190, after previous filtration if necessary. The body belongs to the second class. To make quite sure of the actual nature of the sulphur or hydrated silicic acid filtered off, examine these residuary matters as directed § 188, or as directed § 203.

37   *b.* THERE IS STILL A RESIDUE LEFT. In that case put aside the test-tube containing the specimen which has been boiled with the hydrochloric acid, and try to dissolve another sample of the substance insoluble in water, or already extracted with water, by boiling with nitric acid, and subsequent addition of water. Evolution of nitric oxide, or nitrous acid, by the action of the nitric acid, shows that a process of oxidation is taking place. And for the rest proceed as in (36).

38   *a. The sample is completely dissolved, or leaves no other residue but sulphur or the gelatinous hydrate of silicic acid;* in this case also the body belongs to the second class. Use this solution to test further for bases, as directed § 185, or, as the case may be, § 189, III. (**109.**)

39   *β. After boiling with nitric acid there is still a residue left.*
Pass on to (**40**).

40   3. If the residue insoluble in water will not entirely dissolve in hydrochloric acid nor in nitric acid, try to effect complete solution of it by means of nitro-hydrochloric acid. To this end mix the contents of the tube treated with nitric acid with the contents of the tube treated with concentrated hydrochloric acid; heat the mixture to boiling, and should this fail to effect complete solution, decant the clear fluid off from the undissolved residue, boil the latter for some time with concentrated nitro-hydrochloric acid, and add the decanted solution in dilute aqua regia as well as the solution in dilute hydrochloric acid decanted in (**35**). Heat the entire mixture once more to boiling, and observe whether complete solution has now been effected, or whether the action of the concentrated nitro-hydrochloric acid has still left a residue. In the *latter* case filter the solution—if necessary after addition of some water*—wash the residue with boiling water, and proceed with the filtrate, and the washings added to it, as directed § 185, or as directed § 190;—in the *former* case proceed with the clear solution in the same way.†

41   4. If boiling nitro-hydrochloric acid has left an undissolved residue,

---

* If the fluid turns turbid upon addition of water, this indicates the presence of bismuth or antimony; the turbidity will disappear again upon addition of hydrochloric acid.

† Where the acid solution on cooling deposits acicular crystals, the latter generally consist of chloride of lead; it is in that case often advisable to decant the fluid off the crystals, and to examine the fluid and crystals separately. Where on boiling with aqua regia metastannic chloride has been formed from binoxide of tin, the washing water, dissolving this, becomes turbid on dropping into the strongly acid fluid which has run off first. In that case receive the washing water in a separate vessel, and treat the two solutions separately with hydrosulphuric acid as directed in § 190, but filter afterwards through the same filter.

wash it thoroughly with water, and then proceed as directed § 188, or as directed § 203, according as there is reason to suppose the presence of only one or several bases and acids.

B. The Substance under Examination is a Metal or an Alloy.

§ 181.

The metals are best classed according to their respective behaviour with nitric acid: this gives us,

I. METALS WHICH ARE NOT ATTACKED BY NITRIC ACID: gold, platinum.

II. METALS WHICH ARE OXIDIZED BY NITRIC ACID, BUT OF WHICH THE OXIDES DO NOT DISSOLVE IN AN EXCESS OF THE ACID NOR IN WATER: antimony, tin.

III. METALS WHICH ARE OXIDIZED BY NITRIC ACID AND CONVERTED INTO NITRATES, WHICH DISSOLVE IN AN EXCESS OF THE ACID OR IN WATER: all the other metals.

Pour nitric acid of 1·20 sp. gr. over a small portion of the metal or alloy under examination, and apply heat.

1. COMPLETE SOLUTION TAKES PLACE, EITHER AT ONCE OR UPON ADDITION OF WATER; this proves the absence of platinum,* gold, antimony,† and tin. Proceed either as directed § 185, or as instructed § 189, III. (109), according as there is reason to suppose the presence of only one or of several metals.

2. A RESIDUE IS LEFT.

   a. A metallic residue. Filter, and treat the filtrate as directed § 189, III. (109), after having seen, in the first place, whether anything has really been dissolved. Wash the residue thoroughly, to free it from all dissolved metals, dissolve in nitro-hydrochloric acid, and test the solution for GOLD and PLATINUM, according to the instructions given in § 128.

   b. A white pulverulent residue; this indicates the presence of ANTIMONY and TIN. Filter, ascertain whether anything has been dissolved, then treat the filtrate as directed § 189, III. (109). Wash the residue thoroughly, then test for TEROXIDE OF ANTIMONY, BINOXIDE OF TIN, and ARSENIC ACID, according to the instructions given in § 134, 5. Part, at least, of the arsenic acid is always found in this precipitate, combined with teroxide of antimony and binoxide of tin.

---

* Alloys of silver and platinum, with the latter metal present in small proportion only, dissolve in nitric acid.

† Very minute traces of antimony, however, are often completely dissolved by nitric acid.

III. ACTUAL EXAMINATION.
*Simple Compounds.**

A. SUBSTANCES SOLUBLE IN WATER.

Detection of the Base.†

§ 182.

1. Add some hydrochloric acid to a portion of the aqueous solution.   46

 a. NO PRECIPITATE IS FORMED; this is a positive proof of the absence of silver and suboxide of mercury, and is likewise au indication of the probable absence of lead. Pass on to (50).

 b. A PRECIPITATE IS FORMED. Divide the fluid in which 47 the precipitate is suspended into two portions, and add ammonia in excess to the one.

  a. The precipitate redissolves, and the fluid becomes clear; this shows the precipitate to have consisted of chloride of silver, and is consequently indicative of the presence of SILVER. To arrive at a positive conviction on this point, the original solution must be tested with chromate of potassa, and with hydrosulphuric acid (see § 115, 4, and § 138, 6).

  β. *The precipitate turns black:* this shows the precipitate 48 to have consisted of subchloride of mercury, which has now been converted by the ammonia into ammonio-chloride of mercury; it is consequently indicative of the presence of SUBOXIDE OF MERCURY. To set all doubt on this point at rest, test the original solution with protochloride of tin, and with metallic copper (see § 116).

  γ. *The precipitate remains unaltered;* it consists in that 49 case of chloride of lead, which is not dissolved by ammonia; this reaction is accordingly indicative of the presence of LEAD. Whether the precipitate consists really of chloride of lead or not is conclusively ascertained: 1st, by diluting the second portion of the fluid in which the precipitate produced by hydrochloric acid is suspended, with a large amount of water, and applying heat; the precipitate must dissolve if it consists of chloride of lead; and 2nd, by adding dilute sulphuric acid to the original solution (§ 117, 8).

2. Add to the fluid acidified with hydrochloric acid solution of 50 hydrosulphuric acid until it smells distinctly of that gas, even after shaking, heat the mixture, add some more solution of hydrosulphuric acid, and let it stand a short time.‡

---

* This term is used here, and wherever it happens to occur hereafter in the present work, to designate compounds supposed to contain only *one* base and *one* acid, or *one* metal and *one* non-metallic element. The principal object of this chapter is to facilitate instruction in analysis, as it is advantageous that the examination of complex compounds should be preceded by the analysis of simple compounds. In actual practical analyses, use can be made of this chapter only exceptionally, as there exists no outward sign by which to judge whether a substance contains only one base or acid, or several.

† Arsenious and arsenic acids, and silicic acid, are included here.

‡ If a precipitate forms immediately upon addition of solution of hydrosulphuric acid, it is unnecessary to apply heat, &c.; but if the fluid remains clear, or is rendered only slightly turbid, the above course of proceeding must be strictly followed, to guard against the risk of overlooking arsenic acid and binoxide of tin.

a. THE FLUID REMAINS CLEAR. Pass on to (56), since this is a proof that lead, bismuth, copper, cadmium, oxide of mercury, gold, platinum, tin, antimony, arsenic, and sesquioxide of iron are not present.

b. A PRECIPITATE IS FORMED.

α. THIS PRECIPITATE IS WHITE; it consists in that case of separated sulphur, and is indicative of the presence of SESQUIOXIDE OF IRON (§ 111, 3). However, as the separation of sulphur may also be caused by other substances, it is indispensable that you should satisfy yourself whether the substance present is really sesquioxide of iron or not. For this purpose test this solution with ammonia, and with ferrocyanide of potassium (§ 111, 5 and 6). — **51**

β. THE PRECIPITATE IS YELLOW; in this case it may consist either of sulphide of cadmium, sulphide of arsenic, or bisulphide of tin; it indicates accordingly the presence of either cadmium, arsenic, or binoxide of tin. To distinguish between them, mix a portion of the fluid wherein the precipitate is suspended with ammonia in excess, add some sulphide of ammonium, and heat. — **52**

aa. The precipitate does not dissolve; it consists of CADMIUM; for sulphide of cadmium is insoluble in ammonia and sulphide of ammonium. The blowpipe is resorted to as a confirmatory test (§ 122, 8).

bb. The precipitate dissolves : BINOXIDE OF TIN or ARSENIC: add ammonia to a small portion of the original solution.

αα. *A white precipitate is formed.* BINOXIDE OF TIN is the substance present. Positive conviction is obtained by reducing the precipitate before the blowpipe, with cyanide of potassium and carbonate of soda (§ 130, 8).

ββ. *No precipitate is formed.* This indicates the presence of ARSENIC. Positive conviction may be arrived at by the production of an arsenical mirror, which is effected by reducing the original substance or the precipitated sulphide of arsenic, either with cyanide of potassium and carbonate of soda, or in some other way; and moreover by exposing the original substance in conjunction with carbonate of soda to the inner flame of the blowpipe (§ 132, 12 and 13). If the solution (50) contained *arsenious* acid, the yellow precipitate (52) formed immediately upon the addition of the hydrosulphuric acid; if *arsenic* acid, it formed only upon the application of heat, or after long standing. For further information respecting the means of distinguishing between the two acids see § 134, 9.

γ. THE PRECIPITATE IS ORANGE-COLORED; it that case it consists of tersulphide of antimony, and indicates the presence of TEROXIDE OF ANTIMONY. For confirmation the original solution is tested with zinc in a small platinum dish (§ 131, 8). — **53**

δ. THE PRECIPITATE IS DARK-BROWN. It consists of protosulphide of tin, and indicates the presence of PROTOXIDE OF TIN. To remove all doubt, test a portion of the original solution with solution of chloride of mercury (§ 129, 8). — **54** *ε.* THE PRECIPITATE IS BROWNISH-BLACK OR BLACK. It may in that case consist of sulphide of lead, sulphide of copper, tersulphide of bismuth, tersulphide of gold, bisulphide of platinum, or sulphide of mercury. To distinguish between these different sulphides, the following experiments are resorted to.

aa. Add dilute sulphuric acid to a portion of the original solution; if a white precipitate is formed, this indicates LEAD. To dispel all doubt, test with chromate of potassa (§ 117).

bb. Add solution of soda to a portion of the original solution; if a yellow precipitate is formed, this indicates OXIDE OF MERCURY. The reactions with protochloride of tin and metallic copper afford positive certainty on the point (§ 119). The presence of oxide of mercury is usually sufficiently indicated by the several changes of color through which the precipitate produced by the solution of hydrosulphuric acid in the fluid under examination is observed to pass; this precipitate is white at first, but changes upon the addition of an excess of the precipitant to yellow, then to orange, and finally to black (§ 119, 3).

cc. Add ammonia in excess to a portion of the original solution; if a bluish precipitate is formed which redissolves in an excess of the precipitant, imparting an azure color to the fluid, or even if the ammonia simply colors the solution azure-blue, without producing a precipitate, this indicates COPPER. To remove all doubt, test with ferrocyanide of potassium (§ 120).

dd. If the precipitate produced by ammonia was white, and excess of ammonia has failed to redissolve it, filter the fluid off, wash the precipitate, dissolve it on a watch-glass in 1 or 2 drops of hydrochloric acid, with addition of 2 drops of water, and then add more water. If the solution turns turbid and milky, this is caused by basic terchloride of bismuth.: the reaction consequently indicates BISMUTH. The blowpipe is resorted to as a conclusive test (§ 121).

ee. Add solution of sulphate of protoxide of iron to a portion of the original solution. The formation of a fine black precipitate is indicative of the presence of GOLD. To remove all doubt as to the nature of the precipitate, expose it to the flame of the blowpipe, or test the original solution with protochloride of tin (§ 126).

ff. Add chloride of potassium and alcohol to a portion of the original solution; the formation of a yellow crystalline precipitate is indicative of the presence of PLATINUM. To remove all doubt, heat the precipitate to redness (§ 127).

3. Mix a small portion of the original solution with chloride of ammonium,* add ammonia to alkaline reaction, and then, no matter whether the latter reagent has produced a precipitate or not, a little sulphide of ammonium, and apply heat, if a precipitate fails to separate in the cold.

a. NO PRECIPITATE IS FORMED; pass on to (62); for iron, cobalt,

---

* The chloride of ammonium is used for the purpose of preventing the precipitation by ammonia of any magnesia which might be present.

nickel, manganese, zinc, chromium, alumina, and silicic acid, are not present.

 b. A PRECIPITATE IS FORMED.

  a. The precipitate is black: protoxide of iron, nickel, or **57** cobalt. Mix a portion of the original solution with some solution of potassa or soda.

   aa. A dirty greenish-white precipitate is formed, which soon changes to reddish-brown upon exposure to the air : PROTOXIDE OF IRON. To remove all doubt, test with ferricyanide of potassium (§ 110).

   bb. A precipitate of a light greenish tint is produced, which does not change color : NICKEL. The reaction with ammonia, and the precipitation of the ammoniacal solution by potassa or soda, will afford positive certainty on the point (§ 108).

   cc. A sky-blue precipitate is formed, which turns to a light-red upon boiling, or is discolored and acquires a dark tint : COBALT. The blowpipe is resorted to as a conclusive test (§ 109).

 β. *The precipitate is not black.* **58**

  aa. If the precipitate is distinctly flesh-colored, it consists of sulphide of manganese, and is consequently indicative of the presence of PROTOXIDE OF MANGANESE. To remove all doubt, add soda to the original solution, or try before the blowpipe (§ 107).

  bb. If the precipitate is bluish-green, it consists of hydrated sesquioxide of chromium, and is consequently indicative of the presence of SESQUIOXIDE OF CHROMIUM. To dispel all doubt, test the original solution with soda, and apply the blowpipe tests (§ 102).

  cc. If the precipitate is white, it may consist of hydrate **59** of alumina, or hydrate of silicic acid, or sulphide of zinc, and may accordingly point to the presence of either alumina or oxide of zinc or silicic acid ; the latter, in that case, is generally contained in the original solution as an alkaline silicate. To distinguish between these three bodies, add to a portion of the original solution a drop of solution of soda, and wait to see whether this produces a precipitate ; then add some more solution of soda until the precipitate formed is redissolved.

   aa. If solution of soda fails to produce a precipitate, **60** there is reason to test for SILICIC ACID. For that purpose evaporate a portion of the original solution with hydrochloric acid to dryness, and treat the residue with hydrochloric acid and water (§ 150, 2), when the silicic acid will be left undissolved. Determine the nature of the alkali which has been dissolved, as directed (**66**).

   ββ. If solution of soda produces a precipitate, which redissolves in an excess of the precipitant, add to a portion of this alkaline fluid a solution of hydrosulphuric acid ; the formation of a white precipitate indicates the presence of ZINC. The reaction with solution of nitrate of protoxide of cobalt before the blowpipe will afford conclusive proof (§ 106). If hydrosulphuric acid fails to produce a precipitate, add to the remaining portion of the alkaline fluid chloride of ammonium, and apply heat. The formation of a white precipitate indicates the presence of ALUMINA. The reaction with solution of nitrate of protoxide of cobalt before the blowpipe will afford conclusive proof (§ 101).

Note to (58) and (59).

As very slight contaminations may impair the distinctness of the tints exhibited by the precipitates considered in (58) and (59), it is advisable, in all cases where the least impurity is suspected, to adopt the following method for the detection of manganese, chromium, zinc, alumina, and silicic acid.

Add solution of soda to a portion of the original solution, first 61 in small quantity, then in excess.

aa. No precipitate is formed: SILICIC ACID may be assumed to be present; proceed as directed (60).

bb. A whitish precipitate is formed, which does not redissolve in an excess of the precipitant, and speedily turns blackish-brown upon exposure to the air: MANGANESE. The blowpipe is resorted to as a conclusive test (§ 107).

cc. A precipitate is formed, which redissolves in an excess of the precipitant: SESQUIOXIDE OF CHROMIUM, ALUMINA, OXIDE OF ZINC.

aa. Add hydrosulphuric acid water to a portion of the alkaline solution. The formation of a white precipitate indicates the presence of ZINC.

$\beta\beta$. If the original or the alkaline solution is green, and if the precipitate produced by soda and redissolved by an excess of the precipitant was of a bluish color, SESQUIOXIDE OF CHROMIUM is present. To remove all doubt, heat the alkaline solution to boiling, or try the reaction before the blowpipe (§ 102).

$\gamma\gamma$. Add chloride of ammonium to the alkaline solution. The formation of a white precipitate indicates the presence of ALUMINA. The reaction with solution of nitrate of protoxide of cobalt before the blowpipe will afford conclusive proof (§ 101).

4. Add to a portion of the original solution chloride of ammonium 62 and carbonate of ammonia, mixed with some caustic ammonia, and heat gently.

a. NO PRECIPITATE IS FORMED: absence of baryta, strontia, and lime. Pass on to (64).

b. A PRECIPITATE IS FORMED: presence of baryta, strontia, 63 or lime.

Add solution of sulphate of lime in sufficient quantity to a portion of the original solution.

a. The solution does not become turbid, not even after the lapse of from five to ten minutes: LIME. To remove all doubt, test with oxalate of ammonia (§ 97).

β. *The solution becomes turbid, but only after the lapse of some time:* STRONTIA. It is only in neutral or, at least, but slightly acid solutions that the reaction is sure to make its appearance. The flame-coloration is resorted to as a conclusive test (§ 96, 6 or 7).

γ. *A precipitate is immediately formed:* BARYTA. To remove all doubt, test with hydrofluosilicic acid (§ 95).

5. Mix that portion of the solution of 4 in which carbonate of **64** ammonia has, after previous addition of chloride of ammonium, failed to produce a precipitate (62), with phosphate of soda, add some more ammonia, and rub the sides of the vessel with a glass rod.

    a. NO PRECIPITATE IS FORMED: absence of magnesia. Pass on to (65).

    b. A CRYSTALLINE PRECIPITATE IS FORMED: MAGNESIA.

6. Evaporate a drop of the original solution on perfectly clean **65** platinum-foil as slowly as possible, and gently ignite the residue.

    a. THERE IS NO FIXED RESIDUE LEFT. Test for AMMONIA, by adding to the original solution hydrate of lime, and observing the odor and reaction of the escaping gas, and the fumes which it forms with acetic acid (§ 91).

    b. THERE IS A FIXED RESIDUE LEFT: potassa or soda. Add **66** bichloride of platinum to a portion of the original solution, having first concentrated it by evaporation if dilute, and shake the mixture.

    α. *No precipitate is formed, not even after the lapse of ten or fifteen minutes:* SODA. The flame coloration is selected as a conclusive test, or the reaction with antimonate of potassa is resorted to for the purpose (§ 90).

    β. *A yellow crystalline precipitate is formed:* POTASSA. The reaction with tartaric acid or the flame coloration is selected as a conclusive test (§ 89).

Simple Compounds.

A. SUBSTANCES SOLUBLE IN WATER. DETECTION OF THE ACID.

I. *Detection of Inorganic Acids.*

§ 183.

Reflect in the first place *which* of the inorganic acids form soluble compounds with the detected base (compare Appendix IV.), and bear this in mind in your subsequent operations, giving due regard also to the result of the preliminary examination.

1. ARSENIOUS ACID and ARSENIC ACID have already been consi- **67** dered in the preceding paragraph (detection of the base). These two acids are distinguished from each other by their respective reaction with nitrate of silver, or with potassa and sulphate of copper (see § 134, 9).

2. The presence of CARBONIC ACID, HYDROSULPHURIC ACID, and **68** CHROMIC ACID, is also indicated already in the course of the process pursued for the detection of the bases. The two former betray their presence by effervescing upon the addition of hydrochloric acid; they may be distinguished from one another by the smell. Should additional proof be required, the presence of carbonic acid may be ascertained beyond a doubt by the reaction with lime-water (see § 149), and that of hydrosulphuric acid by the reaction with solution of acetate of lead (§ 156). The presence of chromic acid is invariably indicated by the yellow or red tint of the solution, as well as by the transition of the red or yellow color to

green, accompanied by the separation of sulphur, upon the addition of hydrosulphuric acid water. To remove all doubt, try the reactions with solutions of acetate of lead and of nitrate of silver (§ 138).

3. Acidify a portion of the solution with hydrochloric acid, or —if oxide of silver or suboxide of mercury has been found—with nitric acid, and add chloride of barium or—where nitric acid has been the acidifying agent used—nitrate of baryta. **69**

    a. THE FLUID REMAINS CLEAR. Absence of sulphuric acid. Pass on to (**70**).

    b. A PRECIPITATE IS PRODUCED, IN FORM OF A FINE WHITE POWDER: SULPHURIC ACID. The precipitate must remain undissolved even after further addition of hydrochloric or nitric acid.

4. Add solution of sulphate of lime to another portion of the solution (which, if it has an acid reaction, must first be neutralized, or made slightly alkaline, by means of ammonia). **70**

    a. NO PRECIPITATE IS FORMED: absence of phosphoric acid, silicic acid, oxalic acid, and fluorine. Pass on to (**73**).

    b. A PRECIPITATE IS FORMED. Add acetic acid in excess. **71**

        α. *The precipitate redissolves readily:* PHOSPHORIC ACID or SILICIC ACID. Evaporate a portion of the original solution, after acidifying with hydrochloric acid, to dryness, and treat the residue with some hydrochloric acid and water. If an insoluble residue is left, mix a sample of the original solution with chloride of ammonium, sulphate of magnesia, and ammonia. A crystalline precipitate shows the presence of PHOSPHORIC ACID. (§ 142).

        β. *The precipitate remains undissolved or dissolves with difficulty:* OXALIC ACID or FLUORINE. Oxalate of lime is pulverulent, fluoride of calcium flocculent and gelatinous. The reaction with binoxide of manganese and sulphuric acid (§ 145) will afford conclusive proof of the presence of oxalic acid; the reaction on glass (etching) of the presence of fluorine (§ 146). **72**

5. Acidify a fresh portion of the original solution with nitric acid, and add solution of nitrate of silver. **73**

    a. THE FLUID REMAINS CLEAR. This is a proof of the absence of chlorine, bromine, iodine, ferrocyanogen, and ferricyanogen; the absence of cyanogen (in simple cyanides) is also probable. Of the soluble metallic cyanides, cyanide of mercury is not precipitated by nitrate of silver; if, therefore, in the analytical process for the detection of the bases, mercury has been found, cyanide of mercury may be present. For the manner of detecting the cyanogen in the latter see § 155, 8. Pass on to (**76**).

    b. A PRECIPITATE IS FORMED.

        α. *The precipitate is orange:* FERRICYANOGEN. The reaction with sulphate of protoxide of iron is resorted to as a confirmatory test (§ 155, Supplement). **74**

        β. *The precipitate is white or yellowish-white.* Treat the precipitate with ammonia in excess—immediately, if the base was an alkali or an alkaline earth—after filtering and washing, if the base was an earth proper or the oxide of a heavy metal.

            aa. The precipitate is not dissolved: IODINE or FERROCYANOGEN. In the former case the precipitate is pale yellow, in the latter white and gelatinous. The reaction with starch and hyponitric acid (§ 154) will afford conclusive proof of the presence of iodine, the reaction with sesquichloride of iron of the presence of ferrocyanogen (§ 155, Supplement).

ββ. The precipitate is dissolved: CHLORINE, BROMINE, or **75** CYANOGEN. If the original substance smells of hydrocyanic acid, and the silver precipitate dissolves with some difficulty in the ammonia, the precipitate may be assumed to consist of cyanide of silver, and, consequently, to indicate the presence of CYANOGEN. To remove all doubt on the point, add to the original solution sulphate of protoxide of iron, solution of soda, and hydrochloric acid (§ 155). If addition of chlorine-water imparts a yellow tint to the original solution, the precipitate may be held to consist of bromide of silver, and consequently indicates the presence of BROMINE; if the bromine is present only in very small proportion, chloroform or bisulphide of carbon must be used in conjunction with chlorine-water to make the reaction distinctly apparent (§ 153). In the proved absence of both bromine and cyanogen the precipitate consists of chloride of silver, and consequently shows the presence of CHLORINE.

6. Add to a small portion of the aqueous solution hydrochloric **76** acid, drop by drop, until a distinct acid reaction is just imparted to the fluid, then dip in a slip of turmeric-paper, take it out, and dry it at 212° F. If the dipped portion looks brownish-red, BORACIC ACID is present. To settle all doubt on the point, add sulphuric acid and alcohol, and set fire to the latter (§ 144).

7. With regard to NITRIC ACID and CHLORIC ACID, these are **77** usually discovered already in the course of the preliminary examination (6). The reaction with sulphate of protoxide of iron and sulphuric acid (§ 159) will afford conclusive evidence of the presence of the former, treatment of the solid salt with concentrated sulphuric acid, of the presence of the latter acid (§ 160).

Simple Compounds.

A. SUBSTANCES SOLUBLE IN WATER. DETECTION OF THE ACID.

II. *Detection of Organic Acids.*

§ 184.

Consider, in the first place, *which* of the organic acids form soluble compounds with the detected base (Compare Appendix IV.), and bear this in mind in your subsequent operations, giving due regard also to the results of the preliminary examination.

The following course of proceeding presupposes the organic acid to be present in the free state, or in combination with an alkali or an alkaline earth. If the detected base belongs to another group, therefore, it must first be removed. Where the base belongs to group V. or group VI. the removal is effected by means of hydrosulphuric acid, where it belongs to group IV., by means of sulphide of ammonium. After filtering off the sulphides, and removing the excess of sulphide of ammonium by acidifying with hydrochloric acid, heating, and filtering off the eliminated sulphur, proceed to (**78**). Where the basis is alumina or sesquioxide of chromium, try first to precipitate these substances by boiling with carbonate of soda; should this fail, as it will where the acid is non-volatile, precipitate the latter with neutral acetate of lead, wash the precipitate, diffuse it through water, conduct hydrosulphuric acid into the water in which the precipitate is suspended, filter off the sulphide of lead formed, and treat the filtrate as directed below.—Alumina may also be precipitated from its compounds with non-volatile organic acids by solution of soluble glass, as silicate of alumina.

1. Add ammonia to a portion of the aqueous solution of the compound under examination to slight alkaline reaction, then chloride of calcium. If the solution was neutral, or only slightly acid, add chloride of ammonium before adding the chloride of calcium. **78**

    a. NO PRECIPITATE IS FORMED, NOT EVEN AFTER SHAKING THE FLUID NOR AFTER THE LAPSE OF A FEW MINUTES: absence of oxalic acid and tartaric acid. Pass on to (**80**).

    b. A PRECIPITATE IS FORMED. Add lime-water in excess to a fresh portion of the original solution, and add solution of chloride of ammonium to the precipitate formed. **79**

        α. The precipitate redissolves: TARTARIC ACID. The reaction with acetate of potassa may be resorted to as a confirmatory test; but a still more positive proof will be afforded by the deportment which the precipitate produced by the chloride of calcium, and properly washed, exhibits with solution of soda or with ammonia and nitrate of silver (§ 163).

        β. The precipitate does not redissolve: OXALIC ACID. To remove all doubt, try the reaction with concentrated sulphuric acid (§ 145).

2. Heat the fluid of 1, *a*, to boiling, keep at that temperature for some time, and add some more ammonia to the boiling fluid. **80**

    a. IT REMAINS CLEAR: absence of citric acid. Pass on to (**81**).

    b. IT BECOMES TURBID, AND DEPOSITS A PRECIPITATE: CITRIC ACID. To remove all doubt as to the nature of the acid, add solution of acetate of lead in excess, wash the precipitate formed, and see whether it dissolves readily in ammonia (§ 164).

3. Mix the fluid of 2, *a*, with alcohol. **81**

    a. IT REMAINS CLEAR: absence of malic acid. Pass on to (**82**).

    b. A PRECIPITATE IS FORMED: MALIC ACID. To remove all doubt, it is *invariably* necessary to try the reaction with acetate of lead, to see whether the precipitate produced by that reagent dissolves with difficulty in ammonia, and to examine its deportment when the fluid in which it is suspended is heated to boiling (§ 165).

4. Neutralize a portion of the original solution *completely* (if not already absolutely neutral) with ammonia or with hydrochloric acid, and add solution of sesquichloride of iron. **82**

    a. A BULKY PRECIPITATE FORMS, OF A CINNAMON BROWN, OR DIRTY YELLOW COLOR. Wash the precipitate, heat it with ammonia, filter, *strongly* concentrate the filtrate by evaporation, divide into two parts, and add to the one some hydrochloric acid, to the other alcohol and chloride of barium. The formation of a precipitate in the first portion indicates the presence of BENZOIC ACID, a precipitate in the second denotes the presence of SUCCINIC ACID. Compare § 168 and § 169.

b. THE LIQUID ACQUIRES A RATHER INTENSE DEEP RED TINT, AND, UPON PROTRACTED BOILING, A LIGHT REDDISH-BROWN PRECIPITATE SEPARATES: acetic acid or formic acid. Heat a portion of the solid salt under examination or, if the substance is in the fluid state, of the residue left upon evaporating the fluid (which, if acid, you must neutralize first with soda), with sulphuric acid and alcohol (§ 171). The characteristic odor of acetic ether indicates the presence of ACETIC ACID. **83**

If you do not detect acetic acid in the fluid, you may conclude that the substance under examination contains FORMIC ACID; to remove all doubt, try the reactions with nitrate of silver and chloride of mercury (§ 172).

Simple Compounds.

B. SUBSTANCES INSOLUBLE OR SPARINGLY SOLUBLE IN WATER, BUT SOLUBLE IN HYDROCHLORIC ACID, NITRIC ACID, OR NITRO-HYDROCHLORIC ACID.

Detection of the Base.*

§ 185.

Dilute a portion of the solution in hydrochloric acid, nitric acid, or nitro-hydrochloric acid with water,† and proceed as directed § 182, beginning at (46), in cases where the substance is dissolved in nitric acid, and at 2, (50), if the solution already contains hydrochloric acid. **84**

This course will answer for bases of the *second, fifth,* and *sixth* groups, but in testing for bases of the *third* and *fourth* groups, with sulphide of ammonium, according to (56), the usual course of proceeding is under the circumstances departed from. Particular regard must be had in this to the following observations: we have seen (59), that if in cases where we have A SUBSTANCE SOLUBLE IN WATER we obtain, in the course of the examination, a white precipitate upon adding chloride of ammonium, ammonia, and sulphide of ammonium, this precipitate can consist only of SULPHIDE OF ZINC, or ALUMINA, or HYDRATE OF SILICIC ACID. But the case is different if the body is INSOLUBLE IN WATER, but dissolves in hydrochloric acid; for in that case a white precipitate produced by ammonia, in presence of chloride of ammonium, may consist also of PHOSPHATES, BORATES, OXALATES, SILICATES OF THE ALKALINE EARTHS, or of FLUORIDES OF THEIR METALS, since all these bodies are insoluble in water, but dissolve in hydrochloric acid, and (being only very sparingly soluble also in solution of chloride of ammonium) accordingly separate again upon neutralization of that acid. If, therefore, a white precipitate is produced upon testing an acid solution, under the circumstances stated, and in pursuing the course laid down in § 182, (56), proceed as follows:—

1. If the results of the preliminary examination have given you **85**

---

* Regard is also had here to certain salts of the alkaline earths, as this course of examination leads directly to their detection.

† If upon the addition of water the liquid becomes white and turbid or deposits a white precipitate, this indicates the presence of antimony or bismuth, possibly also of tin. Compare § 121, 9, and § 131, 4. Heat with hydrochloric acid until the fluid has become clear again, then pass on to (50).

reason to suspect the presence of SILICIC ACID (20), evaporate a portion of the hydrochloric acid solution to dryness, moisten the residue with hydrochloric acid and add water. If silicic acid is present, it will remain undissolved. Determine the base in the solution as directed (56) or (62), as the case may be.

2. Add to a portion of the original hydrochloric acid solution some tartaric acid, and after this ammonia in excess. 86

   a. NO PERMANENT PRECIPITATE IS FORMED: absence of the above enumerated salts of the alkaline earths. Mix another portion of the original solution with solution of soda in excess, and add to the one half of the clear fluid chloride of ammonium, to the other half solution of hydrosulphuric acid. The formation of a precipitate in the former indicates the presence of ALUMINA; in the latter, the presence of ZINC.

   b. A PERMANENT PRECIPITATE IS FORMED: presence of a salt of an alkaline earth.

      a. Bring a sample of the original substance, on a watch-glass, in contact with a little binoxide of manganese, a few drops of water, and some concentrated sulphuric acid. If evolution of carbonic acid gas takes place instantly, the salt is an OXALATE. To find the base, ignite a fresh sample, dissolve the residue in dilute hydrochloric acid, and examine the solution as directed (62). 87

      β. Add to a portion of the hydrochloric acid solution ammonia until a precipitate forms; then acetic acid until this is redissolved; lastly, acetate of soda and a drop of solution of sesquichloride of iron: the formation of a white flocculent precipitate indicates the presence of PHOSPHORIC ACID. Add now some more sesquichloride of iron until the fluid has acquired a distinct red color, boil, filter boiling, and test the filtrate, which is now free from phosphoric acid, for the alkaline earth with which the phosphoric acid was combined, as directed (62), after having previously removed, by precipitation with ammonia, the iron which may possibly have been dissolved.' 88

      γ. Test for FLUORINE, by heating a portion of the original substance, or of the precipitate produced in the hydrochloric acid solution by ammonia, with sulphuric acid (§ 146). After removal of the fluorine, ascertain the nature of the alkaline earth which you have now in the residue, in combination with sulphuric acid (§ 188). 89

      δ. BORACIC ACID is detected in the hydrochloric acid solution by means of turmeric-paper (§ 144), and the base combined with it, by boiling a portion of the original substance with water and carbonate of soda, filtering, washing, dissolving the carbonate formed in the least possible amount of dilute hydrochloric acid, and examining the solution as directed (62).

Simple Compounds.

B. SUBSTANCES INSOLUBLE OR SPARINGLY SOLUBLE IN WATER, BUT SOLUBLE IN HYDROCHLORIC ACID, NITRIC ACID, OR NITRO-HYDROCHLORIC ACID.

DETECTION OF THE ACID.

I. *Detection of Inorganic Acids.*

§ 186.

1. CHLORIC ACID cannot be present, since all chlorates are soluble in water; NITRIC ACID, which may be present in form of a basic salt, must have been revealed already by the ignition of the body in a glass tube, and so must CYANOGEN (8). For the analysis of the insoluble metallic CYANIDES insoluble in water see § 204. The results of the test with phosphate of soda and ammonia will have directed attention to the presence of SILICIC ACID. Evaporation of the hydrochloric acid solution to dryness, and treatment of the residue with hydrochloric acid and water, will remove all doubt on this point. **90**

2. The course of examination laid down for the detection of the bases leads likewise to that of ARSENIOUS and ARSENIC ACIDS, CARBONIC ACID, HYDROSULPHURIC ACID, and CHROMIC ACID. With regard to the latter acid, I repeat that its presence is indicated by the yellow or red color of the compound, the evolution of chlorine which ensues upon boiling with hydrochloric acid, and the subsequent presence of sesquioxide of chromium in the solution. Fusion of the compound under examination with carbonate of soda is, however, the most conclusive test for chromic acid (§ 138). **91**

3. Boil a portion of the substance with nitric acid. **92**

  a. If nitric oxide gas is evolved, and sulphur separates, this is confirmative of the presence of a metallic sulphide.

  b. If violet vapors escape, the compound is a metallic IODIDE.

  c. If reddish-brown fumes of a chlorine-like smell are evolved, the compound is a metallic BROMIDE, in which case the fumes will color starch yellow (§ 153).

4. Dilute a small portion of the nitric acid solution—or of the filtrate of this solution, should the nitric acid have left an undissolved residue—with water, and add solution of nitrate of silver to the fluid. The formation of a white precipitate, which, after washing, is soluble in ammonia, and fuses without decomposition when heated, indicates the presence of CHLORINE. **93**

5. Boil a portion of the substance with hydrochloric acid, filter if necessary, dilute with water, and add chloride of barium. The formation of a white precipitate, which does not redissolve even upon addition of a large quantity of water, indicates the presence of SULPHURIC ACID. **94**

6. Test for BORACIC ACID as directed § 144, 6.

7. If none of the acids enumerated from 1 to 6 are present, there is reason to suspect the presence of PHOSPHORIC ACID, OXALIC ACID, or FLUORINE, or the total absence of acids. To the presence of oxalic acid your attention will have been called already in the course of the preliminary examination (8). If the acids named had been combined **95**

with an alkaline earth, they would already have been detected in the course of the examination for these bases **(87)** to **(89)**; they need therefore here be tested for only where the examination has revealed the presence of some other base. To that end precipitate the base, if belonging to Group V. or VI., with hydrosulphuric acid, or if belonging to Group IV., with sulphide of ammonium, and filter. If you have precipitated with sulphide of ammonium, add to the filtrate hydrochloric acid to acid reaction, expel in either case the hydrosulphuric acid by boiling, and filter if necessary. Test a portion of this solution for phosphoric acid, oxalic acid, and fluorine, as directed **(70)**. If the basis was alumina or sesquioxide of chromium, test for phosphoric acid with solution of molybdate of ammonia in nitric acid (§ 142, 10); for oxalic acid with binoxide of manganese and sulphuric acid (§ 145); for fluorine with sulphuric acid (§ 146).

Simple Compounds.

B. SUBSTANCES INSOLUBLE OR SPARINGLY SOLUBLE IN WATER, BUT SOLUBLE IN ACIDS.

DETECTION OF THE ACID.

II. *Detection of Organic Acids.*

§ 187.

1. FORMIC ACID cannot be present, as all the formates are soluble in water. **96**

2. ACETIC ACID has been revealed already in the course of the preliminary examination, by the evolution of acetone. The reaction with sulphuric acid and alcohol (§ 171) will afford conclusive proof.

3. Boil a portion of the substance for some time with solution of carbonate of soda in excess, and filter hot. You have now, in most cases, the organic acid in solution in combination with soda. Acidulate the solution slightly with hydrochloric acid, expel the carbonic acid by heat, and test as directed § 184. With bases of the fourth group and also in presence of oxide of lead, this mode of separation is not completely successful. In exceptional cases of the kind add to the filtrate, after boiling with carbonate of soda, sulphide of ammonium until the whole of the metallic oxide is thrown down. **97**

Simple Compounds.

C. SUBSTANCES INSOLUBLE OR SPARINGLY SOLUBLE IN WATER, HYDROCHLORIC ACID, NITRIC ACID, AND NITRO-HYDROCHLORIC ACID.

DETECTION OF THE BASE AND THE ACID.

§ 188.

Under this head we have to consider here SULPHATE OF BARYTA, SULPHATE OF STRONTIA, SULPHATE OF LIME, FLUORIDE OF CALCIUM, SILICA, SULPHATE OF LEAD, compounds of LEAD with CHLORINE and BROMINE; compounds of SILVER with CHLORINE, BROMINE, IODINE, and CYANOGEN; and lastly SULPHUR and CHARCOAL, as the only bodies belonging to this class which are more frequently met with. For the simple **98** silicates I refer to § 205, for the ferro- and ferricyanides, to § 204. The preliminary examination will have informed you whether you need pay any regard to the possible presence of these compounds.

Sulphate of lime and chloride of lead are not altogether insoluble in water, and sulphate of lead may be dissolved in hydrochloric acid. However, as these compounds are so sparingly soluble that complete solution of them is seldom effected, they are included here also among the class of insoluble substances, to insure their detection, should they have been overlooked in the course of the examination of the aqueous or acid solution of the body to be analyzed.

1. Free SULPHUR must have been detected already in the course of the preliminary examination.

2. CHARCOAL is generally black; it is insoluble in aqua regia; put on platinum foil, with the blowpipe flame playing upon the under side of the foil, it is always consumed; by deflagration with nitrate of potassa it yields carbonate of potassa.

3. Pour sulphide of ammonium over a very small quantity of the substance under examination.

   a. It TURNS BLACK; this indicates the presence of lead or a salt of silver.

     α. *The body fused in the glass tube without decomposition* (3): chloride of lead, bromide of lead; chloride of silver, bromide of silver, iodide of silver. Fuse one part of the compound with 4 parts of carbonate of soda and potassa in a small porcelain crucible, let cool, boil the residue with water, and test the filtrate for CHLORINE, BROMINE, and IODINE, as directed (73). Dissolve the residue, which consists either of metallic SILVER or OXIDE OF LEAD, in nitric acid, and test the solution as directed (46).

     β. *The body evolved cyanogen by ignition in the glass tube, and left metallic silver behind:* CYANIDE OF SILVER.

     γ. *The body remained unaltered by ignition in the glass tube:* SULPHATE OF LEAD. Boil a sample of it with solution of carbonate of soda, filter, acidulate the filtrate with hydrochloric acid, and test with chloride of barium for SULPHURIC ACID; dissolve the washed residue in nitric acid, and test the solution with hydrosulphuric acid and with sulphuric acid for LEAD.

   b. IT REMAINS WHITE: absence of an oxide of a heavy metal. Triturate a small sample together with quartz sand, moisten the mixture on a watch-glass with a few drops of concentrated sulphuric acid, and heat gently.

     α. *White fumes are evolved, which redden litmus paper.* This indicates the presence of FLUORIDE OF CALCIUM. Reduce a portion of the substance to a fine powder, decompose this in a platinum crucible with sulphuric acid, and try the reaction on glass (§ 146), to prove the presence of FLUORINE; boil the residue with hydrochloric acid, filter, neutralize the filtrate with ammonia, and test for LIME with oxalate of ammonia.

     β. *No fumes reddening litmus paper are evolved.* Mix a portion of the very finely pulverized substance with 4 times the quantity of pure carbonate of soda and potassa, and fuse the mixture in a platinum crucible, or on platinum foil. Boil the fused mass with water, filter should a residue be left, and wash the latter. Acidulate a portion of the filtrate with hydrochloric acid, and test with chloride of barium for SULPHURIC ACID; and in case you do not find that acid, test another portion of the filtrate for SILICIC ACID, by evaporating the fluid acidified with hydrochloric acid (§ 150, 2).

If the SILICIC ACID was present in the pure state, the mass resulting from the fusion of the substance with carbonate of soda and potassa must have dissolved in water to a clear fluid; but if silicates also happened to be present, the bases of them are left behind undissolved, and may be further examined.

If, on the other hand, sulphuric acid has been found, the alkaline earth which was combined with it is found on the filter as a carbonate. Wash this, then dissolve in dilute hydrochloric acid, and test the solution for BARYTA, STRONTIA, and LIME, as directed (62).

Complex Compounds.*

A. SUBSTANCES SOLUBLE IN WATER, AND ALSO SUCH AS ARE INSOLUBLE IN WATER, BUT DISSOLVE IN HYDROCHLORIC ACID, NITRIC ACID, OR NITRO-HYDROCHLORIC ACID.

Detection of the Bases.†

§ 189.‡

(*Treatment with Hydrochloric Acid : Detection of Silver, Suboxide of Mercury* [*Lead*].)

The systematic course for the detection of the bases is essentially the same for bodies soluble in water, as for those which are soluble only in acids. Where, from the different nature of the original solution, a departure from the ordinary course is rendered necessary, the fact will be distinctly stated. 101

I. SOLUTION IN WATER.

MIX THE PORTION INTENDED FOR THE DETECTION OF THE BASES WITH SOME HYDROCHLORIC ACID.

1. THE SOLUTION HAD AN ACID OR NEUTRAL REACTION PREVIOUSLY TO THE ADDITION OF THE HYDROCHLORIC ACID. 102

a. NO PRECIPITATE IS FORMED; this indicates the absence of silver and suboxide of mercury. Pass on to § 190.

b. A PRECIPITATE IS FORMED. Add more hydrochloric acid, drop by drop, until the precipitate ceases to increase; then add about six or eight drops more of hydrochloric acid, shake the mixture, and filter.

The precipitate produced by hydrochloric acid may consist of chloride of silver, subchloride of mercury, chloride of lead, a basic

---

* I use this term here and hereafter in the present work to designate compounds in which all the more frequently occurring bases, acids, metals, and metalloids are supposed to be present.

† Consult the explanations in the Third Section, with the contents of which you should make yourself thoroughly acquainted first, before proceeding further. Regard is here had also to the presence of the acids of arsenic, and of those salts of the alkaline earths which dissolve in hydrochloric acid, and separate again from that solution unaltered upon neutralization of the acid by ammonia.

‡ Consult the remarks in the Third Section.

salt of antimony, basic chloride of bismuth, possibly also of benzoic acid. The basic salt of antimony and the basic chloride of bismuth, however, redissolve in the excess of hydrochloric acid ; consequently, if the instructions given have been strictly followed, the precipitate collected upon the filter can consist only of chloride of silver, subchloride of mercury, or chloride of lead—(possibly also of benzoic acid, which, however, is altogether disregarded *here*).

Wash the precipitate collected upon the filter twice with cold water, add the washings to the filtrate, and examine the solution as directed § 190, even though the addition of the washings to the acid filtrate should produce turbidity in the fluid (which indicates the presence of compounds of antimony or bismuth).

Treat the twice-washed precipitate on the filter as follows : **103**

α. Pour hot water over it upon the filter, and test the fluid running off with sulphuric acid for LEAD. The non-formation of a precipitate upon the addition of the sulphuric acid simply proves that the precipitate produced by hydrochloric acid contains no lead, and does not by any means establish the total absence of this metal, as hydrochloric acid fails to precipitate lead from dilute solutions.

β. Pour over the now thrice-washed precipitate upon the filter solution of ammonia. If this changes its color to black or gray, it is a proof of the presence of SUBOXIDE OF MERCURY.

γ. Add to the ammoniacal fluid running off in β nitric acid to strongly acid reaction. The formation of a white, curdy precipitate indicates the presence of SILVER.* (If the precipitate did contain lead, the ammoniacal solution generally appears turbid, owing to the separation of a basic salt of lead. This, however, does not interfere with the testing for silver, since the basic salt of lead redissolves upon the addition of nitric acid.)

2. THE ORIGINAL AQUEOUS SOLUTION HAD AN ALKALINE REACTION. **104**

a. THE ADDITION OF HYDROCHLORIC ACID TO STRONGLY ACID REACTION FAILS TO PRODUCE EVOLUTION OF GAS OR A PRECIPITATE, OR THE PRECIPITATE WHICH FORMS AT FIRST REDISSOLVES UPON FURTHER ADDITION OF HYDROCHLORIC ACID : pass on to § 190.

b. THE ADDITION OF HYDROCHLORIC ACID TO THE ORIGINAL SOLUTION PRODUCES A PRECIPITATE WHICH DOES NOT REDISSOLVE IN AN EXCESS OF THE PRECIPITANT, NOT EVEN UPON BOILING.

a. The formation of the precipitate is attended neither with **105** *evolution of hydrosulphuric acid nor of hydrocyanic acid.*

Filter, and treat the filtrate as directed § 190.

aa. THE PRECIPITATE IS WHITE. It may, in that case, consist of a salt of lead or silver, insoluble in water and hydrochloric acid (CHLORIDE OF LEAD, SULPHATE OF LEAD, CHLORIDE OF SILVER, &c.), or it may be HYDRATE OF SILICIC ACID. Test for the bases and acids of these compounds as directed § 203, bearing in mind that the chloride of lead or chloride of silver

---

* If the quantity of silver is only very small, its presence is indicated by opalescence of the fluid.

which may be present, may possibly only have been formed in the process.

bb. THE PRECIPITATE IS YELLOW OR ORANGE. In that case it may consist of SULPHIDE OF ARSENIC (and if the fluid from which it has separated was not boiled long, or only with very dilute hydrochloric acid, also of SULPHIDE OF ANTIMONY or BISULPHIDE OF TIN), which substances were originally dissolved in solution of ammonia, potassa, soda, phosphate of soda, or some other alkaline fluid, with the exception of solutions of alkaline sulphides and cyanides. Examine the precipitate, which may also contain HYDRATE OF SILICIC ACID, as directed (40).

β. *The formation of the precipitate is attended with evolution of hydrosulphuric acid gas, but not of hydrocyanic acid.** **106**

aa. THE PRECIPITATE IS OF A PURE WHITE COLOR, AND CONSISTS OF SEPARATED SULPHUR. In that case a SULPHURETTED ALKALINE SULPHIDE is present. Boil, filter, and treat the filtrate as directed § 194, the precipitate as directed § 203.

bb. THE PRECIPITATE IS COLORED. In that case you may conclude that a METALLIC SULPHUR SALT is present, *i.e.*, a combination of an alkaline sulphur base with a metallic sulphur acid. The precipitate may accordingly consist of TERSULPHIDE OF GOLD, BISULPHIDE OF PLATINUM, BISULPHIDE OF TIN, SULPHIDE OF ARSENIC, or SULPHIDE OF ANTIMONY. It might, however, consist also of SULPHIDE OF MERCURY or of SULPHIDE OF COPPER or SULPHIDE OF NICKEL, or contain these substances, as the former will dissolve in sulphide of potassium, and the latter are slightly soluble in sulphide of ammonium. Filter, and treat the filtrate as directed § 194, the precipitate as directed (40).

γ. *The formation of the precipitate is attended with evolution of hydrocyanic acid, with or without simultaneous disengagement of hydrosulphuric acid.* This indicates the presence of an ALKALINE CYANIDE, and, if the evolution of the hydrocyanic acid is attended with that of hydrosulphuric acid, also of an alkaline SULPHIDE. In that case the precipitate may, besides the compounds enumerated in *a* and *β*, contain many other substances (*e.g.*, cyanide of nickel, cyanide of silver, &c.). Boil, with further addition of hydrochloric acid, or of nitric acid, until the whole of the hydrocyanic acid is expelled, and treat the solution, or, if an undissolved residue has been left, the filtrate, as directed § 190; and the residue (if any) according to § 203. **107**

c. THE ADDITION OF HYDROCHLORIC ACID FAILS TO PRODUCE A PERMANENT PRECIPITATE, BUT CAUSES EVOLUTION OF GAS. **108**

α. *The escaping gas smells of hydrosulphuric acid;* this indicates the presence of a SIMPLE ALKALINE SULPHIDE. Proceed as directed § 194.

β. *The escaping gas is inodorous;* in that case it is CARBONIC ACID which was combined with an alkali. Pass on to § 190.

---

* Should the odor of the evolved gas leave any doubt regarding the actual presence or absence of hydrocyanic acid, add some chromate of potassa to a portion of the fluid, previously to the addition of the hydrochloric acid.

γ. *The escaping gas smells of hydrocyanic acid* (no matter whether hydrosulphuric acid or carbonic acid is evolved at the same time or not). This indicates the presence of an ALKALINE CYANIDE. Boil until the whole of the hydrocyanic acid is expelled, then pass on to § 190.

II. SOLUTION IN HYDROCHLORIC ACID OR IN NITROHYDROCHLORIC ACID.

Proceed as directed § 190.

III. SOLUTION IN NITRIC ACID.

Dilute a small sample of it with water; should this produce **109** turbidity or a precipitate (indicative of the presence of bismuth), add nitric acid until the fluid is clear again, then hydrochloric acid.

1. NO PRECIPITATE IS FORMED. Absence of silver and suboxide of mercury. Treat the principal solution as directed § 190.

2. A PRECIPITATE IS FORMED. Treat a larger portion of the nitric acid solution the same way as the sample, filter, and examine the precipitate as directed **(103)**, the filtrate as directed § 190.

§ 190.*

(*Treatment with Hydrosulphuric Acid, Precipitation of the Metallic Oxides of Group V., 2nd Division, and of Group VI.*)

ADD TO A *small* PORTION OF THE CLEAR ACID SOLUTION HYDROSULPHURIC ACID WATER, UNTIL THE ODOR OF HYDROSULPHURIC ACID IS DISTINCTLY PERCEPTIBLE AFTER SHAKING THE MIXTURE, AND WARM GENTLY.

1. No PRECIPITATE IS FORMED, even after the lapse of some **110** time. Pass on to § 194, for lead, bismuth, cadmium, copper, mercury, gold, platinum, antimony, tin, and arsenic,† are not present ;‡ the absence of sesquioxide of iron and of chromic acid is also indicated by this negative reaction.

2. A PRECIPITATE IS FORMED.

a. The precipitate is of a pure white color, light, and **111** finely pulverulent, and does not redissolve on addition of hydrochloric acid. It consists of separated sulphur, and indicates the presence of SESQUIOXIDE OF IRON.§ None of the

---

* Consult the remarks in the Third Section.

† Where the preliminary examination has led you to suspect the presence of arsenic acid, you must endeavor to obtain the most conclusive evidence of the absence of this acid; this may be done by allowing the fluid to stand for some time at a gentle heat (about 158° F.), or by heating it with sulphurous acid previous to the addition of the hydrosulphuric acid. (Compare § 133, 3.)

‡ In solutions containing much free acid the precipitates are frequently formed only after dilution with water.

§ Sulphur will precipitate also if sulphurous acid, or iodic acid, or bromic acid is present (which substances are not included in our analytical course), and also if chromic acid, or chloric acid, or free chlorine is present. In presence of chromic acid the separation of the sulphur is attended with reduction of the acid to sesquioxide of chromium, in consequence of which the reddish-yellow color of the solution changes to green. (Compare § 139.) The white sulphur suspended in the green solution looks at first like a green precipitate, which frequently tends to mislead beginners.

other metals enumerated in (**110**) can be present. Treat the principal solution as directed § 194.

b. The precipitate is colored.

112 Add to the larger proportion of the acid or acidified solution, best in a small flask, hydrosulphuric acid water in excess, *i.e.*, until the fluid smells distinctly of it, and the precipitate ceases to increase upon continued addition of the reagent; apply a gentle heat, shake vigorously for some time, filter, keep the filtrate (which contains the oxides present of Groups I.—IV.), for further examination according to the instructions of § 194, and *thoroughly* wash* the precipitate, which contains the sulphides of the metals present of Groups V. and VI.

In many cases, and more particularly where there is any reason to suspect the presence of arsenic, it will be found more convenient to transmit hydrosulphuric acid gas through the solution DILUTED WITH WATER, instead of adding hydrosulphuric acid water.

113 If the precipitate is yellow, it consists principally of sulphide of arsenic, bisulphide of tin, or sulphide of cadmium; if orange-colored, this indicates sulphide of antimony; if brown or black, one at least of the following oxides is present: oxide of lead, teroxide of bismuth, oxide of copper, oxide of mercury, teroxide of gold, binoxide of platinum, protoxide of tin. However, as a yellow precipitate may contain small particles of an orange-colored, a brown, or even a black precipitate, and yet its color not be very perceptibly altered thereby, it will always prove the safest way to assume the presence of all the metals named in (**110**) in any precipitate produced by hydrosulphuric acid, and to proceed accordingly as the next paragraph (§ 191) directs.

§ 191.

(Treatment of the Precipitate produced by Hydrosulphuric Acid with Sulphide of Ammonium; Separation of the 2nd Division of Group V. from Group VI.)

114 INTRODUCE A SMALL PORTION OF THE PRECIPITATE PRODUCED BY HYDROSULPHURIC ACID IN THE ACIDIFIED SOLUTION INTO A TEST-TUBE,† ADD A LITTLE WATER, AND FROM TEN TO TWENTY DROPS OF YELLOWISH SULPHIDE OF AMMONIUM, AND EXPOSE THE MIXTURE FOR A SHORT TIME TO A GENTLE HEAT.‡

---

* Compare § 6.

† If there is a somewhat large precipitate, this may be readily effected by means of a small spatula of platinum or horn; but if you have only a very trifling precipitate, make a hole in the bottom of the filter, insert the perforated point into the mouth of the test-tube, rinse the precipitate into the latter by means of the washing-bottle, wait until the precipitate has subsided, and then decant the water.

‡ If the solution contains copper, which is generally revealed by the color of the fluid, and may be ascertained positively by testing with a clean iron rod (see § 120, 10), use solution of sulphide of sodium instead of sulphide of ammonium (in which sulphide of copper is not absolutely insoluble, see § 120, 5), and boil the mixture. But if the fluid, besides copper, contains also oxide of mercury (the presence of which is generally sufficiently indicated by the several changes of color exhibited by the precipitate forming upon the addition of the hydrosulphuric acid [§ 119, 3], and which, in doubtful cases, may be detected with positive certainty by testing a portion of the original solution acidified with hydrochloric acid with protochloride of tin), sulphide of ammonium must be used, although the separation of the sulphides of the antimony group from the

115 1. THE PRECIPITATE DISSOLVES COMPLETELY IN SULPHIDE OF AMMONIUM (or SULPHIDE OF SODIUM, as the case may be): absence of the metals of Group V.—cadmium, lead, bismuth, copper, mercury. Treat the remainder of the precipitate (of which you have digested a portion with sulphide of ammonium) as directed § 192.—If the precipitate produced by hydrosulphuric acid was so trifling that you have used the whole of it in treating with sulphide of ammonium, precipitate the solution obtained in that process by addition of hydrochloric acid, filter, wash the precipitate, and treat it as directed § 192.

116 2. THE PRECIPITATE IS NOT REDISSOLVED, OR AT LEAST NOT COMPLETELY: presence of metals of Group V.

Dilute with 4 or 5 parts of water, filter, and mix the filtrate with hydrochloric acid in slight excess.

a. The fluid simply turns milky, owing to the separation of sulphur. Absence of the metals of Group VI.—gold, platinum, tin, antimony, and arsenic.* Treat the rest of the precipitate (of which you have digested a portion with sulphide of ammonium) according to the directions of § 193.

117 *b. A colored precipitate is formed:* presence of metals of Group VI. by the side of those of Group V. Treat the entire precipitate produced by hydrosulphuric acid the same as you have treated a portion of it, *i.e.*, digest it with yellow sulphide of ammonium or, as the case may be, sulphide of sodium, let subside, pour the supernatant liquid on a filter, digest the residue in the tube once more with yellow sulphide of ammonium (or sulphide of sodium), and filter. Wash the residue† (containing the sulphides of Group V.), and treat it afterwards as directed § 193. Dilute the filtrate—which contains the metals of Group VI. in the form of sulphur salts—with water, add hydrochloric acid to distinctly acid reaction, heat gently, filter the precipitate formed—which contains the sulphides of the metals of Group VI. mixed with sulphur—wash thoroughly, and proceed as directed next paragraph (§ 192.)

§ 192.

(*Detection of the Metals of Group* VI.: *Arsenic, Antimony, Tin, Gold, Platinum.*)

118 If the precipitate consisting of the sulphides of Group VI. has a PURE YELLOW COLOR, this indicates principally arsenic and tin; if

sulphide of copper is not fully effected in such cases; since, were sulphide of sodium used, the sulphide of mercury would dissolve in this reagent, which would impede the ulterior examination of the sulphides of the antimony group.

* That this inference becomes uncertain if the precipitate produced by hydrosulphuric acid, instead of being digested with a small quantity of sulphide of ammonium, has been treated with a larger quantity of that reagent, is self-evident; for the large quantity of sulphur which separates in that case will of course completely conceal any slight traces of sulphide of arsenic or bisulphide of tin which may have been thrown down.

† If the residue suspended in the fluid containing sulphide of ammonium, and insoluble therein, subsides readily, it is not transferred to the filter, but washed in the tube by decantation. But if its subsidence proceeds slowly and with difficulty, it is transferred to the filter, and washed there; a hole is then made in the bottom of the filter, and the residue rinsed into a small porcelain basin by means of a washing-bottle; the application of a gentle heat will now materially aid the subsidence of the residue, and the supernatant water may then be decanted. The sulphides are occasionally suspended in the fluid in a state of such minute division that the fluid cannot be filtered off clear. In cases of the kind some chloride of ammonium should be added to the fluid.

it is distinctly ORANGE-YELLOW, antimony is sure to be present; if it is BROWN or BLACK, this denotes the presence of platinum or gold.

Beyond these general indications the color of the precipitate affords no safe guidance. It is therefore always advisable to test a yellow precipitate also for antimony, gold, and platinum, since minute quantities of the sulphides of these metals are completely hid by a large quantity of bisulphide of tin or sulphide of arsenic. Proceed accordingly as follows:

Heat a little of the precipitate on the lid of a porcelain crucible, or on a piece of porcelain or glass.*

119 1. *Complete volatilization ensues:* probable presence of ARSENIC, absence of the other metals of Group VI. Reduction of a portion of the precipitate with cyanide of potassium and carbonate of soda (§ 132, 12) † will afford positive proof of the presence or absence of arsenic. Whether that metal was present in the form of arsenious acid or in that of arsenic acid, may be ascertained by the methods described § 134, 9.

120 2. *A fixed residue is left.* In that case all the metals of Group VI. must be sought for. Dry the remainder of the precipitate thoroughly upon the filter, triturate it together with about 1 part of anhydrous carbonate of soda and 1 part of nitrate of soda, and transfer the mixture in small portions at a time to a little porcelain crucible, in which you have previously heated 2 parts of nitrate of soda to fusion.‡ As soon as complete oxidation is effected, pour the mass out on a piece of porcelain.

After cooling soak the fused mass (the portion still sticking to the inside of the crucible as well as the portion poured out on the porcelain) in cold water, filter from the insoluble residue—which will remain if the mass contained antimony, tin, gold, or platinum—and wash thoroughly with a mixture of about equal parts of water and alcohol. (The alcohol is added to prevent the solution of the antimonate of soda. The washings are not added to the filtrate.)

The filtrate and the residue are now examined as follows:

121 *a.* EXAMINATION OF THE FILTRATE FOR ARSENIC (which must be present in it in the form of arsenate of soda).

Add nitric acid to the fluid to distinct acid reaction,§

---

* That this preliminary examination may be omitted if the precipitate has any other color than yellow, and that it can give a decisive result only if the sulphur precipitate submitted to the test has been thoroughly washed, is self-evident.

† In cases where the precipitate contains much free sulphur, dissolve the sulphide of arsenic which may be present, by digestion in the ammonia, filter, evaporate the solution, with addition of a small quantity of carbonate of soda, to dryness, and heat the residue with cyanide of potassium and carbonate of soda.

‡ Should the amount of the precipitate be so minute that this operation cannot be conveniently performed, cut the filter, with the dried precipitate adhering to it, into small pieces, triturate these together with some carbonate of soda and nitrate of soda, and project both the powder and the paper into the fusing nitrate of soda. It is *preferable*, however, in such cases, to procure at once, if practicable, a sufficiently large amount of the precipitate, as otherwise there will be but little hope of effecting the positive detection of all the metals of Group VI. Supposing all the metallic sulphides of the sixth group to have been present, the fused mass would consist of antimonate and arsenate of soda, binoxide of tin, metallic gold and platinum, sulphate, carbonate, nitrate, and some nitrite of soda. Compare also § 134, 1.

§ In some cases where a somewhat larger proportion of carbonate of soda has been used, or a very strong heat applied, a trifling precipitate (hydrated binoxide of tin) may separate upon the acidification of the filtrate with nitric acid. This may be filtered off, and then treated in the same manner as the undissolved residue.

heat, to expel carbonic acid and nitrous acid, then divide the fluid into two portions. Add to the one portion some nitrate of silver (not too little), filter (in case chloride of silver* or nitrite of silver should have separated), pour upon the filtrate, along the side of the tube held slanting, a layer of dilute solution of ammonia—2 parts of water to 1 part of solution of ammonia —and let the mixture stand for some' time without shaking. The formation of a reddish-brown precipitate, which appears hovering cloud-like between the two layers (and may be seen far more readily and distinctly . by reflected than by transmitted light), denotes the presence of ARSENIC.

If the arsenic is present in some quantity, and the free nitric acid of the solution is exactly saturated with ammonia, the fluid being stirred during this process, the precipitate of arsenate of silver which forms imparts a brownish-red tint to the entire fluid.

Add to the other portion of the acidified solution, first **122** ammonia, then a mixture of sulphate of magnesia and chloride of ammonium, and rub the sides of the vessel with a glass rod. A crystalline precipitate of arsenate of magnesia and ammonia, which often forms only after long standing, and deposits the crystalline particles more particularly on the side of the vessel, shows the presence of arsenic. By way of confirmation, the arsenic compound may be reduced to the metallic state (compare § 132 and § 133). Whether the arsenic was present in the form of arsenious acid or in that of arsenic acid, may be ascertained by the methods described § 134, 9.

b. EXAMINATION OF THE RESIDUE FOR ANTIMONY, TIN, **123** GOLD, PLATINUM. (As the antimony, if present in the residue, must exist as white pulverulent antimonate of soda, the tin as white flocculent binoxide, the gold and platinum in the metallic state, the appearance of the residue is in itself indicative of its nature.) Transfer the precipitate to the inverted lid of a platinum crucible, or to a small platinum dish, heat with hydrochloric acid, add a little water, and throw in a small compact lump of pure zinc (more particularly, free from lead), no matter whether the precipitate has completely dissolved or not in the hydrochloric acid. This operation leaves the gold and platinum in the same state in which the fused mass contained them, viz., in the metallic state, to which the tin and antimony are now likewise reduced by the action of the zinc. The antimony reveals its presence at once, or after a short time, by blackening the platinum. As soon as the disengagement of hydrogen has pretty nigh stopped, take out the lump of zinc, remove the solution of chloride of zinc by cautious decantation, treat the metals with hydrochloric acid, and test the solution—which, if tin is present, must contain protochloride of tin—with chloride of mercury (§ 129, 8).

After removing the tin by repeated boiling with hydro- **124**

---

* Chloride of silver will separate if the reagents were not perfectly pure, or the precipitate has not been thoroughly washed.

chloric acid, and all the hydrochloric acid by thoroughly washing with water, examine the insoluble residue (if one is left) as follows: Heat it in the platinum dish with some water, with addition of a few grains of tartaric acid, then add some nitric acid, and heat gently. If the residue dissolves completely, no gold or platinum is present; if a residue is left undissolved, you must test it for these metals. For this purpose remove the acid solution (which may be tested again for ANTIMONY with hydrosulphuric acid) by decantation and washing, heat the residue, transferred to a porcelain dish, with a little aqua regia, evaporate the solution until but little of it is left, and test this small remainder for GOLD and PLATINUM as directed § 128.

§ 193.

(*Detection of the Metallic Oxides of Group V., 2nd Division :—Oxide of Lead, Teroxide of Bismuth, Oxide of Copper, Oxide of Cadmium, Oxide of Mercury.*)

THOROUGHLY WASH THE PRECIPITATE WHICH HAS NOT BEEN DISSOLVED BY SULPHIDE OF AMMONIUM, AND BOIL WITH NITRIC ACID. This operation is performed best in a small porcelain dish: the boiling mass must be constantly stirred with a glass rod during the process. A great excess of acid must be avoided. **125**

1. THE PRECIPITATE DISSOLVES, AND THERE REMAINS FLOATING IN THE FLUID ONLY THE SEPARATED LIGHT FLOCCULENT AND YELLOW SULPHUR; this indicates the absence of mercury. CADMIUM, COPPER, LEAD, and BISMUTH may be present. **126**

Filter the fluid from the separated sulphur, and treat the filtrate as follows (should there be too much nitric acid present, the greater part of this must first be driven off by evaporation): add to a portion of the filtrate dilute sulphuric acid in moderate quantity, heat gently, and let the fluid stand some time.

a. NO PRECIPITATE FORMS; absence of lead. Mix the remainder of the filtrate with ammonia in excess, and gently heat. **127**

a. No precipitate is formed; absence of BISMUTH. If the liquid is blue, COPPER is present; very minute traces of copper, however, might be overlooked if the color of the ammoniated fluid alone were consulted. To be quite safe, and also to test for cadmium, evaporate the ammoniated solution nearly to dryness, add a little acetic acid, and, if necessary, some water, and **128**

aa. Test a small portion of the fluid for copper with ferrocyanide of potassium. The formation of a reddish-brown precipitate, or a light brownish-red turbidity, indicates the presence of COPPER (in the latter case only to a very trifling amount). **129**

bb. Mix the remainder of the fluid with solution of hydrosulphuric acid in excess. The formation of a yellow precipitate denotes CADMIUM. If, on account of the presence of copper, the sulphide of cadmium cannot be distinctly recognised, allow the precipitate produced by the hydrosulphuric acid to subside, decant the supernatant fluid, and add to the precipitate solution of cyanide of potassium until the sulphide **130** of copper is dissolved. If a yellow residue is left undissolved, CADMIUM is present; in the contrary case, not.

β. *A precipitate is formed.* BISMUTH is present. Filter **131** the fluid, and test the filtrate for copper and cadmium as directed in (128). To test the washed precipitate more fully for bismuth, slightly dry the filter containing it between blotting-paper, remove the still moist precipitate with a platinum spatula, dissolve on a watch-glass in the least possible quantity of hydrochloric acid, and then add a proper quantity of water. The appearance of a milky turbidity confirms the presence of bismuth.

b. A PRECIPITATE IS FORMED. Presence of LEAD. Mix **132** the whole of the nitric acid solution in a porcelain dish with a sufficient quantity of dilute sulphuric acid, evaporate on the water-bath until the nitric acid is expelled, dilute the residue with some water containing sulphuric acid, filter off at once the sulphate of lead left undissolved, and test the filtrate for bismuth, copper, and cadmium, as directed in (127).* Test the precipitate, after washing, by one of the methods described in § 123.

2. THE PRECIPITATE OF THE METALLIC SULPHIDES DOES NOT **133** COMPLETELY DISSOLVE IN THE BOILING NITRIC ACID, BUT LEAVES A RESIDUE, BESIDES THE LIGHT FLAKES OF SULPHUR THAT FLOAT IN THE FLUID. Probable presence of OXIDE OF MERCURY (which may be pronounced almost certain if the precipitate is heavy and black). Allow the precipitate to subside, filter off the fluid, which is still to be tested for CADMIUM, COPPER, LEAD, and BISMUTH; mix a small portion of the filtrate with a large amount of solution of hydrosulphuric acid, and should a precipitate form or a coloration become visible, treat the remainder of the filtrate according to the directions of (126).

Wash the residue (which may, besides sulphide of mercury, also contain sulphate of lead, formed by the action of nitric acid upon sulphide of lead, and also binoxide of tin, and possibly sulphide of gold and sulphide of platinum, as the separation of the sulphides of tin, gold, and platinum from the sulphides of the metals of the fifth group is often incomplete), and examine one half of it for mercury,† by dissolving it in some hydrochloric acid, with addition of a very small proportion of chlorate of potassa, and testing the solution with copper or protochloride of tin (§ 119); fuse the other half with cyanide of potassium and carbonate of soda, and treat the fused mass with water. If metallic grains remain, or if a metallic powder is left undissolved, wash this residue, heat with nitric acid, and test the solution obtained with sulphuric acid for lead. Wash the residue which the nitric acid may leave undissolved, and extract from it any hydrate of metastannic acid which it may contain, according to the directions of § 130, 1, as metastannic chloride. Should a metallic powder be left undissolved in the process, heat it with aqua regia, and test the solution for gold and platinum as directed § 128.

---

* For another method of distinguishing cadmium, copper, lead, and bismuth from each other, I refer to the Third Section (additions and remarks to § 193).

† If you have an aqueous solution, or a solution in very dilute hydrochloric acid, the oxide of mercury formed was present in the original substance in that form; but if the solution has been prepared by boiling with concentrated hydrochloric acid, or by heating with nitric acid, the mercury may most likely have been originally present in the form of suboxide, and may have been converted into oxide in the process.

§ 194.

(Precipitation with Sulphide of Ammonium, Separation and Detection of the Oxides of Groups III. and IV: Alumina, Sesquioxide of Chromium; —Oxide of Zinc, Protoxide of Manganese, Protoxide of Nickel, Protoxide of Cobalt, Proto- and Sesquioxide of Iron; and also of those Salts of the Alkaline Earths which are precipitated by Ammonia from their Solution in Hydrochloric Acid: Phosphates, Borates, Oxalates, Silicates, and Fluorides.)

134 PUT A *small portion* OF THE FLUID IN WHICH SOLUTION OF HYDROSULPHURIC ACID HAS FAILED TO PRODUCE A PRECIPITATE (110), OR OF THE FLUID WHICH HAS BEEN FILTERED FROM THE PRECIPITATE FORMED (112), in a test-tube, observe whether it is colored or not,* boil to expel the hydrosulphuric acid which may be present, add a few drops of nitric acid, boil, and observe again the color of the fluid; then cautiously add ammonia to alkaline reaction, observe whether this produces a precipitate, then add some sulphide of ammonium, on matter whether ammonia has produced a precipitate or not.

135 *a.* NEITHER AMMONIA NOR SULPHIDE OF AMMONIUM PRODUCES A PRECIPITATE. Pass on to § 195, for iron, nickel, cobalt, zinc, manganese, sesquioxide of chromium, alumina, are not present, nor are phosphates, borates,† silicates, and oxalates‡ of the alkaline earths; nor fluorides of the metals of the alkaline earths, nor silicic acid—originally in combination with alkalies.

136 *b.* SULPHIDE OF AMMONIUM PRODUCES A PRECIPITATE, AMMONIA HAVING FAILED TO DO SO; absence of phosphates, borates,† silicates, and oxalates‡ of the alkaline earths; of the fluorides of the metals of the alkaline earths; of silicic acid—originally in combination with alkalies; and also, if no organic matters are present, of iron, sesquioxide of chromium, and alumina. Pass on to (138).

137 *c.* AMMONIA PRODUCES A PRECIPITATE before the addition of sulphide of ammonium. The course of proceeding to be pursued now depends upon whether, (α) the original solution is simply aqueous, and has a neutral reaction, or (β) the original solution is acid or alkaline. In the former case pass on to (138), since phosphates, borates, oxalates, and silicates of the alkaline earths cannot be present; nor can fluorides of the metals of the alkaline earths, nor, lastly, silicic acid in combination with alkalies.

---

* If the fluid is colorless, it contains no chromium. If colored, the tint will to some extent act as a guide to the nature of the substance present; thus a green tint, or a violet tint turning green upon boiling, points to the presence of chromium; a light green tint to that of nickel; a reddish color to that of cobalt; the turning yellow of the fluid upon boiling with nitric acid to that of iron. It must, however, be always borne in mind that these tints are perceptible only if the metallic oxides are present in larger quantity, and also that complementary colors, such as, for instance, the green of the nickel solution and the red of the cobalt solution will destroy each other, and that, accordingly, a solution may contain both metals and yet appear colorless.

† Presence of much chloride of ammonium has a great tendency to prevent the precipitation of borates of the alkaline earths.

‡ Oxalate of magnesia is thrown down from hydrochloric acid solution by ammonia after some time only, and never completely; dilute solutions are not precipitated by ammonia.

DETECTION OF BASES.

In the latter case regard must be had to the possible presence of all the bodies enumerated in (135) : pass on to (150).

138 1. DETECTION OF THE BASES OF GROUPS III. AND IV. IF PHOSPHATES, &c., OF THE ALKALINE EARTHS ARE NOT PRESENT.*

Mix the fluid mentioned at the beginning of the paragraph (134), a portion of which you have submitted to a preliminary examination, with some chloride of ammonium, then with ammonia, just to alkaline reaction, lastly with sulphide of ammonium until the fluid, after being shaken, smells distinctly of that reagent; shake the mixture until the precipitate begins to separate in flakes, heat gently for some time, and filter.

Keep the FILTRATE,† which contains, or may contain, the bases of Groups II. and I., for subsequent examination according to the directions of § 195. Wash the PRECIPITATE with water to which a very little sulphide of ammonium has been added, then proceed with it as follows :—

139 *a.* IT HAS A PURE WHITE COLOR ; absence of iron, cobalt, nickel. You must test for all the other bases of Groups III. and IV., as the faint tints of sesquioxide of chromium and sulphide of manganese are imperceptible in a large quantity of a white precipitate. Dissolve the precipitate by heating it in a small dish with the least possible amount of hydrochloric acid ; boil—should hydrosulphuric acid be evolved—until this is completely expelled, concentrate by evaporation to a small residue, add concentrated solution of soda in excess, heat to boiling, and keep the mixture for some time in a state of ebullition.

140 α. *The precipitate formed at first dissolves completely in the excess of solution of soda.* Absence of manganese and chromium, presence of alumina or oxide of zinc. Test a portion of the alkaline solution with solution of hydrosulphuric acid for ZINC ; acidify the remainder with hydrochloric acid, add ammonia *slightly* in excess, and apply heat. The formation of a white flocculent precipitate shows the presence of ALUMINA.

141 β. *The precipitate formed does not dissolve, or dissolves only partially, in the excess of solution of soda.* Filter and test the FILTRATE, as in (140), for ZINC and ALUMINA. With the undissolved PRECIPITATE, which, if containing manganese, looks brown or brownish, proceed as follows :—

aa. If the color of the solution gives you no reason to suspect the presence of chromium, test the precipitate for manganese, by means of the reaction with carbonate of soda in the outer blowpipe flame.

142 *bb.* But where the color of the solution indicates chromium, the examination of the residue insoluble in solution of soda is more complicated, since it may in that case contain also oxide of zinc, possibly even the whole

---

* This simpler method will fully answer the purpose in most cases ; for very accurate analysis the method beginning at (150) is preferable, as this will permit also the detection of minute quantities of alkaline earths, which may have been thrown down together with the alumina and sesquioxide of chromium.

† If the filtrate has a brownish color, this points to the presence of nickel, sulphide of nickel, as is well known, being, under certain circumstances, slightly soluble in sulphide of ammonium; this, however, involves no modification of the analytical course.

quantity present of this metal (§ 112). Dissolve the precipitate therefore in hydrochloric acid, evaporate the solution to a small residue, dilute, nearly neutralize the free acid with carbonate of soda, add carbonate of baryta in slight excess, let the fluid digest in the cold until it has become colorless, filter, and test the precipitate for CHROMIUM, by fusion with carbonate of soda and nitrate of soda (§ 102, 8). Remove the baryta from the filtrate, by precipitating with some sulphuric acid, filter, evaporate to a small residue, add concentrated solution of potassa or soda in excess, and test the filtrate for ZINC with hydrosulphuric acid, the precipitate, if any, for MANGANESE as in *aa*.

b. IT IS NOT WHITE ; this points to the presence of **143** chromium, manganese, iron, cobalt, or nickel. If it is black, or inclines to black, one of the three metals last-mentioned is present. Under any circumstances all the oxides of Groups III. and IV. must be looked for.

Remove the washed precipitate from the filter with a spatula, or by rinsing it with the aid of a washing-bottle, through a hole made in the bottom of the filter, into a test-tube, and pour over it rather dilute cold hydrochloric acid in moderate excess.

a. It dissolves completely (except perhaps a little sulphur, **144** which may separate); absence of cobalt and nickel, at least of notable quantities of these two metals.

Boil until the hydrosulphuric acid is *completely* expelled, filter if particles of sulphur are suspended in the fluid, concentrate by evaporation to a small residue, add concentrated solution of potassa or soda in excess, boil, filter the fluid from the insoluble precipitate which is sure to remain, wash the latter, and proceed first to examine the filtrate, then the precipitate.

aa. Test a small portion of the *filtrate* with hydro- **145** sulphuric acid for *zinc;* acidify the remainder with hydrochloric acid, then test with ammonia for ALUMINA. Compare (140).

bb. Dissolve a small portion of the *precipitate* in hydro- **146** chloric acid, and test the solution with sulphocyanide of potassium for IRON. Test another portion for CHROMIUM, by fusing together with carbonate and nitrate of soda (§ 102, 8). If no chromium has been found, examine the remainder for MANGANESE, by the reaction of carbonate of soda in the oxidizing flame. If chromium is present, on the other hand, test the remainder of the precipitate for manganese and zinc (of which latter metal the precipitate may in that case possibly contain the entire quantity originally present in the compound under examination [§'112]) as directed (**142**).

β. *The precipitate is not completely dissolved, a black re-* **147** *sidue being left;* this indicates the presence of cobalt and nickel. Filter, wash the undissolved precipitate, and test the filtrate as directed (**144**); proceed with the residuary precipitate as follows :—

aa. Test a small portion of it with borax, first in the **148** outer, then in the inner blowpipe-flame. If the bead in the oxidizing flame is violet whilst hot, and of a pale

reddish-brown when cold, and turns in the reducing flame
gray and turbid, NICKEL is present; but if the color of the
bead is and remains blue in both flames, and whether hot or
cold, COBALT is present. As in the latter case the presence of nickel cannot be distinctly recognised, examine

bb. The remainder of the precipitate by incinerating **149**
it together with the filter in a coil of platinum wire,
heating the ash with some hydrochloric acid, filtering
the solution, then evaporating nearly to dryness, and adding
nitrite of potassa and, lastly, acetic acid (§ 109, 10). If a
yellow precipitate forms, after standing for some time at a
gentle heat, this confirms the presence of COBALT. Filter after
about twelve hours, and test the filtrate with solution of
soda for nickel.

2. DETECTION OF THE BASES OF GROUPS III. AND IV. IN CASES **150**
WHERE PHOSPHATES, BORATES, OXALATES, OR SILICATES OF THE AL-
KALINE EARTHS, OR FLUORIDES OF THE METALS OF THE ALKALINE
EARTHS, OR HYDRATE OF SILICIC ACID, MAY POSSIBLY HAVE BEEN THROWN
DOWN ALONG WITH THESE BASES, *i.e.*, in cases where the original solution
was acid or alkaline, and a precipitate was produced by ammonia in the
preliminary examination. See (134).

Mix the fluid mentioned in (134) with some chloride of ammonium,
then with ammonia just to alkaline reaction, lastly with sulphide of
ammonium until the fluid, after being shaken, smells distinctly of the
reagent; shake the mixture until the precipitate begins to separate in
flakes, heat gently for some time, and filter. Keep the FILTRATE, which
contains, or may contain, the bases of Groups II. and I., for subsequent
examination according to the directions of § 195. Wash the precipitate
with water to which a very little sulphide of ammonium has been
added, then proceed with it as directed in (152). To give a clear notion
of the obstacles to be overcome in this analytical process, I must remind
you that it is necessary to examine the precipitate for the following
bodies: Iron, nickel, cobalt (these show their presence to a certain
extent by the black or blackish coloration of the precipitate), manganese,
zinc, sesquioxide of chromium (the latter generally reveals its presence
by the color of the solution), alumina;—baryta, strontia, lime, magnesia,
which latter substances may have fallen down in combination with phosphoric acid, boracic acid, oxalic acid, silicic acid, or in form of fluorides.
Besides these bodies, free silicic acid may also be contained in
the precipitate as hydrate.

As the original substance must, under all circumstances, be **151**
afterwards examined for all acids that might possibly be present,
it is not *indispensable* to test for the above enumerated acids at
this stage of the analytical process; still, as it is often interesting to
know these acids at once, more especially in cases where a somewhat
large proportion of some alkaline earth has been found in the precipitate
produced by sulphide of ammonium, a method for the detection of the
acids in question will be found appended by way of supplement to
the method for the detection of the bases.

Remove the precipitate from the filter with a small spatula, or **152**
by rinsing it off with the washing-bottle, and pour over it cold
dilute hydrochloric acid in moderate excess.

a. A RESIDUE REMAINS. Filter, and treat the filtrate as **153**

242   ACTUAL EXAMINATION—COMPLEX COMPOUNDS.

directed in (154). The residue, if it is black, may contain sulphide of nickel and sulphide of cobalt and, besides these, sulphur and silicic acid. Wash, and examine a sample of it in conjunction with phosphate of soda and ammonia before the blowpipe, in the outer flame. If a silica skeleton remains undissolved (§ 150, 8), this proves the presence of silicic acid. If the color of the bead is blue, COBALT is present; if reddish, turning yellow on cooling, NICKEL. Should the color leave you in doubt, incinerate the filter containing the remainder of the residue, and test for cobalt and nickel by means of nitrite of potassa, as directed (149).

    b. NO RESIDUE IS LEFT (except perhaps a little sulphur, **154** which may separate): absence of nickel and cobalt, at least in any notable proportion.

Boil the solution until the sulphuretted hydrogen is expelled, filter if necessary, and then proceed as follows:

    a. Mix a small portion of the solution with dilute **155** sulphuric acid. If a precipitate forms, this may consist of sulphates of BARYTA and STRONTIA, possibly also of sulphate of lime. Filter, wash the precipitate, and examine it either by the coloration of flame (see § 99, at the end), or decompose it by boiling or fusion with carbonated alkali, wash the carbonates produced, dissolve them in hydrochloric acid, and test the solution as directed § 195. Mix the fluid which has not been precipitated by dilute sulphuric acid, or the fluid filtered from the precipitate produced, with 3 volumes of spirit of wine. If a precipitate forms, this consists of sulphate of LIME. Filter, dissolve in water, and add oxalate of ammonia to the solution, as a confirmatory proof of the presence of lime.

    β. Heat a somewhat larger sample with some nitric **156** acid, and test a small portion of the fluid with sulphocyanide of potassium for IRON;* mix the remainder with sesquichloride of iron in sufficient quantity to make a drop of fluid give a yellowish precipitate† when mixed on a watch-glass with a drop of ammonia; evaporate the fluid now until there is only a small quantity left; add to this some water, then a few drops of solution of carbonate of soda, just sufficient to *nearly* neutralize the free acid, and lastly carbonate of baryta in slight excess; stir the mixture, and let it stand in the cold until the fluid above the precipitate has become colorless. Filter now the precipitate (*aa*) from the solution (*bb*), and wash.

    aa. Boil the *precipitate* for some time with solution of **157** soda, filter, and test the filtrate for ALUMINA,‡ by heating

---

* Whether the iron was present as sesquioxide or as protoxide, must be ascertained by testing the original solution in hydrochloric acid with ferricyanide of potassium and sulphocyanide of potassium.

† The addition of sesquichloride of iron is necessary, to effect the separation of phosphoric acid and silicic acid which may be present.

‡ If the solution contains silicic acid, the precipitate taken for alumina may also contain silicic acid. A simple trial with phosphate of soda and ammonia, on a platinum wire, in the blowpipe flame, will show whether the precipitate really contains silicic acid. Should this be the case, ignite the remainder of the supposed alumina precipitate on the lid of a platinum crucible, add some acid sulphate of potassa, fuse the mixture, and treat the fused mass with hydrochloric acid, which will dissolve the alumina, leaving the silicic acid undissolved; precipitate the alumina from the solution by ammonia.

with chloride of ammonium in excess. The part of the precipitate insoluble in solution of soda is examined for CHROMIUM, by fusion with nitrate of potassa and carbonate of soda (§ 102, 8).

bb. Mix the *solution* first with a few drops of hydrochloric acid, boil to expel the whole of the carbonic acid, then add some ammonia and sulphide of ammonium.

aa. *No precipitate forms:* absence of manganese **158** and zinc. Mix the solution containing chloride of barium with dilute sulphuric acid in slight excess, boil, filter, supersaturate with ammonia, and mix with oxalate of ammonia. If a precipitate of oxalate of LIME forms, filter, and test the filtrate with phosphate of soda for magnesia.

ββ. *A precipitate forms.* Filter, and proceed with **159** the filtrate according to the directions of (**158**). The precipitate may consist of sulphide of manganese and sulphide of zinc, and may contain traces also of sulphide of cobalt and sulphide of nickel. Wash it with water containing some sulphide of ammonium, then treat with acetic acid, which will dissolve the sulphide of MANGANESE, if any is present, leaving the other sulphides undissolved. Filter, boil the filtrate with solution of soda, and test the precipitate, which may form, with carbonate of soda in the outer blowpipe flame for MANGANESE. Free the residuary part of the precipitate which acetic acid has failed to dissolve, by washing, from the acetic acid solution still adhering to it, and then treat it with dilute hydrochloric acid, which will dissolve the *zinc*, if any is present. Filter, add some nitric acid to the filtrate, and concentrate the mixture considerably by boiling; then add to it concentrated solution of soda in excess, boil, filter if necessary, and test the filtrate with sulphide of ammonium for ZINC. Should a precipitate insoluble in solution of soda remain in the last operation, or should the dilute hydrochloric acid have left a black residue, test this precipitate and residue for COBALT and NICKEL, if you have not already previously detected the presence of these bodies; compare (**148** and **149**).

γ. If you have found alkaline earths in α and β, and **160** wish to know the acids in combination with which they have passed into the precipitate produced by sulphide of ammonium, this may be ascertained by making the following experiments with the remainder of the hydrochloric acid solution.

aa. Evaporate a small portion in a small dish or on a **161** watch-glass on the water-bath, dry the residue thoroughly, then treat with hydrochloric acid. If there was any SILICIC ACID in the solution, this will be left undissolved. Test the solution now for PHOSPHORIC ACID, by means of molybdic acid (§ 142, 10).

bb. Mix another portion with carbonate of soda *in excess*, boil for some time, filter, and test one-half of the filtrate for OXALIC ACID, by acidifying with acetic acid and adding solution of sulphate of lime; the other half for BORACIC ACID, by slightly acidifying with hydrochloric acid, and testing with turmeric-paper (§ 144 and § 145.)

cc. Precipitate the remainder with ammonia, filter, wash and dry the precipitate, and examine it for FLUORINE according to § 146, 5.   162

§ 195.

(*Separation and Detection of the Oxides of Group II. which are precipitated by Carbonate of Ammonia in Presence of Chloride of Ammonium, viz., Baryta, Strontia, Lime*).

TO A SMALL PORTION OF THE FLUID IN WHICH AMMONIA AND SULPHIDE OF AMMONIUM HAVE FAILED TO PRODUCE A PRECIPITATE (135), OR OF THE FLUID FILTERED FROM THE PRECIPITATE FORMED, ADD CHLORIDE OF AMMONIUM, IF THE SOLUTION CONTAINS NO AMMONIACAL SALT, THEN CARBONATE OF AMMONIA AND SOME CAUSTIC AMMONIA, AND HEAT FOR SOME TIME VERY GENTLY (not to boiling).

1. NO PRECIPITATE FORMS: absence of any notable quantity of baryta, strontia, and lime. Traces of these alkaline earths may, however, be present: to detect them, add to another portion of the fluid some sulphate of ammonia (prepared by supersaturating dilute sulphuric acid with ammonia): if the fluid becomes turbid, it contains traces of BARYTA; add to a third portion some oxalate of ammonia; if the fluid turns turbid — which reaction may perhaps require some time to manifest itself—traces of lime are present. Treat the remainder of the fluid as directed § 196, after having previously removed the traces of lime and baryta which may have been found, by means of the reagents that have served to effect their detection.   163

2. A PRECIPITATE IS FORMED. Presence of LIME, BARYTA, or STRONTIA. Treat the whole fluid of which a portion has been tested with ammonia, and carbonate of ammonia, the same as the sample, filter off the precipitate formed, after gently heating, as directed above, and test portions of the filtrate with sulphate and oxalate of ammonia for traces of lime and baryta, which it may possibly still contain; remove such traces, should they be found, by means of the said reagents, and examine the fluid, thus perfectly freed from baryta, strontia, and lime, for magnesia according to the directions of § 196. Wash the precipitate produced by carbonate of ammonia, dissolve it in the *least possible* amount of dilute hydrochloric acid, and add to a small portion of the fluid a sufficient quantity of solution of sulphate of lime.   164

a. No precipitate is formed, NOT EVEN AFTER THE LAPSE OF SOME TIME. Absence of baryta and strontia; presence of LIME. To remove all doubt, mix another sample with oxalate of ammonia.

b. A precipitate is formed by solution of sulphate of lime.

α. *It is formed immediately;* this indicates BARYTA. Besides this, strontia and lime may also be present.   165

Evaporate the remainder of the hydrochloric acid solution of the precipitate produced by carbonate of ammonia to dryness, digest the residue with strong alcohol, decant the fluid from the undissolved chloride of barium, dilute with an equal volume of water, mix with a few drops of hydrofluosilicic acid—

which will throw down the small portion of baryta that had dissolved in form of chloride of barium—allow the mixture to stand for some time; filter, and mix the filtrate with dilute sulphuric acid. The formation of a precipitate indicates the presence of strontia or lime, or of both. Filter after some time, wash the precipitate with weak spirit of wine, boil with solution of carbonate of soda, to convert the sulphates into carbonates, filter these off, wash, dissolve in hydrochloric acid, evaporate the solution to dryness, dissolve the residue in water, and test a portion of the solution with dilute solution of sulphate of potassa (§ 96, 3). If a precipitate forms immediately, or in the course of half an hour, the presence of STRONTIA is demonstrated. In that case let the fluid with the precipitate in it stand at rest for some time, then filter, and add ammonia and oxalate of ammonia to the filtrate. The formation of a white precipitate indicates LIME. If sulphate of potassa has failed to produce a precipitate, the remainder of the solution of the residue left upon evaporation is tested at once with ammonia and oxalate of ammonia for lime.

β. *It is formed only after some time.* Absence of baryta, **166** presence of STRONTIA. Mix the remainder of the hydrochloric acid solution with sulphate of potassa, let the mixture stand for some time, then filter, and test the filtrate with ammonia and oxalate of ammonia for LIME.

§ 196.

(Examination for Magnesia.)

TO A PORTION OF THE FLUID IN WHICH CARBONATE, SULPHATE, AND OXALATE OF AMMONIA HAVE FAILED TO PRODUCE A PRECIPITATE (163) OR OF THE FLUID FILTERED FROM THE PRECIPITATES FORMED (164), ADD AMMONIA, THEN SOME PHOSPHATE OF SODA, AND, SHOULD A PRECIPITATE NOT AT ONCE FORM, RUB THE INNER SIDES OF THE GLASS-TUBE WITH A GLASS ROD, AND LET THE MIXTURE STAND FOR SOME TIME.

1. NO PRECIPITATE IS FORMED; absence of magnesia. Evaporate another portion of the fluid to dryness,* and ignite gently. **167** *If a residue remains,* treat the remainder of the fluid the same as the sample, and examine the residue, which by the moderate ignition to which it has been subjected has been freed from ammonia, for potassa and soda, according to the directions of § 197.—*If no residue is left,* this is a proof of the absence of the fixed alkalies; pass on at once to § 198.

2. A PRECIPITATE IS FORMED: presence of MAGNESIA. As testing **168** for alkalies can proceed with certainty only after the removal of magnesia, evaporate the remainder of the fluid to dryness, and ignite until all ammoniacal salts are removed. Warm the residue with some water, add baryta-water (prepared from the crystals) as long as a precipitate continues to form, boil, filter, add to the filtrate a mixture of carbonate of ammonia with some caustic ammonia in slight excess, heat for some time gently, filter, evaporate the filtrate to dryness,

---

* The most convenient way is to conduct the evaporation on the lid of a platinum crucible.

adding some chloride of ammonium during the process (to convert into chlorides the caustic alkalies or alkaline carbonates that may happen to form), ignite the residue gently, then dissolve in a little water, precipitate if necessary once more with ammonia and carbonate of ammonia, evaporate again, and if a residue remains, ignite this gently, and examine it according to the directions of § 197.

§ 197.

(Examination for Potassa and Soda.)

YOU HAVE NOW TO EXAMINE FOR POTASSA AND SODA THE GENTLY IGNITED RESIDUE, FREE FROM SALTS OF AMMONIA AND ALKALINE EARTHS, WHICH HAS BEEN OBTAINED IN (167), OR IN (168).

Dissolve it in a little water, filter if necessary, evaporate until there is only a small quantity of fluid left, and transfer one-half of this to a watch-glass, leaving the other half in the porcelain dish.

1. To the one-half in the porcelain dish add, after cooling, a few drops of solution of *bichloride of platinum*. If a yellow crystalline precipitate forms immediately, or after some time, POTASSA is present. Should no precipitate form evaporate to dryness at a gentle heat, and treat the residue with a very small quantity of water, or, if chlorides alone are present, with a mixture of water and alcohol, when the presence of minute traces of potassa will be revealed by a small quantity of a heavy yellow powder being left undissolved (§ 89, 3). 169

2. To the other half of the fluid (on the watch-glass) add some *antimonate of potassa*. If this produces at once or after some time a crystalline precipitate, SODA is present. If the quantity of soda present is only very trifling, it often takes twelve hours before minute crystals of antimonate of soda will separate; you must therefore always wait full that time for the possible manifestation of the reaction, before deciding from its non-appearance that no soda is present. As regards the form of the crystals, consult § 90, 2. 170

§ 198.

(Examination for Ammonia.)

THERE REMAINS STILL THE EXAMINATION FOR AMMONIA. Triturate some of the body under examination or, if a fluid, a portion of the latter, together with an excess of hydrate of lime, and, if necessary, a little water. If the escaping gas smells of ammonia, if it restores the blue color of reddened litmus-paper, and forms white fumes with hydrochloric acid vapors, brought into contact with it by means of a glass rod, AMMONIA is present. The reaction is the most sensitive if the trituration is made in a small beaker, and the latter covered with a glass plate with a slip of moistened turmeric or moist reddened litmus-paper adhering to the under side. 171

DETECTION OF ACIDS.

Complex Compounds.

A, 1. Substances soluble in Water.

DETECTION OF ACIDS.*

I. *In the Absence of Organic Acids.*

§ 199.

Consider, in the first place, *which* are the acids that form with the bases found compounds soluble in water, and let this guide you in the examination. To students the table given in Appendix IV. will prove of considerable assistance.

1. The ACIDS of ARSENIC, as well as CARBONIC ACID, HYDROSUL- **172** PHURIC ACID, CHROMIC ACID, and SILICIC ACID, have generally been detected already in the course of testing for the bases; see (67) and (68).

2. Add to a portion of the solution chloride of barium or, if lead, silver, or suboxide of mercury are present, nitrate of baryta, and, should the reaction of the fluid be acid, add ammonia to neutral or slightly alkaline reaction.

a. No PRECIPITATE IS FORMED: absence of sulphuric acid, **173** phosphoric acid, chromic acid, silicic acid, oxalic acid, arsenious and arsenic acids, as well as of notable quantities of boracic acid and hydrofluoric acid.† Pass on to (175).

b. A PRECIPITATE IS FORMED. Dilute the fluid, and add **174** hydrochloric acid or, as the case may be, nitric acid; if the precipitate does not redissolve, or at least not completely, SULPHURIC ACID is present.

3. Add nitrate of silver to a portion of the solution. If this **175** fails to produce a precipitate, test the reaction, and if acid, add to the fluid some dilute ammonia, taking care to add the reagent so gently and cautiously that the two fluids do not intermix; if the reaction is alkaline, on the other hand, add with the same care some dilute nitric acid, instead of ammonia, and watch attentively whether a precipitate or a cloud will form in the layer between the two fluids.

a. No PRECIPITATE IS FORMED IN THE LAYER BETWEEN THE **176** TWO FLUIDS, NEITHER IMMEDIATELY NOR AFTER SOME TIME. Pass on to (181); there is neither chlorine, bromine, iodine, cyanogen,‡ ferro- and ferricyanogen present, nor sulphur; nor phosphoric acid, arsenic acid, arsenious acid, chromic acid, silicic acid, oxalic acid; nor boracic acid, if the solution was not too dilute.

b. A PRECIPITATE IS FORMED. Observe the color§ of it, **177** then add nitric acid, and shake the mixture.

---

* Consult also the explanations in Section III.

† If the solution contains an ammoniacal salt in somewhat considerable proportion, the non-formation of a precipitate cannot be considered a conclusive proof of the absence of these acids, since the baryta salts of most of them (not the sulphate) are in presence of ammoniacal salts more or less soluble in water.

‡ That the cyanogen in cyanide of mercury is not indicated by nitrate of silver has been mentioned (73).

§ Chloride, bromide, cyanide, and ferrocyanide of silver, and oxalate, silicate, and borate of silver are white; iodide of silver, tribasic phosphate, and arsenite of silver are yellow; arsenate of silver and ferricyanide of silver are brownish-red; chromate of silver is purple-red; sulphide of silver black.

a. The precipitate redissolves completely: absence of chlorine, bromine, iodine, cyanogen, ferro- and ferricyanogen, and also of sulphur. Pass on to (181).

β. *A residue is left:* chlorine, bromine, iodine, cyanogen, **178** ferro- or ferricyanogen may be present; and if the residue is black or blackish, HYDROSULPHURIC ACID or a soluble METALLIC SULPHIDE.—The presence of sulphur may, if necessary, be readily established beyond doubt, by mixing another portion of the solution with some solution of sulphate of copper.

aa. Test another portion of the fluid for IODINE, and subsequently for BROMINE, by the methods described in § 157.

bb. Test a small portion of the fluid with sesqui- **179** chloride of iron for FERROCYANOGEN; and, if the color of the silver precipitate leads you to suspect the presence of FERRICYANOGEN, test another portion for this latter substance with sulphate of iron.—If the original solution has an alkaline reaction, some hydrochloric acid must be added before the addition of the sesquichloride of iron, or the sulphate of iron.

cc. CYANOGEN, if present in form of a simple cyanide of an alkali metal soluble in water, may usually be readily recognised by the smell of hydrocyanic acid which the body under examination emits, and which is rendered more strongly perceptible by addition of a little dilute sulphuric acid.—If no ferrocyanogen or ferricyanogen is present, the presence of cyanogen may be ascertained by the method given in § 155, 6.

dd. Should bromine, iodine, cyanogen, ferrocyanogen, **180** ferricyanogen, and sulphur not be present, the precipitate which nitric acid has failed to dissolve consists of CHLORIDE of silver.

But where the analytical process has revealed the presence of any of the other bodies, a special examination for chlorine may become necessary, viz. in cases where the *quantity* of the precipitate will not enable the operator to pronounce with positive certainty on the presence or absence of the latter element.*
In such cases, which are of rare occurrence however, the methods given in § 157 are resorted to.

4. Test another portion for NITRIC ACID, by means of sulphate **181** of iron and sulphuric acid (§ 159).

5. To ascertain whether CHLORIC ACID is present, pour a little concentrated sulphuric acid over a small sample of the solid substance on a watch-glass: ensuing yellow coloration of the acid resolves the question in the affirmative (§ 160).

You have still to test for phosphoric acid, boracic acid, silicic acid, oxalic acid, and chromic acid, as well as for hydrofluoric acid.

For the first five acids test only in cases where both chloride of barium and nitrate of silver have produced precipitates in neutral solutions. Compare also foot note to (173).

6. Test for PHOSPHORIC ACID, by adding to a portion of the fluid

---

* Supposing, for instance, the solution of nitrate of silver to have produced a copious precipitate insoluble in nitric acid, and the subsequent examination to have shown mere traces of iodine and bromine, the presence of chlorine may be held to be demonstrated, without requiring additional proof.

ammonia in excess, then chloride of ammonium and sulphate of **182**
magnesia (§ 142, 7). Very minute quantities of phosphoric acid
are detected most readily by means of molybdic acid (§ 142, 10).

7. To effect the detection of OXALIC ACID and HYDROFLUORIC ACID, add
chloride of calcium to a fresh portion of the solution. If the reaction of
the fluid is acid, add ammonia to alkaline reaction. If the chloride of
calcium produces a precipitate which is not redissolved by addition of
acetic acid, one or both bodies are present. Examine therefore now a
sample of the original substance for fluorine according to the directions
of § 146, 5, another sample for oxalic acid by the method given in
§ 145, 7.

8. Acidulate a portion of the fluid slightly with hydrochloric **183**
acid, then test for BORACIC ACID, by means of turmeric-paper
(§ 144, 6).

9. Should SILICIC ACID not yet have been found in the course of
testing for the bases, acidulate a portion of the fluid with hydrochloric
acid, evaporate to dryness, and treat the residue with hydrochloric acid
(§ 150, 3).

10. CHROMIC ACID is readily recognised by the yellow or red color of
the solution, and by the purple-red color of the precipitate produced by
nitrate of silver. If there remains the least doubt on the point, test
for chromic acid with acetate of lead and acetic acid (§ 138, 7).

Complex Compounds.

A, 1. SUBSTANCES SOLUBLE IN WATER.

DETECTION OF ACIDS.

II. *In Presence of Organic Acids.*

§ 200.

1. The examination for the inorganic acids, inclusive of oxalic **184**
acid, is made in the manner described in § 199. As the tartrates
and citrates of baryta and oxide of silver are insoluble in water,
tartaric acid and citric acid can be present only in cases where both
chloride of barium and nitrate of silver have produced precipitates in
the neutral fluid; still, in drawing a conclusion, you must bear in
mind that the said salts are slightly soluble in solutions of salts of
ammonia.

The proper testing for the organic acids requires in the first place the
removal of those bases the presence of which might prove an obstacle,
i.e., all the bases of groups III., IV., V., and VI. Their removal is effected
by the methods described in § 184, at the beginning; and the examination for the organic acids is then conducted as follows :—

2. Make a portion of the fluid feebly alkaline by addition of **185**
ammonia, add some chloride of ammonium, then chloride of calcium, shake vigorously, and let the mixture stand at rest from ten
to twenty minutes.

  a. No PRECIPITATE IS FORMED, NOT EVEN AFTER THE LAPSE OF
SOME TIME. Absence of tartaric acid ; pass on to (**186**).

  b. A PRECIPITATE IS FORMED IMMEDIATELY, OR AFTER SOME
TIME. Filter, wash, and keep the filtrate for further examination
according to the directions of (**186**).

Digest and shake the precipitate with solution of soda, without applying heat, then dilute with a little water, filter, and boil the filtrate some time. If a precipitate separates, TARTARIC ACID may be assumed to be present. Filter hot, and subject the precipitate to the ammonia and nitrate of silver test described in § 163, 8.

3. Mix the fluid in which chloride of calcium has failed to produce a precipitate, or that which has been filtered from the precipitate formed—in which latter case some more chloride of calcium is to be added—with alcohol. **186**

    a. NO PRECIPITATE IS FORMED. Absence of citric acid and malic acid. Pass on to (190). **187**

    b. A PRECIPITATE IS FORMED. Filter and treat the filtrate as directed in (190). As regards the precipitate, treat this as follows:— **188**

After washing with some alcohol, dissolve on the filter in a little dilute hydrochloric acid, add ammonia to the filtrate to alkaline reaction, and boil for some time.

    α. THE FILTRATE REMAINS CLEAR. Absence of citric acid. Probable presence of MALIC ACID. Add alcohol again to the fluid, and test the lime precipitate in the manner directed § 166, to make sure whether malic acid is really present or not.

    β. A HEAVY WHITE PRECIPITATE IS FORMED. Presence of CITRIC ACID. Filter boiling, and test the filtrate for malic acid in the same manner as in α. To remove all doubt as to whether the precipitate is citrate of lime or not, it is advisable to dissolve once more in some hydrochloric acid, to supersaturate again with ammonia, and to boil; if the precipitate really consisted of citrate of lime, it will now be thrown down again. (Compare § 164, 3.) **189**

4. Heat the filtrate of (188) (or the fluid in which addition of alcohol has failed to produce a precipitate (187), to expel the alcohol, neutralize *exactly* with hydrochloric acid, and add sesquichloride of iron. If this fails to produce a light brown flocculent precipitate, neither succinic nor benzoic acid is present. If a precipitate of the kind is formed, filter, digest, and heat the washed precipitate with ammonia in excess; filter, evaporate the filtrate nearly to dryness, and test a portion for SUCCINIC ACID with chloride of barium and alcohol (§ 168); the remainder for BENZOIC ACID with hydrochloric acid (§ 169). Benzoic acid may generally be readily detected also in the original substance, by pouring some dilute hydrochloric acid over a small portion of the latter, which will leave the benzoic acid undissolved; it is then filtered and heated on platinum foil (§ 169, 1). **190**

5. Evaporate a portion of the solution to dryness—if acid, after previous saturation with soda—introduce the residue or a portion of the original dry substance into a small tube, pour some alcohol over it, add about an equal volume of concentrated sulphuric acid, and heat to boiling. Evolution of the odor of acetic ether demonstrates the presence of ACETIC ACID. This odor is rendered more distinctly perceptible by shaking the cooling or cold mixture. **191**

6. To effect the detection of FORMIC ACID, add to a portion of the solution a sufficient quantity of nitrate of silver, then soda **192**

until the fluid is *exactly* neutralized, and boil. If formic acid is present, reduction of the silver to the metallic state ensues (§ 172, 4). The reaction with nitrate of suboxide of mercury may be had recourse to as a conclusive test (§ 172, 5).*

Complex Compounds.

A, 2. Substances insoluble in Water, but soluble in Hydrochloric Acid, Nitric Acid, or Nitrohydrochloric Acid.

DETECTION OF THE ACIDS.

In the Absence of Organic Acids.

§ 201.

In the examination of these compounds attention must be directed to all acids, with the exception of chloric acid. Cyanogen compounds and silicates are not examined by this method. (Compare § 204 and § 205.)

1. Carbonic acid, sulphur (in form of metallic sulphides), arsenious acid, arsenic acid, and chromic acid, if present, have been found already in the course of the examination for bases; nitric acid, if present, has been detected in the course of the preliminary examination,* by the ignition of the powdered substance in a glass tube (8). **193**

2. Mix a sample of the substance with 4 parts of pure carbonate of soda and potassa, and, should a metallic sulphide be present, add some nitrate of soda; fuse the mixture in a platinum crucible if there are no reducible metallic oxides present, in a porcelain crucible if such oxides are present; boil the fused mass with water, and add a little nitric acid, leaving the reaction of the fluid, however, still alkaline; heat again, filter, and proceed with the filtrate according to the directions of § 199, to effect the detection of all the acids which were combined with the bases.† **194**

3. As the phosphates of the alkaline earths are only incompletely decomposed by fusion in conjunction with carbonate of soda and potassa, it is always advisable in cases where alkaline earths are present, and phosphoric acid has not yet been detected, to dissolve a fresh sample of the body under examination in hydrochloric acid or nitric acid, and test the solution for phosphoric acid with solution of molybdic acid, after removal of the arsenic acid and silicic acid, should these be present. (§ 142, 10.) **195**

4. If in the course of the examination for bases alkaline earths have

---

* In presence of chromic acid the reduction of oxide of silver and of suboxide of mercury is not a positive proof of the presence of formic acid. In cases where the two acids are present the following method must be resorted to :—Mix the original solution with some nitric acid, add oxide of lead in excess, shake the mixture, filter, add to the filtrate dilute sulphuric acid in excess, and distil. Test the distillate as directed § 173. In presence of tartaric acid also it is the safest way to distil the formic acid first, with addition of dilute sulphuric acid.

† If the body examined has been found to contain a metallic sulphide, a separate portion of it must be examined for sulphuric acid, by heating it with hydrochloric acid, filtering, adding water to the filtrate, and then testing the fluid with chloride of barium.

been found, it is also advisable to test a separate portion of the body under examination for FLUORINE, by the method described in § 146, 5.

5. That portion of the substance under examination which has been treated as directed in (194), can be tested for SILICIC ACID only in cases where the fusion has been effected in a platinum crucible; in cases where a porcelain crucible has been used it is necessary to examine a separate portion of the body for silicic acid, by evaporating the hydrochloric or nitric acid solution (§ 150, 3).

6. Examine a separate sample of the body for OXALIC ACID as directed in (198).

Complex Compounds.

A, 2. SUBSTANCES INSOLUBLE IN WATER, BUT SOLUBLE IN HYDROCHLORIC ACID, NITRIC ACID, OR NITRO-HYDROCHLORIC ACID.

DETECTION OF THE ACIDS.

II. *In Presence of Organic Acids.*

§ 202.

1. Conduct the examination for inorganic acids according to the directions of § 201.
2. Test for ACETIC ACID as directed § 171, 7,
3. Dissolve a portion of the compound under examination in the least possible amount of hydrochloric acid, filter if necessary, and test the undissolved residue which may be left for BENZOIC ACID, by application of heat; add to the filtrate solution of carbonate of soda in considerable excess, and, besides this, also a little solid carbonate of soda, boil the mixture for a few minutes, then filter the fluid from the precipitate. In the filtrate you have now all the organic acids in solution, combined with soda. Acidify the filtrate with hydrochloric acid, heat, and proceed according to the direction of (185).

Complex Compounds.

B, SUBSTANCES INSOLUBLE OR SPARINGLY SOLUBLE BOTH IN WATER AND IN HYDROCHLORIC ACID, NITRIC ACID, OR NITRO-HYDROCHLORIC ACID.

DETECTION OF THE BASES, ACIDS, AND NON-METALLIC ELEMENTS.

§ 203.

To this class belong the following bodies and compounds.
SULPHATE OF BARYTA, SULPHATE OF STRONTIA, and SULPHATE OF LIME.*
SULPHATE OF LEAD† and CHLORIDE OF LEAD.‡

---

* Sulphate of lime passes partially into the solution effected by water, and often completely into that effected by acids.
† Sulphate of lead may pass completely into the solution effected by acids.
‡ Chloride of lead can here only be found if the precipitate insoluble in acids has not been thoroughly washed with hot water.

CHLORIDE OF SILVER, bromide of silver, iodide of silver, cyanide of silver,* ferro- and ferricyanide of silver.†
SILICIC ACID and many SILICATES.
Native alumina, or alumina which has passed through a process of intense ignition, and many aluminates.
Ignited sesquioxide of chromium and CHROME-IRONSTONE (a compound of sesquioxide of chromium and protoxide of iron).
Ignited and native binoxide of tin (tin-stone).
Some metaphosphates and some arsenates.
FLUORIDE OF CALCIUM and a few other compounds of fluorine.
SULPHUR.
CARBONACEOUS MATTER.
Of these compounds those printed in small capitals are more frequently met with. As the silicates perform a highly important part in mineral analysis, a special chapter (§ 205—§ 208) is devoted to them.

The substance under examination which is insoluble in water and in acids is in the first place subjected to the preliminary experiments here described in *a—e*, if the quantity at the disposal of the operator is not absolutely too small to admit of this proceeding; in cases where the quantity proves insufficient for the purpose, the operator must omit this preliminary examination, and at once pass on to (205,) bearing in mind, however, that the body may contain *all* the aforesaid substances and compounds.

a. Examine attentively and carefully the physical state and condition of the residue, to ascertain whether you have to deal with a homogeneous mass or with a mass composed of dissimilar particles; whether the body is sandy or pulverulent, whether it has the same color throughout, or is made up of variously-colored particles, &c. The microscope, or even a simple magnifying glass, will be found very useful at this stage of the examination. **200**

b. Heat a small sample in a glass tube sealed at one end. If brown fumes arise, and SULPHUR sublimes, this is of course a proof of the presence of that substance. **201**

c. If the substance is black, this indicates, in most cases, the presence of carbonaceous matter. Heat a small sample on platinum-foil over the blowpipe flame; if the substance which blackens the fingers is consumed, this may be held to be a positive proof of the presence of CARBON in some shape or other. Graphite, which may be readily recognised by its property of communicating its blackish-gray color to the fingers, to paper, &c., requires the aid of oxygen for its ready combustion. **202**

d. Heat a small sample, together with a small lump of cyanide of potassium and some water, for some time, filter, and test the filtrate with sulphide of ammonium. The formation of a brownish-black precipitate shows that the residue under examination contains a compound of SILVER. **203**

e. If an undissolved residue has been left in *d*, wash this tho- **204**

---

* Bromide, iodide, and cyanide of silver are decomposed by boiling with nitro-hydrochloric acid, and converted into chloride of silver; they can accordingly be found here only in cases where the operator has to deal with a substance which—as nitrohydrochloric acid has failed to effect its solution—is examined directly by the method described in this paragraph (§ 203).

† With regard to the examination of these compounds, compare also § 204.

roughly with water, and if white, sprinkle a few drops of sulphide of ammonium over it; if it turns black, salts of LEAD are present. If, however, the residue left in *d* is black, heat it with some acetate of ammonia, adding a few drops of acetic acid, filter, and test the filtrate for LEAD, by means of sulphuric acid and hydrosulphuric acid.*

The results obtained by these preliminary experiments serve to guide the operator now in his further course of proceeding.

1, *a*. SALTS OF LEAD ARE NOT PRESENT. Pass on to (206). **205**

    b. SALTS OF LEAD ARE PRESENT. Heat the substance repeatedly with a concentrated solution of acetate of ammonia until the salt of lead is completely dissolved out. Test a portion of the filtrate for CHLORINE, another for SULPHURIC ACID, and the remainder for LEAD, by addition of sulphuric acid in excess, and by hydrosulphuric acid. If acetate of ammonia has left a residue, wash this, and treat it as directed in (206).

2, *a*. SALTS OF SILVER ARE NOT PRESENT. Pass on to (207). **206**

    b. SALTS OF SILVER ARE PRESENT. Digest the substance free from lead, or which has been freed from that metal by acetate of ammonia, repeatedly with cyanide of potassium and water, at a gentle heat (in presence of sulphur, in the cold), until all the salt of silver is removed. If an undissolved residue is left, wash this, and proceed with it according to the directions of (207). Of the *filtrate*, which contains cyanide of potassium, mix the larger portion with sulphide of ammonium, to precipitate the silver. Wash the precipitated sulphide of silver, then dissolve in nitric acid, dilute the solution, and add hydrochloric acid, to ascertain whether the precipitate really consisted of sulphide of silver. Test another small portion of the filtrate for SULPHURIC ACID.†

3, *a*. SULPHUR IS NOT PRESENT. Pass on to (208). **207**

    b. SULPHUR IS PRESENT. Heat the substance free from silver and lead in a covered porcelain crucible until all the sulphur is expelled, and if a residue is left, treat this according to the directions of (208).

4. Mix the substance free from silver, lead, and sulphur with **208** 2 parts of carbonate of soda, 2 parts of carbonate of potassa, and 1 part of nitrate of potassa,‡ heat the mixture in a platinum crucible until the mass is in a state of calm fusion, place the red-hot crucible on a thick cold iron plate, and let it cool. By this means you will generally succeed in removing the fused mass from the crucible in an unbroken lump. Soak the mass now in water, boil, filter, and wash the residue until chloride of barium no longer produces a precipitate in the washings. (Add only the first washings to the filtrate.)

    a. The *solution so obtained* contains the acids which were **209**

---

* The presence of lead in silicates, *e.g.* in glass containing lead, cannot be detected by this method.

† As the carbonate of potassa contained in the cyanide of potassium may have produced a total or partial decomposition of any sulphates of the alkaline earths which happened to be present.

‡ Addition of nitrate of potassa is useful even in the case of white powders, as it counteracts the injurious action of silicate of lead, should any be present, upon the platinum crucible. In the case of black powders the proportion of nitrate of potassa must be correspondingly increased, in order that carbon, if present, may be consumed as completely as possible, and that any chrome-ironstone present in the compound may be more thoroughly decomposed.

present in the substance decomposed by fluxing. But it may, besides these acids, contain also such bases as are soluble in caustic alkalies. Proceed as follows :—

α. Test a small portion of the solution for SULPHURIC ACID.

β. Test another portion with molybdic acid for PHOSPHORIC ACID and ARSENIC ACID. If a yellow precipitate forms, remove the arsenic acid which may be present with hydrosulphuric acid, and then test once more for phosphoric acid, after having removed also any silicic acid that may be present.

γ. Test another portion for FLOURINE (§ 146, 7).

δ. If the solution is yellow, CHROMIC ACID is present. To remove all doubt on the point, acidify a portion of the solution with acetic acid, and test with acetate of lead.

ε. Acidify the remainder of the solution with hydrochloric **210** acid, evaporate to dryness, and treat the residue with hydrochloric acid and water. If a residue is left which refuses to dissolve even in boiling water, this consists of SILICIC ACID. Test the hydrochloric acid solution now in the usual way for those bases which, being soluble in caustic alkalies, may be present.

b. Dissolve the residue left in (208) in hydrochloric acid **211** (effervescence indicates the presence of alkaline earths), and test the solution for the bases as directed in § 190. (If much silicic acid has been found in ε (210), it is advisable to evaporate the solution of the residue to dryness, and to treat the residuary mass with hydrochloric acid and water, in order that the silicic acid remaining may also be removed as completely as possible.)

5. If you have found in 4 that the residue insoluble in acids **212** contains a silicate, treat a separate portion of it according to the directions of (228), to ascertain whether this silicate contains alkalies.

6. If a residue is still left undissolved upon treating the **213** residue left in (208) with hydrochloric acid (211), this may consist either of silicic acid, which has separated, or of an undecomposed portion of sulphate of baryta; it may, however, also be fluoride of calcium, and if it is dark-colored, chrome-ironstone, as the last-named two compounds are only with difficulty decomposed by the method given in (208). I would therefore remind the student that fluoride of calcium may be readily decomposed by means of sulphuric acid; and, as regards the decomposition of chrome-ironstone, I can recommend the following method, first proposed by *Hart:* Project the fine powder into 8 times the quantity of fused borax, stir the mixture frequently, and keep the crucible for half-an-hour at a bright red heat. Add now to the fusing mass carbonate of soda so long as effervescence continues, and finally add 3 times the weight of the chrome-ironstone of a mixture of equal parts of carbonate of soda and nitrate of potassa, whilst actively stirring the mixture with a platinum wire. Let the mass cool, and, when cold, boil it with water.

7. If the residue insoluble in acids contained silver, you have **214** still to ascertain whether that metal was present in the original substance as chloride, bromide, iodide, &c., of silver, or whether it has been converted into the form of chloride of silver by the treatment employed to effect the solution of the original substance. For that

purpose treat a portion of the original substance with boiling water until the soluble part is completely removed; then treat the residuary portion in the same way with dilute nitric acid, wash the undissolved residue with water, and test a small sample of it for silver according to the directions of (**203**). If silver is present, proceed to ascertain the salt-radical with which the metal is combined; this may easily be effected by boiling the remainder of the residue in the first place with rather dilute solution of soda, filtering, and testing the filtrate, after acidifying it, for ferro- and ferricyanogen. Digest the washed residue now with finely granulated zinc and water, with addition of some sulphuric acid, and filter after the lapse of ten minutes. You may now at once test the filtrate for chlorine, bromine, iodine, and cyanogen; or you may first throw down the zinc with carbonate of soda, in order to obtain the salt-radicals in combination with sodium.

SECTION II.

PRACTICAL COURSE

IN PARTICULAR CASES.

I. SPECIAL METHOD OF EFFECTING THE ANALYSIS OF CYANIDES, FERROCYANIDES, ETC., INSOLUBLE IN WATER, AND ALSO OF INSOLUBLE MIXED SUBSTANCES CONTAINING SUCH COMPOUNDS.*

§ 204.

THE analysis of ferrocyanides, ferricyanides, &c., by the common method is often attended by the manifestation of such anomalous reactions as easily to mislead the analyst. Moreover, acids often fail to effect the complete solution of these compounds. For these reasons it is advisable to analyze them, and mixtures containing them, by the following special method:— 215

Treat the substance with water until the soluble parts are entirely removed, and boil the residue with strong solution of potassa or soda; after a few minutes' ebullition add some carbonate of soda, and boil again for some time; filter, should a residue remain, and wash the latter.

1. The *residue*, if any has been left, is now free from cyanogen, unless the substance under examination contains cyanide of silver, in which case the residue would of course still contain cyanogen. Examine the residue now by the common method, beginning at (**35**). 216

2. The *solution* or *filtrate*, which, if combinations of compound cyanogen radicals (Ferrocyanogen, Cobalticyanogen, &c.), were originally present, contains these combined with alkali metals, may also contain other acids, which have been separated from their bases by the process of boiling with car- 217

---

* Before entering upon this course of analysis, consult the special remarks to the paragraph (§ 204) in the Third Section.

bonate of soda, and lastly also, such oxides as are soluble in caustic alkalies.

Treat the solution as follows:—

a. Mix the alkaline fluid with a sufficient quantity of hydrosulphuric acid, to test for metals of the fourth and fifth groups.* **218**

α. *No permanent precipitate is formed.* Absence of zinc and lead. Pass on to (**219**).

β. *A permanent precipitate is formed.* Add to the fluid a little yellow sulphide of sodium, drop by drop, until the metals of the fourth and fifth groups present in the alkaline solution are just thrown down, heat moderately, filter, wash the precipitate, and treat the filtrate as directed in (**219**). Dissolve the washed precipitate in nitric acid, which may leave sulphide of mercury behind, and examine the solution for copper and lead, as well as for zinc and other metals of the fourth group, which may, in the same way as copper, have passed into the alkaline solution, by the agency of organic matters.

b. To test the alkaline fluid, which now also contains some sulphide of an alkali metal, for mercury, which may be present, as its sulphide is soluble in sulphide of potassium, and for the metals of the sixth group, mix with a sufficient quantity of water, then with dilute nitric acid to acid reaction, and if the fluid does not smell strongly of hydrosulphuric acid, add some more of the latter reagent. **219**

α. *No precipitate is formed.* Absence of mercury and of the oxides of the sixth group. Pass on to (**220**).

β. *A precipitate is formed.* Filter, wash the precipitate, then examine it for mercury and the metals of the sixth group according to the directions of § 191.

c. The fluid, acidified with nitric acid, and therefore abundantly supplied with nitrates of alkalies, may still contain those metals which in combination with cyanogen form compound radicals (iron, cobalt, manganese, chromium), and, besides these, also alumina. You have to test it also for cyanogen, respectively ferrocyanogen, cobalticyanogen, &c., and for other acids. Divide it, therefore, into two portions, α and β. Examine α for the acids according to the directions of § 199, or, as the case may be, §. 200. (Cobalticyanogen may be recognised as such by its giving with salts of nickel greenish, with salts of manganese and zinc white precipitates, in which the presence of cobalt is revealed by fusing with borax.) **220**

Evaporate β to dryness, and heat the residue to fusion. Pour the fused mass upon a piece of porcelain, boil with water, filter, and examine the residue for IRON, MANGANESE, COBALT, and ALUMINA. Test a portion of the filtrate (if yellow) for CHROMIC ACID, the remainder for ALUMINA (which may have passed partially or com-

---

* The analyst must, of course, avoid adding solution of hydrosulphuric acid, or conducting hydrosulphuric acid gas into the fluid, until the mixture smells of the reagent (accordingly, until all the alkali present has been converted into hydrosulphate of sulphide of alkali metal), since this might lead to the precipitation also of the alumina which may be present in the alkaline solution, and even of sulphides of metals of the sixth group—a precipitation which is not intended here.

pletely into the solution, through the agency of the caustic alkalies formed in the process of fusion from the nitrates of the alkalies present.)

II. ANALYSIS OF SILICATES.

§ 205.

Whether the body to be analyzed is a silicate or contains one, is ascertained by the preliminary examination with phosphate of soda and ammonia before the blowpipe; since in the process of fusion the metallic oxides dissolve, whilst the separated silicic acid floats about in the liquid bead as a transparent swollen mass (§ 150, 8).

The analysis of the silicates differs, strictly speaking, from the common course only in so far as the preliminary treatment is concerned, which is required to effect the separation of the silicic acid from the bases, and to obtain the latter in solution.

The silicates and double silicates are divided into two distinct classes, which require respectively a different method of analysis; viz., (1) silicates readily decomposable by acids (hydrochloric acid, nitric acid, sulphuric acid), and (2) silicates which are not, or only with difficulty, decomposed by acids. Many minerals consist of mixtures of the two classes of silicates.

To ascertain to which of these two classes a silicate belongs, reduce it to a very fine powder, and digest a portion with hydrochloric acid at a temperature near the boiling-point. If this fails to decompose it, try another portion by long-continued heating with a mixture of three parts of concentrated sulphuric acid and 1 part of water. If this also fails, after some time, to produce the desired effect, the silicate belongs to the second class. Whether decomposition has been effected by the acid or not, may generally be learned from external indications, as a colored solution forms almost invariably, and the separated gelatinous, flocculent, or finely-pulverulent hydrate of silicic acid takes the place of the original heavy powder which grated under the glass rod with which it was stirred. But whether the decomposition is complete, or extends only to one of the components of this mineral, may be ascertained by boiling the separated hydrate of silicic acid in a solution of carbonate of soda. If perfect solution ensues, complete decomposition has been effected; if not, the decomposition is only partial. The result of these preliminary tests will show whether the silicate should be examined according to § 206, or to § 207, or to § 208.

Before proceeding further, examine a portion of the pulverized compound also for water, by heating it in a perfectly dry glass tube. If the substance contains hygroscopic moisture, it must first be dried by protracted exposure to a temperature of 212° F. Apply a gentle heat at first, but ultimately an intense heat, by means of the blowpipe; you may also conveniently combine with this a preliminary examination for fluorine (§ 146, 8).

A. SILICATES DECOMPOSABLE BY ACIDS.

§ 206.

a. Silicates decomposable by hydrochloric acid or by nitric acid.*

1. Digest the finely pulverized silicate with hydrochloric acid at

---

* Nitric acid is preferable to hydrochloric acid in cases where compounds of silver or lead are present.

a temperature near the boiling-point, until complete decomposition is effected, filter off a small portion of the fluid, evaporate the remainder, together with the silicic acid suspended therein, to dryness, and expose the residue to a temperature somewhat exceeding 212° F., with constant stirring, until no more, or very few, hydrochloric acid fumes escape; allow it to cool, moisten the residue with hydrochloric acid or, as the case may be, with nitric acid, afterwards add a little water, and heat gently for some time. **222**

This operation effects the separation of the silicic acid, and the solution of the bases in the form of chlorides, or, as the case may be, nitrates. Filter, wash the residue thoroughly, and examine the solution by the common method, beginning at § 189, II. or III.* To be quite safe, the residuary silicic acid may be digested with ammonia, filtered, and the filtrate tested for silver, by supersaturation with nitric acid.

2. As in silicates, and more particularly in those decomposed by hydrochloric acid, there are often found other acids, as well as metalloids, the following observations and instructions must be attended to, that none of these substances may be overlooked:— **223**

α. SULPHIDES of METALS and CARBONATES are detected in the process of treating with hydrochloric acid.

β. If the separated silicic acid is black, and turns subsequently white upon ignition in the air, this indicates the presence of CARBON or of ORGANIC SUBSTANCES. In presence of the latter, the silicates emit an empyreumatic odor upon being heated in a glass tube.

γ. Test the portion of the hydrochloric acid solution filtered off before evaporating for SULPHURIC ACID, PHOSPHORIC ACID, and ARSENIC ACID—for sulphuric acid with chloride of barium, after diluting with water; for arsenic acid by heating the solution to 158° Fah., conducting sulphuretted hydrogen into it, and examining the precipitate formed; for phosphoric acid with solution of molybdate of ammonia in nitric acid. Where arsenic is found, the fluid filtered off the sulphide of arsenic is tested for the phosphoric acid, after the removal of the hydrosulphuric acid.

δ. BORACIC ACID is best detected by fusing a portion of the substance in a platinum spoon with carbonate of soda and **224**

---

* Minute traces of titanic acid are occasionally met with in silicates. If the separation of the silicic acid has been effected on the water-bath, the greater part of the titanic acid passes into the hydrochloric acid solution, whilst a small portion separates along with the silicic acid. The titanic acid is found in the following way: *a*. Heat the silicic acid repeatedly in a platinum dish with hydrofluoric acid and sulphuric acid until the silica is completely removed as fluoride of silicon, then evaporate to dryness. *b*. Precipitate the solution with ammonia, filter, wash the precipitate (which contains alumina, sesquioxide of iron, &c., and the rest of the titanic acid), add to it the residue left in *a*, and fuse the whole with a quantity of acid sulphate of potassa sufficient to effect solution; continue heating until the greater part of the excess of sulphuric acid is removed. Let the fused mass cool, dissolve in cold water (which, if the operation has been successful, and no more silicic acid remains, will effect complete solution to a clear fluid); filter if necessary, dilute freely, transmit sulphuretted hydrogen gas through the fluid until the whole of the sesquioxide of iron present is reduced to protoxide, and keep the fluid boiling half an hour, with constant transmission of carbonic acid through it (without previously filtering off the sulphur). The titanic acid will gradually separate, the bases remaining in solution (TH. SCHEERER). Filter, wash, ignite, and examine the residue finally as directed in § 104, 10.

potassa, boiling the fused mass with water, and examining the solution for boracic acid by the method given in § 144, 6.

ε. With many silicates, boiling with water is sufficient to dissolve the metallic CHLORIDES present, which may then be readily detected in the filtrate by means of solution of nitrate of silver; the safest way, however, is to dissolve the mineral in dilute nitric acid, and test the solution with nitrate of silver.

ζ. Metallic FLUORIDES, which often occur in silicates in greater or smaller proportion, are detected by the method described § 146, 6.

b. Silicates which resist the action of hydrochloric acid, but are decomposed by concentrated sulphuric acid.

Heat the finely pulverized mineral with a mixture of 3 parts of concentrated pure sulphuric acid and 1 part of water (best in a platinum dish), finally drive off the greater portion of the sulphuric acid, boil the residue with hydrochloric acid, dilute, filter, and treat the filtrate as directed § 190; and the residue, which, besides the separated silicic acid, may contain also sulphates of the alkaline earths, &c., according to the directions of § 203. If you wish to examine silicates of this class for acids and salt radicals, treat a separate portion of the substance according to the directions of § 207. 225

B. SILICATES WHICH ARE NOT DECOMPOSED BY ACIDS.*

§ 207.

As the silicates of this class are most conveniently decomposed by fusion with carbonate of soda and potassa, the portion so treated cannot, of course, be examined for alkalies. The analytical process is therefore properly divided into two principal parts, viz., a portion of the mineral is examined for the silicic acid and the bases, with the exception of the alkalies, whilst another portion is specially examined for the latter.—Besides these, there are some other experiments required, to obtain information as to the presence or absence of other acids. 226

1. *Detection of the silicic acid and the bases, with the exception of the alkalies.*

Reduce the mineral to a very fine powder, mix this with 4 parts of carbonate of soda and potassa, and heat the mixture in a platinum crucible over a gas or *Berzelius* spirit-lamp until the mass is in a state of calm fusion. Place the red-hot crucible on a thick cold iron plate, and let it cool there; this will generally enable you to remove the fused cake from the crucible, in which case break the mass to pieces, and keep a portion for subsequent examination for acids. Put the remainder, or, if the mass still adheres to the crucible, the latter, with its contents, into a porcelain dish, pour water over it, add hydrochloric acid, and heat gently until the mass is dissolved, with the exception of the silicic acid, which separates in flakes. Remove the crucible from 227

---

* It will be understood, from what has been stated § 205, that these are not decomposed by heating with hydrochloric acid and sulphuric acid in open vessels; but by heating them, reduced to a fine powder, in a sealed glass tube, with a mixture of 3 parts of concentrated sulphuric acid and 1 part of water, or with hydrochloric acid, to 392° or 410° Fah., most of them are decomposed, and may accordingly be analyzed also in this manner (AL. MITSCHERLICH).

the dish if necessary, evaporate the contents of the latter to dryness, and treat the residue as directed (222).

2. *Detection of the alkalies.*

To effect this, the silicates under examination must be decomposed by means of a substance free from alkalies. Hydrofluoric acid or a metallic fluoride answers this purpose best; but fusion with hydrate of baryta will also accomplish the end in view. **228**

a. DECOMPOSITION BY MEANS OF A METALLIC FLUORIDE.—Mix 1 part of the very finely pulverized mineral with 5 parts of fluoride of barium, or pure, finely pulverized fluoride of calcium, stir the mixture in a platinum crucible with concentrated sulphuric acid to a thickish paste, and heat gently for some time in a place affording a free escape to the vapors; finally heat a little more strongly, until the excess of sulphuric acid is completely expelled. Boil the residue now with water, add chloride of barium cautiously as long as a precipitate continues to form, then baryta-water to alkaline reaction, boil, filter, mix with carbonate of ammonia and some ammonia as long as a precipitate forms, and proceed exactly as directed (168).

b. DECOMPOSITION BY MEANS OF HYDRATE OF BARYTA.—Mix 1 part of the very finely pulverized substance with 4 parts of hydrate of baryta, expose the mixture for half an hour in a platinum crucible to the strongest possible heat of a good *Berzelius* or gas-lamp, and treat the fused or agglutinated mass with hydrochloric acid and water until it is dissolved; precipitate the solution with ammonia and carbonate of ammonia, filter, evaporate to dryness, ignite, dissolve the residue in water, precipitate again with ammonia and carbonate of ammonia, filter, evaporate, ignite, and test the residue for potassa and soda as directed § 197. If the residue still contains magnesia, this may be readily removed, by adding to the aqueous solution of the residue a little pure oxalic acid, evaporating to dryness, igniting the dry mass, then treating it with water, which will leave the magnesia undissolved. Filter, acidify the filtrate with hydrochloric acid, evaporate to dryness, and examine the residue for potassa and soda. **229**

3. *Examination for fluorine, chlorine, boracic acid, phosphoric acid, arsenic acid, and sulphuric acid.*

Use for this purpose the portion of the fused mass reserved in (227), or, if necessary, fuse a separate portion of the finely pulverized substance with 4 parts of pure carbonate of soda and potassa until the mass flows calmly; boil the fused mass with water, filter the solution, which contains all the fluorine as fluoride of sodium, all the chlorine as chloride of sodium, all the boracic acid as borate, all the sulphuric acid as sulphate, all the arsenic acid as arsenate, and at least part of the phosphoric acid as phosphate of soda, and treat the filtrate as follows:— **230**

a. Acidify a small portion of it with nitric acid, and test for CHLORINE with nitrate of silver.

b. Test another portion for BORACIC ACID as directed § 144, 6.

c. To effect the detection of the FLUORINE, treat a third portion of the filtrate as directed § 146, 7.

d. Acidify the remainder with hydrochloric acid and test a small portion with chloride of barium for SULPHURIC ACID; heat the remainder to 158° Fah., and test with hydrosulphuric acid for ARSENIC ACID. If no precipitate forms, evaporate the fluid, if a precipitate forms, the filtrate, to dryness, test the residue with hydrochloric acid and water, and examine the solution for PHOSPHORIC ACID with sulphate of magnesia, or with solution of molybdate of ammonia in nitric acid (§ 142).

C. SILICATES WHICH ARE PARTIALLY DECOMPOSED BY ACIDS.

§ 208.

231 Most of the native rocks and minerals are mixtures of several silicates, of which some are often decomposable by acids, others not. If such minerals were analyzed by the same method as the absolutely insoluble silicates, the analyst would indeed detect all the elements present, but the analysis would afford no satisfactory insight into the actual composition of the mineral.

It is therefore advisable to examine separately those parts of the mineral which show a different deportment with acids. For this purpose digest the very finely pulverized mineral for some time with hydrochloric acid at a gentle heat, filter off a small portion of the solution, evaporate the remainder to dryness, and expose to a temperature somewhat exceeding 212° Fah., with stirring, until no more, or very little hydrochloric acid vapor is evolved; let the residue cool, moisten it when cold with hydrochloric acid, heat gently with water, and filter.

The filtrate contains the bases of that part of the mixed mineral which has been decomposed by the hydrochloric acid; examine this as directed (**222**). Examine the portion first filtered off as directed in (**223**, γ). Test portions of the original substance for other acids as directed (**224**). Boil the residue—which, besides the silicic acid separated from the decomposed portion of the silicate, contains that part of the mixed mineral which has resisted the action of the hydrochloric acid—with an excess of solution of carbonate of soda, filter hot, and wash, first with hot solution of carbonate of soda, finally with boiling water. Treat the residuary undecomposed part of the mineral, thus freed from the admixed separated silicic acid, according to the instructions given in § 207. In cases where it is of no consequence or interest to effect the separation of the silicic acid of the part decomposed by acids, you may omit the troublesome operation with carbonate of soda, and may proceed at once to the decomposition of the residue.

III. ANALYSIS OF NATURAL WATERS.

§ 209.

232 In the examination of natural waters the analytical process is simplified by the circumstance that we know from experience the elements and compounds which are usually found in them. Now, although a quantitative analysis alone can properly inform us of the true nature and character of a water, since the differences between the various waters are principally caused by the different proportions in which the several constituents are respectively present, a qualitative analysis may yet render very good service, especially if the analyst notes with proper care whether a reagent produces a faint or a distinctly marked turbidity, a slight or a copious precipitate; since these circumstances will enable him to make an approximate estimation of the relative proportions in which the several constituents are present.

I separate here the analysis of the common fresh waters (spring-water, well-water, brook-water, river-water, &c.) from that of the mineral waters, in which latter we may also include sea-water; for, although no well-defined limit can be drawn between the two classes, still the analytical examination of the former is necessarily far more simple than that of the latter, as the number of substances to be looked for is much more limited than in the case of mineral waters.

A. ANALYSIS OF FRESH WATERS (SPRING-WATER, WELL-WATER, BROOK-WATER, RIVER-WATER, &c.).

§ 210.

We know from experience that the substances to be had regard to in the analysis of such waters are the following:— 233

a. BASES: Potassa, soda, ammonia, lime, magnesia, protoxide of iron.

b. ACIDS, &c.: Sulphuric acid, phosphoric acid, silicic acid, carbonic acid, nitric acid, nitrous acid, chlorine.

c. ORGANIC MATTERS.

d. MECHANICALLY SUSPENDED SUBSTANCES: Clay, &c.

The fresh waters contain indeed also other constituents besides those enumerated here, as may be inferred from the origin and formation of springs, &c., and as has, moreover, been fully established by the results of analytical investigations;* but the quantity of such constituents is so trifling that they escape detection, unless hundreds of pounds of the water are subjected to the analytical process. I therefore omit here the mode of their detection, and refer to § 211.

1. Boil the carefully collected water (1000 to 2000 grms.) in a 234 glass flask or retort to one half. This generally produces a precipitate. Pass the fluid through a perfectly clean filter (free from iron and lime), wash the precipitate well, after having removed the filtrate, then examine both as follows:

a. Examination of the precipitate.

The precipitate contains those constituents of the water 235 which were only kept in solution through the agency of free carbonic acid, or, as the case may be, in the form of bicarbonates, viz., carbonate of lime, carbonate of magnesia, hydrated sesquioxide of iron (which was in solution as bicarbonate of protoxide of iron, and precipitates upon boiling as sesquioxide—if phosphoric acid is present, also in combination with that acid), phosphate of lime—also silicic acid, and sometimes sulphate of lime, if that substance is present in large proportion; and clay which was mechanically suspended in the water.

---

* *Chatin* ("Journ. de Pharm. et de Chim.," 3 Sér. t. xxvii. p. 418) found iodine in all fresh-water plants, but not in land plants, a proof that the water of rivers, brooks, ponds, &c., contains traces, even though extremely minute, of metallic iodides. According to *Marchand* ("Comp. Rend.," t. xxxi. p. 495), all natural waters contain iodine, bromine, and lithia. *Van Ankum* has demonstrated the presence of iodine in almost all the potable waters of Holland. And it may be affirmed with the same certainty that all, or at all events most, natural waters contain compounds of strontia, baryta, fluorine, &c.

Dissolve the precipitate on the filter in the least possible quantity of dilute hydrochloric acid (effervescence indicates the presence of CARBONIC ACID), and mix separate portions of the solution :—

a. With sulphocyanide of potassium : red coloration indicates the presence of IRON ;

β. After previous boiling, with ammonia ; filter if necessary, mix the filtrate with oxalate of ammonia, and let the mixture stand for some time in a warm place. The formation of a white precipitate indicates the presence of LIME—in the form of carbonate, or also in that of sulphate if sulphuric acid is detected in γ. Filter, mix the filtrate again with ammonia, add some phosphate of soda, stir with a glass rod, and let the mixture stand for twelve hours. The formation of a white crystalline precipitate, which is often visible only on the sides of the vessel when the fluid is poured out, indicates the presence of MAGNESIA (carbonate of) ; **236**

γ. With chloride of barium, and let the mixture stand for twelve hours in a warm place. The formation of a precipitate—which, if very inconsiderable, is seen best if the supernatant clear fluid is cautiously decanted, and the small quantity remaining shaken about in the glass—indicates the presence of SULPHURIC ACID.

δ. Evaporate another portion of the solution to dryness, treat the residue with hydrochloric acid and water, filter, and test the filtrate for phosphoric acid with solution of molybdate of ammonia in nitric acid (§ 142, 10), or with acetate of soda and sesquichloride of iron (§ 142, 9). **237**

b. *Examination of the filtrate.*

a. Mix a portion of the filtrate with a little hydrochloric acid and chloride of barium. The formation of a white precipitate, which makes its appearance at once, or perhaps only after standing some time, indicates SULPHURIC ACID. **238**

β. Mix another portion with nitric acid, and add nitrate of silver. A white precipitate or a white turbidity indicates the presence of CHLORINE.

γ. Test a portion of the filtrate for PHOSPHORIC ACID, by acidifying with hydrochloric acid, and proceeding as in (**237**).

δ. Evaporate another, larger portion of the filtrate until highly concentrated, and test the reaction of the fluid. If it is alkaline, and a drop of the concentrated clear solution effervesces when mixed on a watch-glass with a drop of acid, a CARBONATE of an alkali is present. Should this be the case, evaporate the fluid to perfect dryness, boil the residue with spirit of wine, filter, evaporate the alcoholic solution to dryness, dissolve the residue in a little water, and test the solution for NITRIC ACID as directed § 159, 7, 8 or 9.*

ε. Mix the remainder of the filtrate with some chloride of ammonium, ammonia, and oxalate of ammonia, and let the mixture stand some time. The formation of a precipitate indicates the presence of LIME. Filter, and test,—

---

* The nitric acid may often be found without trouble, by evaporating the water to a small residue, and testing this at once for it; in which case the above course of proceeding, which, though very accurate, is somewhat roundabout, may be dispensed with.

aa. A small portion with ammonia and phosphate of soda for MAGNESIA.

bb. Evaporate the remainder to dryness, heat the residue to redness, remove the magnesia which may be present (168), and test for POTASSA and SODA, according to the directions of § 197.

2. Acidify a tolerably large portion of the filtered water with **239** pure hydrochloric acid, and evaporate nearly to dryness; divide the residue into 2 parts, and,—

a. Test the one part with hydrate of lime for AMMONIA (§ 91, 3).

b. Evaporate the other part to dryness, moisten the residue with hydrochloric acid, add water, warm, and filter if a residue remains. The residue may consist of SILICIC ACID and, if the water has not been filtered quite clear, also of CLAY mechanically suspended in the water; these two substances may be separated by boiling with solution of carbonate of soda. The precipitate is often dark-colored from the presence of organic substances; but it becomes perfectly white upon ignition.

3. Mix another portion of the water, fresh taken from the well, **240** &c., with lime-water. If a precipitate is thereby produced, FREE CARBONIC ACID or BICARBONATES are present. If the former is present (free carbonic acid), no permanent precipitate is obtained when a larger portion of the water is mixed with only a small amount of lime-water, since in that case soluble bicarbonate of lime is formed.

4. Test for NITROUS ACID,* by mixing a portion of the water **241** with some iodide of potassium starch-paste (made of 1 part of the purest iodide of potassium, 20 parts of starch, and 500 parts of water) and pure dilute sulphuric acid, and observe whether a blue coloration makes its appearance, either at once or, at least, after a few minutes (§ 158, 1).

5. To detect the presence of ORGANIC MATTERS, evaporate a portion **242** of the water to dryness, and gently ignite the residue: blackening of the mass denotes the presence of organic substances. If this experiment is to give conclusive results, the evaporation of the water, as well as the ignition of the residue, must be conducted in a glass flask or a retort.

6. Offensive substances, decaying organic matters, are detected best by filling a bottle two-thirds with the water, covering it with the hand, shaking, and smelling.—If the smell is of sulphuretted hydrogen, proceed as directed § 212, 3. Whether there are other smelling organic matters present besides, may be ascertained by adding a little sulphate of copper to the water, before trying it by the sense of smell.

7. If you wish to examine the MATTERS MECHANICALLY SUSPENDED **243** in a water (in muddy brook or river-water, for instance), fill a large glass bottle with the water, cork securely, and let it stand at rest for several days, until the suspended matter has subsided; remove now the clear supernatant fluid with the aid of a syphon, filter the remainder, and examine the sediment remaining on the filter. As this sediment may consist of the finest dust of various minerals, treat it first with dilute hydrochloric acid, then examine the part insoluble in that menstruum in the manner directed § 205 (Analysis of Silicates).

---

* SCHÖNBEIN found this acid in rain- and snow-water.

8. As oxide of lead may be present, arising from leaden pipes, treat a larger quantity of the water with sulphuretted hydrogen, let stand for some time, and, should a black precipitate form, examine this as directed in § 193. To detect very minute traces of lead, acidify 6 or 8 quarts of the water with acetic acid; add a little acetate of ammonia, to prevent the lead precipitating as sulphate; evaporate to a small residue, filter, and conduct hydrosulphuric acid into the filtrate.

B. ANALYSIS OF MINERAL WATERS.
§ 211.

The analysis of mineral waters embraces a larger number of constituents than that of fresh water. The following are the principal of the additional elements to be looked for:— 244

OXIDE OF CÆSIUM, OXIDE OF RUBIDIUM, LITHIA, BARYTA, STRONTIA, ALUMINA, PROTOXIDE OF MANGANESE, BORACIC ACID, BROMINE, IODINE, FLUORINE, HYDROSULPHURIC ACID, [hyposulphurous acid]*, CRENIC ACID and APOCRENIC ACID, [formic acid, propionic acid, &c., nitrogen gas, oxygen gas, light carburetted hydrogen gas*].

The analyst has moreover to examine the muddy ochreous or hard sinter-deposits of the spring, or also the residue left upon the evaporation of very large quantities of water, for ARSENIOUS ACID, ARSENIC ACID, TEROXIDE OF ANTIMONY, OXIDE OF COPPER, OXIDE OF LEAD, PROTOXIDE OF COBALT, PROTOXIDE OF NICKEL, and the oxides of other heavy metals. The greatest care is required in this examination, to ascertain whether these oxides come really from the water, and do not perhaps proceed from metal pipes, stopcocks, &c.† The absolute purity of the reagents employed in these delicate investigations must also be ascertained with the greatest care.

I. EXAMINATION OF THE WATER.
a. OPERATIONS AT THE SPRING.
§ 212.

1. Filter the water at the spring, if not perfectly clear, through Swedish filter-paper, and collect the filtrate in large bottles with glass stoppers. The sediment remaining on the filter, which possibly contains, besides the flocculent matter suspended in the water, also those constituents which separate at once upon coming in contact with the air (hydrate of sesquioxide of iron, and compounds of sesquioxide of iron with phosphoric acid, silicic acid, arsenic acid), is taken to the laboratory, to be examined afterwards according to the directions of § 214. 245

2. The presence of FREE CARBONIC ACID is usually sufficiently visible to the eye. However, to convince yourself by positive reactions, test the water with fresh-prepared solution of litmus, and 246

---

* Respecting the constituents in brackets, I refer to the corresponding chapter in my "Quantitative Analysis" (§ 206, &c.), as the detection of these matters generally comprises also their quantitative estimation.

† Compare "Chemische Untersuchung der wichtigsten Mineralwasser des Herzogthums Nassau," von Professor Dr. Fresenius; I. Der Kochbrunnen zu Wiesbaden; II. Die Mineralquellen zu Ems; III. die Quellen zu Schlangenbad; IV. die Quellen zu Langenschwalbach; V. die Schwefelquelle zu Weilbach; VI. die Mineralquelle zu Geilnau; VII. die neue Natronquelle zu Weilbach; published at Wiesbaden, by Kriedelund Niedner. 1850—1860.

with lime-water. If carbonic acid is present, the former acquires a wine-red color; the latter produces turbidity, which must disappear again upon addition of the mineral water in excess.

3. Free HYDROSULPHURIC ACID is most readily detected by the smell. For this purpose half fill a bottle with the mineral water, cover with the hand, shake, take off the hand, and smell the bottle. In this way distinct traces of hydrosulphuric acid are often found which would escape detection by reagents. However, if you wish to have some visible reactions, fill a large white bottle with the water, add a few drops of solution of acetate of lead in solution of soda, place the bottle on a white surface, and look in at the top, to see whether the water acquires a brownish color or deposits a blackish precipitate;—or half fill a large bottle with the water, and close with a cork to which is attached a small slip of paper, previously steeped in solution of acetate of lead and then moistened with a little solution of carbonate of ammonia; shake the bottle gently from time to time, and observe whether the paper slip acquires a brownish tint in the course of a few hours. If the addition of the solution of acetate of lead to the water has imparted a brown color to the fluid, or produced a precipitate in it, whilst the reaction with the paper slip gives no result, this indicates that the water contains an alkaline sulphide, but no free hydrosulphuric acid. **247**

4. Mix a wineglass-full of the water with some tannic acid, another wineglass-full with some gallic acid. If the former imparts a red-violet, the latter a blue-violet color to the water, PROTOXIDE OF IRON is present. Instead of the two acids, you may employ infusion of galls, which contains them both. The colorations make their appearance only after some time, and increase in intensity from the top—where the air acts on the fluid—towards the bottom part of the vessel. **248**

b. OPERATIONS IN THE LABORATORY.

§ 213.

As it is always desirable to obtain even in the qualitative examination some information as to the quantitative composition of a mineral water, *i. e.* as to the proportions in which the several constituents are contained in it, it is advisable to analyze a comparatively small portion for the principal constituents, and to ascertain, as far as may be practicable, the relative proportions in which these constituents exist, and thus to determine the character of the water; and then to examine a very large amount of the water for those elements which are present only in minute quantities. For this purpose proceed as follows:—

1. EXAMINATION FOR THOSE CONSTITUENTS OF THE WATER WHICH ARE PRESENT IN LARGER QUANTITIES. **249**

    a. Boil about 3 lbs. of the clear water, or of the filtrate, brought from the spring, in a glass flask for 1 hour, taking care, however, to add from time to time some distilled water, that the quantity of liquid may remain undiminished, and thus the separation of any but *those* salts be prevented which owe their solution to the presence and agency of carbonic acid. Filter after an hour's ebullition, and examine the precipitate and the filtrate as directed § 210.

    b. Test for AMMONIA, SILICIC ACID, ORGANIC MATTERS, &c., by the methods given in § 210.

2. EXAMINATION FOR THOSE FIXED CONSTITUENTS OF THE WATER 250
WHICH ARE PRESENT IN MINUTE QUANTITIES ONLY.—Evaporate a
large quantity (at least 20 lbs.) of the water in a silver or porcelain
dish to dryness; conduct this operation with the most scrupulous cleanliness in a place as free as possible from dust. If the water contains no carbonate of an alkali, add pure carbonate of potassa to slight predominance. The process of evaporation may be conducted at first over a gaslamp, but ultimately the sand-bath must be employed. Heat the dry mass to very faint redness; if in a silver dish, you may at once proceed to ignite it; but if you have it in a porcelain dish, first transfer it to a silver or platinum vessel before proceeding to ignition. If the mass turns black in this process, ORGANIC MATTERS may be assumed to be present.*

Mix the residue thoroughly, that it may have the same composition throughout, and then divide it into 3 portions, one (c) amounting to about one-half, and each of the other two (a and b) to one-fourth.

a. EXAMINATION FOR IRON AND PHOSPHORIC ACID.

Warm the portion a with some water, add perfectly pure 251
hydrochloric acid in moderate excess, digest for some time at
a temperature near the boiling-point, filter through paper
washed with hydrochloric acid and water, and test.

α. A sample for IRON, by means of sulphocyanide of potassium.
β. The remainder for PHOSPHORIC ACID, by means of solution of molybdate of ammonia in nitric acid (§ 142, 10).

b. EXAMINATION FOR FLUORINE.

Heat the portion b with water, add chloride of calcium as 252
long as a precipitate continues to form, let deposit and filter
the fluid from the precipitate, which consists chiefly of carbonate of lime and carbonate of magnesia. After having washed and dried the precipitate, ignite it, then pour water over it in a small dish, add acetic acid in slight excess, evaporate on the waterbath to dryness, heat until all smell of acetic acid has disappeared, add water, heat again, filter the solution of the acetates of the alkaline earths, wash, dry or ignite the residue, and test it for FLUORINE as directed § 146, 5.

c. EXAMINATION FOR THE REMAINING CONSTITUENTS PRESENT IN MINUTE QUANTITIES.

Boil the portion c repeatedly with water, filter, and wash 253
the undissolved residue with boiling water. You have now
a residue (α), and a solution (β).

α. The *residue* consists chiefly of carbonate of lime, carbonate of magnesia, silicic acid, and—in the case of chalybeate springs—hydrate of sesquioxide of iron. But it may contain also minute quantities of BARYTA, STRONTIA, ALUMINA, and PROTOXIDE OF MANGANESE, and must accordingly be examined for these substances.

Pour water over a portion of the insoluble residue, in a platinum or porcelain dish, add hydrochloric acid to slightly acid reaction,

---

* This inference is, however, correct only if the water has been effectually protected from dust during the process of evaporation; if this has not been the case, and you yet wish to ascertain beyond doubt whether organic matters are present, evaporate a separate portion of the water in a retort. If you find organic matter, and wish to know whether it consists of crenic acid or of apocrenic acid, treat a portion of the residue as directed § 214, 3.

then 4 or 5 drops of dilute sulphuric acid, evaporate to dryness, moisten with hydrochloric acid, then add water, warm gently, filter, and wash the residue which is left undissolved.

aa. EXAMINATION OF THE RESIDUE INSOLUBLE IN HY- **254** DROCHLORIC ACID FOR BARYTA AND STRONTIA.

This residue will generally consist of silicic acid ; but it may contain also sulphates of the alkaline earths and carbon. If there is much silicic acid present, remove this in the first place, as far as practicable, by boiling with dilute solution of soda ; filter, wash the residue, if any has been left, dry, incinerate the filter in a platinum crucible, add some carbonate of soda and potassa and, in presence of carbon, some nitrate of potassa, and heat for some time to fusion. If the residue contains but little silicic acid, the treatment with solution of soda may be omitted, and the fusion with carbonate of potassa and soda, &c., at once proceeded with. Boil the fused mass with water, filter, wash thoroughly, dissolve the residue (which must have been left, if sulphates of the alkaline earths are present) on the filter in the least possible quantity of dilute hydrochloric acid, add an equal volume of spirit of wine, then some pure hydrofluosilicic acid, and let the mixture stand 12 hours. If in the course or at the end of the 12 hours a precipitate makes its appearance, this denotes the presence of BARYTA. Filter, and warm the filtrate in a platinum dish, adding from time to time some water, until the spirit of wine is quite driven off. Mix the strongly concentrated fluid now with saturated solution of sulphate of lime. If this produces a precipitate, whether after a short time or after several hours' standing, this precipitate consists of sulphate of STRONTIA. To make quite sure, examine it before the blowpipe (see § 96, 7).

The examination for baryta and strontia may be much facilitated and shortened by the use of the spectrum apparatus. If you are in possession of an apparatus of the kind, ignite the remainder of the residue insoluble in water in a platinum crucible intensely over the blast, extract the ignited mass with a little water, filter, evaporate the filtrate with hydrochloric acid to dryness, and examine the residue by spectrum analysis (compare § 99).

bb. EXAMINATION OF THE HYDROCHLORIC ACID SOLU- **255** TION FOR PROTOXIDE OF MANGANESE AND ALUMINA.

Mix the solution in a flask with some pure chloride of ammonium, add ammonia until the fluid is just turning alkaline, then some yellow sulphide of ammonium, close the flask, filled to the neck, and let it stand for 24 hours in a moderately warm place. If a precipitate has formed at the end of that time, filter, dissolve the precipitate in hydrochloric acid, boil, add solution of potassa (§ 32 *c*) in excess, boil again, filter, and test the filtrate for ALUMINA, with chloride of ammonium ;* the residue with carbonate of soda before the blowpipe for MANGANESE.

---

* You are not justified in regarding this substance as an ingredient of the water, except in cases where the process of evaporation has been conducted in a platinum or silver dish, but not in a porcelain dish.

β. The *alkaline solution* contains the salts of the alkalies, **256** and usually also magnesia and traces of lime. You have to examine it now for NITRIC ACID, BORACIC ACID, IODINE, BROMINE, and LITHIA. Evaporate the fluid until *highly* concentrated, let it cool, and place the dish in a slanting position, that the small quantity of liquid may separate from the saline mass; transfer a few drops of the concentrated solution to a watch-glass, by means of a glass rod, acidify very slightly with hydrochloric acid, and test with turmeric-paper for BORACIC ACID. Evaporate the whole contents of the dish, with stirring, to perfect dryness, and divide the residuary powder into 2 portions, one (*aa*) of two-thirds, the other (*bb*) of one-third.

aa. EXAMINE THE LARGER PORTION FOR NITRIC ACID, **257** IODINE, AND BROMINE.

Put the powder into a flask, pour spirits of wine of 90 per cent. over it, boil on the water-bath, and filter hot; repeat the same operation a second and a third time. Mix the alcoholic extract with a few drops of solution of potassa, distil the spirit of wine off until but little of it remains, and let cool. If minute crystals separate, these may consist of nitrate of potassa; pour off the fluid, wash the crystals with some spirit of wine, dissolve them in a very little water, and test the solution for nitric acid, best by means of indigo, or with brucia, or iodide of potassium, starch-paste and zinc (§ 159). Evaporate the alcoholic solution now to dryness. If you have not yet found nitric acid, dissolve a small portion of the residue in a very little water, and examine the solution for that acid. Treat the remainder of the residue or, if it has been unnecessary to search for nitric acid, the entire residue, three times with warm alcohol, filter, evaporate the filtrate to dryness, with addition of a drop of solution of potassa, dissolve the residue in a very little water, add some starch-paste, acidify slightly with sulphuric acid, and test for iodine by adding some nitrite of potassa in solution, or a drop of solution of hyponitric acid in sulphuric acid.* After having carefully observed the reactions, test the same fluid for bromine with chloroform or sulphide of carbon, and chlorine water in the manner described in § 157.

bb. EXAMINE THE SMALLER PORTION FOR LITHIA.

Warm the smaller portion of the residue, which, if **258** lithia is present, must contain that alkali as carbonate or phosphate, with water, add hydrochloric acid to distinctly acid reaction, evaporate *nearly* to dryness, then mix with pure spirit of wine of 90 per cent., which will separate the greater portion of the chloride of sodium, and give all the lithia in the alcoholic solution. Drive off the alcohol by evaporation and, if you have a spectrum apparatus, examine the residue with this for LITHIA (§ 93, 3). If you have no spectrum apparatus, dissolve the residue in water mixed with a few drops of hydrochloric acid,

---

* The nitric acid originally present may have been destroyed by the ignition of the residue in (**250**), if the latter contained organic matter. If you have reason to fear that such has been the case, and you have not already found nitric acid in (**249**), examine a larger portion of non-ignited residue for that acid, according to the directions of (**257**).

add a little sesquichloride of iron, then ammonia in *slight* excess, and a small quantity of oxalate of ammonia; let the mixture stand for some time, then filter off the fluid, which is now entirely free from phosphoric acid and lime; evaporate the filtrate to dryness, and gently ignite the residue until the salts of ammonia are expelled; treat the residue with some chlorine water (to remove the iodine and bromine) and a few drops of hydrochloric acid, and evaporate to dryness; add a little water and (to remove the magnesia) some finely divided oxide of mercury, evaporate to dryness, and gently ignite the residue until the chloride of mercury is just driven off; treat the residue now with a mixture of absolute alcohol and anhydrous ether, filter the solution obtained, concentrate the filtrate by evaporation, and set fire to the alcohol. If it burns with a carmine flame, LITHIA is present. By way of confirmation convert the lithia found into phosphate of LITHIA (§ 93, 3).

3. EXAMINATION FOR THOSE CONSTITUENTS OF THE WATER WHICH ARE PRESENT IN MOST MINUTE QUANTITIES ONLY.

1. Evaporate 200 or 300 lbs. of the water in a large perfectly 259
clean iron vessel until the salts soluble in water begin to separate. If the mineral water contains no carbonate of soda, add sufficient of that substance to impart a perceptibly alkaline reaction to the fluid. After evaporation filter the solution off, wash the precipitate, without adding the washings to the first filtrate, and

    a Examine the precipitate by the method given in § 214 for sinter deposits;

    b. Mix the solution with hydrochloric acid to acid reaction, heat, just precipitate the sulphuric acid which may be present with chloride of barium, filter, evaporate the filtrate to dryness, digest the residue with alcohol of 90 per cent., and examine the solution for CÆSIUM and RUBIDIUM according to the directions of § 93, last paragraph.

2. Test a portion of the original water for NITROUS ACID accord- 260
ing to the directions of (241). If the water contains hydrosulphuric acid, this is removed first by very cautious addition of some sulphate of silver (under no circumstances must silver salt be allowed to remain in solution).

II. EXAMINATION OF THE SINTER-DEPOSIT.

§ 214.

1. Free the ochreous or sinter-deposit from impurities, by picking, 261
sifting, elutriation, &c., and from the soluble salts adhering to it, by washing with water; digest a large quantity (about 200 grammes) of the residue with water and hydrochloric acid (effervescence: CARBONIC ACID) until the soluble part is completely dissolved; dilute, let cool, filter, and wash the residue.

    a. Examination of the filtrate.

      a. Heat the larger portion of the filtrate nearly to boiling, 262
and add gradually a solution of pure hyposulphite of soda until the whole of the sesquichloride of iron is converted to

protochloride; heat and conduct carbonic acid into the fluid until the mixture smells no longer, or only very faintly, of sulphurous acid. Then conduct hydrosulphuric acid into the unfiltered fluid, which, if necessary, should previously be diluted. Let the fluid, now stand in a moderately warm place until it retains only a faint smell of sulphuretted hydrogen, then filter and wash.

263 Displace the washing water by strong alcohol, and remove the greater part of the free sulphur by digestion and washing with sulphide of carbon; then warm the precipitate gently with some yellowish sulphide of ammonium, filter, wash with water containing sulphide of ammonium, and evaporate the filtrate and washings in a small porcelain dish to dryness. Pour pure red fuming nitric acid over the residue, warm until the greater part of the nitric acid is driven off, add carbonate of soda in slight excess, then a little nitrate of soda, heat to fusion, treat the fused mass with cold water, filter, wash with a mixture of spirit of wine and water, and test the aqueous solution for ARSENIC ACID (121) and (122), the residue for ANTIMONY, TIN, and COPPER, by dissolving in dilute hydrochloric acid, and testing one half of the solution in the platinum dish with zinc for antimony and tin (123), the other half with ferrocyanide of potassium for copper.

264 If a residue has been left upon treating the precipitate produced by hydrosulphuric acid with sulphide of ammonium, wash, and remove from the filter by means of the washing bottle; boil with a little dilute nitric acid, filter, wash, and pour solution of hydrosulphuric acid over the contents of the filter—that any sulphate of lead present may not be overlooked—then test for BARYTA and STRONTIA as in (254). Mix the filtrate (the nitric acid solution) with some pure sulphuric acid, evaporate on the water-bath to dryness, and treat the residue with water. If this leaves an undissolved residue, the latter consists of sulphate of LEAD. To make quite sure, filter, wash the residue, treat it with hydrosulphuric acid water, and observe whether that reagent imparts a black color to it. Test the fluid filtered from the sulphate of lead which may have separated, *a* with ammonia, *b* with ferrocyanide of potassium, for COPPER.

265 Evaporate a portion of the fluid filtered from the precipitate produced by hydrosulphuric acid to dryness, treat the residue with hydrochloric acid and water, filter, and test the filtrate for PHOSPHORIC ACID with solution of molybdate of ammonia in nitric acid. Mix the remainder in a flask with chloride of ammonium, ammonia, and yellowish sulphide of ammonium, close the flask, filled up to the neck, and let it stand in a moderately warm place until the fluid above the precipitate looks no longer greenish, but yellow; filter, and wash the precipitate with water to which some sulphide of ammonium has been added. Dissolve the washed precipitate in hydrochloric acid, and examine for COBALT, NICKEL, IRON, MANGANESE, ZINC, ALUMINA, and SILICIC ACID, according to the directions of (152) to (160).—Examine now the fluid filtered from the precipitate produced by sulphide of ammonium for LIME and MAGNESIA in the usual way.

β. Mix a portion of the hydrochloric acid solution, considerably diluted, with chloride of barium, and let the mixture stand 12 hours in a warm place. The formation of a white precipitate indicates the presence of SULPHURIC ACID.

b. *Examination of the residue.*

This consists usually of silicic acid, clay, and organic matters, but it may also contain sulphate of baryta and sulphate of strontia. Boil in the first place with solution of soda or potassa, to dissolve the SILICIC ACID ; then fuse the residue with carbonate of soda and potassa, and a little nitrate of potassa. Boil the mass with water, wash the residue thoroughly, and dissolve it in some hydrochloric acid ; the silicic acid still present then separates, add ammonia to the filtrate, filter again from the ALUMINA, &c., which may precipitate, evaporate the filtrate to dryness, gently ignite the residue, redissolve it in very little water, with addition of a drop of hydrochloric acid, and test for BARYTA and STRONTIA as directed (254). **266**

2. As regards the examination for FLUORINE, the best way is to take for this purpose a separate portion of the ochreous or sinter-deposit. Ignite (which operation will also reveal the presence of organic matters), stir with water, add acetic acid to acid reaction, evaporate until the acetic acid is completely driven off, and proceed as directed (252). **267**

3. Boil the ochreous or sinter-deposit for a considerable time with concentrated solution of potassa or soda, and filter. **268**

a. Acidify a portion of the filtrate with acetic acid, add ammonia, let the mixture stand 12 hours, and then filter the fluid from the precipitate of alumina and hydrated silicic acid, which usually forms ; again add acetic acid to acid reaction, then a solution of neutral acetate of copper. If a brownish precipitate is formed, this consists of APOCRENATE of copper. Mix the fluid filtered from the precipitate with carbonate of ammonia, until the green color has changed to blue, and warm. If a bluish-green precipitate is produced, this consists of CRENATE of copper.

b. If you have detected arsenic, use the remainder of the alkaline fluid to ascertain whether the arsenic existed in the sinter as ARSENIOUS ACID or as ARSENIC ACID. Compare § 134, 9.

IV. ANALYSIS OF SOILS.

§ 215.

Soils must necessarily contain all the constituents which are found in the plants growing upon them, with the exception of those supplied by the atmosphere and the rain. When we find, therefore, a plant the constituent elements of which are known, growing in a certain soil, the mere fact of its growing there gives us some insight into the composition of that soil, and may accordingly save us, to some extent, the trouble of a qualitative analysis.

Viewed in this light, it would appear quite superfluous to make a qualitative analysis of soils still capable of producing plants ; for it is well known that the ashes of plants contain almost invariably the same constituents, and the differences between them are caused principally by differences in the relative proportions in which the several constituents

I.   T

are present. But if, in the qualitative analysis of a soil, regard is had also—in so far as may be done by a simple estimation—to the quantities and proportions of the several constituent ingredients, and to the state and condition in which they are found to be present in the soil, an analysis of the kind, if combined with an examination of the physical properties of the soil, and a mechanical separation of its component parts,* may give most useful results, enabling the analyst to judge sufficiently of the condition of the soil, to supersede the necessity of a *quantitative* analysis, which would require much time, and is a far more difficult task.

As plants can only absorb substances in a state of solution, it is a matter of especial importance, in the qualitative analysis of a soil, to know which are the constituents that are soluble in water;† which those that require an acid for their solution (in nature principally carbonic acid); and, finally, which those that are neither soluble in water nor in acids, and are not, accordingly, in a position for the time being to afford nutriment to the plant. With regard to the insoluble substances, another interesting question to answer is whether they suffer disintegration readily, or slowly and with difficulty, or whether they altogether resist the action of disintegrating agencies; and also what are the products which they yield upon their disintegration.‡

In the analysis of soils, the constituents soluble in water, those soluble in acids, and the insoluble constituents, must be examined separately. The examination of the organic portion also demands a separate process.

The analysis is therefore properly divided into the following four parts:

1. *Preparation and Examination of the Aqueous Extract.*

§ 216.

About two pounds (1000 grammes) of the air-dried soil are used for the preparation of the aqueous extract. To prepare this extract

---

* With regard to the mechanical separation of the component parts of a soil, and the examination of its physical properties and chemical condition, compare Fr. Schulze's paper, "Anleitung zur Untersuchung der Ackererden auf ihre wichtigsten physikalischen Eigenschaften und Bestandtheile."—Journal f. prakt. Chemie, Vol. 47, p. 241; also Fresenius' "Quantitative Analysis," § 258.

† It was formerly universally assumed that substances soluble in water, or in water containing carbonic acid, circulated freely in the soil so long as there existed agents for their solution; but since it has been discovered that arable soil possesses in a similar manner to porous charcoal, the property of withdrawing from dilute solutions the bodies dissolved in them, this notion is exploded, and we now know that arable soil will bind and retain with a certain force bodies otherwise soluble—from which we conclude accordingly that the aqueous extract of a soil cannot be expected to contain the whole of the substances present in that soil in a state immediately available for the plant. Neither can we expect to find these matters in the aqueous extract in the same proportion in which they are present in the soil, since the latter will readily give up to water those substances in regard to which its power of absorption has been satisfied, whilst it will more or less strongly retain others. But although, for this reason, the examination of the aqueous extract of a soil has no longer the same value as it was formerly considered to have, yet it is still useful to ascertain what substances a soil will actually give up to water. It is for this reason that I have retained the chapter on the preparation and examination of the aqueous extract.

‡ For more ample information on this subject, I refer the reader to Fresenius' "Chemie für Landwirthe, Forstmänner und Cameralisten;" published at Brunswick, by F. Vieweg and Son, 1847, p. 485.

quite clear is a matter of some difficulty; in following the usual course, viz., digesting or boiling the earth with water, and filtering, the fine particles of clay are speedily found to impede the operation, by choking up the pores of the filter; they also almost invariably render the filtrate turbid, at least the portion which passes through first. I have found the following method the most practical.* Close the neck of several middle-sized funnels with small filters of coarse blotting-paper, moisten the paper, press it close to the sides of the funnel, and then introduce the air-dried soil, in small lumps ranging from the size of a pea to that of a walnut, but not pulverized or even crushed; fill the funnels with the soil to the extent of about two-thirds. Pour distilled water into them, in sufficient quantity to cover the soil; if the first portion of the filtrate is turbid, pour it back into the funnel. Let the operation proceed quietly. When the first quantity of the fluid has passed, fill the funnels a second, and after this a third time with water, and continue this process of lixiviation until the filtrates weigh twice or three times as much as the soil used. Collect the several filtrates in one vessel, and mix them intimately together. Keep a portion of the lixiviated soil.

a. Strongly concentrate two-thirds of the aqueous solution **270** by cautiously evaporating in a porcelain dish, filter off a portion, and test its reaction; put aside a portion of the filtrate for the subsequent examination for organic matters, according to the directions of (**280**). Warm the remainder, and add nitric acid. Evolution of gas indicates the presence of an ALKALINE CARBONATE. Then test with nitrate of silver for CHLORINE. *b.* Transfer the remainder of the concentrated fluid, together with the precipitate which usually forms in the process of concentration, to a small porcelain, or, which is preferable, a small platinum dish, evaporate to dryness, and cautiously heat the brownish residue over the lamp until complete destruction of the organic matter is effected. In presence of NITRATES this operation is attended with deflagration, which is more or less violent according to the greater or smaller proportion in which these salts are present. *c.* Test a small portion of the gently ignited residue with carbonate of soda before the blowpipe for MANGANESE. *d.* Warm the remainder with water, add some hydrochloric acid (effervescence indicates the presence of CARBONIC ACID), evaporate to dryness, heat a little more strongly, to effect the complete separation of the silicic acid, moisten with hydrochloric acid, add water, warm, and filter. The washed residue generally contains some carbonaceous matter, also a little clay—if the aqueous extract was not perfectly clear—and lastly SILICIC ACID. To detect the latter, make a hole in the point of the filter, rinse the residue through, boil with solution of carbonate of soda, filter, saturate with hydrochloric acid, evaporate to dryness, and treat the residue with water, which will leave the silicic acid undissolved.

e. Test a small portion of the hydrochloric acid solution **271** with chloride of barium for SULPHURIC ACID; another portion with solution of molybdate of ammonia in nitric acid for PHOSPHORIC ACID; a third portion with sulphocyanide of potassium

---

* Recommended by Fr. Schulze, "Anleitung zur Untersuchung der Ackererden auf ihre wichtigsten physikalischen Eigenschaften und Bestandtheile."—Journ. f. prakt. Chemie, Vol. 47, p. 241.

for SESQUIOXIDE OF IRON. Add to the remainder a few drops of sesquichloride of iron (to remove the phosphoric acid), then ammonia cautiously until the fluid is slightly alkaline, warm a little, filter, throw down the LIME from the filtrate by means of oxalate of ammonia, and proceed for the detection of MAGNESIA, POTASSA, and SODA, in the usual way, strictly according to the directions of § 196.

f. Alumina is not likely to be found in the aqueous extract. (Fr. Schulze never found any.) However, if you wish to test for it, boil the ammonia precipitate obtained in *e* (271) with pure solution of soda or potassa, filter, and test the filtrate with chloride of ammonium. **272**

g. If you have detected iron, test a portion of the remaining third of the aqueous extract with ferricyanide of potassium, another with sulphocyanide of potassium, both after previous addition of some hydrochloric acid : this will indicate the degree of oxidation in which the iron is present. Mix the remainder of the aqueous extract with a little sulphuric acid, evaporate on the water-bath nearly to dryness, and test the residue for AMMONIA, by adding hydrate of lime. **273**

2. *Preparation and Examination of the Acid Extract.*

§ 217.

Heat about 50 grammes of the soil from which the part soluble in water has been removed as far as practicable* (see § 216), with moderately strong hydrochloric acid (effervescence indicates CARBONIC ACID) for several hours on the water-bath, filter, and make the following experiments with the filtrate, which, owing to the presence of sesquichloride of iron, has in most cases a reddish-yellow color :— **274**

1. Test a small portion of it with sulphocyanide of potassium for SESQUIOXIDE OF IRON, another with ferrocyanide of potassium for PROTOXIDE OF IRON. **275**

2. Test a small portion with chloride of barium for SULPHURIC ACID, another, after separation of the silicic acid with solution of molybdate of ammonia in nitric acid, for PHOSPHORIC ACID.

3. Mix a larger portion of the filtrate with ammonia to neutralize the free acid, then with yellowish sulphide of ammonium ; let the mixture stand in a warm place, in a flask filled up to the neck, until the fluid looks yellow ; then filter, and test the filtrate in the usual way for LIME, MAGNESIA, POTASSIA, and SODA. **276**

4. Dissolve the precipitate obtained in 3 (**276**) in hydrochloric acid, evaporate the solution to dryness, moisten the residue with hydrochloric acid, add water, warm, filter, and examine the filtrate according to the directions of (150), for IRON, MANGANESE, ALUMINA, and, if necessary, also for lime and magnesia, which may have been thrown down by the sulphide of ammonium, in combination with phosphoric acid. **277**

5. The separated SILICIC ACID obtained in 4 is usually colored by organic matter. It must, therefore, be ignited to obtain it pure.

---

* Complete lixiviation is generally impracticable.

6. If it is a matter of interest to ascertain whether the hydrochloric acid extract contains ARSENIC ACID, OXIDE OF COPPER, &c., 278 treat the remainder of the solution first with sulphite of soda, then with hydrosulphuric acid, as directed in (262) to (264).

7. Should you wish to look for FLUORINE, ignite a fresh portion of the earth, and then proceed according to the directions of (430).

3. *Examination of the Inorganic Constituents insoluble in Water and Acids.*

§ 218.

The operation of heating the lixiviated soil with hydrochloric 279 acid (274) leaves still the greater portion of it undissolved. If you wish to subject this undissolved residue to a chemical examination, wash, dry, and sift, to separate the large and small stones from the clay and sand; moreover, separate the two latter from each other by elutriation. Subject the several portions to the analytical process given for the silicates (§ 205).

4. *Examination of the Organic Constituents of the Soil.*[*]

§ 219.

The organic constituents of the soil, which exercise so great an influence upon its fertility, both by their physical and chemical action, are partly portions of plants in which the structure may still be recognised (fragments of straw, roots, seeds of weeds, &c.), partly products of vegetable decomposition, which are usually called by the general name of HUMUS, but differ in their constituent elements and properties, according to whether they result from the decay of the nitrogenous or non-nitrogenous parts of plants—whether alkalies or alkaline earths have or have not had a share in their formation—whether they are in the incipient or in a more advanced stage of decomposition. To separate these several component parts of humus would be an exceedingly difficult task, which, moreover, would hardly repay the trouble; the following operations are amply sufficient to answer all the purposes of a qualitative analysis of the organic constituents of a soil.

a. *Examination of the Organic Substances soluble in Water.*

Evaporate the portion of the filtrate of (270) which has been put 280 aside for the purpose of examining the organic constituents, on the water-bath to perfect dryness, and treat the residue with water. The ulmic, humic, and geic acids, which were present in the solution in combination with bases, remain undissolved, whilst crenic acid and apocrenic acid are dissolved in combination with ammonia; for the manner of detecting the latter acids, see (268).

b. *Treatment with an Alkaline Carbonate.*

Dry a portion of the lixiviated soil, and sift to separate the 281 fragments of straw, roots, &c., together with the small stones, from the finer parts; digest the latter for several hours at a tempera-

---

[*] Compare Fresenius' "Chemie für Landwirthe, Forstmänner und Cameralisten;" published at Brunswick, by F. Vieweg and Son, 1847, §§ 282—285.

ture of 176°—194° F., with solution of carbonate of soda, and filter. Mix the filtrate with hydrochloric acid to acid reaction. If brown flakes separate, these proceed from ulmic acid, humic acid, or geic acid. The larger the quantity of ulmic acid present the lighter, the larger that of humic acid or geic acid, the darker the brown color of the flakes.

c. Treatment with Caustic Alkali.

282 Wash the soil boiled with solution of carbonate of soda (*b*) with water, boil several hours with solution of potassa, replacing the water in proportion as it evaporates, dilute, filter, and wash. Treat the brown fluid as in *b*. The ulmic and humic acids which separate now, are new products resulting from the action of boiling solution of potassa upon the ulmine and humine originally present.

V. DETECTION OF INORGANIC SUBSTANCES IN PRESENCE OF ORGANIC SUBSTANCES.

§ 220.

The impediments which the presence of coloring, slimy, and other organic substances throw in the way of the detection of inorganic bodies, and that the latter can often be effected only after the total destruction of the organic admixture, will be readily conceived, if we reflect that in dark colored fluids changes of color or the formation of precipitates escape the eye, that slimy fluids cannot be filtered, &c. Now, as these difficulties are very often met with in the analysis of medicinal substances, and more especially in the detection of inorganic poisons in articles of food or in the contents of the stomach, and, lastly, also in the examination of plants and animals, or parts of them, for their inorganic constituents, I will here point out the processes best adapted to lead to the attainment of the object in view, both in the general way and in special cases.

1. *General Rules for the Detection of Inorganic Substances in Presence of Organic Matters, which by their Color, Consistence, &c., impede the Application of the Reagents, or obscure the Reactions produced.*

§ 221.

We confine ourselves here, of course, to the description of the most generally applicable methods, leaving the adaptation of the modifications which circumstances may require in special cases to the discretion of the analyst.

283 1. THE SUBSTANCE UNDER EXAMINATION DISSOLVES IN WATER, BUT THE SOLUTION IS DARK-COLORED OR OF SLIMY CONSISTENCE.

   a. Heat a portion of the solution with hydrochloric acid on the water-bath, and gradually add chlorate of potassa until the mixture is decolorized and perfectly fluid; heat until it exhales no longer the odor of chlorine, then dilute with water, and filter. Examine the filtrate in the usual way, commencing at § 190. Compare also § 225.

   b. Boil another portion of the solution for some time with nitric acid, filter, and test the filtrate for SILVER, POTASSA, and HYDROCHLORIC ACID. If the nitric acid succeeds in effecting the ready and complete destruction of the coloring and slimy matters, &c., this method is often altogether preferable to all others.

c. ALUMINA and SESQUIOXIDE OF CHROMIUM might escape detection by this method, because ammonia and sulphide of ammonium fail to precipitate these oxides from fluids containing non-volatile organic substances. Should you have reason to suspect the presence of these oxides, mix a third portion of the substance with carbonate of soda and chlorate of potassa, and throw the mixture gradually into a red-hot crucible. Let the mass cool, then treat it with water, and examine the solution for chromic acid and alumina, the residue for alumina (§ 103).

2. BOILING WATER FAILS TO DISSOLVE THE SUBSTANCE, OR EFFECTS 284 ONLY PARTIAL SOLUTION; THE FLUID ADMITS OF FILTRATION.

Filter, and treat the filtrate either as directed § 189, or, should it require decoloration, according to the directions of (283). The residue may be of various kinds.

a. IT IS FATTY. Remove the fatty matter by means of ether, and should a residue be left, treat this as directed § 175.

b. IT IS RESINOUS. Use alcohol instead of ether, or apply both liquids successively.

c. IT IS OF A DIFFERENT NATURE, *e.g.*, woody fibre, &c.

α. Dry, and ignite a portion of the dried residue in a porcelain or platinum vessel until total or partial incineration is effected; boil the residue with nitric acid and water, and examine the solution as directed (109); if a residue has been left, treat this according to the directions of § 203.

β. Examine another portion for the heavy metals, and for acids, as directed in (283 and 284)—since with the method given in *a* arsenic, cadmium, zinc, &c., may volatilize, besides the compounds of mercury which may be present.

γ. Test the remainder for ammonia, by triturating it together with hydrate of lime.

3. THE SUBSTANCE DOES NOT ADMIT OF FILTRATION OR ANY OTHER 285 MEANS OF SEPARATING THE DISSOLVED FROM THE UNDISSOLVED PART.

Treat the substance in the same manner as the residue in (284). As regards the charred mass (284) *c, a*, it is often advisable to boil the mass, carbonized at a gentle heart, with water, filter, examine the filtrate, wash the residue, incinerate it, and examine the ash.

2. *Detection of Inorganic Poisons in Articles of Food, in Dead Bodies, &c., in Chemico-legal Cases.**

§ 222.

The chemist is sometimes called upon to examine an article of 286 food, the contents of the stomach of an individual, a dead body, &c., with a view to detect the presence of some poison, and thus to establish the fact of a wilful or accidental poisoning; but it is more frequently the case that the question put to him is of a less general nature,

---

* Compare : *a.* Fresenius, "die Stellung des Chemikers bei gerichtlich-chemischen Untersuchungen," &c. (Annal. der Chemie und Pharm. 49, 275); and *b.* Fresenius and v. Babo's "Abhandlung über ein neues, unter allen Umständen sicheres Verfahren zur Ausmittelung und quantitativen Bestimmung des Arsens bei Vergiftungsfällen."—Annal. der Chemie und Pharmacie, 49, 287.

and that he is called upon to determine whether a certain substance placed before him contains a *metallic* poison ; or, more pointedly still, whether it contains arsenic or hydrocyanic acid, or some other particular poison— as it may be that the symptoms point clearly in the direction of that poison, or that the examining magistrate has, or believes he has, some other reason to put this question.

It is obvious that the task of the chemist will be the easier, the more special and pointed the question which is put to him. However, the analyst will always act most wisely, even in cases where he is simply requested to state whether *a certain poison, e.g.*, arsenic, is present or not, if he adopts a course of proceeding which will not only permit the detection of the *one* poison specially named, the presence of which may perhaps be suspected on insufficient grounds, but will moreover inform him as to the presence or absence of other similar poisons.

But we must not go too far in this direction either ; if we were to attempt to devise a method that should embrace *all* poisons, we might unquestionably succeed in elaborating such a method at the writing-desk ; but practical experience would but too speedily convince us that the intricate complexity inseparable from such a course, must necessarily impede the easy execution of the process, and impair the certainty of the results, to such an extent indeed, that the drawbacks would be greater than the advantages derivable from it.

Moreover, the attendant circumstances permit usually at least a tolerably safe inference as to the group to which the poison belongs. Acting on these views, I give here,—

1. A method which ensures the detection of the minutest traces of arsenic, allows of its quantitative determination, and permits at the same time the detection of all other metallic poisons.

2. A method to effect the detection of hydrocyanic acid, which leaves the substance still fit to be examined both for metallic poisons and for vegeto-alkalies.

3. A method to effect the detection of phosphorus, which does not interfere with the examination for other poisons.

This Section does not, therefore, profess to supply a complete guide in every possible case or contingency of chemico-legal investigations. But the instructions given in it are the tried and proved results of my own practice and experience. Moreover, they will generally be found sufficient, the more so as in the Section on the vegeto-alkalies, I give the description of the best processes by which the detection of these latter poisons in criminal cases may be effected.

I. METHOD FOR THE DETECTION OF ARSENIC (WITH DUE REGARD TO THE POSSIBLE PRESENCE OF OTHER METALLIC POISONS).

§ 223.

Of all metallic poisons arsenic is the most dangerous, and at the same time the one most frequently used, more particularly for the wilful poisoning of others. And again, among the compounds of arsenic, arsenious acid (white arsenic) occupies the first place, because— (1) It kills even in small doses ; (2) It does not betray itself, or at least very slightly, by the taste : and (3) It is but too readily procurable.

As arsenious acid dissolves in water only sparingly and—on account of the difficulty with which moisture adheres to it—very slowly, the greater portion of the quantity swallowed exists usually in the body still in the undissolved state; as, moreover, the smallest grains of it may be readily detected by means of an exceedingly simple experiment; and lastly, as—no matter what opinion may be entertained about the normal presence of arsenic in the bones, &c.—this much is certain, that at all events *arsenious acid in grains or powder* is never normally present in the body, the particular care and efforts of the analyst ought always to be directed to the detection of the arsenious acid in substance—and this end may indeed usually be attained.

A. *Method for the Detection of undissolved Arsenious Acid.*

1. If you have to examine some article of food, substances rejected from the stomach, or some other matter of the kind, mix the whole as uniformly as may be practicable, reserve one-third for unforeseen contingencies, and mix the other two-thirds in a porcelain dish with distilled water, with a stirring rod; let the mixture stand a little, then pour off the fluid, together with the lighter suspended particles, into another porcelain dish. Repeat this latter operation several times, if possible with the same fluid, pouring it from the second dish back into the first, &c. Finally, wash once more with pure water, remove the fluid, as far as practicable, and try whether you can find in the dish small, white, hard grains which feel gritty and grate under the glass rod. If not, proceed as directed § 224 or § 225. But if so, put the grains, or part of them, on blotting-paper, removing them from the dish with the aid of pincers, and try the deportment of one or several grains upon heating in a glass tube, and of some others upon ignition with a splinter of charcoal (compare § 132, 2 and 11). If you obtain in the former experiment a white crystalline sublimate, in the latter a lustrous arsenical mirror, the fact is clearly demonstrated that the grains selected and examined consisted really of arsenious acid. If you wish to determine the quantity of the poison, or to test for other metallic poisons, unite the contents of both dishes, and proceed as directed § 224 or § 225.

2. If a stomach is submitted to you for analysis, empty the contents into a porcelain dish, turn the stomach inside out, and (*a*), search the inside coat for small, white, hard, sandy grains. The spots occupied by such grains are often reddened; the grains are also frequently found firmly imbedded in the membrane. (*b*) Mix the contents in the dish uniformly, put aside one-third for unforeseen contingencies, and treat the other two-thirds as in 1. The same course is pursued also with the intestines. In other parts of the body—with the exception perhaps of the pharynx and œsophagus—arsenious acid cannot be found in grains, if the poison has been introduced through the mouth. If you have found grains of the kind described, examine them as directed in 1; if not, or if you wish to test also for other metallic poisons, proceed according to the instructions of § 224 or § 225.

B. *Method of detecting soluble Arsenical and other Metallic Compounds by means of Dialysis.*

§ 224.

If method *A* has failed to show the presence of arsenious acid in the solid state, and the process described in § 225, in which the organic substances are totally destroyed by chlorate of potassa and hydrochloric acid, is at once resorted to, the operator must, of course, in the event of the presence of arsenic being revealed, give up all notion of ascertaining, as far as the portion operated upon is concerned, in what form or combination the poison detected has been administered; as the process will always simply give a solution containing arsenic acid, no matter whether the poison was originally present in that form, or as arsenious acid, or as a sulphide, or in the metallic state, &c. This defect may be remedied, however, by making after A, and before proceeding to B, first a dialytic experiment.*

Fig. 34.

The experiment requires the apparatus shown in Fig. 34 (Hoop Dialyser).

The hoop is made of wood or, better, of sheet gutta percha; it is 2 inches in depth, and 8 or 10 inches in diameter. The disk of parchment-paper used should measure 3 or 4 inches more in diameter than the hoop to be covered. It is moistened, and then bound to the hoop by string, or by an elastic band; but it should not be firmly secured. The parchment-paper must not be porous. Its soundness is tried by sponging the upper surface with pure water, and then observing whether wet spots show on the opposite side, in which case the defects are remedied by applying liquid albumen and coagulating this by heat. When the dialyser has thus been got ready, free from all defects, the mass to be examined (residue and fluid of § 223, A) is, after previous addition, where the circumstances of the case require it, of two-thirds of the stomach, intestinal canal, &c., cut small, and after digestion of the whole mixture for 24 hours at about 89 to 90° Fahrenheit, poured into the hoop, upon the surface of the parchment-paper, but only to a

---

* Dialysis is a means of chemical analysis only quite recently introduced into the science by GRAHAM (Phil. Mag., Fourth Series, Nos. 153—155). It is based upon the different deportment of bodies dissolved in water when placed in contact with moist membranes. A certain class of bodies, to which the name "*crystalloids*" is given, have the power of penetrating suitable membranes placed in contact with their solution, whilst another class of bodies, to which the name "*colloids*" is given, do not possess that property. To the former class belong all crystallizable bodies, to the latter all non-crystallizable bodies, such as gelatine, gum, dextrin, caramel, tannic acid, albumen, extractive matters, hydrated silicic acid, &c. The septum must consist of a colloid matter, such as an animal membrane, or, which is the best and most suitable, of parchment-paper, and must on the other side be in contact with water. GRAHAM explains the action which ensues upon the assumption that the crystalloids appropriate to them the water absorbed by the colloid septum, acquiring thereby a medium for diffusion, whilst the dissolved colloids, possessing only a most feeble affinity for water, are unable to separate that liquid from the septum, and fail therefore also to penetrate the latter, and thus to open the door for their own passage outward by diffusion.

depth of half-an-inch at the most. The dialyser is then floated in a basin containing about 4 times as much water as the fluid to be dialysed amounts to. After 24 hours the one-half or three-fourths of the crystalloid substances will be found in the external water, which generally appears colorless. Concentrate this by evaporation on the water-bath, acidify with hydrochloric acid, treat with sulphuretted hydrogen, and proceed generally as directed in § 291, &c. If an arsenical compound soluble in water (or some other soluble metallic salt) was present, the corresponding sulphide is obtained almost pure. By floating the dialyser successively on fresh supplies of water, the whole of the crystalloid substances present may finally be withdrawn from the mass. When this has been accomplished, examine the residue for metallic compounds insoluble in water, as directed in § 225.

*C. Method for the Detection of Arsenic in whatever Form of Combination it may exist, which allows also a Quantitative Determination of that Poison, and permits at the same time the Detection of other Metallic Poisons which may be present.**

§ 225.

If you have found no arsenious acid in substance by the method described in *A*, nor a soluble arsenic compound by dialysis, evaporate the mass in the porcelain dish, which has been diluted by washing with water (see *A*, 1), on the water-bath, to a pasty consistence. If you have to analyze a stomach, intestinal tube, &c., cut this into pieces, and add two-thirds to the mass in the dish, if this has not been done already in the process of dialysis.

In examining other parts of the body (the lungs, liver, &c.), cut them also into small pieces, and use two-thirds for the analysis. The process is divided into the following parts.†

1. *Decoloration and Solution.*

Add to the matters in the porcelain dish, which, by way of illustration, may amount to from 4 to 8 ounces (120-240 grm.), an amount of pure hydrochloric acid of 1·12 sp. gr., about equal to or somewhat exceeding the weight of the dry substances present, and sufficient water to give to the entire mass the consistence of a thin paste. The quantity of hydrochloric acid added should never exceed one-third of the entire liquid present. Heat the dish now on the water-bath, adding every five minutes about two grammes (half a drachm) of chlorate of potassa to the hot fluid, with stirring, until the contents of the dish show a light-yellow color and a perfectly homogeneous appearance, and are quite fluid; replace the evaporating water from time to time in this process. When this point is attained, add again a portion of chlorate of potassa, and then remove the dish from the water-bath. When the contents are quite cold, transfer them cautiously to a linen strainer or

290

---

* This method is essentially the same as that which I have elaborated and published in 1844, jointly with L. v. Babo; compare "Annal. der Chemie und Pharmacie," Bd. 49, p. 308. I have since that time had frequent occasion to apply it; I have also had it tried by others, under my own inspection, and *I have invariably found it to answer the purpose perfectly.*

† I think I need hardly observe that in such extremely delicate experiments the vessels and reagents used in the process must be perfectly free from arsenic, from heavy metals in general, and indeed from every impurity.

to a white filter, according to the greater or less quantity of substance; allow the whole of the fluid to pass through, and heat the filtrate on the water-bath with renewal of the evaporating water, until the smell of chlorine has quite or nearly gone off. Wash the residue well with hot water, and dry it; then mark it I., and reserve for further examination, according to the instructions of (303). Evaporate the washings on the water-bath to about 3 or 4 oz. (about 100 grm.), add this, together with any precipitate that may have formed therein, to the principal filtrate.

2. *Treatment of the Solution with Hydrosulphuric Acid* (Separation 291
of the Arsenic as Tersulphide, and, as the case may be, of all the Metals of Groups V. and VI. in form of Sulphides).

Transfer the fluid obtained in 1, which amounts to about three or four times the quantity of the hydrochloric acid used, to a flask, heat this on the water-bath to 158° Fah., and transmit through it, for about 12 hours, a slow stream of washed hydrosulphuric acid gas, then let the mixture cool, continuing the transmission of the gas all the while; rinse the delivery pipe with some ammonia, add the ammoniated solution thus obtained, after acidifying, to the principal fluid, cover the flask lightly with unsized paper, and put it in a moderately warm place (about 86° Fah.) until the odor of hydrosulphuric acid has nearly disappeared. Collect the precipitate obtained in this manner on a moderately sized filter, and wash until the washings are quite free from chlorine. Concentrate the filtrate and washings somewhat, mix the fluid in a proper-sized flask with ammonia to alkaline reaction, then with sulphide of ammonium, closely cork the flask, which must now be nearly full, and reserve for further examination according to the instructions of (307).

3. *Purification of the Precipitate produced by Hydrosulphuric* 292
Acid.

Thoroughly dry the precipitate obtained in 2—which, besides organic matters and free sulphur, must contain, in form of tersulphide, the whole of the arsenic present in the analyzed substance, as well as, in form of sulphides, all the metals of Groups V. and VI. which may happen to be present—together with the filter, in a small porcelain dish, heated on the water-bath; add pure fuming nitric acid (perfectly free from chlorine), drop by drop, until the mass is completely moistened, then evaporate on the water-bath to dryness. Moisten the residue uniformly all over with pure concentrated sulphuric acid, previously warmed; then heat for two or three hours on the water-bath, and finally on the air-sand- or oil-bath at a somewhat higher, though still moderate temperature (338° Fah.), until the charred mass becomes friable, and a small sample of it—to be returned afterwards to the mass—when mixed with water and then allowed to subside, gives a colorless fluid; should the fluid standing over the sediment show a brownish tint, or the residue, instead of being friable, consist of a brown oily liquid, add to the mass some cuttings of pure Swedish filtering-paper, and continue the application of heat. By attending to these rules you will always completely attain the object in view, viz., the destruction of the organic substances, without loss of any of the metals. Warm the residue on the water-bath with a mixture of 8 parts of water and 1 part of hydrochloric acid, filter, wash the undissolved part thoroughly with hot, distilled water, with addition of a little hydrochloric acid, and add the washings, which must be concentrated if necessary, to the filtrate.

Dry the washed carbonaceous residue, then mark it II., and reserve for further examination according to the instructions given in (304).

4. PRELIMINARY EXAMINATION FOR ARSENIC AND OTHER METALLIC **293** POISONS OF GROUPS V. AND VI. (Second Precipitation with Hydrosulphuric Acid.)

The clear and colorless or, at the most, somewhat yellowish fluid obtained in 3 contains all the arsenic which may have been present, in form of arsenious acid, and may contain also tin, antimony, mercury, copper, bismuth, and cadmium. Supersaturate a small portion of it cautiously and gradually with a mixture of carbonate of ammonia and some ammonia, and observe whether a precipitate is produced. Acidify the supersaturated sample of the fluid with hydrochloric acid, which will redissolve the precipitate that may have been produced by ammonia; then return the sample to the fluid, and treat the latter with hydrosulphuric acid in strict accordance with the directions of (291).

This process may lead to three different results, which are to be carefully distinguished.

a. The hydrosulphuric acid fails to produce a precipitate; **294** but subsequently, after the fluid treated as directed in (291) has stood for some time, a trifling white or yellowish-white precipitate separates. In this case probably no metals of Groups V. and VI. are present. Nevertheless, treat the filtered and washed precipitate as directed in (297), to guard against overlooking even the minutest traces of arsenic, &c.

b. A precipitate is formed, of a pure yellow color like that **295** of tersulphide of arsenic. Take a small portion of the fluid, together with the precipitate suspended therein, add some ammonia, and shake the mixture for some time, without application of heat. If the precipitate dissolves readily and, with the exception of a trace of sulphur, completely, and if, in the preliminary examination (293), carbonate of ammonia has failed to produce a precipitate, arsenic alone is present, and no other metal (at all events, no quantity worth mentioning —tin or antimony). Mix the solution of the small sample in ammonia with hydrochloric acid to acid reaction, return the acidulated sample to the fluid from which it was taken, and which contains the yellow precipitate produced by the hydrosulphuric acid, and proceed as directed in (297). If, on the other hand, the addition of ammonia to the sample completely or partially fails to redissolve the precipitate, or if, in the preliminary examination (293), carbonate of ammonia has produced a precipitate, there is reason to suppose that another metal is present, perhaps with arsenic. In this latter case also, add to the sample in the test-tube hydrochloric acid to acid reaction, return the acidulated sample to the fluid from which it was taken, and which contains the yellow precipitate produced by the hydrosulphuric acid, and proceed as directed in (298).

c. A precipitate is formed of another color. In that case you have to assume that other metals are present, perhaps with arsenic. Proceed as directed in (298).   **296**

5. *Treatment of the Yellow Precipitate produced by Hydrosulphuric Acid, in Cases where the Results of the Examination in* (295) *lead to the Assumption that Arsenic alone is present.* Determination of the Weight of the Arsenic.   **297**

As soon as the fluid precipitated according to the directions of (293) has nearly lost the smell of sulphuretted hydrogen, transfer the yellow precipitate to a small filter, wash thoroughly, pour upon the still moist precipitate solution of ammonia, and wash the filter—on which, in this case, nothing must remain undissolved, except some sulphur—thoroughly with dilute ammonia; evaporate the ammonical fluid in a small accurately tared porcelain dish, on the water-bath, dry the residue at 212° Fah. until its weight suffers no further diminution, and weigh. If it is found, upon reduction, that the residue consisted of perfectly pure tersulphide of arsenic, calculate for every part of it 0·8049 of arsenious acid, or 0·6098 of arsenic. Treat the residue in the dish according to the instructions given in (300).

6. *Treatment of the Yellow Precipitate produced by Hydrosulphuric Acid, in Cases where the Results of the Examination in* (295), *or in* (296) *lead to the Assumption that another Metal is present—perhaps with Arsenic.* Separation of the Metals from each other. Determination of the Weight of the Arsenic.   **298**

If you have reason to suppose that the fluid precipitate by hydrosulphuric acid (293) contains other metals, perhaps with arsenic, proceed as follows :—As soon as the precipitation is thoroughly accomplished, and the smell of sulphuretted hydrogen has nearly gone off, transfer the precipitate to a small filter, wash thoroughly, perforate the point of the filter, and rinse the contents with the washing-bottle into a small flask, using the least possible quantity of water for the purpose; add to the fluid in which the precipitate is now suspended, first ammonia, then some yellowish sulphide of ammonium, and let the mixture digest for some time at a gentle heat. Should part of the precipitate remain undissolved, filter this off, wash, perforate the filter, rinse off the residuary precipitate, mark it III., and reserve for further examination according to the instructions given in (305). Evaporate the filtrate, together with the washings, in a small porcelain dish, to dryness. Treat the residue with some pure fuming nitric acid (free from chlorine), nearly drive off the acid by evaporation, then add, as C. Meyer was the first to recommend, gradually and in small portions at a time, a solution of pure carbonate of soda until it predominates. Add now a mixture of 1 part of carbonate and 2 parts of nitrate of soda in sufficient, yet not excessive quantity, evaporate to dryness, and heat the residue very gradually to fusion. Let the fused mass cool, and, when cold, extract it with cold water. If a residue remains undissolved, filter, wash with a mixture of equal parts of spirit of wine and water, mark it IV., and reserve for further examination, according to the directions of (306). Mix the solution, which must contain all the arsenic as arsenate of soda, with the washings, previously freed from alcohol by evaporation, add gradually and cautiously pure dilute sulphuric acid to strongly acid   **299** reaction, evaporate in a small porcelain dish, and, when the fluid is strongly concentrated, add again sulphuric acid, to see whether the quantity first added has been sufficient to expel all nitric acid and nitrous acid; heat now very cautiously until heavy fumes of hydrated sulphuric acid begin to escape; then let the liquid cool, and, when cold, add water, transfer the solution to a small flask, heat to 158° F., and conduct into it, for at least 6 hours, a slow stream of washed hydrosulphuric acid gas. Let the mixture finally cool, continuing the transmission of the gas all the while. If arsenic is present, a yellow precipitate will form. When the precipitate has completely subsided, and the fluid has nearly lost the smell of sulphuretted hydrogen, filter, wash the precipitate, dissolve it in ammonia, and proceed with the solution as directed in (297), to determine the weight of the arsenic.

7. *Reduction of the Sulphide of Arsenic.*

300 The production of metallic arsenic from the sulphide, which may be regarded as the keystone of the whole process, demands the greatest care and attention. The method recommended in § 132, 12, viz., to fuse the arsenical compound, mixed with cyanide of potassium and carbonate of soda, in a slow stream of carbonic acid gas, is the best and safest, affording, besides the advantage of great accuracy, also a positive guarantee against the chance of confounding the arsenic with any other body, more particularly antimony; on which account it is more especially adopted for medico-legal investigations.

Take care to have the whole apparatus filled with carbonic acid, and to give the proper degree of force to the gaseous stream, before applying heat. It is advisable to substitute for the evolution flask in Fig. 31 (§ 132, 12), a flask which will allow the operator to regulate the current of gas. The arrangement shown in Fig. 35, which is a matter of easy contrivance, will fully answer the purpose. Caoutchouc stoppers should be used; the compression stopcock is furnished with an adjusting screw.

Fig. 35.

As regards the process of reduction, either proceed at once with the sulphide of arsenic, or previously convert the latter into arsenic acid (see 301). In the former case take care, if possible, not to use the whole of the residue in the dish, obtained by the evaporation of the ammoniacal solution, but use only a portion of it, so that the process may be repeated several times if necessary. Should the residue be too trifling to admit of being divided into several portions, dissolve it in a few drops of ammonia, add a little carbonate of soda, and evaporate on the water-bath to dryness, taking care to stir the mixture during the process; divide the dry mass into several portions, and proceed to reduction.

Otto* recommends to convert the sulphide first into arsenic acid, then to reduce the latter with cyanide of potassium. The following is the process given by him to effect the conversion: Pour concentrated nitric acid over the sulphide of arsenic in the dish, evaporate, and repeat the same operation several times if necessary, then remove every trace of nitric acid by repeatedly moistening the residue with water, and drying again; when the nitric acid is *completely* expelled, treat the residue with a few drops of water, add carbonate of soda in powder, to form an alkaline mass, and thoroughly dry this in the dish, with frequent stirring, taking care to collect the mass within the least possible space in the middle of the dish. The dry mass thus obtained is admirably adapted for reduction. I can, from the results of my own experience, fully confirm this statement of Otto; but I must once more repeat that it is indispensable for the success of the operation that the residue should be perfectly free from *every trace* of nitric acid or nitrate, since otherwise deflagration is sure to take place during the process of fusion with cyanide of potassium, and, of course, the experiment will fail.

When the operation is finished, cut off the reduction-tube at c (see Fig. 36), set aside the fore part, which contains the arsenical mirror, put the other part of the tube into a cylinder, pour water over it, and let it stand some time; then filter the solution obtained, add

Fig. 36.

to the filtrate hydrochloric acid to acid reaction; then conduct some hydrosulphuric acid into it, and observe whether this produces a precipitate. In cases where the reduction of the sulphide of arsenic has been effected in the direct way, without previous conversion to arsenic acid, a trifling yellow precipitate will usually form; had traces of antimony been present, the precipitate would be orange-colored and insoluble in carbonate of ammonia. After all the soluble salts of the fused mass have been dissolved out, examine the metallic residue which may be left, for traces of tin and antimony; these being the only metals that can possibly be present if the instructions here given have been strictly followed. Should *appreciable* traces of these metals, or either of them, be found, proper allowance must be made for this in calculating the weight of the arsenic.

8. *Examination of the reserved Residues, marked severally* I., II., III., *and* IV., *for other Metals of the Fifth and Sixth Groups.*

   a. Residue I. Compare (290).
   This may contain more particularly chloride of silver and sulphate of lead, possibly also binoxide of tin. Incinerate the

---

* "Anleitung zur Ausmittelung der Gifte," von Dr. Fr. Jul. Otto, p. 36.

residue (I.) in a porcelain dish, burn the carbon with the aid of some nitrate of ammonia, extract the residue with water, dry the part left undissolved, then fuse it with cyanide of potassium in a porcelain crucible. When the fused mass is cold, treat it with water until all that is soluble in it is completely removed; warm the residue with nitric acid, and proceed as directed in § 181.

b. *Residue II.* Compare (292). **304**

The carbonaceous residue which is obtained in the purification of the crude sulphide by means of nitric acid and sulphuric acid, may more particularly contain lead, mercury, and tin; antimony and bismuth may also be present.

Heat the residue for some time with nitrohydrochloric acid, and filter the solution; wash the undissolved residue with water, at first mixed with some hydrochloric acid, add the washings to the filtrate, and treat the dilute fluid thus obtained with hydrosulphuric acid. Should a precipitate form, examine this according to the instructions given in § 191. Incinerate the residue insoluble in nitrohydrochloric acid, fuse the ash in conjunction with cyanide of potassium, and proceed with the fused mass as directed in (303).

c. *Residue III.* Compare (298). **305**

Examine the precipitate insoluble in sulphide of ammonium for the metals of the fifth group according to the instructions given in § 193.

d. *Residue IV.* Compare (299). **306**

This may contain tin and antimony, perhaps also copper. Proceed as directed (123). If the color of the residue was black (oxide of copper), treat the reduced metals according to the instructions given in § 181.

9. *Examination of the filtrate reserved in* (291), *for Metals of the Fourth and Third Groups, especially for Zinc and Chromium.*

a. As we have seen in (291), the fluid filtered from the precipitate produced by hydrosulphuric acid, and temporarily reserved for further examination, has already been mixed with sulphide of ammonium. The addition of this reagent to the filtrate is usually attended with the formation of a precipitate consisting of sulphide of iron and phosphate of lime, but which may possibly also contain sulphide of zinc. Filter the fluid from this precipitate, and treat the filtrate as directed in (388); wash the precipitate with water mixed with some sulphide of ammonium, dissolve by warming with hydrochloric acid, and boil the solution with nitric acid, to convert the protoxide of iron into sesquioxide; add, if necessary sufficient sesquichloride of iron for carbonate of soda to produce a brownish-yellow precipitate in a sample of the fluid; neutralize almost completely with carbonate of soda, precipitate with carbonate of baryta, and filter; the precipitate contains all the sesquioxide of iron and all the phosphoric acid. Concentrate the filtrate, precipitate the baryta with dilute sulphuric acid, filter, add to the filtrate ammonia to alkaline reaction, and precipitate with sulphide of ammonium the zinc which may be present. For the further examination of the precipitate see § 106. **307** *b.* The fluid filtered from the precipitate produced by sulphide of ammonium (307) will usually contain all the chromium that may be present, as sulphide of ammonium fails to precipitate sesquioxide of chromium from solutions containing organic matters. If you wish to ascertain whether chromium is really present, evaporate the filtrate to dryness, ignite, mix the fixed residue with 3 parts of nitrate of potassa and 1 part of carbonate of soda, and project the mixture into a crucible heated to moderate redness. Allow the fused mass to cool, and, when cold, boil with water: yellow coloration of the fluid shows the presence of alkaline chromate, and accordingly of chromium. For confirmatory tests see § 138. 308

II. Method for the Detection of Hydrocyanic Acid.

§ 226.

In cases of actual or suspected poisoning with hydrocyanic acid, where it is required to separate that acid from articles of food or from the contents of the stomach, and thus to prove its presence, it is highly necessary to act with the greatest expedition, as the hydrocyanic acid speedily undergoes decomposition. Still this decomposition is not quite so rapid as is generally supposed, and indeed it requires some time before the *complete* decomposition of the *whole* of the acid present is effected.* 309

Although hydrocyanic acid betrays its presence, even in minute quantities, by its peculiar odor, still this sign must never be looked upon as conclusive. On the contrary, to adduce positive proof of the presence of the acid, it is always indispensable to separate it, and to convert it into certain known compounds.

The method of accomplishing this, which I am about to describe, is based upon distillation of the acidified mass, and examination of the distillate for hydrocyanic acid. Now, as the non-poisonous salts, ferro- and ferricyanide of potassium give by distillation likewise a product containing hydrocyanic acid, it is, of course, indispensable—as Otto very properly observes—first to ascertain whether one of these salts may not be present. To this end, stir a small portion of the mass to be examined with water, filter, acidify the filtrate with hydrochloric acid, and test a portion of it with sesquichloride of iron, another with sulphate of protoxide of iron. If no blue precipitate or coloration forms in either, soluble ferro- and ferricyanides are not present, and you may safely proceed as follows;

Test, in the first place, the reaction of the mass under examination; if necessary, after mixing and stirring it with water. If it is not already strongly acid, add solution of tartaric acid until the fluid strongly reddens litmus-paper; introduce the mixture into a retort, and place the body of the retort, with the neck pointing upwards, in an iron or copper vessel, but so that it does not touch the bottom, which 310

---

* Thus I succeeded in separating a notable quantity of hydrocyanic acid from the stomach of a man who had poisoned himself with that acid in very hot weather, and whose intestines were handed to me full 36 hours after death.—A dog was poisoned with hydrocyanic acid, and the contents of the stomach, mixed with the blood, were left for 24 hours exposed to an intense summer heat, and then examined: the acid was still detected.

should, moreover, by way of precaution, be covered with a cloth; fill the vessel with a solution of chloride of calcium, and apply heat, so as to cause gentle ebullition of the contents of the retort. Conduct the vapors passing over, with the aid of a tight-fitting tube, bent at a very obtuse angle, through a *Liebig's* condensing apparatus, and receive the distillate in a small weighed flask. When about half-an-ounce of distillate has passed over, remove the receiver, and replace it by a somewhat larger flask, also previously tared. Weigh the contents of the first receiver now, and proceed as follows :

a. Mix one-fourth of the distillate with solution of potassa **311** or soda to strongly alkaline reaction, then add a small quantity of solution of sulphate of protoxide of iron, mixed with a little sesquichloride of iron; digest a few minutes at a very gentle heat, and supersaturate finally with hydrochloric acid. If a blue precipitate forms, this shows the distillate to contain a relatively large, if a blue-greenish fluid is obtained, from which blue flakes separate after long standing, a relatively small, quantity of hydrocyanic acid.

b. Treat another fourth as directed § 155, 7, to convert the **312** hydrocyanic acid into sulphocyanide of iron. As the distillate might, however, contain acetic acid, do not neglect to add towards the end of the process a little more hydrochloric acid, in order to neutralize the adverse influence of the acetate of ammonia.

c. If the experiments *a* and *b* have demonstrated the pre- **313** sence of hydrocyanic acid, and you wish now also to approximately determine its quantity, continue the distillation until the fluid passing over contains no longer the least trace of hydrocyanic acid; add one-half of the contents of the second receiver to the remaining half of the contents of the first, mix the fluid with nitrate of silver, then with ammonia until it predominates, and finally with nitric acid to strongly acid reaction. Allow the precipitate which forms to subside, filter on a tared filter, dried at 212° Fah., wash the precipitate, dry it thoroughly at 212° Fah., and weigh. Ignite the weighed precipitate in a small porcelain crucible, to destroy the cyanide of silver, fuse the residue with carbonate of soda and potassa—to effect the decomposition of the chloride of silver which it may contain—boil the mass with water, filter, acidify the filtrate with nitric acid, and precipitate with nitrate of silver; determine the weight of the chloride of silver which may precipitate, and deduct the amount found from the total weight of the chloride and cyanide of silver: the difference gives the quantity of the latter; by multiplying the quantity found of the cyanide of silver by 0·2017, you find the corresponding amount of anhydrous hydrocyanic acid; and by multiplying this again by ·2—as only one-half of the distillate has been used—you find the total quantity of hydrocyanic acid which was present in the examined mass. Instead of decomposing the fused silver precipitate by fusion with carbonate of soda and potassa, it may be reduced also by means of zinc, with addition of dilute sulphuric acid, and the chlorine determined in the filtrate.

Instead of pursuing this indirect method, you may also **314** determine the quantity of the hydrocyanic acid by the following direct method: introduce half of the distillate into a retort, together with powdered borax; distil to a small residue, and determine the hydrocyanic acid in the distillate as cyanide of silver. Hydrochloric acid can no longer be present in this distillate, as the soda of the borax retains it in the retort (*Wackenroder.*)

III. METHOD FOR THE DETECTION OF PHOSPHORUS.

§ 227.

Since phosphorus paste has been employed to poison mice, &c., 315 and the poisonous action of lucifer matches has become more extensively known, phosphorus has not unfrequently been resorted to as an agent for committing murder. The chemist is therefore occasionally called upon to examine some article of food, or the contents of a stomach, for this substance. It is obvious that, in cases of the kind, his whole attention must be directed to the separation of the phosphorus in the *free state,* or to the production of such reactions as will enable him to infer the presence of *free phosphorus;* since the mere finding of phosphorus in form of phosphates would prove nothing, as phosphates invariably form constituents of animal and vegetable bodies.

A. Detection of Unoxidized Phosphorus.

1. Ascertain in the first place whether the substance under ex- 316 amination does not betray the presence of phosphorus to the sense of smell, by the peculiar odor of that element, or to the sense of sight, by luminosity in the dark. To this end take care to increase the contact of the phosphorus in the substance with the air, by rubbing, stirring, and shaking.

2. Put a little of the substance into a flask, fasten to the loosely 317 inserted cork a strip of filtering-paper moistened with neutral solution of nitrate of silver, and heat to from 86° to 104° Fah. If the paper does not turn black, not even after some time, no unoxidized phosphorus is present, and there is consequently no need to try 3 and 4, but the operator may at once pass on to B (**324**). If, on the other hand, the paper strip turns black, this is no positive proof of the presence of unoxidized phosphorus, as hydrosulphuric acid, formic acid, putrifying matters, &c., will also cause blackening of the paper. Treat therefore the principal mass of the substance now by the methods 3 and 4. (To ascertain whether the blackening proceeds from the presence of hydrosulphuric acid, try the reaction with a strip of paper moistened with solution of lead or with terchloride of antimony.)—T. SCHERER.—Annal d. Chem. u. Pharm., 112, 214.

3. As the luminosity of phosphorus is always one of the most striking proofs of the presence of that element in the unoxidized -318 state, examine a larger portion of the substance by the following excellent and approved method, recommended by E. MITSCHERLICH.*

Mix the substance under examination with water and some sulphuric acid, and subject the mixture to distillation in a flask, A (see Fig. 37). This flask is connected with an evolution tube, *b,* and the latter again with a glass cooling or condensing tube, *c c c,* which passes through the

---

* "Journal für prakt. Chemie," vol. 66, p. 238.

bottom of a cylinder, B, in which it is fastened by means of a cork, *a*, and opens into a glass vessel, C. Cold water is made to run from D through a stopcock, into a funnel, *i*, the lower end of which rests upon the bottom of B; the cooling water flows off through *g*.

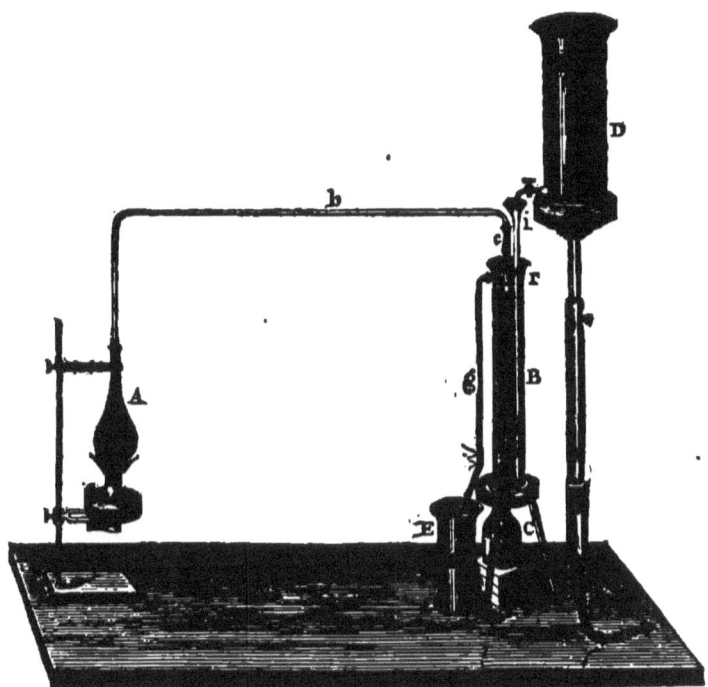

Fig. 37.

Now, if the substance in A contains phosphorus, there will appear, in the dark, in the refrigerated part of the condensing tube at the point *r*, where the aqueous vapors distilling over enter that part of the tube, a strong luminosity, usually a luminous ring. If you take for distillation 5 oz. of a mixture containing only $\frac{1}{10}$th of a grain of phosphorus, and accordingly only 1 part of phosphorus in 100,000 parts of mixture, you may distil over 3 oz. of it—which will take at least half-an-hour—without the luminosity ceasing; MITSCHERLICH, in one of his experiments, stopped the distillation after half-an-hour, allowed the flask to stand uncorked a fortnight, and then recommenced the distillation: the luminosity was as strong as at first.

If the fluid contains substances which prevent the luminosity of phosphorus in general, such as ether, alcohol, or oil of turpentine, no luminosity is observed so long as these substances continue to distil over. In the case of ether and alcohol, however, this is soon effected, and the luminosity accordingly very speedily makes its appearance; but it is quantity of putrid blood mixed with the composition scraped from the tip of a common lucifer match; and this even in presence of substances which prevent the luminosity of the phosphorus in experiments made by MITSCHERLICH's method.

323 5. If there is sufficient phosphorus present to permit a quantitative determination, this may be effected most conveniently by SCHERER's modification of MITSCHERLICH's method, viz., by distilling the mass, acidified with sulphuric acid, in an atmosphere of carbonic acid. I would suggest, with respect to this, to have the distilling flask furnished with a double perforated cork, and to transmit pure carbonic acid gas until the apparatus is filled with it, then to shut off the carbonic acid stream. A flask with double perforated cork serves for receiver; the mouth of the condensing tube passes into one of the openings; into the other is inserted a bent glass-tube, which leads into a U-shaped tube containing a solution of pure nitrate of silver.

When the distillation is over, minute globules of phosphorus are found in the receiver. A moderate stream of carbonic acid is now once more transmitted through the apparatus, and a gentle heat applied, with a view to effect the formation of larger globules by aggregation. These are then washed and weighed the same as in MITSCHERLICH's method. The fluid poured off the phosphorus globules is luminous in the dark when shaken. It requires, however, a larger proportion of phosphorus to obtain distinct luminosity in this way than is the case with MITSCHERLICH's method. The phosphorus in the fluid may, after oxidation by nitric acid or chlorine, be determined as phosphoric acid. However, the result is reliable only if the operation has been conducted with the requisite care and caution to guard against portions of the boiling fluid, which often contains phosphoric acid, spirting out of the vessel, and being thus lost. To obtain the remainder of the phosphorus originally present in the examined substance, treat the contents of the U-shaped tube finally with nitric acid, throw down the silver by hydrochloric acid, filter through a washed filter, concentrate in a porcelain dish, and precipitate the phosphoric acid as phosphate of magnesia and ammonia, finally weigh it as pyrophosphate of magnesia.

B. *Detection of Phosphorous Acid.*

324 Should all attempts to show the presence of unoxidized phosphorus fail, try whether it may not be practicable to find the first product of the oxidation of phosphorus in the air, *i.e.*, phosphorous acid. For this purpose transfer the residue left in the distilling flask in (318) or in (323), or the residue left in (320), to the apparatus illustrated by Fig. 38, having previously tested the purity of the zinc and sulphuric acid, as directed (321), then proceed according to the instruction of (321), and observe whether the coloration of the hydrogen flame reveals the presence of phosphorus. Should this be the case, the end in view is attained; if not, the presence of organic substances may be the preventive cause. If, therefore, the flame remains uncolored, shut the compression stopcock at once, connect with the apparatus a U-shaped tube containing a solution of neutral nitrate of silver, open the cock again, and let the gas pass for many hours, in a slow stream, through the silver solution. The formation of a precipitate in the silver solution shows the presence of phosphorous acid. Examine the precipitate of phosphide of silver as directed in (321).

3. *Examination of the Inorganic Constituents of Plants, Animals, or Parts of the same, of Manures, &c.* (*Analysis of Ashes.*)

§ 228.

A. PREPARATION OF THE ASH.

It is sufficient for the purposes of a qualitative analysis to incinerate a comparatively small quantity of the substance which it is intended to examine for its inorganic constituents; the substance must previously be most carefully cleaned. The incineration is effected best in a small clay muffle, but it may be conducted also in a Hessian crucible placed in a slanting position, or, under certain circumstances, even in a small porcelain or.platinum dish, with the aid of a wide glass-tube or lamp-glass, to increase the draught. The heat must always be moderate, to guard against the volatilization of certain constituents, more especially of metallic chlorides. It is not always necessary to continue the combustion until all the carbon is consumed. With ashes containing a large proportion of fusible salts, as, *e. g.* the ash of beetroot molasses, it is even advisable to effect, in the first place, complete carbonization, then to boil the charred mass with water, and finally to incinerate the washed and dried residue. For further particulars see *Quantitative Analysis*, 3rd Edition, § 250. 325

B. EXAMINATION OF THE ASH.

As the qualitative analysis of the ash of a vegetable substance is generally undertaken, either as a practical exercise, or for the purpose of determining its general character, and the state or condition in which any given constituent may happen to be present, or also with a view to make, as far as practicable, an approximate estimation of the respective quantities of the several constituents, it is usually the best way to examine separately, (1) the part soluble in water, (2) the part soluble in hydrochloric acid, and (3) the residue which is insoluble in either menstruum. This can be done the more readily, as the number of bodies to which regard must be had in the analysis is only small, and the several processes may accordingly be expeditiously performed. 326

a. Examination of the Part soluble in Water.

Boil the ash with water, filter, and whilst the residue is being washed, examine the solution as follows :—

1. Add to a portion, after heating it, hydrochloric acid in excess, warm, and let the fluid stand at rest. Effervescence indicates CARBONIC ACID combined with alkalies; smell of hydrosulphuric acid indicates the SULPHIDE of an ALKALI METAL, formed from an alkaline sulphate by the reducing action of the carbon. Turbidity from separation of sulphur, with smell of sulphurous acid, denotes a HYPOSULPHITE (which occurs occasionally in the ash of pit-coal). Filter if necessary, and add to the filtrate—or to the fluid where no filtration is required—some chloride of barium; the formation of a white precipitate indicates the presence of SULPHURIC ACID. 327

2. Evaporate another portion of the solution until it is reduced 328

to a small volume, add hydrochloric acid to acid reaction—effervescence indicates the presence of CARBONIC ACID. Evaporate now to dryness, and treat the residue with hydrochloric acid and water. The portion left undissolved consists of SILICIC ACID. Filter, add ammonia, chloride of ammonium, and sulphate of magnesia ; the formation of a white precipitate indicates the presence of PHOSPHORIC ACID. Instead of this reaction, you may also mix the fluid filtered from the silicic acid with acetate of soda, and then cautiously add, drop by drop, sesquichloride of iron, or you may test with nitric acid solution of molybdate of ammonia (§ 142).

3. Add to another portion of the solution nitrate of silver as long as a precipitate continues to form; warm gently, and then cautiously add ammonia ; if a black residue is left, this consists of sulphide of silver, proceeding from the sulphide of an alkali metal, or from a hyposulphite. Mix the ammoniacal solution now—after previous filtration if necessary—cautiously with nitric acid in slight excess, to effect the solution of the phosphate of silver precipitate formed, leaving thus only CHLORIDE (iodide,* bromide) OF SILVER undissolved. Filter, and examine the precipitate as directed (178) ; neutralize the filtrate exactly with ammonia. If this produces a *bright yellow* precipitate, the phosphoric acid found in (328) was present in the tribasic, if a *white* precipitate, it was present in the bibasic form. **329**

4. Heat a portion of the solution with hydrochloric acid, then make it alkaline with ammonia ; mix the alkaline fluid with oxalate of ammonia, and let it stand at rest. The formation of a white precipitate indicates LIME. Filter, and mix the filtrate with ammonia and phosphate of soda ; the formation of a crystalline precipitate, which often becomes visible only after long standing, indicates MAGNESIA. Magnesia is often found in distinctly appreciable, lime only in exceedingly minute quantity, even where alkaline carbonates and phosphates are present. **330**

5. For POTASSA and SODA examine as directed § 197. If magnesia is present, neutralize with hydrochloric acid the portion of the filtrate intended to be tested for the alkalies, then remove the magnesia as directed § 196, 2, before proceeding to the examination for the alkalies.

6. LITHIA, which is much more frequently found in ashes than has hitherto been believed, and OXIDE OF RUBIDIUM, which is almost constantly present with potassa, may be most readily and conveniently detected by spectrum analysis (§ 93) in the residue consisting of the alkali salts.

b. *Examination of the Part soluble in Hydrochloric Acid.*

Warm the residue left undissolved by water with hydrochloric acid†—effervescence indicates CARBONIC ACID, combined with alkaline earths ; evolution of chlorine denotes OXIDES OF MANGANESE. Evaporate to dryness, and heat a little more strongly, to effect the **331**

---

* To detect the iodine in aquatic plants, dip the plant in weak solution of potassa (*Chatin*), dry, incinerate, treat with water, and examine the aqueous solution as directed (257).

† If the residue still contains much carbon, after further incineration.

separation of the silicic acid; moisten the residue with hydrochloric acid and some nitric acid, add water, warm, and filter.

1. Test a portion of the solution with hydrosulphuric acid. If this produces any other than a perfectly white precipitate, you must examine it in the usual way. The ashes of plants occasionally contain COPPER; if the plant has been manured with excrements deodorized by nitrate of lead, they may contain lead; other metals are also occasionally found.

2. Mix a portion of the original solution with carbonate of soda, **332** as long as the precipitate formed redissolves upon stirring; then add acetate of soda, and some acetic acid. This produces, in most cases, a white precipitate of PHOSPHATE OF SESQUIOXIDE OF IRON. If the fluid in which this precipitate is suspended is reddish, there is more sesquioxide of iron present than corresponds to the phosphoric acid; if it is colorless, add sesquichloride of iron, drop by drop, until the fluid looks reddish. (From the quantity of the precipitate of phosphate of sesquioxide of iron formed, you may estimate the PHOSPHORIC ACID present.) Heat to boiling,* filter hot, and mix the filtrate, after addition of ammonia, with yellowish sulphide of ammonium, in a stoppered flask, filled almost to the mouth; should a precipitate form, after long standing, examine this before the blowpipe for MANGANESE and ZINC (which latter metal may exceptionally be present) according to the direction of **(141)**, and the fluid filtered from it for LIME and MAGNESIA, in the usual way **(330)**.

3. To test for BARYTA and STRONTIA, mix a somewhat larger quantity of the hydrochloric acid solution with some dilute sulphuric acid, let the mixture stand some time, and, if a precipitate forms, examine this according to the instructions of **(254)**.

c. Examination of the Residue insoluble in Hydrochloric Acid.

The residue insoluble in hydrochloric acid contains,

1. The silicic acid, which has separated on treating with hydro- **333** chloric acid.

2. Those ingredients of the ash which are insoluble in hydrochloric acid. These are, in most ashes, sand, clay, carbon; substances, therefore, which are present in consequence of defective cleaning or imperfect combustion of the plants, or matter derived from the crucible. It is only the ashes of the stalk of cereals and others abounding in silicic acid that are not completely decomposed by hydrochloric acid.

Boil the washed residue with solution of carbonate of soda in **334** excess, filter hot, wash with boiling water, and test for silicic acid in the filtrate by evaporation with hydrochloric acid (§ 150, 2). If the ash was of a kind to be completely decomposed by hydrochloric acid, the analysis may be considered as finished—for the accidental admixture of clay and sand will rarely interest the analyst sufficiently to warrant a more minute examination by fluxing. But if the ash abounded in silicic acid, and it may therefore be supposed that the hydrochloric acid has failed to effect complete decomposition, evaporate half of the residue insoluble in solution of carbonate of soda with pure solution of soda in excess, in a silver or platinum dish, to dryness. This decomposes the silicates of the ash, whilst but little affecting the sand. Acidify now with

---

* If this should fail to decolorize the fluid, add some more acetate of soda.

hydrochloric acid, evaporate to dryness, &c., and proceed as in (331). For the detection of the alkalies use the other half of the residue, treating this according to the instructions of (228).

SECTION III.

EXPLANATORY NOTES AND ADDITIONS TO THE SYSTEMATIC COURSE OF ANALYSIS.

I. Additional Remarks to the Preliminary Examination.

To §§ 175—178.

The inspection of the physical properties of a body may, as already stated § 175, in many cases enable the analyst to draw certain general inferences as to its nature. Thus, for instance, if the analyst has a white substance before him, he may at once conclude that it is not cinnabar, or if a light substance, that it is not a compound of lead, &c.

Inferences of this kind are quite admissible to a certain extent; but if carried too far, they are apt to mislead the operator, by blinding him to every reaction not exactly in accordance with his preconceived notions.

As regards the examination of substances at a high temperature, platinum foil or small iron spoons may also be used in the process; however, the experiment in the glass tube gives, in most cases, results more clearly evident, and affords moreover the advantage that volatile bodies are less likely to escape detection, and that a more correct and precise notion can be formed of the nature of the heated substance, than exposure on platinum foil or in an iron spoon will permit. To ascertain the products of oxidation of a body, it is sometimes advisable also to heat it in a short glass-tube, open at both ends, and held in a slanting position; small quantities of a metallic sulphide, for instance, may be readily detected by this means. (Compare § 156, 6.)

With respect to the preliminary examination by means of the blowpipe, I have to remark that the student must avoid drawing positive conclusions from pyrochemical experiments, until he has acquired some practice in this branch of analytical chemistry. A slight incrustation of the charcoal support, which may *seem* to denote the presence of a certain metal, is not always a *conclusive* proof of the presence of that metal; nor would it be safe to assume the absence of a substance simply because the blowpipe flame fails to effect reduction, or solution of nitrate of protoxide of cobalt fails to impart a color to the ignited mass, &c. The blowpipe reactions are, indeed, in most cases, unerring, but it is not always easy to produce them, and they are moreover liable to suffer modification by accidental circumstances.

The student should never omit the preliminary examination; the notion that this omission will save time and trouble is very erroneous.

II. Additional Remarks to the Solution, etc., of Substances.

To §§ 179—181.

It is a task of some difficulty to fix the exact limit between substances which are soluble in water and those that are insoluble in that menstruum, since the number of bodies which are sparingly soluble in water is very considerable, and the transition from *sparingly* soluble to *insoluble* is very gradual. Sulphate of lime, which is soluble in 430 parts of water, might perhaps serve as a limit between the two classes, since this salt may still be positively detected in aqueous solution by the delicate reagents which we possess for lime and sulphuric acid.

When examining an aqueous fluid by evaporating a few drops of it upon platinum foil, to see whether it holds a solid body in solution, a very minute residue sometimes remains, which leaves the analyst in doubt respecting the nature of the substance. In cases of the kind test, in the first place, the reaction of the fluid with litmus-papers; in the second place, add to a portion of it a drop of solution of chloride of barium; and lastly, to another portion some carbonate of soda. Should the fluid be neutral, and remain unaltered upon the addition of these reagents, the analyst need not, as a general rule, examine it any further for bases or acids; since if the fluid contained any of those bases or acids which principally form sparingly soluble compounds, the chloride of barium and the carbonate of soda would have revealed their presence. The analyst may therefore feel assured that the detection of the substance of which the residue left upon evaporation consists will be more readily effected in the class of bodies insoluble in water.

If water has dissolved any part of the substance under examination, the student will always do well to examine the solution both for acids and bases, since this will lead more readily to a correct apprehension of the nature of the compound—an advantage which will amply counterbalance the drawback of sometimes meeting with the same substance both in the aqueous and in the acid solution.

The following substances (with few exceptions) are insoluble in water, but soluble in hydrochloric acid or in nitric acid: the phosphates, arsenates, arsenites, borates, carbonates, and oxalates of the earths and metals; also several tartrates, citrates, malates, benzoates, and succinates; the oxides and sulphides of the heavy metals; alumina, magnesia; many of the metallic iodides and cyanides, &c. Nearly the whole of these compounds are, indeed, decomposed, if not by dilute, by boiling concentrated hydrochloric acid;* but this decomposition gives rise to the formation of insoluble compounds where oxide of silver is present, and of sparingly soluble compounds in the presence of suboxide of mercury and lead. This is not the case with nitric acid, and accordingly the latter effects complete solution in many cases where hydrochloric acid leaves a residue. On the other hand, however, nitric acid leaves, besides the bodies insoluble in any simple acid, teroxide of antimony, binoxide of tin, binoxide of lead, &c., undissolved, and dissolves many other substances less readily than hydrochloric acid, *e. g.* sesquioxide of iron and alumina.

Substances not soluble in water are therefore, briefly expressed, to be treated as follows: try to dissolve them in dilute or concentrated, cold

* For the exceptions see § 203.

or boiling hydrochloric acid; if this fails to effect complete solution, try to dissolve a fresh portion in nitric acid; if this also fails, treat the body with aqua regia, which is an excellent solvent, more particularly for metallic sulphides. To examine separately the solution in hydrochloric acid or in nitric acid, on the one hand, and that in nitrohydrochloric acid on the other, is, in most cases, neither necessary nor desirable. To prepare a solution in nitric acid or in aqua regia, where the nature of the substance does not absolutely demand it, is not advisable, as a solution in hydrochloric acid is much better suited for precipitation by hydrosulphuric acid. Nor is it advisable to concentrate a solution in aqua regia by evaporation, to drive off the excess of the acids, as the operation might lead to the escape of volatile chlorides, more particularly of chloride of arsenic. It is therefore always best to use no more aqua regia than is just necessary to effect solution.

With regard to the solution of metals and alloys, I have to remark that, upon boiling them with nitric acid, white precipitates will frequently form, although neither tin nor antimony be present. Inexperienced students often confound such precipitates with the oxides of these two metals, although their appearance is quite different. These precipitates consist simply of nitrates sparingly soluble in the nitric acid present, but readily soluble in water. Consequently the analyst should ascertain whether these white precipitates will dissolve in water or not, before he concludes them to consist of tin or antimony.

III. Additional Remarks to the Actual Examination.

To §§ 182—204.

A. General Review and Explanation of the Analytical Course.

a. Detection of the Bases.

The classification of the bases into groups, and the methods which serve to detect and isolate them individually, have been fully explained in Part I., Section III. The systematic course of analysis, from § 189 to § 198, is founded upon this classification of the bases; and as a correct apprehension of it is of primary importance, I will here subjoin a brief explanation of the grounds upon which this division rests. Respecting the detection of the several bases individually, I refer the student to the recapitulations and remarks in §§ 88—125.

The general reagents which serve to divide the bases into principal groups are—HYDROCHLORIC ACID, HYDROSULPHURIC ACID, SULPHIDE OF AMMONIUM, and CARBONATE OF AMMONIA: this is likewise the order of succession in which they are applied. Sulphide of ammonium performs a *double* part.

Let us suppose we have in solution the whole of the bases, together with arsenious and arsenic acids, and also phosphate of lime—which latter may serve as a type for the salts of the alkaline earths soluble in acids and reprecipitated unaltered by ammonia.

Chlorine forms insoluble compounds only with silver and mercury; chloride of lead is sparingly soluble in water. The insoluble subchloride of mercury corresponds to the suboxide of that metal. If, therefore, we add to our solution:

1. *Hydrochloric Acid,*

we remove from it the metallic oxides of the first division of the fifth group, viz., the whole of the OXIDE OF SILVER and the whole of the SUB-OXIDE OF MERCURY. From concentrated solutions a portion of the LEAD may likewise precipitate as chloride; this is, however, immaterial, as a sufficient quantity of the lead remains in the solution to permit the subsequent detection of this metal.

Hydrosulphuric acid completely precipitates the oxides of the fifth and sixth groups from solutions containing a free mineral acid, since the affinity of the metallic radicals of these oxides for sulphur, and that of the hydrogen for oxygen, is sufficiently powerful to overcome the affinity between the metal and the oxygen, and that between the oxide and a strong acid, EVEN THOUGH THE ACID BE PRESENT IN EXCESS. But none of the other bases are precipitated under these circumstances, since those of the first and second groups form no sulphur compounds insoluble in water, besides that their sulphides cannot possibly form in acid solutions; as regards those of the third group, sulphide of aluminium and sulphide of chromium cannot possibly be formed in the humid way; and the affinity which the metallic radicals of the oxides of the fourth group possess for sulphur, combined with that manifested by hydrogen for oxygen, is not sufficiently powerful to overcome the affinity of the metal for oxygen and of the oxide for a strong acid, IF THE LATTER IS PRESENT IN EXCESS.

If, therefore, after the removal of the oxide of silver and suboxide of mercury, by means of hydrochloric acid, we add to the solution, which still contains free hydrochloric acid,

2. *Hydrosulphuric Acid,*

we remove from it the remainder of the oxides of the fifth, together with those of the sixth group, viz., OXIDE OF LEAD, OXIDE OF MERCURY, OXIDE OF COPPER, TEROXIDE OF BISMUTH, OXIDE OF CADMIUM, TEROXIDE OF GOLD, BINOXIDE OF PLATINUM, PROTOXIDE OF TIN, BINOXIDE OF TIN, TEROXIDE OF ANTIMONY, ARSENIOUS ACID, and ARSENIC ACID. All the other oxides remain in solution, either unaltered, or reduced to a lower degree of oxidation, *e. g.*, sesquioxide of iron to protoxide; chromic acid to sesquioxide of chromium, &c.

The sulphides corresponding to the oxides of the sixth group combine with basic metallic sulphides (the sulphides of the alkali metals), and form with them sulphur salts soluble in water; while the sulphides corresponding to the oxides of the fifth group do not possess this property, or possess it only to a limited extent.* If, therefore, we treat the whole of the sulphides precipitated by hydrosulphuric acid from an acid solution, with—

3. *Sulphide of Ammonium* (or, in certain cases, *Sulphide of Sodium*),

with addition, if necessary, of some sulphur or yellow sulphide of ammonium, the sulphides of mercury, lead, copper, bismuth, and cadmium remain undissolved, whilst the other sulphides dissolve as double compounds of sulphide of GOLD, PLATINUM, ANTIMONY, TIN, ARSENIC, with SULPHIDE OF AMMONIUM (or, as the case may be, SULPHIDE OF SODIUM),

---

* Sulphide of mercury combines with sulphide of potassium and sulphide of sodium, but not with sulphide of ammonium; sulphide of copper dissolves a little in sulphide of ammonium, but not in sulphide of potassium or sulphide of sodium.

and precipitate again from this solution upon the addition of hydrochloric acid, either unaltered, or in a state of higher sulphuration (they take up sulphur from the yellow sulphide of ammonium). The rationale of this precipitation is as follows:—The acid decomposes the sulphur salt formed. The sulphur base (sulphide of ammonium or sulphide of sodium) is decomposed by the hydrochloric acid into chloride and hydrosulphuric acid; and the liberated electro-negative sulphide (sulphur acid) precipitates. Sulphur precipitates at the same time if the sulphide of ammonium contains an excess of that element. The analyst must bear in mind that this eliminated sulphur makes the precipitated sulphides appear of a lighter color than they are naturally.

The sulphides corresponding to the oxides still remaining in solution are part of them—as those of the alkalies and alkaline earths—soluble in water; part—as those of alumina and sesquioxide of chromium—decomposed by water into hydrated oxides and hydrosulphuric acid; part— as those of the fourth group—insoluble in water. These latter would accordingly have been precipitated by hydrosulphuric acid, but for the free acid present. If, therefore, this free acid is removed, *i.e.*, if the solution is made alkaline, and then treated with more hydrosulphuric acid, if required, or, what will answer both purposes at once, if

4. *Sulphide of Ammonium,*

is added to the solution,* the sulphides corresponding to the oxides of the fourth group will precipitate: viz., the SULPHIDES OF IRON, MANGANESE, COBALT, NICKEL, and ZINC. But in conjunction with them, HYDRATE OF ALUMINA, HYDRATED SESQUIOXIDE OF CHROMIUM, and PHOSPHATE OF LIME are thrown down, because the affinity which the oxide of ammonium possesses for the acid of the salt of alumina or of sesquioxide of chromium, or for that which keeps the phosphate of lime in solution, causes the elements of the sulphide of ammonium to transpose with those of the water, thus giving rise to the formation of oxide of ammonium and of hydrosulphuric acid. The former combines with the acid, the latter escapes, being incapable of entering into combination with the liberated oxides or with the phosphate of lime,—the oxides and the lime-salt precipitate.

There remain now in solution only the alkaline earths and the alkalies. The neutral carbonates of the former are insoluble in water, whilst those of the latter are soluble in that menstruum. If, therefore, we now add

5. *Carbonate of Ammonia,*

together with a little pure ammonia, to guard against the possible formation of bicarbonates, the whole of the alkaline earths ought to precipitate. This is, however, the case only as regards BARYTA, STRONTIA, and LIME; of magnesia we know that, owing to its disposition to form double compounds with salts of ammonia, it precipitates only in part; and that the presence of an additional salt of ammonia will altogether prevent its precipitation, at least within a reasonable space of time. To guard against any uncertainty arising from this cause, chloride of ammonium is added previously to the addition of the carbonate of ammonia,

* After previous neutralization of the free acid by ammonia, to prevent unnecessary evolution of hydrosulphuric acid; and after previous addition also, if necessary, of chloride of ammonium to prevent the precipitation of magnesia by ammonia.

the mixture soon after filtered, and thus the precipitation of the magnesia is altogether prevented.

We have now still in solution MAGNESIA and the ALKALIES. The detection of magnesia may be effected by means of phosphate of soda and ammonia; but its separation requires a different method, since the presence of phosphoric acid would impede the further progress of the analysis. The process which serves to effect the removal of the magnesia is based upon the insolubility of that earth in the pure state. The substance under examination is accordingly ignited in order to expel the salts of ammonia, and the magnesia is then precipitated by means of baryta, the alkalies, together with the newly formed salt of baryta and the excess of the caustic baryta added, remaining in solution. By the addition of carbonate of ammonia the compounds of baryta are removed from the solution, which now only contains the fixed alkalies, the salt of ammonia formed, and the excess of the salt of ammonia added. If the salts of ammonia are then removed by ignition, the residue consists of the fixed alkalies alone. This method of separating the baryta affords the advantage over that of effecting the removal of that earth by means of sulphuric acid, that the alkalies are obtained in the most convenient form for their subsequent individual detection and isolation, viz., as chlorides. But as carbonate of baryta is slightly soluble in salts of ammonia, and gives, upon evaporation with chloride of ammonium, carbonate of ammonia and chloride of barium, it is usually necessary, after the expulsion of the salts of ammonia by ignition, to precipitate it once more with carbonate of ammonia, in order to obtain a solution perfectly free from baryta.

Lastly, to effect the detection of the AMMONIA, a fresh portion of the substance must of course be taken.

b. DETECTION OF THE ACIDS.

Before passing on to the examination for acids and salt-radicals, the analyst should first ask himself *which* of these substances may be expected to be present, to judge from the nature of the detected bases and the class to which the substance under examination belongs with respect to its solubility in water or acids, since this will save him the trouble of unnecessary experiments. Upon this point I refer the student to the table in Appendix IV., in which the various compounds are arranged according to their several degrees of solubility in water and acids.

The general reagents applied for the detection of the acids are, for the inorganic acids, CHLORIDE OF BARIUM and NITRATE OF SILVER, for the organic acids CHLORIDE OF CALCIUM and SESQUICHLORIDE OF IRON. It is therefore indispensable that the analyst should first assure himself whether the substance under examination contains only inorganic acids, or whether the presence of organic acids must also be looked for. The latter is invariably the case if the body, when ignited, turns black, owing to separation of carbon.—In the examination for bases the different reagents serve to effect the actual separation of the several groups of bases from each other; but in the examination for acids they serve simply to demonstrate the presence or absence of the acids belonging to the different groups.

Let us suppose we have an aqueous solution containing the whole of the acids, in combination with soda, for instance.

Baryta forms insoluble compounds with sulphuric acid, phosphoric acid, arsenious acid, arsenic acid, carbonic acid, silicic acid, boracic acid, chromic acid, oxalic acid, tartaric acid, and citric acid; fluoride of barium also is insoluble, or at least only sparingly soluble; all these compounds are soluble in hydrochloric acid, with the exception of sulphate of baryta. If, therefore, to a portion of our neutral or, if necessary, neutralized solution, we add,

1. *Chloride of Barium,*

the formation of a precipitate will denote the presence of at least one of these acids. By treating the precipitate with hydrochloric acid we learn at once whether sulphuric acid is present or not, as all the salts of baryta being soluble in this menstruum, with the exception of the sulphate, a residue left undissolved by the hydrochloric acid can consist only of the latter salt. Where sulphate of baryta is present, the reaction with chloride of barium fails to lead to the positive detection of the whole of the other acids enumerated. For upon filtering the hydrochloric solution of the precipitates and supersaturating the filtrate with ammonia, the borate, tartrate, citrate, &c., of baryta do not always fall down again, being kept in solution by the chloride of ammonium formed. For this reason chloride of barium cannot serve to effect the actual separation of the whole of the acids named, and, except as regards sulphuric acid, we set no value upon this reagent as a means of effecting their individual detection. Still it is of great importance as a reagent, since the non-formation of a precipitate upon its application in neutral or alkaline solutions proves at once the absence of so considerable a number of acids.

The compounds of silver with sulphur, chlorine, iodine, bromine, cyanogen, ferro- and ferricyanogen, and of the oxide of silver with phosphoric acid, arsenious acid, arsenic acid, boracic acid, chromic acid, silicic acid, oxalic acid, tartaric acid, and citric acid, are insoluble in water. The whole of these compounds are soluble in dilute nitric acid, with the exception of the chloride, iodide, bromide, cyanide, ferrocyanide, ferricyanide, and sulphide of silver. If, therefore, we add to our solution, which, for the reason just now stated, must be perfectly neutral,

2. *Nitrate of Silver,*

and precipitation ensues, this shows at once the presence of one or several of the acids enumerated: chromic acid, arsenic acid, and several others, which form colored salts with silver, may be individually recognised with tolerable certainty by the mere color of the precipitate. By treating the precipitate now with nitric acid, we see whether it contains sulphide of silver or any of the haloid compounds of silver, as these remain undissolved, whilst all the oxide salts dissolve.—Nitrate of silver fails to effect the complete separation of those acids which form with oxide of silver compounds insoluble in water, from the same cause which renders the separation of acids by chloride of barium uncertain, viz., the ammoniacal salt formed prevents the reprecipitation, by ammonia, of several of the salts of silver from the acid solution. Nitrate of silver, besides effecting the separation of chlorine, iodine, bromine, cyanogen, &c., and indicating the presence of chromic acid, &c., serves, like the chloride of barium, to demonstrate at once the absence of a great many acids, where it produces no precipitate in neutral solutions.

The deportment which the solution under examination exhibits with chloride of barium and with nitrate of silver, indicates therefore at once the further course of the investigation. Thus, for instance, where chloride of barium has produced a precipitate, whilst nitrate of silver has failed to do so, it is not necessary to test for phosphoric acid, chromic acid, boracic acid, silicic acid, arsenious acid, arsenic acid, oxalic acid, tartaric acid, and citric acid, provided always the solution did not already contain salts of ammonia. The same is the case if we obtain a precipitate by nitrate of silver, but none by chloride of barium.

Returning now to the supposition which we have assumed here, viz., that the whole of the acids are present in the solution under examination, the reactions with chloride of barium and nitrate of silver would accordingly have demonstrated already the presence of SULPHURIC ACID, and led to the application of the special tests for CHLORINE, BROMINE, IODINE, CYANOGEN, FERROCYANOGEN, FERRICYANOGEN, and SULPHUR;* and there would be reason to test for all the other acids precipitable by these two reagents. The detection of these acids is based upon the results of a series of special experiments, which have already been fully described and explained in the course of the present work: the same remark applies to the rest of the inorganic acids, accordingly to nitric acid and chloric acid.

Of the organic acids oxalic acid, paratartaric acid, and tartaric acid, are precipitated by chloride of calcium in the cold, in presence of chloride of ammonium; the two former immediately, the latter often only after some time; but the precipitation of citrate of lime is prevented by the presence of salts of ammonia, and ensues only upon ebullition or upon mixing the solution with alcohol; the latter agent serves also to effect the separation of malate of lime from aqueous solutions. If, therefore, we add to our fluid,—

3. *Chloride of Calcium* and Chloride of Ammonium,

OXALIC ACID, PARATARTARIC ACID, and TARTARIC ACID are precipitated, but the lime-salts of several inorganic acids, which have not yet been separated, phosphate of lime for instance, precipitate along with them. We must therefore select for the individual detection of the precipitated organic acids such reactions only as preclude the possibility of confounding the organic acids with the inorganic acids that have been thrown down along with them. For the detection of oxalic acid we select accordingly solution of sulphate of lime, with acetic acid (§ 145); to effect the detection of the tartaric and paratartaric acids, we treat the precipitate produced by chloride of calcium with solution of soda, since the lime-salts of these two acids only are soluble in this menstruum in the cold, but insoluble upon ebullition.

Of the organic acids we have now still in solution citric acid and malic acid, succinic acid and benzoic acid, acetic acid and formic acid. CITRIC ACID and MALIC ACID precipitate upon addition of alcohol to the fluid filtered from the oxalate, tartrate, &c., of lime, and which still contains an excess of chloride of calcium. Sulphate and borate of lime invariably precipitate along with the malate and citrate of lime, if sulphuric acid and boracic acid happen to be present; the analyst must therefore carefully guard against confounding the lime precipitates of these acids with

---

* For the separation and special detection of these substances, I refer to § 157.

those of citric acid and malic acid. The alcohol is now removed by evaporation, and,—

4. *Sesquichloride of Iron*

added to the perfectly neutral fluid. This reagent precipitates SUCCINIC ACID and BENZOIC ACID in combination with sesquioxide of iron, whilst FORMIC ACID and ACETIC ACID remain in solution. The methods which serve to effect the separation of the several groups from each other, and the reactions on which the individual detection of the various acids is based, have been fully described and explained in the former part of this work.

B. SPECIAL REMARKS AND ADDITIONS TO THE SYSTEMATIC COURSE OF ANALYSIS.

To § 189.

At the commencement of § 189 the analyst is directed to mix neutral or acid aqueous solutions with hydrochloric acid. This should be done drop by drop. If no precipitate forms, a few drops are sufficient, since the only object in that case is to acidify the fluid in order to prevent the subsequent precipitation of the metals of the iron group by hydrosulphuric acid. In the case of the formation of a precipitate, some chemists recommend that a fresh portion of the solution should be acidified with nitric acid. However, even leaving the fact out of consideration that nitric acid also produces precipitates in many cases—in a solution of potassio-tartrate of antimony, for instance—I prefer the use of hydrochloric acid, *i.e.*, the complete precipitation by that acid of all that is precipitable by it, for the following reasons:—1. Metals are more readily precipitated by hydrosulphuric acid from solutions acidified with hydrochloric acid, than from those acidified with nitric acid;—2. In cases where the solution contains silver, suboxide of mercury, or lead, the further analysis is materially facilitated by the total or partial precipitation of these three metals in the form of chlorides;—and 3. This latter form is the best adapted for the individual detection of these three metals when present in the same solution. Besides, the application of hydrochloric acid saves the necessity of examining whether the mercury, which may be subsequently detected with the other metals of the fifth group, was originally present in the form of oxide or in that of suboxide. That the lead, if present in large proportion, is obtained partly in the form of a chloride, and partly in the precipitate produced by hydrosulphuric acid in the acid solution, can hardly be thought an objection to the application of this method, as the removal of the larger portion of the lead from the solution, effected at the commencement, will only serve to facilitate the examination for other metals of the fifth and sixth groups.

As already remarked, a basic salt of teroxide of antimony may separate from potassio-tartrate of antimony, for instance, or from some other analogous compound, and precipitate along with the insoluble chloride of silver and subchloride of mercury, and the sparingly soluble chloride of lead. This precipitate, however, is readily soluble in the excess of hydrochloric acid which is subsequently added, and exercises therefore no influence whatever upon the further process. The application of heat to the fluid mixed with hydrochloric acid is neither necessary nor even advisable, since it might cause the conversion of a little of the precipitated subchloride of mercury into chloride.

Should bismuth or chloride of antimony be present, the addition of the washings of the precipitate produced by hydrochloric acid to the first filtrate will cause turbidity, if the amount of free hydrochloric acid present is not sufficient to prevent the separation of the basic salt. This turbidity exercises, however, no influence upon the further process, since hydrosulphuric acid as readily converts these finely-divided precipitates into sulphides, as if the metals were in actual solution.

In the case of alkaline solutions, the addition of hydrochloric acid must be continued until the fluid shows a strongly acid reaction. The substance which causes the alkaline reaction of the fluid combines with the hydrochloric acid, and the bodies originally dissolved in that acid separate. Thus, if the alkali is present in the free state, oxide of zinc, for instance, or alumina, &c., may precipitate. But these oxides will redissolve in an excess of hydrochloric acid, whereas chloride of silver will not redissolve, and chloride of lead only with difficulty. If a metallic sulphur salt is the cause of the alkaline reaction, the sulphur acid, *e. g.*, tersulphide of antimony, precipitates upon the addition of the hydrochloric acid, whilst the sulphur base, *e. g.*, sulphide of sodium, transposes with the constituents of the hydrochloric acid, forming chloride of sodium and hydrosulphuric acid. If an alkaline carbonate, a cyanide, or the sulphide of an alkali metal is the cause of the alkaline reaction, carbonic acid, or hydrocyanic acid, or hydrosulphuric acid escapes. All these phenomena should be carefully observed by the analyst, since they not only indicate the presence of certain substances, but demonstrate also the absence of entire groups of bodies.

Precipitates are produced also by hydrochloric acid in solutions containing alkali salts of antimonic acid, tantalic acid, hyponiobic acid, molybdic acid, and tungstic acid. The antimonic, tantalic, and molybdic acids precipitates dissolve (the tantalic acid precipitate to an opalescent fluid), whilst the *hyponiobic* and *tungstic acid* precipitates are not dissolved by an excess of the hydrochloric acid. The two latter acids are therefore eventually left in the precipitate, which may contain also chloride of silver, subchloride of mercury, chloride of lead, and silicic acid. Separation of sulphur ensuing some time after addition of hydrochloric acid, and attended by a smell of sulphurous acid, indicates presence of *hyposulphites*.

To §§ 190 and 191.

A judicious distribution and economy of time is especially to be studied in the practice of analysis; many of the operations may be carried on simultaneously, which the student may readily perceive and arrange for himself.

In cases where the analyst has simply to deal with metallic oxides of the sixth group—*e. g.*, teroxide of antimony—and of the fourth or fifth group—*e. g.*, iron or bismuth,—he need not precipitate the acidified solution with hydrosulphuric acid, but may, after neutralization, at once add sulphide of ammonium in excess. The sulphide of iron, &c., will in that case precipitate, whilst the antimony, &c., will remain in solution, from which they will, by addition of an acid, at once be thrown down as tersulphide of antimony, &c. This method has the advantage that the fluid is diluted less than is the case where solution of hydrosulphuric acid is employed, and that the operation is performed more expeditiously and conveniently than is the case where hydrosulphuric acid gas is conducted into the fluid. Finally, I must once more remind the student that the perfect purity of the reagents, and their proper application, rank amongst the most indispensable conditions of successful analysis. This applies more particularly to hydrosulphuric acid, especially when used in the gaseous form. In such cases students often lose sight of the circumstance that hydrosulphuric acid gas fails to precipitate highly acid solutions unless they be previously diluted with water. Arsenic acid also is apt to escape detection if the analyst omits to properly support the action of the hydrosulphuric acid by application of heat.

It happens occasionally that in treating acid solutions with hydrosulphuric acid, or in decomposing by hydrochloric acid the sulphide of ammonium used to effect the solution of sulphides of the sixth group that may be present, precipitates are obtained which look almost like pure sulphur, and thus leave the analyst in doubt whether it is really requisite to examine them for metals. In such cases the precipitate may be washed, first with water, then with alcohol, and treated finally with sulphide of carbon, to remove the sulphur; this will show whether or not a trifling quantity of a sulphide is mixed with the sulphur.

The following sulphides of the rarer elements pass into the precipitate produced by hydrosulphuric acid in an acid solution; the sulphides of palladium, rhodium, osmium, ruthenium, iridium,* molybdenum, tellurium, selenium.†

The following rarer compounds cause separation of sulphur, by decomposing the hydrosulphuric acid:

The higher oxides and chlorides of manganese and cobalt, vanadic acid (with blue coloration of the fluid), nitrous acid, sulphurous acid, hyposulphurous acid, hypochlorous and chlorous acids, bromic acid, and iodic acid.

On treating the precipitate with sulphide of ammonium (sulphide of sodium), the sulphides of iridium, molybdenum, tellurium, and selenium dissolve with the sulphides of arsenic, antimony, &c., whilst the sulphides of palladium, rhodium, osmium, and ruthenium remain undissolved with the sulphides of lead, bismuth, &c.

To § 192.

If a precipitate containing all the sulphides of the sixth group (of tin, antimony, arsenic, tellurium, selenium, molybdenum, gold, platinum, and iridium) is fused, according to the instructions of § 192, with carbonate and nitrate of soda, and the fused mass treated with cold water, the *telluric acid, selenic acid,* and *molybdic acid* dissolve with the arsenic acid, whilst the *iridium* is left undissolved with the binoxide of tin, the antimonate of soda, the gold, and the platinum.

For the way of detecting the rarer elements in the solution and in the precipitate, see § 135.

To § 193.

Besides the methods described in the systematic course, to distinguish between cadmium, copper, lead, and bismuth, the following process will also be found to give highly satisfactory results. Add carbonate of soda to the nitric acid solution as long as a precipitate continues to form, then solution of cyanide of potassium in excess, and heat gently. This effects the complete separation of lead and bismuth in the form of carbonates, whilst copper and cadmium are obtained in solution in the form of

---

* The metals of the platinum ores are precipitated with difficulty by hydrosulphuric acid. To attain the end in view, hydrosulphuric acid gas must be perseveringly conducted into the fluid, and heat applied to promote the action of the acid.

† Tungsten and vanadium are not found in the precipitate thrown down from an acid solution by hydrosulphuric acid. They can be present only where the fluid has first been mixed with sulphide of ammonium, then with acid in excess; but in that case the sulphides of nickel and cobalt will also be found with those of the fifth and sixth groups.

cyanide of copper and potassium, and cyanide of cadmium and potassium. Lead and bismuth may now be readily separated from one another by means of sulphuric acid. The separation of the copper from the cadmium is effected by adding to the solution of the cyanides of these two metals in cyanide of potassium, hydrosulphuric acid in excess, gently heating, and then adding some more cyanide of potassium, in order to redissolve the sulphide of copper which may have precipitated along with the sulphide of cadmium. A residuary yellow precipitate (sulphide of cadmium), insoluble in the cyanide of potassium, demonstrates the presence of cadmium. Filter the fluid from this precipitate, and add hydrochloric acid to the filtrate, when the formation of a black precipitat (sulphide of copper) will demonstrate the presence of copper.

Where there is reason to suppose that the precipitate containing the sulphides of the fifth group (of copper, bismuth, &c.), contains also the sulphides of palladium, rhodium, osmium, and ruthenium, proceed as follows:

Fuse the precipitate with hydrate of potassa and chlorate of potassa, heat, ultimately to redness, let cool, then treat the fused mass with water. The solution contains osmate and ruthenate of potassa, which latter imparts a deep yellow color to it. If the fluid is cautiously neutralized with nitric acid, *black sesquioxide of ruthenium* separates; if more nitric acid is added to the filtrate,- and the fluid then distilled, *osmic acid* passes over. If the residue left upon the extraction of the fused mass with water is gently ignited in hydrogen gas,* then cautiously treated with dilute nitric acid, the copper, lead, &c., are dissolved, whilst the rhodium and palladium are left undissolved. The *palladium* may then be dissolved out of the residue by means of aqua regia, leaving the *rhodium* undissolved. For the further examination of the isolated metals, I refer to § 124. A separate portion of the precipitate of the sulphides must be examined for mercury, in the event of the above process being adopted.

To § 194.

Assuming all elements not yet precipitated to be present in the fluid filtered from the precipitate produced in an acid solution by hydrosulphuric acid, the precipitate produced by addition of chloride of ammonium to this filtrate, neutralization with ammonia, and addition of sulphide of ammonium in excess, will contain the following elements:

a. In the form of sulphides: cobalt, nickel, manganese, iron, zinc, uranium;

b. In the form of oxides: aluminium, beryllium, thorium, zirconium, yttrium, terbium, erbium, cerium, lanthanium, didymium, chromium, titanium, tantalum, niobium.†

Where there is reason to suspect the presence of some of the rarer elements in the precipitate, the following method may be recommended as the most suitable in many cases:

Dry the washed precipitate, ignite in a porcelain crucible, then fuse perseveringly in a platinum crucible with *acid sulphate of potassa;* let the fused mass cool, soak in cold water, and digest for some time, without application of heat. Filter the solution from the residue.

The RESIDUE, which contains the acids of tantalum and niobium, and may contain also silicic acid and a little undissolved sesquioxide of iron and sesquioxide of chromium, gives, on fusion with hydrate of soda and some nitrate of soda, a mass out of which dilute solution of soda will dissolve chromate and silicate of potassa, leaving undissolved, with the sesquioxide of iron, tantalate and hyponiobate of soda (being insoluble in solution of soda). After removing the excess of soda, treat repeatedly with a very dilute solution of carbonate of soda, in which the HYPONIOBATE of soda dissolves much more readily than the TANTALATE. For further examination compare § 104, 11 and 12.

Treat the SOLUTION, which contains all the other bases, &c., of the third and fourth groups, with hydrosulphuric acid, to reduce the sesquioxide of iron, dilute considerably, heat to boiling, and keep boiling for some time, whilst conducting carbonic acid into

---

* Cadmium may escape in this operation.

† Of hyponiobic acid only the trifling traces redissolved on the precipitation by hydrochloric acid can be present here.

the fluid. If a precipitate is formed, examine this for TITANIC ACID; it may possibly contain also a little ZIRCONIA.

Concentrate the filtrate by evaporation, with addition of some nitric acid, precipitate with *ammonia*, filter, and wash; redissolve the washed precipitate in hydrochloric acid, and precipitate again with ammonia. This will give almost the whole of the ZINC, MANGANESE, NICKEL, and COBALT in solution, whilst the earths are left undissolved with the sesquioxides of iron, uranium, and chromium. Redissolve the precipitate in hydrochloric acid, and add to the solution *concentrated solution of potassa*, without applying heat. This will leave in solution the sesquioxide of chromium, the alumina, and the berylla whilst precipitating the other earths with the sesquioxides of iron and uranium. Dilute the alkaline solution, and boil some time; this will throw down the berylla and the sesquioxide of chromium, leaving the ALUMINA in solution. The latter earth may then be precipitated by chloride of ammonium. Fuse the precipitate of berylla and sesquioxide of chromium with carbonate of soda and nitrate of potassa, and separate the BERYLLA from the CHROMIC ACID in the same way in which the separation of alumina from chromic acid is effected (§ 103).

The precipitate, which contains the sesquioxides of iron and uranium and the earths insoluble in potassa, may under circumstances, *e.g.*, in presence of yttria and sesquioxide of cerium, also contain alumina and berylla. Dissolve it in hydrochloric acid, remove an over-large excess of the acid by evaporation, dilute, add *carbonate of baryta*, and let the mixture stand from four to six hours in the cold.

The precipitate produced contains the SESQUIOXIDE OF IRON, the SESQUIOXIDE OF URANIUM, and the ALUMINA which may still be present. Redissolve in hydrochloric acid, and separate the sesquioxide of uranium from the alumina and the sesquioxide of iron, by means of carbonate of ammonia added in excess. From the fluid filtered off the precipitate produced by carbonate of baryta, remove the baryta by sulphuric acid, concentrate strongly by evaporation, neutralize exactly with potassa (leaving the reaction rather acid than alkaline), add *neutral sulphate of potassa* in crystals, boil, and let the fluid stand twelve hours. Then filter, and wash with a solution of neutral sulphate of potassa. The filtrate contains that portion of the berylla which may have escaped solution by potassa, also yttria (together with oxide of erbium and oxide of terbium). These substances are precipitated by ammonia, and may then easily be separated by treating with a concentrated warm solution of *oxalic acid*, in which the BERYLLA is soluble, whilst the oxalates of YTTRIA and of OXIDE OF ERBIUM and OXIDE OF TERBIUM are left undissolved. Now boil the precipitate of the double sulphates of zirconia, &c. and potassa repeatedly in water, with addition of some hydrochloric acid, which will dissolve the THORIA and the OXIDES OF CERIUM, leaving the sulphate of ZIRCONIA and potassa undissolved. The thoria and the oxides of cerium may then be precipitated from the solution by ammonia, and tested by the reactions described in § 104.

To §§ 195—198.

The fluid filtered from the precipitate produced by sulphide of ammonium may not only contain the alkaline earths and the alkalies, but some nickel, and also vanadic acid and that portion of the tungstic acid which has been left unprecipitated by hydrochloric acid. The nickel, the vanadic acid, and the tungstic acid, are present as sulphides dissolved in the excess of sulphide of ammonium; they are thrown down in that form by just acidifying the fluid with hydrochloric acid. Filter the precipitate, wash, dry, fuse with carbonate of soda and nitrate of potassa, and treat the fused mass with water; this will dissolve the vanadate and tungstate of potassa, leaving the protoxide of nickel undissolved. From this solution the vanadic acid may be separated by means of solid chloride of ammonium, the tungstic acid by evaporating with hydrochloric acid and treating the residue with water. The two acids may then be examined as directed § 113, *b*, and § 135, *c*.

For the detection of lithium, cæsium, and rubidium, I refer to the analysis of mineral waters (258) and (259).

To § 203.

If the rarer elements are taken into account, the number of bodies which are left undissolved by treating a substance under examination with water, hydrochloric acid, nitric acid, and aqua regia, is much enlarged. The following bodies, more especially, are either altogether, or in the ignited state, or in certain combinations, insoluble or slowly and sparingly soluble in acids:

Berylla, thoria, and zirconia; the oxides of cerium; titanic acid and tantalic acid; hyponiobic acid and niobic acid; molybdic acid and tungstic acid; rhodium, iridium, osmio-iridium, ruthenium.

When you have, in the systematic course of analysis, arrived at (208), fuse the substance, free from silver, lead, and sulpbur, with carbonate of soda and some nitrate of potassa, extract the fused mass repeatedly with hot water, and, if a residue is left, fuse this some time, in a silver crucible, with hydrate of potassa and nitrate of potassa, and again treat the fused mass repeatedly with water. The alkaline solutions, which may be examined separately or together, may contain berylla, a portion of the titanic acid, the tantalic acid, the niobic acid, the molybdic acid, the tungstic acid, the osmic and ruthenic acids, and a portion of the iridium present.

If the residue left undissolved by the preceding operation is fused with acid sulphate of potassa, and the fused mass treated with water, the thoria and zirconia, the oxides of cerium, the remainder of the titanic acid, and the rhodium may dissolve. A residue left by this operation may consist of platinum ore metals that have escaped decomposition by fluxing, and had best be mixed with chloride of sodium, and ignited in a stream of chlorine.

With respect to the separation and individual detection of the several elements that have passed into the different solutions, the requisite directions and instructions have been given in the third section of Part I., and in the additional remarks to §§ 189-198.

To § 204.

The analysis of cyanogen compounds is not very easy in certain cases, and it is sometimes a difficult task even to ascertain whether we have really a cyanide before us or not. However, if the reactions of the substance under examination upon ignition (8) be carefully observed, and also whether upon boiling with hydrochloric acid any odor of hydrocyanic acid is emitted (35), the presence or absence of a cyanide will generally not long remain a matter of doubt.

It must above all be borne in mind that the insoluble cyanogen compounds occurring in pharmacy, &c., belong to two distinct classes. Viz., they are either SIMPLE CYANIDES, or COMPOUNDS OF METALS WITH FERROCYANOGEN or some other analogous compound radical.

All the simple cyanides are decomposed by boiling with concentrated hydrochloric acid into metallic chlorides and hydrocyanic acid. Their analysis is therefore never difficult. But the ferrocyanides, &c., to which indeed the method described § 204 more exclusively refers, suffer by acids such complicated decompositions that their analysis by means of acids is a task not so easily accomplished. Their decomposition by potassa or soda is far more simple. The alkali yields its oxygen to the metal combined with the ferrocyanogen, &c., the oxide thus formed precipitates, and the reduced potassium or sodium forms with the liberated radical soluble ferrocyanide, &c., of potassium or sodium. But several oxides are soluble in an excess of potassa, as, *e.g.*, oxide of lead, oxide of zinc, &c. If, therefore, the double ferrocyanide of zinc and potassium, for instance, is boiled with solution of caustic potassa, it dissolves completely in that menstruum, and we may assume that the solution contains ferrocyanide of potassium and oxide of zinc dissolved in potassa. Were we to add an acid to this solution, we should of course simply re-obtain the original precipitate of the double ferrocyanide of zinc and potassium, and the experiment would consequently be of no avail. To prevent this failure, we conduct hydrosulphuric acid into the solution in potassa, but only until the precipitable oxides are completely thrown down, and not until the solution smells of sulphuretted hydrogen. This serves to convert into sulphides all the heavy metals which the potassa holds in solution as oxides. Those sulphides which are insoluble in potassa, such as sulphide of lead, sulphide of zinc, &c., precipitate, whilst those which are soluble in alkaline sulphides, such as bisulphide of tin, tersulphide of antimony, &c., remain in solution. To effect the detection of these also, the fluid is now acidified with nitric acid, and, if necessary, more hydrosulphuric acid conducted into it. The fluid must be diluted before the nitric acid is added, and a large excess of the latter avoided, as otherwise the fluid will show a blue coloration caused by the decomposition of the liberated hydroferrocyanic acid.

The fluid filtered from the precipitated oxides and sulphides accordingly always contains the cyanogen as ferrocyanide, &c., of potassium—provided, of course, the analyzed compound is really a double ferrocyanide, &c. From most of these compounds—ferrocyanide, ferricyanide, chromicyanide, and manganocyanide of potassium—the cyanogen partly separates as hydrocyanic acid, upon boiling the solutions with moderately diluted sulphuric acid, and may thus be readily detected by this means, should the direct way of detecting the radicals fail to give the desired result. But the cobalticyanide of potassium is not decomposed by dilute sulphuric acid, and the analyst is accordingly directed to effect the detection of the compound radical in that salt by means of solution of nickel, manganese, zinc, &c. By fusion with nitrate of potassa all these double compounds suffer decomposition, cobalticyanide of potassium not excepted. The reason why the fusion of these double compounds with nitrate of potassa should be preceded by evaporation with an excess of nitric acid, is simply to prevent the occurrence of explosions. Caution is always highly advisable in this operation.

If you simply wish to examine for bases in simple or compound cyanides, and for that purpose to destroy the cyanogen compound, reduce the substance under examination to powder, transfer this to a platinum vessel, pour over it concentrated sulphuric acid, diluted with a little water, and heat long and intensely until the whole of the free sulphuric acid is driven off. The residuary mass consists of sulphates, which are then dissolved in hydrochloric acid and water (H. ROSE—Zeitschrift f. Analyt. Chem. I, 194).

APPENDIX.

I.

DEPORTMENT OF THE MOST IMPORTANT MEDICINAL ALKALOIDS WITH REAGENTS, AND SYSTEMATIC METHOD OF EFFECTING THE DETECTION OF THESE SUBSTANCES.

§ 229.

THE detection and separation of the vegeto-alkalies, or alkaloids, is a task of far greater difficulty than that of most of the inorganic bases. Although this difficulty is in some measure owing to the circumstance that scarcely one of the compounds which the alkaloids form with other substances is absolutely insoluble or particularly characterized by its color or any striking property, yet the principal cause of it must be ascribed to the want of accurate and minute investigations of the salts and other compounds of the alkaloids, and of the products of their decomposition. We consequently generally see and apprehend the reactions only in their external manifestation, but without being able to connect them with the causes producing them, which makes it impossible to understand all the conditions which may exercise a modifying influence.

Although therefore, in the present imperfect state of our knowledge of these bodies, an attempt to define their deportment with reagents, and base thereon a method of effecting their separation or, at least, their individual detection in presence of each other, must of necessity fall very short of perfection, yet, having made a great many experiments on the nature and deportment of these substances, I will attempt here, for the benefit of young chemists, and more particularly pharmaceutists, to describe in some measure the reactions which the most important of the alkaloids manifest with other bodies, and to lay down a systematic method of effecting their individual detection.

The classification of the alkaloids into groups which I have adopted is based upon their deportment with certain general reagents. I have verified by numerous experiments the whole of the reactions described in the succeeding paragraphs.

Recent investigations have shown that several of the alkaloids, in the state in which they are obtained by the usual methods of preparing them, and in which they occur in commerce or are used in pharmacy, must not be looked upon as simple organic bodies, but as consisting of several, often even of many, very closely allied alkaloids. Thus, for instance, common narcotine contains three or four homologous bases; thus SCHÜTZENBERGER has produced from so-called brucia, by fractional crystallization, ten alkaloids, analogous in form of crystallization and in deportment with nitric acid, but differing in composition and in solubility in water. As these varieties have not as yet been studied with regard to their reactions, I have of course been obliged to disregard them altogether in the subjoined course of analysis, which therefore must be held to refer simply to the vegeto-alkalies in the pure state, such as they have hitherto been assumed to exist.

I. Volatile Alkaloids.

The volatile alkaloids are fluid at the common temperature, and may be volatilized in the pure state as well as when mixed with water. They are accordingly obtained in the distillate when their salts are distilled with strong fixed bases and water. Their vapors, when brought in contact with those of volatile acids, form a white cloud.

1. Nicotina, or Nicotine (C_{10} H, N).

§ 230.

1. Nicotina, in its pure state, form a colorless, oily liquid, of 1·048 sp. gr.; the action of air imparts a yellowish or brownish tint to it. It boils at 482° F., suffering, however, partial decomposition in the process; but, when heated in a stream of hydrogen gas, it distils over unaltered, between 212° and 392° F. It is miscible in all proportions with water, alcohol, and ether.

Nicotina has a peculiar, disagreeable, somewhat ethereal, tobacco-like odor, an acrid, pungent taste, and very poisonous properties. Dropped on paper, it makes a transparent stain, which slowly disappears; it turns turmeric-paper brown, and reddened litmus-paper blue. Concentrated aqueous solution of nicotina shows these reactions more distinctly than the alkaloid in the pure state.

2. Nicotina has the character of a pretty strong base; it precipitates metallic oxides from their solutions, and forms salts with acids. The salts of nicotina are freely soluble in water and alcohol, insoluble in ether; they are inodorous, but taste strongly of tobacco; part of them are crystallizable. Their solutions, when distilled with solution of potassa, give a distillate containing nicotina. By neutralizing this with oxalic acid, and evaporating, oxalate of nicotina is produced, which may be freed from any admixture of oxalate of ammonia, by means of spirit of wine, in which the former salt is soluble, the latter insoluble.

3. If an aqueous solution of nicotina, or a solution of a salt of nicotina mixed with solution of soda or potassa, is shaken with *ether*, the nicotina is dissolved by the ether; if the latter is then allowed to evaporate on a watch-glass, the nicotina remains behind in drops and streaks; on warming the watch-glass, it volatilizes in white fumes of strong odor.

4. *Bichloride of platinum* produces in aqueous solutions of nicotina whitish-yellow flocculent precipitates. On heating the fluid containing the precipitate, the latter dissolves, but upon continued application of heat it very speedily separates again in form of an orange-yellow, crystalline, heavy powder, which, under the microscope, appears to be composed of roundish crystalline grains. If a rather dilute solution of nicotina, supersaturated with hydrochloric acid, is mixed with bichloride of platinum, the fluid at first remains clear; after some time, however, the double salt separates in small crystals (oblique, four-sided prisms), clearly discernible with the naked eye.

5. *Terchloride of gold* produces a reddish-yellow flocculent precipitate, sparingly soluble in hydrochloric acid.

6. Solution of *iodine in iodide of potassium* and water, when added in small quantity to an aqueous solution of nicotina, produces a yellow precipitate, which after a time disappears. Upon further addition of iodine solution, a copious kermes-colored precipitate separates; but this also disappears again after a time.

7. Solution of *tannic acid* produces a copious white precipitate, which redissolves upon addition of hydrochloric acid.

8. If an aqueous solution of nicotina is added to a solution of *chloride of mercury* in excess, an abundant, flocculent, white precipitate is formed. If solution of chloride of ammonium is now added to the mixture in sufficient quantity, the entire precipitate, or the greater part of it, redissolves. But the fluid very soon turns turbid, and deposits a heavy white precipitate.

2. CONIA, or CONINE ($C_{16} H_{15} N$).

§ 231.

1. Conia forms a colorless oily liquid, of 0·87 sp. gr.; the action of the air imparts to it a brown tint. In the pure state it boils at about 392° F.; when heated in a stream of hydrogen gas, it distils over unaltered; but when distilled in vessels containing air, it turns brown and suffers partial decomposition; with aqueous vapors it distils over freely. It dissolves sparingly in water, 100 parts of water of the common temperature dissolving 1 part of conia. The solution turns turbid on warming. Conia is miscible in all proportions with alcohol and ether. The aqueous and alcoholic solutions manifest strong alkaline reaction. Conia has a very strong, pungent, repulsive odor, which affects the head, a most acrid and disagreeable taste, and very poisonous properties.

2. Conia is a strong base; it accordingly precipitates metallic oxides from their solutions, in a similar way to ammonia, and forms salts with acids. The salts of conia are soluble in water and in spirit of wine, but nearly insoluble in ether. Hydrochlorate of conia crystallizes readily; the smallest quantity of this base, brought in contact with a trace of hydrochloric acid, yields almost immediately a corresponding quantity of non-deliquescent rhombic crystals (TH. WERTHEIM). The solutions of the salts of conia turn brownish upon evaporation, with partial decomposition of the conia. The dry salts of conia do not smell of the alkaloid; when moistened, they smell only feebly of it; but upon addition of solution of soda, they at once emit a strong conia odor. When salts of conia are distilled with solution of soda, the distillate contains conia. On neutralizing this with oxalic acid, evaporating to dryness, and treating the residue with spirit of wine, the oxalate of conia formed is dissolved, whilst any oxalate of ammonia that may be present is left undissolved. As conia is only sparingly soluble in water, and dissolves with still greater difficulty in solutions of alkalies, a concentrated solution of a salt of conia turns milky upon addition of solution of soda. The minute drops which separate unite gradually, and collect on the surface.

3. If an aqueous solution of a salt of conia is shaken with *solution of soda* and *ether*, the conia is dissolved by the ether. If the latter is then allowed to evaporate on a watch-glass, the conia is left in yellowish-colored oily drops.

4. *Concentrated nitric acid* imparts a fine blood-red tint to conia; *sulphuric acid*, a purple-red color, which subsequently turns to olive-green.

5. *Terchloride of gold* produces a yellowish-white precipitate, insoluble in hydrochloric acid; *chloride of mercury*, a copious white precipitate, soluble in hydrochloric acid. *Bichloride of platinum* does not precipitate aqueous solutions of salts of conia, the conia compound corresponding to ammonio-bichloride of platinum being insoluble in spirit of wine and ether, but soluble in water.

6. To solution of *iodine in iodide of potassium* and water, and to solution of *tannic acid*, conia comports itself the same as nicotina.

7. *Chlorine water* produces in a mixture of water and conia a strong, white turbidity.

8. If an aqueous solution of conia is mixed with a solution of *albumen*, the albumen coagulates. Aniline is the only other volatile vegeto-alkali which shows this reaction.

The volatile alkaloids are easily recognised when pure; the great object of the analyst must accordingly always be to obtain them in that state. The way of effecting this is the same for nicotia as for conia, and has already been given in the foregoing paragraphs, viz., to distil with addition of solution of soda, neutralize with oxalic acid, evaporate, dissolve in alcohol, evaporate the solution, treat the residue with water, add solution of soda, shake the mixture with ether, and let the latter evaporate spontaneously. Conia is distinguished from nicotina chiefly by its odor, its sparing solubility in water, and its comportment with chlorine water, and bichloride of platinum.

II. Non-volatile Alkaloids.

The non-volatile alkaloids are solid, and cannot be distilled over with water.

FIRST GROUP.

Non-volatile Alkaloids which are precipitated by Potassa or Soda from the Solutions of their Salts, and redissolve readily in an Excess of the Precipitant.

Of the alkaloids of which I purpose to treat here, *one* only belongs to this group, viz.,

Morphia, or Morphine $(C_{34}H_{19}\overset{+}{N}O_{6}=\mathrm{Mo})$.

§ 232.

1. Crystallized morphia $(\overset{+}{\mathrm{Mo}} + 2 \text{ aq.})$ usually appears in the form of colorless, brilliant four-sided prisms, or, when obtained by precipitation, as a white crystalline powder. It has a bitter taste, and dissolves very sparingly in cold, but somewhat more readily in boiling water. Of cold alcohol it requires about 90 parts by weight for solution; of boiling alcohol from 20 to 30 parts. The solutions of morphia in alcohol as well as in hot water manifest distinctly alkaline reaction. This alkaloid is nearly insoluble in ether; it dissolves in amyl-alcohol in the cold, but more freely with the aid of heat. At a moderate heat the crystallized morphia loses the two equivalents of water.

2. Morphia neutralizes acids completely, and forms with them the SALTS OF MORPHIA. These salts are readily soluble in water and in spirit of wine, but insoluble in ether and amyl-alcohol; their taste is disagreeably bitter. Most of them are crystallizable.

3. *Potassa* and *ammonia* precipitate from the solutions of salts of morphia—generally only after some time—$\overset{+}{\text{Mo}} + 2$ aq., in the form of a white crystalline powder. Stirring and friction on the sides of the vessel promote the separation of the precipitate, which redissolves with great readiness in an excess of potassa, but more sparingly in ammonia. It dissolves also in chloride of ammonium and, though with difficulty only, in carbonate of ammonia.

4. *Carbonate of potassa and carbonate of soda* produce the same precipitate as potassa and ammonia, but fail to redissolve it upon addition in excess. Consequently if a fixed alkaline bicarbonate is added to a solution of morphia in caustic potassa, or if carbonic acid is conducted into the solution, $\overset{+}{\text{Mo}} + 2$ aq. separates,—especially after previous ebullition—in the form of a crystalline powder. A more minute inspection, particularly through a magnifying glass, shows this powder to consist of small acicular crystals; seen through a glass which magnifies 100 times, these crystals present the form of four-sided prisms.

5. *Bicarbonate of soda and bicarbonate of potassa* speedily produce in solutions of neutral salts of morphia a precipitate of hydrated morphia, in the form of a crystalline powder. The precipitate is insoluble in an excess of the precipitants. These reagents fail to precipitate acidified solutions of salts of morphia in the cold.

6. The action of strong *nitric acid* upon morphia or one of its salts, in the solid state or in concentrated solutions, produces a fluid varying from red to yellowish-red. Dilute solutions do not change their color upon addition of nitric acid in the cold, but upon heating they acquire a yellow tint.

7. If morphia or a compound of morphia is treated with pure concentrated *sulphuric acid*, and heat applied, a colorless solution is obtained; if, after cooling, 8 to 20 drops of *sulphuric acid mixed with some nitric acid** are added, and 2 or 3 drops of water, the fluid acquires a violet-red coloration (gentle heating promotes the reaction); and if from 4 to 6 clean lentil-sized fragments of *binoxide of manganese* are now added, or a fragment of *chromate of potassa* (OTTO), the fluid acquires an intensely mahogany-brown color. If the fluid is then diluted with 4 parts of water, properly cooled in a test tube, and ammonia added until the reaction is *almost* neutral, a dirty-yellow color makes its appearance, which turns brownish red upon supersaturation with ammonia, without depositing an appreciable precipitate (I. ERDMANN).

8. *Neutral sesquichloride of iron* imparts to neutral solutions of salts of morphia a beautiful dark blue color, which disappears upon the addition of an acid. If the solution contains an admixture of animal or vegetable extractive matters, or of acetates, the color will appear clouded and less distinct.

9. If *iodic acid* is added to a solution of morphia or of a salt of

---

* Mix 6 drops of nitric acid of 1·25 sp. gr. with 100 cubic centimetres of water, and add 10 drops of this mixture to 20 grammes of pure concentrated sulphuric acid.

morphia, IODINE separates. In concentrated aqueous solutions the separated iodine appears as a kermes-brown precipitate, whilst to alcoholic and dilute aqueous solutions it imparts a brown or yellowish-brown color. The addition of starch-paste to the fluid, no matter whether made before or after that of the iodic acid, considerably heightens the delicacy of the reaction, since the blue tint of the iodide of starch remains still perceptible in exceedingly dilute solutions, which is not the case with the brown color imparted by iodine. As other nitrogenous bodies (albumen, caseïne, fibrine, &c.) likewise reduce iodic acid, this reaction has only a relative value; however, if *ammonia* is added after the iodic acid, the fluid becomes colorless if the separation of iodine has been caused by other substances, whilst the coloration becomes much more intense if it is owing to the presence of morphia (LEFORT).*

SECOND GROUP.

NON-VOLATILE ALKALOIDS WHICH ARE PRECIPITATED BY POTASSA FROM THE SOLUTIONS OF THEIR SALTS, BUT DO NOT REDISSOLVE TO A PERCEPTIBLE EXTENT IN AN EXCESS OF THE PRECIPITANT, AND ARE PRECIPITATED BY BICARBONATE OF SODA EVEN FROM ACID SOLUTIONS, if the latter are not diluted in a larger proportion than 1 : 100 ; Narcotina, Quina, Cinchonia.

a. NARCOTINA, or NARCOTINE ($C_{46}H_{25}\overset{+}{N}O_{14}=Na$).

§ 233.

1. Crystallized narcotina ($\overset{+}{Na}$ + aq.) appears usually in the form of colorless, brilliant, straight rhombic prisms, or, when precipitated by alkalies, as a white, loose, crystalline powder. It is insoluble in water. Alcohol and ether dissolve it sparingly in the cold, but somewhat more readily upon heating. Solid narcotina is tasteless, but the alcoholic and ethereal solutions are intensely bitter. Narcotina does not alter vegetable colors. At 338° F. it fuses, with loss of 1 eq. of water.

2. Narcotina dissolves readily in acids, combining with them to salts. These salts have invariably an acid reaction. Those with weak acids are decomposed by a large amount of water, and, if the acid is volatile, even upon simple evaporation. Most of the salts of narcotine are amorphous, and soluble in water, alcohol, and ether; they have a bitter taste.

3. *Pure alkalies,* and *alkaline carbonates* and *bicarbonates,* immediately precipitate from the solutions of salts of narcotine $\overset{+}{Na}$ + aq. in the form of a white powder, which, seen through a lens magnifying 100 times, appears an aggregate of small crystalline needles. The precipitate is insoluble in an excess of the precipitants. If solution of narcotina is mixed with ammonia, and ether added in sufficient quantity, the nar-

---

* LEFORT (Zeitschrift f. anal. Chem. I., 134) recommends the following method for the detection of small quantities of morphia : moisten strips of very white unsized paper with the morphia solution, dry, and repeat the operation several times, so as to ensure absorption by the paper of a tolerably large quantity of the fluid ; the dried paper contains the morphia in the solid state, most finely divided. Nitric acid, sesquichloride of iron, and iodic acid and ammonia will readily and with positive distinctness show the characteristic reactions on paper so prepared.

cotine which has separated upon the addition of the ammonia, redissolves in the ether, and the clear fluid presents two distinct layers. If a drop of the ethereal solution is evaporated on a watch-glass, the residue is seen, upon inspection through a lens magnifying a hundred times, to consist of small, distinct, elongated, and lance-shaped crystals.

4. *Concentrated nitric acid* dissolves narcotine to a colorless fluid, which acquires a pure yellow tint upon application of heat.

5. If a small quantity of pure narcotine is treated with from 4 to 6 drops of pure *concentrated sulphuric acid*, no coloration is observed, not even on the application of a gentle heat, or, at least, the heated fluid acquires only a barely perceptible yellow tint, but upon adding now, after cooling, from 8 to 20 drops of sulphuric acid mixed with a little *nitric acid* (see foot-note to § 232, 7, page 319), and 2 or 3 drops of water, the fluid acquires an intense red color. Addition of binoxide of manganese does not materially change the color. If, after dilution, *ammonia* is added to *nearly* neutral reaction, the color becomes less intense, in consequence of the dilution. Addition of ammonia in excess produces a copious dark brown precipitate (J. ERDMANN).

6. If the solution of a salt of narcotine is mixed with *chlorine water*, it acquires a yellow color, slightly inclining to green; if ammonia is then added, a much more intensely colored yellowish-red fluid is obtained.

7. If narcotine or one of its salts is dissolved in an excess of dilute *sulphuric acid*, some finely levigated *binoxide of manganese* added, the mixture heated to boiling, and kept in ebullition for the space of several minutes, the narcotine absorbs oxygen and is converted into opianic acid, cotarnine (a base soluble in water), and carbonic acid. Ammonia will now of course fail to precipitate narcotine from the filtrate.

b. QUINA, or QUININE ($C_{40}H_{24}N_2O_4 = \overset{+}{Q}$).

§ 234.

1. Crystallized quina ($\overset{+}{Q}$ + 6 aq.) appears either in the form of fine crystalline needles of silky lustre, which are frequently aggregated into tufts, or as a loose white powder. It is sparingly soluble in cold, but somewhat more readily in hot water. It is readily soluble in spirit of wine, both cold and hot, but less so in ether. The taste of quina is intensely bitter; the solutions of quina manifest alkaline reaction. Upon exposure to heat it loses the 6 eq. of water.

2. Quina neutralizes acids completely. The salts taste intensely bitter; most of them are crystallizable, difficultly soluble in cold, readily soluble in hot water and in spirit of wine. The acid salts dissolve very freely in water; the solutions reflect a bluish tint. If a cone of light is thrown into them, by means of a lens either horizontally or vertically, a blue cone of light is seen even in highly dilute solution.

3. *Potassa, ammonia,* and the neutral *carbonates of the alkalies* produce in solutions of salts of quina (if they are not too dilute) a white, loose, pulverulent precipitate of hydrated quina, which immediately after precipitation appears opaque and amorphous under the microscope, but assumes, after the lapse of some time, the appearance of an aggregate of crystalline needles. The precipitate redissolves only to a scarcely per-

ceptible extent in an excess of potassa, but more so in ammonia. It is hardly more soluble in fixed alkaline carbonates than in pure water. If a solution of quina is mixed with ammonia, ether added, and the mixture shaken, the quina which has separated upon the addition of the ammonia, redissolves in the ether, and the clear fluid presents two distinct layers. In this point quina differs essentially from cinchonia ; by means of this reaction the former may therefore be readily detected in presence of the latter, and separated from it.

4. *Bicarbonate of soda* also produces both in neutral and acid solutions of salts of quina a white precipitate. In acidified solutions containing 1 part of quina to 100 parts of acid and water, the precipitate forms immediately ;—if the proportion of the quina to the acid and water is as 1 : 150, the precipitate separates only after an hour or two, in the form of distinct needles, aggregated into groups. If the proportion is as 1 : 200, the fluid remains clear, and it is only after from twelve to twenty-four hours' standing that a slight precipitate makes its appearance. The precipitate is not altogether insoluble in the precipitant, and the separation is accordingly the more complete the less the excess of the precipitant; the precipitate contains carbonic acid.

5. *Concentrated nitric acid* dissolves quina to a colorless fluid, turning yellowish upon application of heat.

6. The addition of *chlorine water* to the solution of a salt of quina fails to impart a color to the fluid, or, at least, imparts to it only a very faint tint ; but if ammonia is now added, the fluid acquires an intense emerald-green color. If, after the addition of the chlorine water, some solution of *ferrocyanide of potassium* is added, then a few drops of *ammonia* or some other alkali, the fluid acquires a magnificent deep red tint, which, however, speedily changes to a dirty brown. This reaction is delicate and characteristic. Upon addition of an acid* to the red fluid, the color vanishes, but reappears afterwards upon cautious addition of ammonia. (O. LIVONIUS, communicated in a letter to the author ; A. VOGEL.)

7. *Concentrated sulphuric acid* likewise dissolves pure quina and pure salts of quina to a colorless or very faint yellowish fluid ; application of a gentle heat turns the fluid yellow, application of a stronger heat brown. Sulphuric acid containing an admixture of nitric acid dissolves quina to a colorless or very faint yellowish fluid.

8. As regards HERAPATH's quinine reaction, based upon the polarizing properties of sulphate of iodide of quinine, I refer to *Phil. Mag.* vi. 171.

c. CINCHONIA, or CINCHONINE $(C_{40}H_{24}N_2O_2 = \overset{+}{Ci})$.

§ 235.

1. Cinchonia appears either in the form of transparent, brilliant, four-sided prisms, or fine white crystalline needles, or, if precipitated from concentrated solutions, as a loose white powder. At first it is tasteless, but after some time the bitter taste of the bark becomes perceptible. It is nearly insoluble in cold water, and dissolves only with extreme difficulty in hot water ; it dissolves sparingly in cold dilute spirit of wine, more readily in hot spirit of wine, and the most freely in absolute alcohol. From hot alcoholic solutions the greater portion of the dissolved cinchonia separates upon cooling in a crystalline form. Solu-

---

* Acetic acid answers the purpose best.

tions of cinchonia taste bitter, and manifest alkaline reaction. Cinchonia is insoluble in ether.*

2. Cinchonia neutralizes acids completely. The salts have the bitter taste of the bark; most of them are crystallizable: they are generally more readily soluble in water and in spirit of wine than the corresponding quina compounds. Ether fails to dissolve them.

3. Cinchonia, when heated cautiously, fuses at first without loss of water; subsequently white fumes arise which, like benzoic acid, condense upon cold substances, in the form of small brilliant needles, or as a loose sublimate, a peculiar aromatic odor being exhaled at the same time. If the operation is conducted in a stream of hydrogen gas, long brilliant prisms are obtained (HLASIWETZ).

4. *Potassa, ammonia*, and the *neutral carbonates of the alkalies* produce in solutions of salts of cinchonia a white loose precipitate of CINCHONIA, which does not redissolve in an excess of the precipitants. If the solution was concentrated, the precipitate does not exhibit a distinctly crystalline appearance, even though viewed through a lens magnifying 200 times; but if the solution was so dilute that the precipitate formed only after some time, it appears under the microscope to consist of distinct crystalline needles aggregated into star-shaped tufts.

5. *Bicarbonate of soda* and *bicarbonate of potassa* precipitate cinchonia in the same form as in 4, both from neutral and acidified solutions of cinchonia salts, but not so completely as the simple carbonates of the alkalies. Even in solutions containing 1 part of cinchonia to 200 of water and acid, the precipitate forms immediately; its quantity increases after standing some time.

6. *Concentrated sulphuric acid* dissolves cinchonia to a colorless fluid, which upon application of heat first acquires a brown, and finally a black color. Addition of some nitric acid leaves the solution colorless in the cold, but upon application of heat the fluid, after passing through the intermediate tints of yellowish-brown and brown, turns finally black.

7. The addition of *chlorine water* to the solution of a salt of cinchonia fails to impart a color to the fluid; if ammonia is now added, a yellowish-white precipitate is formed.

8. If the solution of a cinchonia salt containing only very little or no free acid, is mixed with ferrocyanide of potassium, a flocculent precipitant of ferrocyanide of cinchonia is formed. If an excess of the precipitant is added, and a gentle heat very slowly applied, the precipitate dissolves, but separates again upon cooling, in brilliant gold-yellow scales, or in long needles, often aggregated in the shape of a fan. With the aid of the microscope, this reaction is as delicate as it is characteristic (CH. DOLLFUS: BILL; SELIGSOHN).

Recapitulation and Remarks.

§ 236.

The non-volatile alkaloids of the second group are altered or precipitated by various other reagents besides those mentioned above; the reactions are, however, not adapted to effect their individual detection and separation. Thus, for instance, bichloride of platinum produces in solutions of the salts of the three alkaloids belonging to this group a

* The cinchonia of commerce usually contains in admixture another alkaloid, called cinchotina, which is soluble in ether. This alkaloid crystallizes in large rhomboidal crystals of brilliant lustre, which fuse at a high temperature, and cannot be sublimed, even in a stream of hydrogen gas (HLASIWETZ).

yellowish-white precipitate, chloride of mercury a white precipitate, tincture of galls a yellowish-white flocculent precipitate, solution of iodine in iodide of potassium a reddish-brown, phospho-molybdic acid a yellow precipitate, &c.

Narcotina and quina being soluble in ether, whilst cinchonia is insoluble in that menstruum, the two former alkaloids may be most readily separated by this means from the latter. For this purpose the analyst need simply mix the aqueous solution of the three alkaloids with ammonia in excess, then add ether, and separate the solution of quina and narcotine from the undissolved cinchonia. If the ethereal solution is now evaporated, the residue dissolved in hydrochloric acid and a sufficient amount of water to make the dilution as 1 : 200, and bicarbonate of soda is then added, the narcotina precipitates, whilst the quina remains in solution. By evaporating the solution, and treating the residue with water, the quina is obtained in the free state.*

THIRD GROUP.

NON-VOLATILE ALKALOIDS WHICH ARE PRECIPITATED BY POTASSA FROM THE SOLUTIONS OF THEIR SALTS, AND DO NOT REDISSOLVE TO A PERCEPTIBLE EXTENT IN AN EXCESS OF THE PRECIPITANT; BUT ARE NOT PRECIPITATED FROM (even somewhat concentrated) ACID SOLUTIONS BY THE BICARBONATES OF THE FIXED ALKALIES: Strychnia, Brucia, Veratria.

a. STRYCHNIA, or STRYCHNINE ($C_{42}H_{22}N_2O_4 = \overset{+}{Sr}$).

§ 237.

1. Strychnia appears either in the form of white brilliant rhombic prisms, or, when produced by precipitation or rapid evaporation, as a white powder. It has an exceedingly bitter taste. It is nearly insoluble in cold, and barely soluble in hot water. It is almost insoluble in absolute alcohol and ether, and only sparingly soluble in dilute spirit of wine. It dissolves freely in amyl-alcohol, more especially with the aid of heat. It does not fuse when heated. It is exceedingly poisonous.

2. Strychnia neutralizes acids completely. The salts of strychnia are, for the most part, crystallizable; they are soluble in water. All the salts of strychnia have an intolerably bitter taste and are exceedingly poisonous.

3. *Potassa* and *carbonate of soda* produce in solutions of salts of strychnia white precipitates of STRYCHNIA, which are insoluble in an excess of the precipitants. Viewed under a microscope magnifying one hundred times the precipitate appears as an aggregate of small crystalline needles. From dilute solutions the strychnia separates only after the lapse of some time, in the form of crystalline needles, which are distinctly visible even to the naked eye.

4. *Ammonia* produces the same precipitate as potassa. The preci-

---

* The reaction with ammonia and ether, though well adapted to effect the separation of quina from cinchonine, fails to effect the separation of the former vegeto-alkali from the other bases found in bark, which are not as yet officinal, viz., α quinidine, β quinidine, γ quinidine, and cinchonidine; since, as G. KERNER (Zeitschrift f. analyt. Chem., 1, 150) has shown, several of these other vegeto-alkalies are pretty freely soluble in ether. In fact, no qualitative reaction will enable the analyst to fully effect this purpose; but it may be accomplished by means of a simple volumetrical method, based upon the circumstance that the quina thrown down by ammonia from a solution of the sulphate, requires less ammonia to redissolve it than all the other vegeto-alkalies of the bark. For further particulars I refer to KERNER'S paper on the subject.

pitate redissolves in an excess of ammonia; but after a short time—or if the solution is highly dilute, after a more considerable lapse of time—the strychnia crystallizes from the ammoniacal solution in the form of needles, which are distinctly visible to the naked eye.

5. *Bicarbonate of soda* produces in neutral solutions of salts of strychnia a precipitate of strychnia, which separates in fine needles shortly after the addition of the reagent, and is insoluble in an excess of the precipitant. But upon adding one drop of acid (so as to leave the fluid still alkaline), the precipitate dissolves readily in the liberated carbonic acid. The addition of bicarbonate of soda to an acid solution of strychnia causes no precipitation, and it is only after the lapse of twenty-four hours, or even a longer period, that strychnia crystallizes from the fluid in distinct prisms, in proportion as the free carbonic acid escapes. If a concentrated solution of strychnia, supersaturated with bicarbonate of soda, is boiled for some time, a precipitate forms at once; from dilute solutions this precipitate separates only after concentration.

6. *Sulphocyanide of potassium* produces in concentrated solutions of salts of strychnia immediately, in dilute solutions after the lapse of some time, a white crystalline precipitate, which appears under the microscope as an aggregate of flat needles, truncated or pointed at an acute angle, and is but little soluble in an excess of the precipitant.

7. *Chloride of mercury* produces in solutions of salts of strychnia a white precipitate, which changes after some time to crystalline needles, aggregated into stars, and distinctly visible through a lens. Upon heating the fluid these crystals redissolve, and upon subsequent cooling of the solution the double compound recrystallizes in larger needles.

8. If a few drops of pure *concentrated sulphuric acid* are added to a little strychnia in a porcelain dish, solution ensues, without coloration of the fluid. If small quantities of oxidizing agents (chromate of potassa, permanganate of potassa, ferricyanide of potassium, peroxide of lead, binoxide of manganese) are now added—best in the solid form, as dilution is prejudicial to the reaction—the fluid acquires a magnificent blue-violet color, which, after some time, changes to wine-red, then to reddish-yellow. With chromate of potassa and permanganate of potassa the reaction is immediate; on inclining the dish, blue violet streaks are seen to flow from the salt fragment, and by pushing the latter about, the coloration is soon imparted to the entire fluid. With ferricyanide of potassium the reaction is less rapid; but it is slowest with peroxides. The more speedy the manifestation of the reaction the more rapid is also the change of color from one tint to another. I prefer chromate of potassa, recommended by OTTO, or permanganate of potassa, recommended by GUY, as the most sensitive, to all other oxidizing agents. JORDAN succeeded, with chromate of potassa, in distinctly showing the presence of $\frac{1}{500000}$th grain of strychnia. J. ERDMANN prefers binoxide of manganese in lentil-sized fragments. Metallic chlorides and considerable quantities of nitrates, also large quantities of organic substances, prevent the manifestation of the reaction or impair its delicacy. It is therefore always advisable to free the strychnia first, as far as practicable, from all foreign matters before proceeding to try this reaction. If the solution colored red (by binoxide of manganese) is mixed with from 4 to 6 times its volume of water, and ammonia is then added until the reaction is *nearly* neutral, the fluid shows a magnificent violet-purple tint; upon addition of more ammonia the color becomes yellowish-green to yellow

(J. ERDMANN). I have found, however, that this reaction is seen only where larger, though still very minute, quantities of strychnia are present.

9. Strong *chlorine water* produces in solutions of salts of strychnia a white precipitate, which dissolves in ammonia to a colorless fluid.

10. *Concentrated nitric acid* dissolves strychnia and its salts to a colorless fluid, which turns yellow upon the application of heat.

b. BRUCIA, or BRUCINE ($C_{46} H_{26} N_2 O_6 = \overset{+}{Br}$).

§ 238.

1. Crystallized brucia ($\overset{+}{Br}$ + 8 aq.) appears either in the form of transparent straight rhombic prisms, or in that of crystalline needles aggregated into stars, or as a white powder composed of minute crystalline scales. Brucia is difficultly soluble in cold, but somewhat more readily in hot water. It dissolves freely in alcohol, both in absolute and dilute, also in cold, but more readily still in hot, amyl-alcohol; but it is almost insoluble in ether. Its taste is intensely bitter. When heated, it fuses with loss of its water of crystallization.

2. Brucia neutralizes acids completely. The salts of brucia are readily soluble in water, and of an intensely bitter taste. Most of them are crystallizable.

3. *Potassa* and *carbonate of soda* throw down from solutions of salts of brucia a white precipitate of brucia, which is insoluble in an excess of the precipitant. Viewed under the microscope, immediately after precipitation, it appears to consist of very minute grains; but upon further inspection, these grains are seen—with absorption of water—to suddenly form into needles, which latter subsequently arrange themselves without exception into concentric groups. These successive changes of the precipitate may be traced distinctly even with the naked eye.

4. *Ammonia* produces in solutions of salts of brucia a whitish precipitate, which appears at first like a number of minute drops of oil, but changes subsequently—with absorption of water—to small needles. The precipitate redissolves, immediately after separation, very readily in an excess of the precipitant; but after a very short time—or, in dilute solutions, after a more considerable lapse of time—the brucia, combined with crystallization water, crystallizes from the ammoniacal fluid in small concentrically grouped needles, which addition of ammonia fails to redissolve.

5. *Bicarbonate of soda* produces in neutral solutions of salts of brucia a precipitate of brucia, combined with crystallization water; this precipitate separates after the lapse of a short time, in form of concentrically aggregated needles of silky lustre, which are insoluble in an excess of the precipitant, but dissolve in free carbonic acid (compare strychnia). Bicarbonate of soda fails to precipitate acid solutions of salts of brucia; and it is only after the lapse of a considerable time, and with the escape of the carbonic acid, that the alkaloid separates from the fluid in regular and comparatively large crystals.

6. *Concentrated nitric acid* dissolves brucia and its salts to intensely red fluids, which subsequently acquire a yellowish-red tint, and turn yellow upon application of heat. Upon addition of protochloride of tin or sulphide of ammonium to the heated fluid, no matter whether

concentrated or after dilution with water, the faint yellow color changes to a most intense violet.

7. If a little brucia is treated with from 4 to 6 drops of pure *concentrated sulphuric acid*, a solution of a faint rose color is obtained, which afterwards turns yellow. If from 8 to 20 drops of sulphuric acid mixed with some *nitric acid* (see foot-note to § 232, 7, page 319) are added, the fluid transiently acquires a red, afterwards a yellow color. Addition of binoxide of manganese transiently imparts a red, then a gamboge tint to the fluid. If the fluid is then, with proper cooling, diluted with 4 parts of water, *ammonia* added to nearly neutral reaction, or even to alkaline reaction, the solution acquires a gold-yellow color (J. ERDMANN).

8. Addition of *chlorine water* to the solution of a salt of brucia imparts to the fluid a fine bright red tint; if ammonia is then added, the red color changes to yellowish-brown.

9. *Sulphocyanide of potassium* produces in concentrated solutions of salts of brucia immediately, in dilute solutions after some time, a granular crystalline precipitate, which, when viewed under the microscope, appears composed of variously aggregated polyhedral crystalline grains. Friction applied to the sides of the vessel promotes the separation of the precipitate.

10. *Chloride of mercury* also produces a white granular precipitate, which, when viewed under the microscope, appears composed of small roundish crystalline grains.

c. VERATRIA, or VERATRINE ($C_{44} H_{34} N_2 O_{14}$) $\overset{+}{\text{Ve}}$.

§ 239.

1. Veratria appears in the form of small prismatic crystals, which acquire a porcelain-like look in the air, or as a white or yellowish-white powder of acrid and burning, but not bitter taste; it is exceedingly poisonous. Veratria acts with great energy upon the membranes of the nose; even the most minute quantity of the powder excites the most violent sneezing. It is insoluble in water; in alcohol it dissolves readily, but more sparingly in ether. At 239° Fah. it fuses like wax, and solidifies upon cooling to a transparent yellow mass.

2. Veratria neutralizes acids completely. Some salts of veratria are crystallizable, others dry up to a gummy mass. They are soluble in water, and have an acrid and burning taste.

3. *Potassa, ammonia,* and the *mono-carbonates of the alkalies* produce in solutions of salts of veratria a flocculent white precipitate, which, viewed under the microscope, immediately after precipitation, does not appear crystalline. After the lapse of a few minutes, however, it alters its appearance, and small scattered clusters of short prismatic crystals are observed, instead of the original coagulated flakes. The precipitate does not redissolve in an excess of potassa or of carbonate of potassa. It is slightly soluble in ammonia in the cold, but the dissolved portion separates again upon application of heat.

4. With *bicarbonate of soda* and *bicarbonate of potassa* the salts of veratria comport themselves like those of strychnia and brucia. However, the veratria separates readily upon boiling, even from dilute solutions.

5. If veratria is acted upon by *concentrated nitric acid*, it agglutinates into small resinous lumps, which afterwards dissolve slowly in the acid. If the veratria is pure the solution is colorless.

6. If veratria is treated with *concentrated sulphuric acid*, it also agglutinates at first into small resinous lumps; but these dissolve with great readiness to a faint yellow fluid, the color of which gradually increases in depth and intensity, and changes afterwards to a reddish-yellow, then to an intense blood-red. The color persists 2 or 3 hours, then disappears gradually. Addition of sulphuric acid, containing nitric acid, or of binoxide of manganese causes no great change of color. If the fluid is then diluted with water, and ammonia added until the reaction is nearly neutral, a yellowish solution is obtained, in which ammonia added in excess produces a greenish light-brown precipitate (J. ERDMANN).

7. *Sulphocyanide of potassium* produces only in concentrated solutions of salts of veratria flocculent-gelatinous precipitates.

8. Addition of *chlorine-water* to the solution of a salt of veratria imparts to the fluid a yellowish tint, which, upon addition of ammonia, changes to a faint brownish color. In concentrated solutions chlorine produces a white precipitate.

Recapitulation and Remarks.

§ 240.

The alkaloids of the third group also are precipitated by many other reagents besides those above-mentioned, as, for instance, by tincture of galls, bichloride of platinum, solution of iodine in iodide of potassium, phosphomolybdic acid, &c. But as these reactions are common to all, they are of little importance in an analytical point of view.[*]

Strychnia may be separated from brucia and veratria by means of absolute alcohol, since it is insoluble in that menstruum, whilst the two latter alkaloids readily dissolve in it. The identity of strychnia is best established by the reaction with sulphuric acid and the above-mentioned oxidizing agents;[†] also by the form of its crystals—when thrown down by alkalies—viewed under the microscope; and lastly, by the form of the precipitate which sulphocyanide of potassium and chloride of mercury produce in solutions of its salts. Brucia and veratria are not readily separated from one another, but may be detected in presence of each other. The identity of brucia is best established by the reactions with nitric acid and protochloride of tin or sulphide of ammonium, or by the form of the crystalline precipitate which ammonia produces in solutions of salts of brucia. Veratria is sufficiently distinguished from brucia and the other alkaloids which we have treated of, by its characteristic deportment at a gentle heat, and also by the form of the precipitate which alkalies produce in solutions of its salts. To distinguish veratria in presence of brucia, the reaction with concentrated sulphuric acid is selected.

---

[*] If the precipitate produced in the solution of a salt of strychnia by iodide of potassium containing iodine is dissolved in spirit of wine mixed with some sulphuric acid, and the solution is evaporated, strongly polarizing prismatic crystals of sulphate of iodide of strychnia are obtained. DE VRIJ and VAN DER BURG (Jahresber v. LIEBIG and KOPP, 1857, 602). Whether this reaction is characteristic for strychnia, can be known only after the optical properties of analogous compounds of the other alkaloids shall have been studied.

[†] The only substance which shows somewhat analogous reactions in this respect, is aniline. A. GUY has, however, called attention to the fact that aniline, treated with sulphuric acid and oxidizing agents, acquires a pale green color at first, which gradually deepens, and only then changes to a magnificent blue, which, after persisting some time, turns finally black.

To these alkaloids I will add *salicine*, although this substance does not properly belong to the same class of chemical compounds.

§ 241.

SALICINE $(C_{26}H_{18}O_{14})$.

1. Salicine appears either in the form of white crystalline needles and scales of silky lustre, or, where the crystals are very small, as a powder of silky lustre. It has a bitter taste, is readily soluble in water and in alcohol, but insoluble in ether.

2. No reagent precipitates salicine as such.

3. If salicine is treated with *concentrated sulphuric acid*, it agglutinates into a resinous lump, and acquires an intensely blood-red color, without dissolving in the acid; the color of the sulphuric acid is at first unaltered.

4. If an aqueous solution of salicine is mixed with *hydrochloric acid* or *dilute sulphuric acid*, and the mixture boiled for a short time, the fluid suddenly becomes turbid, and deposits a fine granular crystalline precipitate (saliretine).

SYSTEMATIC COURSE FOR THE DETECTION OF THE ALKALOIDS TREATED OF IN THE PRECEDING PARAGRAPHS, AND OF SALICINE.

The analytical course which I am now about to describe is based upon the supposition that the analyst has to examine a concentrated aqueous solution—effected by the agency of an acid—of one or several of the non-volatile alkaloids, which solution is free from any admixture of substances that might tend to obscure or modify the reactions. For the modifications which the presence of coloring or extractive matters, &c., requires, I refer to § 244.

I. DETECTION OF THE ALKALOIDS, AND OF SALICINE, IN SOLUTIONS SUPPOSED TO CONTAIN ONLY ONE OF THESE SUBSTANCES.[*]

§ 242.

1. Add dilute solution of potassa or soda, drop by drop, to a portion of the aqueous solution until the fluid acquires a scarcely perceptible alkaline reaction; stir, and let the fluid stand for some time.

a. NO PRECIPITATE IS FORMED; this proves the total absence of the alkaloids, and indicates the presence of SALICINE. To set all doubt at rest, test the original substance with concentrated sulphuric acid, and also with hydrochloric acid. Compare § 241.

b. A PRECIPITATE IS FORMED. Add solution of potassa or soda, drop by drop, until the fluid manifests a strongly alkaline reaction.

---

[*] Where the detection of one of the five more frequently occurring poisonous alkaloids alone is the object, the following simple method, devised by J. ERDMANN, will fully answer the purpose.

In this method, which is more especially applicable in cases where the disposable quantity of substance is very small, the alkaloids are supposed to be present in the pure state and in the solid form.

1. Treat the substance under examination with from 4 to 6 drops of pure concentrated sulphuric acid.

Yellow color, speedily changing to red: VERATRIA.

Rose color, changing afterwards to yellow: BRUCIA.

The other alkaloids, if pure, impart no color to the sulphuric acid.

2. No matter whether there is color or not; add to the fluid obtained in 1 from 8 to

a. The precipitate redissolves: MORPHIA. To arrive at a positive conclusion on this point, test another portion of the solution with iodic acid (§ 232, 9), and a portion of the original substance with sulphuric acid, &c. (§ 232, 7).

β. *The precipitate remains undissolved:* Presence of an alkaloid of the second or third group. Pass on to 2.

2. Add to a second portion of the original solution two or three drops of dilute sulphuric acid, then a saturated solution of bicarbonate of soda until the acid reaction is just neutralized; vigorously rub the inside of the vessel, and allow the mixture to stand for half an hour.

a. NO PRECIPITATE IS PRODUCED: Absence of narcotine and cinchonia. Pass on to 3.

b. A PRECIPITATE IS FORMED: Narcotina, cinchonia, and perhaps also quina, as the precipitation of the latter substance by bicarbonate of soda depends entirely upon the degree of dilution of the fluid. Add to a portion of the original solution ammonia in excess, then a sufficient quantity of ether, and shake the mixture.

a. The precipitate which forms at first upon the addition of the ammonia redissolves in the ether, and the clear fluid presents two distinct layers: narcotina or quina. To distinguish between the two, test a fresh portion of the original solution with chlorine water and ammonia. If the solution turns green, QUINA, if yellowish-red, NARCOTINE is present. The reaction with a mixture of sulphuric acid and nitric acid (§ 233, 5) is resorted to as a conclusive test.

β. *The precipitate which forms upon the addition of ammonia does not redissolve in the ether*—CINCHONIA. The deportment of cinchonia at a high temperature (235, 3), or the reaction with ferricyanide of potassium (§ 235, 8), may serve as a conclusive test.

3. Put a portion of the original substance, or of the residue remaining 20 drops of concentrated sulphuric acid mixed with nitric acid (see foot-note to § 232, 7, page 319), then 2 or 3 drops of water. After a quarter or half-hour the fluid shows:

a. a violet-red color: MORPHIA;
b. an onion-red color: NARCOTINA;
c. a transient-red tint, changing to yellow: BRUCIA;
d. the red color of the sulphuric acid solution of VERATRIA is not materially altered;
e. with STRYCHNIA no coloration is observed.

3. Put into the fluid obtained in 2, no matter whether colored or not, from 4 to 6 clean fragments of binoxide of manganese, of the size of a lentil. After an hour the fluid shows:

a. a mahogany-brown color: MORPHIA;
b. a yellowish-red to blood-red color: NARCOTINA;
c. a transient purple-violet tint, changing to deep onion-red: STRYCHNIA;
d. a transient red tint, changing to gamboge-yellow: BRUCIA;
e. a dark dirty cherry-red color: VERATRIA.

4. Pour the colored fluid obtained in 3, into a test tube containing 4 times the volume of water, and add ammonia until the neutralization point is *almost* attained. Heat must be as much as possible avoided in these operations.

a. dirty-yellow color, changing to brownish-red upon supersaturation with ammonia, without immediate deposition of a notable precipitate: MORPHIA;
b. reddish coloration, more or less intense according to the degree of dilution; upon supersaturation with ammonia, copious dark-brown precipitate: NARCOTINA;
c. violet-purple colored solution, becoming yellowish-green to yellow upon addition of ammonia in excess: STRYCHNIA.
d. gold-yellow solution, not materially changed by excess of ammonia: BRUCIA;
e. faint brownish solution, turning yellowish upon further addition of ammonia, and depositing a greenish light-brown precipitate: VERATRIA.

upon the evaporation of the solution, on a watch-glass, and treat with concentrated sulphuric acid.

a. A rose-coloured solution is obtained, which becomes intensely red upon addition of nitric acid: BRUCIA. The reaction with nitric acid and protochloride of tin is resorted to as a conclusive test (§ 238, 6).

b. A yellow solution is obtained, the color of which gradually changes to yellowish-red, then to blood-red, and turns finally crimson: VERATRIA.

c. A colorless fluid is obtained, which remains colorless after standing for some time.

Add to the fluid a fragment of chromate of potassa; if this imparts to it a deep blue color, STRYCHNIA is present; if it leaves the fluid unaltered, QUINA is present. The reaction with chlorine water and ammonia is resorted to as a conclusive test.

II. DETECTION OF THE ALKALOIDS, AND OF SALICINE, IN SOLUTIONS SUPPOSED TO CONTAIN SEVERAL OR ALL OF THESE SUBSTANCES.

§ 243.

1. Add to a portion of the aqueous solution dilute solution of potassa or soda, drop by drop, until the fluid acquires a scarcely perceptible alkaline reaction; stir, and let the fluid stand for some time.

a. NO PRECIPITATE IS FORMED; this proves the total absence of the alkaloids, and indicates the presence of SALICINE. To remove all doubt on the point, test the original substance with concentrated sulphuric acid, and with hydrochloric acid. Compare § 242, 1, *a.*

b. A PRECIPITATE IS FORMED: add solution of potassa or soda, drop by drop, until the fluid manifests a strongly alkaline reaction.

α. *The precipitate redissolves.* Absence of the alkaloids of the second and third groups. Presence of MORPHIA is indicated. The reactions with iodic acid (§ 232, 9), and with sulphuric acid, &c., (§ 232, 7), are resorted to as conclusive tests. Examination for salicine, see 4.

β. *The precipitate does not redissolve, or at least not completely.* Filter, and treat the precipitate as directed in 2. Saturate the filtrate with carbonic acid, or mix it with bicarbonate of soda or bicarbonate of potassa, and boil nearly to dryness. Treat the residue with water; if it dissolves completely, this is a sign that no morphia is present; but if there is an insoluble residue left, this indicates the presence of morphia. The reactions with iodic acid (§ 232, 9), and with sulphuric acid, &c. (§ 232, 7), are resorted to as conclusive tests.

2. Wash the filtered precipitate of 1, *b*, β, with cold water, dissolve in a slight excess of dilute sulphuric acid, and add solution of bicarbonate of soda to the fluid until the acid reaction is neutralized; stir the mixture, vigorously rubbing the sides of the vessel, and allow the fluid to stand for an hour.

a. NO PRECIPITATE IS FORMED. Absence of narcotina and cinchonia. Boil the solution nearly to dryness, and treat the residue with cold water. If it dissolves completely, pass on to 4; but if an insoluble residue is left, examine this for quina—of which a minute quantity might be present—and for strychnia, brucia, and veratria, according to the directions of 3.

b. A PRECIPITATE IS FORMED. This may contain narcotina, cinchonia, and also quina, compare § 242, 2, *b.* Filter, and treat the filtrate as directed § 243, 2, *a*. Wash the precipitate with cold water, dissolve in a little hydrochloric acid, add ammonia in excess, then a sufficient quantity of ether.

α. *The precipitate which forms at first upon the addition of the ammonia redissolves completely in the ether, and the clear fluid presents two distinct layers.* Absence of cinchonia ; presence of quina or narcotina. Evaporate the ethereal solution, dissolve the residue in a little hydrochloric acid and a sufficient amount of water to make the dilution at least as 1 : 200 ; add bicarbonate of soda to neutralization, and allow the fluid to stand for some time. The formation of a precipitate indicates the presence of NARCOTINA. Filter, and test the precipitate with chlorine water and ammonia, and with a mixture of sulphuric acid and nitric acid (§ 233). Evaporate the filtrate, or the fluid if no precipitate has formed, to dryness, and treat the residue with water. If part of it remains undissolved, wash this, dissolve in hydrochloric acid, and add chlorine water and ammonia. Green color : QUINA.

β. *The precipitate produced by the ammonia does not redissolve in the ether, or at least not completely* : CINCHONIA. Quina or narcotina may also be present. Filter, and examine the filtrate for quina and narcotina as in α. The precipitate consists of cinchonia, and may be further examined according to § 235, 3, or 8.

3. Wash the insoluble residue of § 243, 2, *a*, with water, dry on the water-bath, and digest with absolute alcohol.

a. IT DISSOLVES COMPLETELY : absence of strychnia ; presence of (quina) brucia or veratria. Evaporate the alcoholic solution on the water-bath to dryness, and, if quina has already been detected, divide the residue into two portions, and test one part for BRUCIA, with nitric acid and protochloride of tin (§ 238, 6), the other for VERATRIA, by means of concentrated sulphuric acid (§ 239, 6) ; but if no quina has as yet been detected, divide the residue into three portions, *a*, *b*, and *c ;* examine *a* and *b* for BRUCIA and VERATRIA, in the manner just stated, and *c* for quina, with chlorine-water and ammonia. However, if brucia is present, dissolve *c* in hydrochloric acid, add ammonia and ether, let the mixture stand for some time, evaporate the ethereal solution, and examine the residue for quina.

b. It does not dissolve, or at least not completely : presence of STRYCHNIA ; perhaps also of (quina) brucia and veratria. Filter, and examine the filtrate for (QUINA) BRUCIA and VERATRIA as directed § 243, 3, *a*. The identity of the precipitate with strychnia is demonstrated by the reaction with sulphuric acid and chromate of potassa (§ 237, 8).

4. Mix a portion of the original solution with hydrochloric acid, and boil the mixture for some time. The formation of a precipitate indicates the presence of SALICINE. To set all doubt on this point at rest, test the original substance with concentrated sulphuric acid (§ 241, 3).

III. DETECTION OF THE ALKALOIDS, IN PRESENCE OF COLORING AND EXTRACTIVE VEGETABLE OR ANIMAL MATTERS.

§ 244.

The presence of mucilaginous, extractive, and coloring matters renders the detection of the alkaloids a task of considerable difficulty. These matters obscure the reactions so much that we are even unable to determine by a preliminary experiment, whether the substance under examination contains one of the alkaloids we have treated of in the foregoing paragraphs, or not. I will now give several methods by means of which the separation of the alkaloids from such extraneous matters may be effected, and their detection made practicable. Which of these methods to select will, of course, always depend upon the particular circumstances of the case.

1. STAS'S METHOD FOR EFFECTING THE DETECTION OF POISONOUS ALKALOIDS.*

This method is based, *a*, upon the solubility of the acid salts of the alkaloids in water and in spirit of wine ; *b*, upon the fact that the alkaloids, even those sparingly soluble in ether, will pass into the ethereal solution upon mixing the aqueous solution of any of their salts with a fixed alkali or with the carbonate of a fixed alkali in excess, and shaking the mixture repeatedly with ether ;—lastly, *c*, the removal from the alkaloids of other matters soluble in ether, is based upon the insolubility of the salts of the alkaloids in ether, owing to which an aqueous solution of the acid sulphate of the alkaloid can be obtained, by shaking the ethereal solution of the pure alkaloid with dilute sulphuric acid. I will give here, first the original method of STAS, then, in 2, OTTO's modifications of that method.

a. If you have to look for the suspected organic bases in the contents of the stomach or intestines, or in articles of food, or in pappy matters in general, heat the suspected substance with double its weight of strong alcohol, acidified with from 0·5 grm. to 2 grm. of tartaric acid or oxalic acid, to from 158° F. to 167° F. When quite cold, filter, and wash the undissolved part with strong alcohol, adding the washings to the filtrate.

If you have to deal with the heart, liver, lungs, or similar organs, cut them into fine shreds, moisten with the acidified alcohol, press, and repeat the same operation, until the soluble parts are completely extracted ; collect the fluids obtained, and filter.

b. Concentrate the alcoholic fluid at a temperature not exceeding 95° F., and, if no insoluble matter separates, continue to evaporate nearly to dryness. Conduct this process either under a bell-glass over sulphuric acid, with or without rarefaction of the air, or in a tubular retort, through which a current of air is passed. If fatty or other insoluble matters separate in the process of concentration, pass the concentrated fluid through a moistened filter, and evaporate the filtrate nearly to dryness, conducting the process, as above, either under a bell-glass or in a retort.

c. Digest the residue with cold absolute alcohol, filter, wash the insoluble residue thoroughly with alcohol, and let the alcoholic solution evaporate in the air or in vacuo ; dissolve the acid residue in a little water, and add bicarbonate of soda as long as effervescence ensues.

d. Add to the mixture four or five times its volume of pure ether, free from oil of wine, and shake ; then allow it to stand at rest, and let a little of the supernatant ether evaporate spontaneously on a watch-glass. If this leaves oily streaks upon the glass, which gradually collect into a drop, and emit, upon the application of a gentle heat, a disagreeable, pungent, and stifling odor, there is reason to infer the presence of a liquid volatile base ; whilst a solid residue or a turbid fluid, with solid particles suspended in it, indicates a non-volatile, solid base. In the latter case the base may emit a disagreeable animal smell, but not a pungent

* "Bulletin de l'Académie de Médecine de Belgique," IX. 304. "Jahrb. f. prakt. Pharm.," XXIV. 313. "Jahresbericht" von Liebig und Kopp, 1851, p. 640.

odor, as is the case with volatile bases. The blue color of reddened litmus-paper is permanently restored. If no residue is left, add to the fluid some solution of soda or potassa, and shake with repeatedly renewed ether, which will now dissolve the base. It follows from the assumption that the bases present will pass into the ethereal solution, that Stas's method is principally calculated for the detection of the poisonous alkaloids which are soluble in ether, though some of them only sparingly. The following are the vegeto-alkalies which Stas enumerates as discoverable by his method: Conia, Nicotina, Aniline, Picoline, Petinine, Morphia, Codeia, Brucia, Strychnia, Veratria, Colchicia, Delphia, Emetine, Solania, Aconitina, Atropia, and Hyoscyamia.

a. There is reason to infer the presence of a volatile base.

Add to the contents of the vessel from which you have taken the small portion of ether for evaporation on the watch-glass, one or two cubic centimetres of strong solution of potassa or soda, shake the mixture, let it stand at rest, pour the supernatant fluid into a flask, and treat the residue again three or four times with ether, until the last portion poured off leaves no longer a residue upon evaporation. Mix the ethereal fluid now with some dilute sulphuric acid (1 part of acid to 5 parts of water) until the well-shaken fluid manifests acid reaction; allow the mixture to stand at rest, decant the supernatant ether from the acid aqueous fluid, and treat the latter once more with ether in the same way.

aa. Mix the residual acid solution (which may contain sulphates of ammonia, nicotina, aniline, picoline, and petinine—indeed which must contain these bases, if they are present in the examined substance, since their compounds with sulphuric acid are quite insoluble in ether; and in which, if conia is present, the greater part of the latter alkaloid is also found) with concentrated solution of soda or potassa in excess, and treat with ether, which will again dissolve the liberated bases; decant the ether, and leave it to spontaneous evaporation, at the lowest possible temperature; place the dish with the residue in vacuo over sulphuric acid. In this process the ether and ammonia escape, leaving the volatile organic base in the pure state. The nature of the organic base is then finally ascertained.

bb. The ether decanted from the acid solution contains the animal matters which it has removed from the alkaline fluid. It leaves, therefore, upon spontaneous evaporation, a trifling faint yellow residue of nauseous odor, which contains also some sulphate of conia, if that base was present in the examined matter.

β. There is reason to infer the presence of a solid base.

Add a few drops of alcohol to the ethereal solution obtained by treating with ether the previously acid residue mixed either simply with bicarbonate of soda, or first with that reagent, then with solution of soda or potassa (see *c* and *d*), and leave the mixture to spontaneous evaporation. If this fails to give the base in a distinctly crystalline form and sufficiently pure, add a few drops of water slightly acidified with sulphuric acid, which will usually serve to separate the mass into a fatty portion, adhering to the dish, and an acid aqueous solution, which contains the base as an acid sulphate. Decant or filter, wash with a little slightly acidified water, and evaporate the solution to a considerable extent, under a bell-glass over sulphuric acid. Mix the residue with a highly concentrated solution of pure carbonate of potassa, treat the mixture with absolute alcohol, decant, and let the alcoholic fluid evaporate, which will generally leave the base in a state of perfect purity, or nearly so.

2. Otto's Modifications of Stas's Method.*

a. In the method just described, the morphia which may be present will only pass into the ethereal solution if the solution obtained in 1, *c*, is shaken with ether immediately after the addition of the bicarbonate of soda, and the ether then quickly decanted. But if the operation of shaking with ether is delayed, so as to afford the morphia time to crystallize, the crystals will deposit, being almost absolutely insoluble in ether (P. Pöllnitz); and if the ethereal solution is allowed to stand some time, the dissolved morphia will separate in small crystals, on the sides of the vessel.—As it is therefore, under the circumstances stated, always likely to happen that the morphia may remain, wholly or in part, undissolved by the ether, it is of the highest importance never to neglect mixing the alkaline fluid obtained in 1, *d*—after repeated extraction with ether, and subsequent addition of some solution of soda, to dissolve the morphia, which may have separated, and after evaporating the ether still present—with a concentrated solution of chloride of ammonium, and letting the mixture stand exposed to the open air, to allow the morphia to crystallize.

b. Instead of the process described in 1, β, to effect the detection of non-volatile alkaloids, Otto recommends the following method, which is in principle the same as that recommended by Stas for the detection of the volatile bases.

Let the ethereal solution evaporate, dissolve the residuary impure alkaloid in a little water mixed with sulphuric acid, and shake the solution repeatedly with ether, which will remove the foreign organic matters present, and leave the acid vegeto-alkaline sulphate unaffected. Mix now the acid aqueous solution with carbonate of soda in excess, shake repeatedly with ether (to dissolve the liberated alkaloids), and let the ethereal solution evaporate, when the alkaloids held in solution by the ether will be left in a very pure state and, to a great extent, in the crystalline form. This method has stood the test of numerous experiments.

c. But what Otto recommends most, is the treatment with ether of the alkaloid in the form of salt, before its separation, by means of an alkali, and its solution in ether.—If, therefore, you wish to follow this method, shake the acid aqueous fluid of 1, *c*, which contains the alkaloid in combination with tartaric acid or oxalic acid, repeatedly with ether, so long as the ether becomes colored and leaves a residue upon evaporation; then, and not before, add carbonate of soda, dissolve the alkaloid by means of ether, and proceed generally as directed in 1, *d*. Upon evaporating the ether, the alkaloid is now left at once in a very pure state.

3. Method of L. v. Uslar and J. Erdmann.†

This may be considered an improvement upon Stas's method, as regards non-volatile alkaloids, more especially morphia, whilst Stas's method deserves the preference for volatile alkaloids. The new method is the same in principle as that of Stas's, simply substituting amyl-alcohol for ether.

Mix the matters to be examined with water, if necessary, to the consistence of a thin paste, acidify slightly with hydrochloric acid, digest

---

* Annal. d. Chem. u. Pharm., 100, 44.
† Annal. d. Chem. u. Pharm., 120, page 121; and 122, page 360.

one or two hours at from 140° to 176° Fah., and pass through a linen cloth moistened with water. Extract the residue with water acidified with hydrochloric acid, add the solution obtained to the first fluid, supersaturate with ammonia, and evaporate to dryness, with addition of pure quartz sand, which will enable you to reduce the residue to powder. Boil the powder repeatedly with amyl-alcohol, to extract the whole of the alkaloid from it. Filter the extracts hot through felt, moistened with amyl-alcohol. The filtrate, which is mostly colored yellow, holds, besides the alkaloid, fatty and coloring matters in solution. To remove these latter, transfer the filtrate to a cylindrical vessel, mix it with from ten to twelve times its volume of almost boiling water, acidified with hydrochloric acid, and vigorously shake the mixture for some time. The hydrochlorate of the alkaloid passes into the aqueous solution, whilst the fatty and coloring matters remain dissolved in the amyl-alcohol.* Remove the latter by means of an india-rubber pipette, then shake the acid solution repeatedly with fresh quantities of amyl-alcohol, until all the fatty and coloring matters are completely removed. Concentrate now by evaporation, mix with ammonia in slight excess, add hot amyl-alcohol, and shake vigorously. When the liquid has separated into two distinct layers, draw off, by means of a pipette, the upper layer, which contains the solution of the alkaloid in amyl-alcohol, treat the fluid once more with hot amyl-alcohol, then completely drive off the latter by heating on the water-bath, which will often leave the alkaloid sufficiently pure for examination by the usual reactions. Should it, however, still look yellowish or brownish, dissolve it once more in dilute hydrochloric acid, shake the solution with amyl-alcohol, remove the latter with the pipette, then supersaturate with ammonia, shake again with amyl-alcohol, draw off the latter with the pipette, and evaporate it on the water-bath. It is only in very rare cases that the alkaloid left by this evaporation requires a repetition of this process of purification. The last evaporation of the pure alkaloid is best conducted in a small porcelain crucible, placed obliquely. Before proceeding to the decisive reactions, pour a few drops of concentrated sulphuric acid over the alkaloid, and observe whether it still turns brown on the application of this test; in which case the process of purification must be repeated.—USLAR and ERDMANN have detected and isolated by this method very minute traces of alkaloids, e.g., 5 milligrammes of hydrochlorate of morphia, 1 drop of nicotine, 9 milligrammes of strychnine, mixed with from 2 to 3 pounds of contents of the stomach. In his second paper on the subject ERDMANN calls particular attention to the fact that he succeeded in separating by this method the poisonous alkaloids from quite putrid intestines of poisoned animals, from a fortnight to a month after death. The latter experiments referred to strychnia and morphia. Of course only those portions of the alkaloids can be detected which have not yet suffered decomposition. With regard to morphia, there would seem to exist no doubt but that this alkaloid suffers decomposition in the organism. A rabbit had given it 0.1 grm. of hydrochlorate of morphia, and was killed three and a half hours after: no morphia was found in the urine, brain, and spinal marrow; only very little of it in the blood, a little in the stomach and the small intestines, more in the other intestines.

---

* If less water is used, traces of the hydrochlorate of the alkali are apt to remain in the amyl-alcoholic solution. J. ERDMANN recommends always to put aside the first amyl-alcohol, as well as that used to effect the removal of the fatty matters, that they may, if necessary, be shaken once more with the stated quantity of acidulated water.

4. Methods of Detecting Strychnia, based upon the Use of Chloroform.*

a. Rodgers and Girdwood's Method.†

Digest the substance under examination with dilute hydrochloric acid (1 part of acid to 10 parts of water) and filter; evaporate the filtrate on the water-bath to dryness, extract the residue with spirit of wine, evaporate the solution, treat the residue with water, filter, supersaturate the filtrate with ammonia, add ½ oz. (15 grammes) of chloroform, shake, transfer the chloroform to a dish, by means of a pipette, evaporate on the water-bath, moisten the residue with concentrated sulphuric acid, to effect carbonization of foreign organic matters, treat with water, after the lapse of several hours, then filter. Supersaturate the filtrate again with ammonia, and shake it with about 1 drachm (4 grammes) of chloroform. Repeat the same operation until the residue left upon the evaporation of the chloroform is no longer charred by sulphuric acid. Transfer the chloroform solution which leaves a pure residue, no longer affected by sulphuric acid, drop by drop, by means of a capillary tube, to the same spot on a heated porcelain dish, letting it evaporate, then test the residue with sulphuric acid and chromate of potassa. Rodgers and Girdwood succeeded in detecting by this method so small a quantity of strychnia as the $\frac{1}{30000}$th part of a grain.

b. Method recommended by E. Prollius.)‡

Boil twice with spirit of wine, mixed with some tartaric acid, evaporate at a gentle heat, filter the residuary acid aqueous solution through a moistened filter, add ammonia in slight excess, then from 20 to 25 grains (about 1½ grm.) of chloroform, shake, free the deposited chloroform thoroughly from the ley, by decanting and shaking with water, mix the chloroform so purified with 3 parts of spirit of wine, and let the fluid evaporate. If there is any notable quantity of strychnia present, it is obtained in crystals.

5. Method of Effecting the Detection of Strychnia in Beer, by *Graham* and *A. W. Hofmann*.§

This method, which is based on the known fact that a solution of a salt of strychnia, when mixed and shaken with animal charcoal, yields its strychnia to the charcoal, will undoubtedly be found applicable also for the detection of other alkaloids. The process is conducted as follows:—

Shake 2 ounces of animal charcoal in half-a-gallon of the aqueous neutral or feebly acid fluid under examination; let the mixture stand for from 12 to 24 hours, with occasional shaking, filter, wash the charcoal twice with water, then boil for half-an-hour with 8 ounces of spirit of wine of 80—90 per cent., avoiding loss of alcohol by evaporation. Filter the spirit of wine hot from the charcoal, and distil the filtrate; add a few drops of solution of potassa to the residual watery fluid, shake with

---

* These methods are no doubt useful also for effecting the separation of other alkaloids; however, the deportment of the latter with chloroform has not yet been sufficiently studied.
† Liebig and Kopp's "Jahresbericht," 1857, 603.—Pharm. Journ. Trans., xvi. 497.
‡ Chem. Centrabl., 1857, 231.
§ Chem. Soc. Quart. Journ., v. 173.

ether, let the mixture stand at rest, then decant the supernatant ether. The ethereal fluid leaves, upon spontaneous evaporation, the strychnia in a state of sufficient purity to admit of its further examination by reagents (see § 237).

MACADAM* employed the same method in his numerous experiments to detect strychnia in the bodies of dead animals. He treated the comminuted matters with a dilute aqueous solution of oxalic acid in the cold, filtered through muslin, washed with water, heated to boiling, filtered still warm, from the coagulated albuminous matters, shook with charcoal, and proceeded in the manner just described. According to his statements, the residue left by the evaporation of the alcoholic solution was generally at once fit to be tested for strychnia. Where it was not so, he treated the residue again with solution of oxalic acid, and repeated the process with animal charcoal.

6. SEPARATION BY DIALYSIS.

The dialytic method devised by GRAHAM, and described in § 224, may also be advantageously employed to effect the separation of alkaloids from the contents of the stomach, intestines, &c. Acidify with hydrochloric acid, and place the matter in the dialyser. The alkaloids, being crystalline bodies, penetrate the membrane, and are found, for the greater part, after 24 hours, in the outer fluid; from this they may, then, according to circumstances, either be thrown down at once, after concentration by evaporation; or they may be purified by one of the above described methods.

II.

GENERAL PLAN OF THE ORDER AND SUCCESSION IN WHICH SUBSTANCES SHOULD BE ANALYZED FOR PRACTICE.

§ 245.

It is not a matter of indifference whether the student, in analyzing for the sake of practice, follows no rule or order whatever in the selection of the substances which he intends to analyze, or whether, on the contrary, his investigations and experiments proceed systematically. Many ways, indeed, may lead to the desired end, but one of them will invariably prove the shortest. I will, therefore, here point out a course which experience has shown to lead safely and speedily to the attainment of the object in view.

Let the student take 100 compounds, systematically arranged (*see below*), and let him analyze these compounds successively in the order in which they are placed. A careful and diligent examination of these will be amply sufficient to impart to him the necessary degree of skill in practical analysis. When analyzing for the sake of practice only, the student must above all things possess the means of verifying the results obtained by his experiments. The compounds to be examined ought, therefore, to be mixed for him by a friend who knows their exact composition.

A. *From* 1 *to* 20.

AQUEOUS SOLUTIONS OF SIMPLE SALTS : *e. g.*, sulphate of soda, nitrate of lime, chloride of copper, &c. These investigations will serve to teach the student the method of analyzing substances soluble in water which

---

* Pharm. Journ. Trans., xvi. 120, 160.

contain but one base. In these investigations it is only intended to ascertain which base is present in the fluid under examination; but neither the detection of the acid, nor the proof of the absence of all other bases besides the one detected, is required.

B. *From* 21 *to* 50.

SALTS, ETC., CONTAINING ONE BASE AND ONE ACID, OR ONE METAL AND ONE METALLOID (in form of powder): *e.g.*, carbonate of baryta, borate of soda, phosphate of lime, arsenious acid, chloride of sodium, bitartrate of potassa, acetate of copper, sulphate of baryta, chloride of lead, &c. These investigations will serve to teach the student how to make a preliminary examination of a solid substance, by heating in a tube or before the blow-pipe; how to convert it into a proper form for analysis, *i. e.*, how to dissolve or decompose it; how to detect *one* metallic oxide, even in substances *insoluble* in water; and how to demonstrate the presence of *one* acid. The detection of both the base and the acid is required, but it is not necessary to prove that no other bodies are present.

C. *From* 51 *to* 65.

AQUEOUS OR ACID SOLUTIONS OF SEVERAL BASES. These investigations will serve to teach the student the method of separating and distinguishing several metallic oxides from each other. The proof is required that no other bases are present besides those detected. No regard is paid to the acids.

D. *From* 66 *to* 80.

DRY MIXTURES OF EVERY DESCRIPTION. A portion of the salts should be organic, another inorganic; a portion of the compounds soluble in water or hydrochloric acid, another insoluble; *e. g.*, mixtures of chloride of sodium, carbonate of lime, and oxide of copper;—of phosphate of magnesia and ammonia, and arsenious acid;—of tartrate of lime, oxalate of lime, and sulphate of baryta;—of phosphate of soda, nitrate of ammonia, and acetate of potassa, &c.

These investigations will serve to teach the student how to treat mixtures of different substances with solvents; how to detect several acids in presence of each other; how to detect the bases in presence of phosphates of the alkaline earths;—and they will serve as a general introduction to scientific and practical analysis. All the component parts must be detected, and the nature of the substance ascertained.

E. *From* 81 *to* 100.

NATIVE COMPOUNDS, ARTICLES OF COMMERCE, &c. Mineral and other waters, minerals of every description, soils, potash, soda, alloys, colors, &c.

III.

ARRANGEMENT OF THE RESULTS OF THE ANALYSES PERFORMED FOR PRACTICE.

§ 246.

The manner in which the results of analytical investigations ought to be arranged is not a matter of indifference. The following examples will serve to illustrate the method which I have found the most suitable in this respect.

PLAN OF ARRANGING THE RESULTS OF EXPERIMENTS, Nos. 1—20.

Colorless fluid of neutral reaction.

H Cl *no precipitate,* consequently no Ag O Hg_2 O	H S *no precipitate,* no Pb O „ Hg O „ Cu O „ Bi_2O_3 „ Cd O „ $As O_3$ „ $As O_5$ „ $Sb O_3$ „ $Sn O_2$ „ Sn O „ Au O^3 „ Pt O_2 „ $Fe_2 O_3$	$NH_4 S$ *no precipitate,* no Fe O „ Mn O „ Ni O „ Co O „ Zn O „ Al_2O_3 „ Cr_2O_3	$NH_4 O, CO_2$ and $NH_4 Cl$ *a white precipitate,* consequently either Ba O, Sr O, or Ca O, no precipitate by solution of sulphate of lime, consequently LIME. Confirmation by means of $\overline{O}$

PLAN OF ARRANGING THE RESULTS OF EXPERIMENTS, Nos. 21—50.

White powder, fusing in the water of crystallization upon application of heat, then remaining unaltered—soluble in water—reaction neutral.

H Cl *no precipitate.*	H S *no precipitate.*	$NH_4 S$ *no precipitate.*	$NH_4 O, CO_2$ and $NH_4 Cl$ *no precipitate.*	$2 NaO, HO, PO_5$ and $NH_4 O$ *a white precipitate,* consequently MAGNESIA.

The detected base being Mg O, and the analyzed substance being soluble in water, the acid can only be Cl, I, Br, SO_3, NO_5, $\overline{A}$, &c. The preliminary examination has proved the absence of the organic acids and of nitric acid.

Ba Cl produces a white precipitate which H Cl fails to dissolve; consequently SULPHURIC ACID.

APPENDIX. 341

PLAN OF ARRANGING THE RESULTS OF EXPERIMENTS, NOS. 51—100.

A white powder, acquiring a permanent yellow tint upon application of heat, without forming a sublimate, and without emitting visible fumes marked by acid or alkaline reaction. Before the blowpipe, a ductile metallic globule, and yellow incrustation, with white border upon cooling. Insoluble in water, effervescing with hydrochloric acid, incompletely soluble in that acid, readily soluble in nitric acid to a colorless fluid.

H Cl	H S	N H$_4$ S	N H$_4$ O, CO$_2$	No fixed residue upon evaporation.	Hydrate of lime has failed to evolve ammonia.
White precipitate, insoluble in an excess of the precipitant, unaltered by ammonia: quite soluble in hot water; SO$_3$ producing a white precipitate in the solution: LEAD.	Black precipitate, insoluble in sulphide of ammonium, readily soluble in nitric acid. SO$_3$ produces a white precipitate : LEAD. Examination for Cu, Bi, and Cd : results negative.	White precipitate; ammonia, applied by itself, produces no precipitate; solution of precipitate in hydrochloric acid remains clear upon addition of soda in excess. N H$_4$, Cl — no precipitate. H S — white precipitate : ZINC.	White precipitate; upon dissolving this in hydrochloric acid, and adding solution of sulphate of lime to the fluid, a white precipitate forms after some time: STRONTIA. Precipitation with sulphate of potassa, filtrate tested for lime with $\bar{O}$: results negative.		

Of the acids CARBONIC ACID has already been found. Of the remaining acids the following cannot be present:

The preliminary examination has proved the absence of organic acids with nitric acid.

Cl O$_3$ cannot be present, because the analyzed substance is insoluble in water.

S and SO$_3$ not, because the analyzed substance is readily soluble in nitric acid

Cr O$_3$ not, as the nitric acid solution is colorless.

P O$_5$, Si O$_3$, H F, and $\bar{O}$ not, because the solution filtered from the sulphide of lead was not precipitated by simple addition of ammonia.

B O$_3$ might be present in trifling quantity; the examination for it gave a negative result.

Cl, I, Br might be present in the form of basic compounds of lead. However, nitrate of silver has produced no precipitate in the nitric acid solution; accordingly, they cannot be present.

The analyzed compound contains, therefore $\begin{cases} \text{bases : oxide of lead, oxide of zinc, strontia.} \\ \text{acids : carbonic acid.} \end{cases}$

IV.

TABLE

OF THE

MORE FREQUENTLY OCCURRING FORMS AND COMBINATIONS OF THE SUBSTANCES TREATED OF IN THE PRESENT WORK,

ARRANGED

WITH ESPECIAL REGARD TO THE CLASS TO WHICH THEY RESPECTIVELY BELONG ACCORDING TO THEIR SOLUBILITY

IN WATER, IN HYDROCHLORIC ACID, IN NITRIC ACID, OR IN NITROHYDROCHLORIC ACID.

§ 247.

PRELIMINARY REMARKS.

The class to which the several compounds respectively belong according to their solubility in water or acids (see § 179), is expressed by figures. Thus 1 or I means a substance soluble in water; 2 or II a substance insoluble in water, but soluble in hydrochloric acid, nitric acid, or nitrohydrochloric acid; 3 or III a substance insoluble in water, in hydrochloric acid, and in nitric acid. For those substances which stand as it were on the limits between the various classes, the figures of the classes in question are jointly expressed: thus 1—2 signifies a substance sparingly soluble in water, but soluble in hydrochloric acid or nitric acid; 1—3 a body sparingly soluble in water, and of which the solubility is not notably increased by the addition of acids; and 2—3 a substance insoluble in water, and sparingly soluble in acids. Wherever the deportment of a substance with hydrochloric acid differs materially from that which it exhibits with nitric acid, this is stated in the notes.

The Roman figures denote official and more commonly occurring compounds.

The haloid salts and sulphur compounds are placed in the columns of the corresponding oxides. The salts given are, as a general rule, the neutral salts; the basic, acid, and double salts, if official, are mentioned in the notes; the small figures placed near the corresponding neutral or simple salts refer to these.

Cyanogen, chloric acid, citric acid, malic acid, benzoic acid, succinic acid, and formic acid, are of more common occurrence in combination with a few bases only, and have therefore been omitted from the table. The most frequently occurring compounds of these substances are: cyanide of potassium I, ferrocyanide of potassium I, ferricyanide of potassium I, sesqui-ferricyanide of iron (Prussian blue) III, ferrocyanide of zinc and potassium II—III, chlorate of potassa I, the citrates of the alkalies I, the malates of the alkalies I, malate of sesquioxide of iron I, the benzoates of the alkalies I, the succinates of the alkalies I, and the formates of the alkalies I.

INDEX OF THE SOLUBILITY OF

	KO	NaO	NH_4O	BaO	SrO	CaO	MgO	Al_2O_3	MnO	FeO	Fe_2O_3	CoO	NiO	ZnO
	I	I	I	I	1	I-II	II	II	2	2	II	II	II	II
S	I	I	I	I	1	I-II	2		II	II	2	2_{15}	2_{15}	2_{17}
Cl	I	I	I_{12}	I	I	I	1	1	I	I	I_{12}	I	I	1
I	I	1	1	1	1	1	1		1	I	1			1
SO_3	I_1	I	I_{13}	III	III	I-III	I	$I_{1\cdot 13}$	I	I	I	I	I	I
NO_5	I	I	I	I	I	1	1	1	1	1	1	I	1	1
PO_5	1	I_{10}	1_{10}	2	2	II_{14}	2	2	2	2	II	2	2	2
CO_2	I_2	I_{11}	I	II	II	II	II		II	II		II	2	II
$\bar{O}$	I_3	1	I	2	2	II	2	2	2	1-2	1-2	2	2	2
BO_3	1_4	I_4	1	2	2	2	2	2	2	2	2	2	2	2
$\bar{A}$	I	I	I	I	1	I	1	1	1	1	I	1	1	I
$\bar{T}$	$I_{4\cdot 9}$	I_7	1_6	2	2	II	1-2	1	1-2	1-2	I_8	1	2	2
AsO_5	I	I	1	2	2	2	2	2	2	2	2	2	2	2
AsO_3	I	1	1	2	2	2	2			2	2	2	2	
CrO_3	I	1	1	2	2	2	1	2	1		1		2	1

NOTES.

1. SULPHATE of potassa and alumina I.
2. Bicarbonate of potassa I.
3. Binoxalate of potassa I.
4. Tartarized borax (bitartrate of potassa and borate of soda) I.
5. Bitartrate of potassa I-II.
6. Tartrate of potassa and ammonia I.
7. Tartrate of potassa and soda I.
8. Tartrate of potassa and sesquioxide of iron I.
9. Tartrate of antimony and potassa I.
10. Phosphate of soda and ammonia I.
11. Bicarbonate of soda I.
12. Sesquichloride of iron and chloride of ammonium I.
13. Sulphate of alumina and ammonia I.
14. Basic phosphate of lime II.
15. Sulphide of cobalt is pretty readily decomposed by nitric acid, bu very difficultly by hydrochloric acid. This substance is no officinal.

SUBSTANCES IN WATER OR ACIDS.

	CdO	PbO	SnO	SnO₂	BiO₃	CuO	Hg₂O	HgO	AgO	PtO₂	AuO₃	SbO₃	Cr₂O₃
	2	II₁₈	2	2 & 3	2	II₂₃	II	II	2	2		II₃₅	II & III
S	2	II	2₂₀	2₂₀	2	2₂₃	II	II	2₃₀	2₃₁		II₃₆	
Cl	1	I-III	I	I	I	I₂₄	II-III	I₂₆	III	I₃₂·₃₃	I₃₄	I₃₇	I & III
I	1	I-II	1	1			II	II	3				
SO₃	I	II-III	1		1	I₂₅	1-2	1₂₉	I-II	1		2	I-II
NO₅	1	I		1	I₂₁	I	I₂₇	I	I	1			I
PO₅	2	2	2			2	2	2	2				2
CO₂	2	II			2	II	2	2	2				
Ō	2	2	2	1	2	2	2	2	2			1-2	1
BO₃	1-2	2	2		2	2	1						2
A	1	I₁₉	1	1	1	I₃₆	1-2	1	1				1
T	1-2	2	1-2		2	1	1-2	2	2			I₃₃	1
AsO₃		2			2·	2	2	2	2			2	2
AsO₅		2				II	2	2	2			2	
CrO₃		II-III	2		2	1	2	1-2	2			2	2

16. The same applies to sulphide of nickel.
17. Sulphide of zinc is readily soluble in nitric acid, somewhat more sparingly soluble in hydrochloric acid.
18. Minium is converted by hydrochloric acid into chloride of lead; by nitric acid into oxide, which redissolves in an excess of the acid, and into brown binoxide of lead, which is insoluble in nitric acid.
19. Trisacetate of lead I.
20. Proto- and bisulphide of tin are decomposed and dissolved by hydrochloric acid; by nitric acid they are converted into binoxide, which is insoluble in an excess of the acid. Sublimed bisulphide of tin dissolves only in nitrohydrochloric acid.
21. Basic nitrate of teroxide of bismuth II.
22. Ammoniated oxide of copper 1.
23. Sulphide of copper is difficultly decomposed by hydrochloric acid, but with facility by nitric acid.
24. Chloride of copper and ammonium I.
25. Sulphate of copper and ammonia I.

26. Basic acetate of copper, partially soluble in water, and completely in acids.
27. Basic nitrate of suboxide of mercury and ammonia II.
28. Ammonio-chloride of mercury II.
29. Basic sulphate of oxide of mercury II.
30. Sulphide of silver soluble only in nitric acid.
31. Bisulphide of platinum is not affected by hydrochloric acid, and but little by boiling nitric acid; it dissolves in hot nitrohydrochloric acid.
32. Bichloride of platinum and chloride of potassium 1—3.
33. Bichloride of platinum and chloride of ammonium 1—3.
34. Terchloride of gold and chloride of sodium I.
35. Teroxide of antimony is soluble in hydrochloric acid, but not in nitric acid.
36. Tersulphide of antimony and sulphide of calcium I—II.
37. Basic terchloride of antimony II.
38. Tartrate of teroxide of antimony and potassa I.

Note to Page 111.

For a full account of THALLIUM, the Editor refers to the original papers of its discoverer, Mr. CROOKES, published in the *Chemical News*, Philosophical Transactions, and Proceedings during the years 1861, 2, and 3.

V.

TABLE OF WEIGHTS AND MEASURES.

GRAMMES.		GRAINS.	DECIGRAMMES.		GRAINS.
1	=	15·4323	1	=	1·5432
2	...	30·8646	2	...	3·0864
3	...	46·2969	3	...	4·6296
4	...	61·7292	4	...	6·1728
5	...	77·1615	5	...	7·7160
6	...	92·5938	6	...	9·2592
7	...	108·0261	7	...	10·8024
8	...	123·4584	8	...	12·3456
9	...	138·8907	9	...	13·8888

CENTIGRAMMES.		GRAINS.	MILLIGRAMMES.		GRAINS.
1	=	·1543	1	=	·0154
2	...	·3086	2	...	·0308
3	...	·4630	3	...	·0463
4	...	·6173	4	...	·0617
5	...	·7717	5	...	·0771
6	...	·9260	6	...	·0926
7	...	1·0804	7	...	·1080
8	...	1·2347	8	...	·1234
9	...	1·3891	9	...	·1389

METRES.		INCHES.	DECIMETRES.		INCHES.
1	=	39·37	1	=	3·937
2	...	78·74	2	...	7·874
3	...	118·11	3	...	11·811
4	...	157·48	4	...	15·748
5	...	196·85	5	...	19·685
6	...	236·22	6	...	23·622
7	...	275·59	7	...	27·559
8	...	314·96	8	...	31·496
9	...	354·33	9	...	35·433

CENTIMETRES.		INCHES.	MILLIMETRES.		INCHES.
1	=	·3937	1	=	·03937
2	...	·7874	2	...	·07874
3	...	1·1811	3	...	·11811
4	...	1·5748	4	...	·15748
5	...	1·9685	5	...	·19685
6	...	2·3622	6	...	·23622
7	...	2·7559	7	...	·27559
8	...	3·1496	8	...	·31496
9	...	3·5433	9	...	·35433

One kilogramme	=	15432 grains.
One cubic centimetre	=	0·0610 cubic inch.
One litre	=	61·0270 cubic inches.

ALPHABETICAL INDEX.

A.

	PAGE
Acetic acid (as reagent)	33
deportment with reagents	193
detection of, in simple compounds	223
in complex compounds	250, 252
Acids, as reagents	30
Actual examination	214
Alcohol (as reagent)	30
Alkaloids, detection of	329
in presence of coloring and extractive vegetable or animal matter	332
Alkaline solutions, examination of	229
Alloys, examination of	209, 213
Alumina, deportment with reagents	90
detection of, in soluble simple compounds	218, 224
in soluble complex compounds	239, 240, 242
in insoluble complex compounds	253
phosphate (see phosphate of alumina).	
Ammonia (as reagent)	44
deportment with reagents	78
detection of, in simple compounds	219
in complex compounds	246
in soils	276
in fresh waters	265
in mineral waters	267
carbonate of (as reagent)	52
molybdate of (as reagent)	54
oxalate of (as reagent)	50
Antimonic acid, detection of	309
Antimony, detection of, in alloys	213
properties of	131
teroxide of, detection of, in simple compounds	215
in complex compounds	235
in sinter deposits	272
in food, &c.	289

	PAGE
Antimony, teroxide of, deportment with reagents	131
Apocrenic acid, detection of, in soils	277
in mineral waters	273
Apparatus and utensils	25
Arsenic, properties of	134
acid, deportment with reagents	143
produced from arsenious acid	140
the tersulphide	140
Arsenious acid, deportment with reagents	134
Arsenious and arsenic acids, detection of, in simple compounds	215
in complex compounds	234, 255
in mineral waters	266
in food, &c.	285
in sinter deposits	272
Arsenious from arsenic acid, how to distinguish	147
Ashes of plants, animals, manures, &c., examination of	297

B.

	PAGE
Baryta, deportment of, with reagents	82
detection of, in soluble simple compounds	219
in insoluble simple compounds	228
in soluble complex compounds	244
in insoluble complex compounds	252
in mineral waters	269
in sinter deposits	273
carbonate of (as reagent)	59
hydrate of (as reagent)	68
nitrate of (as reagent)	58
water (as reagent)	45
Bases (as reagents)	42
Beaker glasses	26
Benzoic acid, detection of, in simple compounds	222
in complex compounds	252
deportment with reagents	192

ALPHABETICAL INDEX.

	PAGE
Berylla, deportment with reagents	74
detection of	311, 312
Bismuth, detection of, in alloys	209
in articles of food, &c.	289
properties of	119
teroxide, deportment of, with reagents	119
detection of, in simple compounds	216
in complex compounds	236, 237
hydrated (as reagent)	47
Blowpipe	13
flame	14, 15
Boracic acid, deportment with reagents	160
detection of, in simple compounds 221, 224, 225	
in complex compounds	244, 249
in silicates	259, 261
in mineral waters	270
Borax (as reagent)	71
Bromic acid, detection of	310
Bromine, properties and deportment with reagents	172
detection of 221, 225, 227, 248	
in mineral waters	270
Brucia, deportment with reagents	326
detection of, in simple compounds 329, 330, 331	
in complex compounds	332
Butyric acid, deportment with reagents	196

C.

	PAGE
Cadmium, properties of	121
oxide, detection of, in simple compounds	215
in complex compounds	236, 237
deportment with reagents	121
Cæsium, oxide, deportment with reagents	80
detection of	271
Carbon, detection of, in compound bodies	227, 253
in silicates	259
properties of	167
Carbonic acid, deportment with reagents	167
detection of, in simple compounds	219
in complex compounds	230
in soils	275, 276
in well and mineral waters	264, 265, 266
Cerium, oxides, deportment with reagents	95
detection of	312
Charcoal for blowpipe experiments	15

	PAGE
Chloric acid, detection of	221, 248
deportment with reagents	184
Chloride of ammonium (as reagent)	55
of barium (as reagent)	58
of calcium (as reagent)	59
of mercury (as reagent)	63
Chlorine (as reagent)	35
properties and deportment with reagents	171
detection of, in soluble simple compounds	221, 225
in insoluble simple compounds	227
in soluble complex compounds	248
in insoluble complex compounds	254
in soils	275
in fresh and mineral waters	264
in silicates	260, 261
Chloroform (as reagent)	30
Chlorous acid, deportment with reagents	182
detection of	310
Chrome-ironstone, analysis of	255
Chromic acid, deportment with reagents	152
detection of, in simple compounds	219
in complex compounds	249
in insoluble compounds	255
Chromium, sesquioxide, deportment with reagents	92
detection of, in soluble simple compounds	217, 218
in complex compounds	240, 243
Cinchonia, deportment with reagents	322
detection of, in simple compounds	330
in complex compounds	332
Citric acid, deportment with reagents	187
detection of, in simple compounds	222
in complex compounds	250
Cobalt, properties of	104
protoxide, deportment with reagents	104
detection of, in simple compounds	217
in complex compounds	241, 242
nitrate (as reagent)	72
Coloration of flame	21
Conia, deportment with reagents	317
Copper (as reagent)	47
properties of	117
oxide, deportment with reagents	117
detection of, in simple compounds	216
in complex compounds	236, 237

ALPHABETICAL INDEX. 351

Copper, oxide, detection of, in sinter deposits . 272
— sulphate (as reagent) . . 62
Crenic acid, detection of, in soils . 277
— in mineral waters 273
Crystallization 5
Cyanide of potassium (as reagent) . 55
— in the moist way . 55
— in the dry way . 70
Cyanides, insoluble in water, analysis of 256
Cyanogen, detection of, in simple compounds 220, 227
— in complex compounds 231, 248
— properties of . . . 176

D.

Decantation 8
Deflagration 12
Dialysis 282
Didymium, oxide, deportment with reagents 96
— detection of . . 311
Distillation 10
Distilling apparatus . . . 10

E.

Edulcoration 8
Erbium, oxide, deportment with reagents . 95
— detection of . . . 311, 312
Ether (as reagent) 30
Evaporation 9

F.

Ferricyanide of potassium (as reagent) 57
Ferricyanogen, detection of, in simple compounds . 220
— in complex compounds 248, 256, 257
Ferrocyanide of potassium (as reagent) 56
Ferrocyanogen, detection of, in simple compounds . 220
— in complex compounds 248, 256, 257
Filtering paper 7
— stands 7
Filtration 7
Flame, coloration of . . . 21
— parts of 13
Fluoride of calcium (as reagent) . 69
Fluorine, detection of, in simple compounds 220, 224, 225, 227
— in complex compounds 244, 252
— in insoluble compounds 255
— in mineral waters 268
— in sinter deposits 273
— in silicates 260, 261
Fluxing 11
Formic acid, deportment with reagents 194

Formic acid, detection of, in simple compounds . 223
— in complex compounds 250
Funnels 7, 26
Fusion 11

G.

Gas-lamp 18, 25
Geic acid, detection of, in soils . 277
Georgina paper 66
Gold, properties of . . . 125
— detection of, in alloys . 213
— terchloride of (as reagent) . 65
— teroxide, deportment with reagents . . 125
— detection of, in simple compounds 216
— in complex compounds 235

H.

Halogens (as reagents) . . . 31
Humic acid, detection of, in soils . 277
Hydriodic acid, deportment with reagents . 174
Hydrobromic acid, deportment with reagents . 172
Hydrochloric acid (as reagent) . 34
— deportment with reagents . 171
Hydrocyanic acid, deportment with reagents . 176
— in organic matters . 290
Hydroferricyanic acid, deportment with reagents . 178
Hydroferrocyanic acid, deportment with reagents . 178
Hydrofluoric acid, properties and deportment with reagents . 163
— detection of . 249
Hydrofluosilicic acid (as reagent) . 37
— deportment with reagents . 156
Hydrogen acids (as reagents) . 34
Hydrosulphuric acid (as reagent) . 37
— deportment with reagents . 178
— detection of, in simple compounds 219
— in complex compounds 248
— in mineral waters 267
Hypochlorous acid, deportment with reagents . 182
— detection of . 310
Hyponiobic acid, deportment with reagents . 98
— detection of . 309
Hypophosphorous acid, deportment with reagents . 182
Hyposulphurous acid, deportment with reagents . 154
— detection of . 309, 310

ALPHABETICAL INDEX.

I.

	PAGE
Ignition	10
Indigo solution (as reagent)	67
Inorganic bodies, detection of, in presence of organic bodies	278
Iodic acid, deportment with reagents	154
detection of	310
Iodine, detection of, in simple compounds	220, 225, 227
in complex compounds	248
in mineral waters	277
properties of	174
Iron (as reagent)	47
properties of	105
protoxide, deportment with reagents	105
detection of, in simple compounds	217
in complex compounds	242
in well and mineral waters	264, 267, 268
sulphate of protoxide (as reagent)	60
sesquichloride (as reagent)	61
sesquioxide, deportment with reagents	107
detection of, in simple compounds	215
in complex compounds	231, 242
in soils	276
in well and mineral waters	264, 268
Iridium, oxide, deportment with reagents	147
detection	310, 312

L.

Lactic acid, deportment with reagents	196
Lamps, use of	17
Lanthanium, oxide, deportment with reagents	96
detection of	311
Lead, properties of, and deportment of oxide with reagents	114
oxide, detection of, in soluble simple compounds	214, 216
in insoluble simple compounds	227
in soluble complex compounds	229, 237
in insoluble complex compounds	254
in organic matters	289
in sinter deposits	272
acetate (as reagent)	62
Lime, deportment with reagents	85
detection of, in soluble simple compounds	218
in soluble complex compounds	242, 243, 245
Lime, detection of, in insoluble simple compounds	227, 228
in insoluble complex compounds	252
in soils	276
in well and mineral waters	264
sulphate (as reagent)	59
water (as reagent)	46
Lithia, deportment with reagents	80
detection of, in mineral waters	270
Litmus-paper	65

M.

Magnesia, deportment with reagents	86
detection of, in simple compounds	219
in complex compounds	245
in soils	276
in well and mineral waters	264, 265
sulphate of (as reagent)	60
Malic acid, detection of, in simple compounds	222
in complex compounds	250
deportment with reagents	188
Manganese, properties of	101
protoxide, detection of, in simple compounds	217
in complex compounds	240, 243
in soils	275, 276
in mineral waters	269
protoxide, deportment with reagents	101
Marsh's apparatus	138
Mercury, detection of, in articles of food, &c.	289
properties of	113
chloride (as reagent)	63
oxide, deportment with reagents	116
detection of, in soluble simple compounds	216
oxide, detection of, in soluble complex compounds	237
suboxide, deportment with reagents	113
detection of, in simple compounds	24
in complex compounds	229
nitrate of, (as reagent)	62
Metallic poisons, detection of, in articles of food, &c.	279
Metals (as reagents)	12
Mineral waters, analysis of	266
Molybdenum, deportment of oxide of, with reagents	148
detection of	309, 312

ALPHABETICAL INDEX. 353

Morphia, deportment with reagents . 318
   detection of, in simple compounds 329, 330
   in complex compounds . 331

N.

Narcotina, deportment with reagents . 320
   detection of, in simple compounds 329, 330
   in complex compounds . 332
Nickel, properties of . . . 102
   protoxide, deportment with reagents . . 102
   detection of, in simple compounds . 217
   in complex compounds, 241, 242
Nicotina, deportment with reagents . 316
Niobic acid, detection . . . 312
Nitric acid (as reagent) . . . 33
   deportment with reagents . 183
   detection of, in simple compounds . 221
   in complex compounds . 248
   in soils . . 275
   in well and mineral waters, 264, 270
Nitrohydrochloric acid (as reagent) . 36
Nitrous acid, deportment with reagents 181
   detection of . . . 310
   in fresh waters 265
   in mineral waters 271

O.

Osmium, oxides, deportment with reagents 123
   detection of 310, 311, 312
Oxalic acid, properties of . . . 162
   deportment with reagents . 162
   detection of, in simple compounds 220, 224, 225
   in complex compounds 243, 249, 252
Oxidizing flame 14
Oxygen acids (as reagents) . . 31
   bases (as reagents) . . . 42

P.

Palladium, properties of . . . 123
   protoxide of, deportment with reagents . . 123
   detection of . . . 310
   sodio-chloride as reagent . 65
Paratartaric acid, deportment with reagents 190
Perchloric acid, deportment with reagents 186
Phosphate of soda and ammonia (as reagent) 71
Phosphates of alkaline earths, detection of, in simple compounds . 224

Phosphates in complex compounds . 241
Phosphoric acid, deportment with reagents . . 156
   monobasic . . 160
   bibasic . . . 160
   detection of, in simple compounds, 220, 224, 225
   in complex compounds, 243, 248, 251, 255
   in soils 275, 276
   in mineral waters, 264, 268
   in silicates, 259, 261, 262
   in articles of food, &c. . 292
Phosphorous acid, deportment with reagents 167
Phosphorus, properties of . . 156
Pincers 26
Platinum, detection of, in alloys . 213
   properties of . . . 126
   bichloride of (as reagent) . 64
   binoxide of, deportment with reagents . 126
   detection of in simple compounds 216
   in complex compounds 235
   crucibles and their use 11, 25
   foil and wire . . 16, 25
Porcelain dishes and crucibles . . 26
Potassa (as reagent) 42
   antimonate (as reagent) . . 54
   bichromate (as reagent) . . 53
   nitrite (as reagent) . . 53
   sulphate (as reagent) . . 50
   deportment with reagents . 75
   detection of, in simple compounds . . 219
   in complex compounds . 246
   in well and mineral waters . 265
   in silicates . 261
   in soils . . 276
Potassium, ferricyanide of (as reagent) 57
   ferrocyanide of (as reagent) 56
   sulphocyanide of (as reagent) 57
Precipitation 6
Preliminary examination of solid bodies 204
   of fluids . 209
Propionic acid, deportment with reagents 196

Q.

Quina, detection of, in simple compounds 331
Quina, detection of, in complex compounds . 332
   deportment with reagents . 321

R.

Racemic acid, deportment with reagents 190

A A

LONDON:
SAVILL AND EDWARDS, PRINTERS, CHANDOS STREET,
COVENT GARDEN.

London, New Burlington Street,
June, 1867.

MESSRS. CHURCHILL & SONS'

Publications,

IN

MEDICINE

AND THE VARIOUS BRANCHES OF

NATURAL SCIENCE.

"It would be unjust to conclude this notice without saying a few words in favour of Mr. Churchill, from whom the profession is receiving, it may be truly said, the most beautiful series of Illustrated Medical Works which has ever been published."—*Lancet.*

"All the publications of Mr. Churchill are prepared with so much taste and neatness, that it is superfluous to speak of them in terms of commendation."—*Edinburgh Medical and Surgical Journal.*

"No one is more distinguished for the elegance and *recherché* style of his publications than Mr. Churchill."—*Provincial Medical Journal.*

"Mr. Churchill's publications are very handsomely got up: the engravings are remarkably well executed."—*Dublin Medical Press.*

"The typography, illustrations, and getting up are, in all Mr. Churchill's publications, most beautiful."—*Monthly Journal of Medical Science.*

"Mr. Churchill's illustrated works are among the best that emanate from the Medical Press."—*Medical Times.*

"We have before called the attention of both students and practitioners to the great advantage which Mr. Churchill has conferred on the profession, in the issue, at such a moderate cost, of works so highly creditable in point of artistic execution and scientific merit."—*Dublin Quarterly Journal.*

Messrs. Churchill & Sons are the Publishers of the following Periodicals, offering to Authors a wide extent of Literary Announcement, and a Medium of Advertisement, addressed to all Classes of the Profession.

THE BRITISH AND FOREIGN MEDICO-CHIRURGICAL REVIEW,
AND
QUARTERLY JOURNAL OF PRACTICAL MEDICINE AND SURGERY.
Price Six Shillings. Nos. I. to LXXVIII.

THE QUARTERLY JOURNAL OF SCIENCE.
Price Five Shillings. Nos. I. to XIV.

THE QUARTERLY JOURNAL OF MICROSCOPICAL SCIENCE,
INCLUDING THE TRANSACTIONS OF THE ROYAL MICROSCOPICAL SOCIETY OF LONDON.
Edited by Dr. LANKESTER, F.R.S., and GEORGE BUSK, F.R.S. Price 4s. Nos. I. to XXVI. *New Series.*

THE JOURNAL OF MENTAL SCIENCE.
By authority of the Medico-Psychological Association.
Edited by C. L. ROBERTSON, M.D., and HENRY MAUDSLEY, M.D.
Published Quarterly, price Half-a-Crown. *New Series.* Nos. I. to XXV.

JOURNAL OF CUTANEOUS MEDICINE.
Edited by ERASMUS WILSON, F.R.S.
Published Quarterly, price 2s. 6d. No. I.

ARCHIVES OF MEDICINE:
A Record of Practical Observations and Anatomical and Chemical Researches, connected with the Investigation and Treatment of Disease. Edited by Dr. LIONEL S. BEALE, F.R.S. Published Quarterly; Nos. I. to VIII., 3s. 6d.; IX. to XII., 2s. 6d., XIII. to XVI., 3s.

ARCHIVES OF DENTISTRY:
Edited by EDWIN TRUMAN. Published Quarterly, price 4s. Nos. I. to IV.

THE YEAR-BOOK OF PHARMACY
AND
CHEMISTS' DESK COMPANION FOR 1866.
BEING A PRACTICAL SUMMARY OF RESEARCHES IN PHARMACY, MATERIA MEDICA, AND PHARMACEUTICAL CHEMISTRY, DURING THE YEAR 1865.
Edited by CHARLES H. WOOD, F.C.S., and CHAS. SHARP. Price 2s. 6d.

THE ROYAL LONDON OPHTHALMIC HOSPITAL REPORTS, AND JOURNAL OF OPHTHALMIC MEDICINE AND SURGERY.
Vol. V., Part 4, 2s. 6d.

THE MEDICAL TIMES & GAZETTE.
Published Weekly, price Sixpence, or Stamped, Sevenpence.
Annual Subscription, £1. 6s., or Stamped, £1. 10s. 4d., and regularly forwarded to all parts of the Kingdom.

THE HALF-YEARLY ABSTRACT OF THE MEDICAL SCIENCES.
Being a Digest of the Contents of the principal British and Continental Medical Works; together with a Critical Report of the Progress of Medicine and the Collateral Sciences. Post 8vo. cloth, 6s. 6d. Vols. I. to XLIV.

THE PHARMACEUTICAL JOURNAL,
CONTAINING THE TRANSACTIONS OF THE PHARMACEUTICAL SOCIETY.
Published Monthly, price One Shilling.
⁎ Vols. I. to XXVI., bound in cloth, price 12s. 6d. each.

THE BRITISH JOURNAL OF DENTAL SCIENCE.
Published Monthly, price One Shilling. Nos. I. to CXXXI.

THE MEDICAL DIRECTORY FOR THE UNITED KINGDOM.
Published Annually. 8vo. cloth, 10s. 6d.

ANNALS OF MILITARY AND NAVAL SURGERY AND TROPICAL MEDICINE AND HYGIENE,
Embracing the experience of the Medical Officers of Her Majesty's Armies and Fleets in all parts of the World.
Vol. I., price 7s.

A CLASSIFIED INDEX

TO

MESSRS. CHURCHILL & SONS' CATALOGUE.

ANATOMY.

	PAGE
Anatomical Remembrancer	7
Flower on Nerves	16
Hassall's Micros. Anatomy	19
Heale's Anatomy of the Lungs	19
Heath's Practical Anatomy	20
Holden's Human Osteology	20
Do. on Dissections	20
Huxley's Comparative Anatomy	21
Jones' and Sieveking's Pathological Anatomy	22
MacDougal—Hirschfeld on the Nervous System	v
Maclise's Surgical Anatomy	25
St. Bartholomew's Hospital Catalogue	31
Sibson's Medical Anatomy	32
Waters' Anatomy of Lung	37
Wheeler's Anatomy for Artists	38
Wilson's Anatomy	39

CHEMISTRY.

Bernays' Notes for Students	9
Bloxam's Chemistry	10
Bowman's Practical Chemistry	10
Do. Medical do.	10
Fownes' Manual of Chemistry	16
Do. Actonian Prize	16
Do. Qualitative Analysis	16
Fresenius' Chemical Analysis	16
Galloway's First Step	17
Do. Second Step	17
Do. Analysis	17
Do. Tables	17
Griffiths' Four Seasons	18
Horsley's Chem. Philosophy	21
Mulder on the Chemistry of Wine 27	
Plattner & Muspratt on Blowpipe	28
Speer's Pathol. Chemistry	33
Sutton's Volumetric Analysis	34

CLIMATE.

Aspinall on San Remo	7
Bennet's Winter in the South of Europe	9
Chambers on Italy	12
Dalrymple on Egypt	14
Francis on Change of Climate	16
Hall on Torquay	19
Haviland on Climate	19
Lee on Climate	24
Do. Watering Places of England	24
McClelland on Bengal	25
McNicoll on Southport	25
Martin on Tropical Climates	26
Moore's Diseases of India	26
Scoresby-Jackson's Climatology	31
Shapter on South Devon	32
Siordet on Mentone	32
Taylor on Pau and Pyrenees	34

DEFORMITIES, &c.

	PAGE
Adams on Spinal Curvature	6
Do. on Clubfoot	6
Bigg's Orthopraxy	10
Bishop on Deformities	10
Do. Articulate Sounds	10
Brodhurst on Spine	11
Do. on Clubfoot	11
Godfrey on Spine	17
Hugman on Hip Joint	21
Salt on Lower Extremities	31
Tamplin on Spine	34

DISEASES OF WOMEN AND CHILDREN.

Ballard on Infants and Mothers	7
Bennet on Uterus	9
Bird on Children	10
Bryant's Surg. Diseases of Child.	11
Eyre's Practical Remarks	15
Harrison on Children	19
Hood on Scarlet Fever, &c.	21
Kiwisch (ed. by Clay) on Ovaries	15
Lee's Ovarian & Uterine Diseases	24
Do. on Speculum	24
Ritchie on Ovaries	30
Seymour on Ovaria	32
Tilt on Uterine Inflammation	35
Do. Uterine Therapeutics	35
Do. on Change of Life	35
Underwood on Children	36
Wells on the Ovaries	39
West on Women	38
Do. (Uvedale) on Puerp. Diseases	38

GENERATIVE ORGANS, Diseases of, and SYPHILIS.

Acton on Reproductive Organs	6
Coote on Syphilis	14
Gant on Bladder	17
Hutchinson on Inherited Syphilis	22
Judd on Syphilis	23
Lee on Syphilis	24
Parker on Syphilis	27
Wilson on Syphilis	39

HYGIENE.

Armstrong on Naval Hygiene	7
Beale's Laws of Health	8
Do. Health and Disease	8

HYGIENE—continued.

	PAGE
Carter on Training	12
Chavasse's Advice to a Mother	13
Do. Advice to a Wife	13
Dobell's Germs and Vestiges of Disease	15
Fife & Urquhart on Turkish Bath	16
Gordon on Army Hygiene	17
Granville on Vichy	18
Hartwig on Sea Bathing	19
Do. Physical Education	19
Hufeland's Art of prolonging Life	21
Hunter on Body and Mind	21
Lee's Baths of France, Germany, and Switzerland	24
Moore's Health in Tropics	26
Parkes on Hygiene	27
Parkin on Disease	28
Pearse's Notes on Health	28
Pickford on Hygiene	29
Robertson on Diet	30
Routh on Infant Feeding	30
Wells' Seamen's Medicine Chest	38
Wife's Domain	38
Wilson on Healthy Skin	39
Do. on Mineral Waters	39
Do. on Turkish Bath	39

MATERIA MEDICA and PHARMACY.

Bateman's Magnacopia	8
Beasley's Formulary	9
Do. Receipt Book	9
Do. Book of Prescriptions	9
Frazer's Materia Medica	16
Nevins' Analysis of Pharmacop.	27
Pereira's Selecta è Præscriptis	28
Prescriber's Pharmacopœia	29
Royle's Materia Medica	31
Squire's Hospital Pharmacopœias	33
Do. Companion to the Pharmacopœia	33
Stoggall's First Lines for Chemists and Druggists	34
Stowe's Toxicological Chart	34
Taylor on Poisons	35
Waring's Therapeutics	37
Wittstein's Pharmacy	39

MEDICINE.

Adams on Rheumatic Gout	6
Addison on Cell Therapeutics	6
Do. on Healthy and Diseased Structure	6
Aldis's Hospital Practice	6
Anderson (Andrew) on Fever	7
Do. (Thos.) on Yellow Fever	7

a 2

MEDICINE—continued.

	PAGE
Austin on Paralysis	7
Barclay on Medical Diagnosis	8
Do. on Gout	8
Barlow's Practice of Medicine	8
Basham on Dropsy	8
Brinton on Stomach	11
Do. on Intestinal Obstruction	11
Budd on the Liver	11
Do. on Stomach	11
Camplin on Diabetes	12
Catlow on Æsthetic Medicine	12
Chambers on the Indigestions	12
Do. Lectures	12
Cockle on Cancer	13
Davey's Ganglionic Nervous Syst.	15
Day's Clinical Histories	15
Eyre on Stomach	15
Foster on the Sphygmograph	16
Fuller on Rheumatism	17
Gairdner on Gout	17
Gibb on Throat	17
Do. on Laryngoscope	17
Granville on Sudden Death	18
Griffith on the Skin	18
Gully's Simple Treatment	18
Habershon on the Abdomen	18
Do. on Mercury	18
Hall (Marshall) on Apnœa	18
Do. Observations	18
Headland—Action of Medicines	19
Do. Medical Handbook	19
Hooper's Physician's Vade-Mecum	18
Inman's New Theory	22
Do. Myalgia	22
James on Laryngoscope	22
Jones (Bence) on Pathology and Therapeutics	22
Maclachlan on Advanced Life	25
MacLeod on Acholic Diseases	25
Marcet on Chronic Alcoholism	25
Macpherson on Cholera	26
Markham on Bleeding	26
Meryon on Paralysis	26
Mushet on Apoplexy	27
Nicholson on Yellow Fever	27
Parkin on Cholera	28
Pavy on Diabetes	28
Peet's Principles and Practice of Medicine	28
Roberts on Palsy	30
Robertson on Gout	30
Sansom on Cholera	31
Savory's Compendium	31
Semple on Cough	32
Seymour on Dropsy	32
Shaw's Remembrancer	32
Shrimpton on Cholera	32
Smee on Debility	32
Thomas' Practice of Physic	35
Thudichum on Gall Stones	35
Todd's Clinical Lectures	36
Tweedie on Continued Fevers	36
Walker on Diphtheria	37
What to Observe at the Bedside	25
Williams' Principles	38
Wright on Headaches	39

MICROSCOPE.

Beale on Microscope in Medicine	8
Carpenter on Microscope	12
Schacht on do.	31

MISCELLANEOUS.

Acton on Prostitution	6
Barclay's Medical Errors	8

MISCELLANEOUS—cont.d.

	PAGE
Barker & Edwards' Photographs	8
Bascome on Epidemics	8
Buckle's Hospital Statistics	11
Cooley's Cyclopædia	13
Gordon on China	17
Graves' Physiology and Medicine	17
Guy's Hospital Reports	17
Harrison on Lead in Water	19
Hingeston's Topics of the Day	20
Howe on Epidemics	21
Lane's Hydropathy	23
Lee on Homœop. and Hydrop.	24
London Hospital Reports	25
Marcet on Food	25
Massy on Recruits	26
Mayne's Medical Vocabulary	26
Oppert on Hospitals	27
Part's Case Book	28
Redwood's Supplement to Pharmacopœia	30
Ryan on Infanticide	31
St. George's Hospital Reports	31
Simms' Winter in Paris	32
Snow on Chloroform	33
Steggall's Medical Manual	34
Do. Gregory's Conspectus	34
Do. Celsus	34
Waring's Tropical Resident at Home	37
Whitehead on Transmission	38

NERVOUS DISORDERS AND INDIGESTION.

Althaus on Epilepsy, Hysteria	7
Birch on Constipation	10
Carter on Hysteria	12
Downing on Neuralgia	15
Hunt on Heartburn	21
Jones (Handfield) on Functional Nervous Disorders	22
Leared on Imperfect Digestion	23
Lobb on Nervous Affections	24
Radcliffe on Epilepsy	29
Reynolds on the Brain	30
Do. on Epilepsy	30
Rowe on Nervous Diseases	31
Sieveking on Epilepsy	32
Turnbull on Stomach	36

OBSTETRICS.

Barnes on Placenta Prævia	8
Hodges on Puerperal Convulsions	20
Lee's Clinical Midwifery	24
Do. Consultations	24
Leishman's Mechanism of Parturition	24
Pretty's Aids during Labour	29
Priestley on Gravid Uterus	29
Ramsbotham's Obstetrics	29
Do. Midwifery	30
Sinclair & Johnston's Midwifery	32
Smellie's Obstetric Plates	33
Smith's Manual of Obstetrics	33
Swayne's Aphorisms	34
Waller's Midwifery	37

OPHTHALMOLOGY.

Cooper on Injuries of Eye	13
Do. on Near Sight	13
Dalrymple on Eye	14

OPHTHALMOLOGY—cont.d.

	PAGE
Dixon on the Eye	15
Hogg on Ophthalmoscope	20
Hulke on the Ophthalmoscope	21
Jago on Entoptics	22
Jones' Ophthalmic Medicine	23
Do. Defects of Sight	23
Do. Eye and Ear	23
Macnamara on the Eye	25
Nunneley on the Organs of Vision	27
Solomon on Glaucoma	33
Walton on the Eye	37
Wells on Spectacles	37

PHYSIOLOGY.

Carpenter's Human	12
Do. Manual	12
Heale on Vital Causes	19
Richardson on Coagulation	30
Shea's Animal Physiology	32
Virchow's (ed. by Chance) Cellular Pathology	12

PSYCHOLOGY.

Arlidge on the State of Lunacy	7
Bucknill and Tuke's Psychological Medicine	11
Conolly on Asylums	13
Davey on Nature of Insanity	15
Dunn on Psychology	15
Hood on Criminal Lunatics	21
Millingen on Treatment of Insane	26
Murray on Emotional Diseases	27
Noble on Mind	27
Sankey on Mental Diseases	31
Williams (J. H.) Unsoundness of Mind	38

PULMONARY and CHEST DISEASES, &c.

Alison on Pulmonary Consumption	6
Barker on the Lungs	8
Billing on Lungs and Heart	10
Bright on the Chest	11
Cotton on Consumption	14
Do. on Stethoscope	14
Davies on Lungs and Heart	15
Dobell on the Chest	15
Do. on Tuberculosis	15
Do. on Winter Cough	15
Do. First Stage of Consumption	15
Fenwick on Consumption	16
Fuller on the Lungs	17
Do. on Heart	17
Jones (Jas.) on Consumption	22
Laennec on Auscultation	23
Markham on Heart	26
Peacock on the Heart	28
Richardson on Consumption	30
Salter on Asthma	31
Skoda on Auscultation	26
Thompson on Consumption	35
Timms on Consumption	35
Turnbull on Consumption	36
Waters on Emphysema	37
Weber on Auscultation	37

CLASSIFIED INDEX.

RENAL and URINARY DISEASES.

	PAGE
Acton on Urinary Organs	6
Beale on Urine	8
Bird's Urinary Deposits	10
Coulson on Bladder	14
Hassall on Urine	19
Parkes on Urine	27
Thudichum on Urine	35
Todd on Urinary Organs	36

SCIENCE.

	PAGE
Baxter on Organic Polarity	8
Bentley's Manual of Botany	9
Bird's Natural Philosophy	10
Craig on Electric Tension	14
Hardwich's Photography	19
Hinds' Harmonies	20
Howard on the Clouds	21
Jones on Vision	23
Do. on Body, Sense, and Mind	23
Mayne's Lexicon	26
Noad on the Inductorium	27
Pratt's Genealogy of Creation	29
Do. Eccentric & Centric Force	29
Do. on Orbital Motion	29
Do. Astronomical Investigations	29
Do. Oracles of God	29
Rainey on Shells	29
Reymond's Animal Electricity	30
Taylor's Medical Jurisprudence	34
Unger's Botanical Letters	36
Vestiges of Creation	36

SURGERY.

	PAGE
Adams on Reparation of Tendons	6
Do. Subcutaneous Surgery	6
Anderson on the Skin	7
Ashton on Rectum	7
Brodhurst on Anchylosis	11
Bryant on Diseases of Joints	11
Do. Clinical Surgery	11
Callender on Rupture	12
Chapman on Ulcers	12
Do. Varicose Veins	12
Clark's Outlines of Surgery	13
Collis on Cancer	13
Cooper (Sir A.) on Testis	14
Do. (S.) Surg. Dictionary	14
Coulson on Lithotomy	14
Curling on Rectum	14
Do. on Testis	14
Druitt's Surgeon's Vade-Mecum	15
Fayrer's Clinical Surgery	16
Fergusson's Surgery	16
Gamgee's Amputation at Hip-joint	17
Gant's Principles of Surgery	17
Heath's Minor Surgery and Bandaging	20
Higginbottom on Nitrate of Silver	20
Hodgson on Prostate	20
Holt on Stricture	21
James on Hernia	22
Jordan's Clinical Surgery	23
Lawrence's Surgery	23
Do. Ruptures	23
Lee on the Rectum, &c.	24
Liston's Surgery	24
Logan on Skin Diseases	24
Macleod's Surgical Diagnosis	25
Do. Surgery of the Crimea	25

SURGERY—continued.

	PAGE
Maclise on Fractures	25
Maunder's Operative Surgery	26
Nayler on Skin Diseases	27
Nunneley on Erysipelas	27
Pirrie's Surgery	28
Pirrie & Keith on Acupressure	28
Price on Excision of Knee-Joint	29
Salt on Rupture	31
Sansom on Chloroform	31
Smith (Hy.) on Stricture	33
Do. on Hæmorrhoids	33
Do. on the Surgery of the Rectum	33
Do. (Dr. J.) Dental Anatomy and Surgery	33
Squire on Skin Diseases	33
Steggall's Surgical Manual	34
Teale on Amputation	35
Thompson on Stricture	35
Do. on Prostate	35
Do. Lithotomy and Lithotrity	35
Tomes' Dental Surgery	36
Wade on Stricture	36
Webb's Surgeon's Ready Rules	37
Wilson on Skin Diseases	39
Do. Portraits of Skin Diseases	39
Yearsley on Deafness	39
Do. on Throat	39

VETERINARY MEDICINE.

	PAGE
Blaine's Veterinary Art	10
Bourguignon on the Cattle Plague	10

TO BE COMPLETED IN TWELVE PARTS, 4TO., AT 7s. 6d. PER PART.

PARTS I. & II. NOW READY.

A DESCRIPTIVE TREATISE
ON THE
NERVOUS SYSTEM OF MAN,

WITH THE MANNER OF DISSECTING IT.

By LUDOVIC HIRSCHFELD,

DOCTOR OF MEDICINE OF THE UNIVERSITIES OF PARIS AND WARSAW, PROFESSOR OF ANATOMY TO THE FACULTY OF MEDICINE OF WARSAW;

Edited in English (from the French Edition of 1866)

By ALEXANDER MASON MACDOUGAL, F.R.C.S.,

WITH

AN ATLAS OF ARTISTICALLY-COLOURED ILLUSTRATIONS,

Embracing the Anatomy of the entire Cerebro-Spinal and Sympathetic Nervous Centres and Distributions in their accurate relations with all the important Constituent Parts of the Human Economy, and embodied in a series of 56 Single and 9 Double Plates, comprising 197 Illustrations,

Designed from Dissections prepared by the Author, and Drawn on Stone by
J. B. LÉVEILLÉ.

MESSRS. CHURCHILL & SONS' PUBLICATIONS.

MR. ACTON, M.R.C.S.

I.
A PRACTICAL TREATISE ON DISEASES OF THE URINARY AND GENERATIVE ORGANS IN BOTH SEXES. Third Edition. 8vo. cloth, £1. 1s. With Plates, £1. 11s. 6d. The Plates alone, limp cloth, 10s. 6d.

II.
THE FUNCTIONS AND DISORDERS OF THE REPRODUC-TIVE ORGANS IN CHILDHOOD, YOUTH, ADULT AGE, AND ADVANCED LIFE, considered in their Physiological, Social, and Moral Relations. Fourth Edition. 8vo. cloth, 10s. 6d.

III.
PROSTITUTION: Considered in its Moral, Social, and Sanitary Bearings, with a View to its Amelioration and Regulation. 8vo. cloth, 10s. 6d.

DR. ADAMS, A.M.

A TREATISE ON RHEUMATIC GOUT; OR, CHRONIC RHEUMATIC ARTHRITIS. 8vo. cloth, with a Quarto Atlas of Plates, 21s.

MR. WILLIAM ADAMS, F.R.C.S.

I.
ON THE PATHOLOGY AND TREATMENT OF LATERAL AND OTHER FORMS OF CURVATURE OF THE SPINE. With Plates. 8vo. cloth, 10s. 6d.

II.
CLUBFOOT: its Causes, Pathology, and Treatment. Jacksonian Prize Essay for 1864. With 100 Engravings. 8vo. cloth, 12s.

III.
ON THE REPARATIVE PROCESS IN HUMAN TENDONS AFTER SUBCUTANEOUS DIVISION FOR THE CURE OF DEFORMITIES. With Plates. 8vo. cloth, 6s.

IV.
SKETCH OF THE PRINCIPLES AND PRACTICE OF SUBCUTANEOUS SURGERY. 8vo. cloth, 2s. 6d.

DR. WILLIAM ADDISON, F.R.S.

I.
CELL THERAPEUTICS. 8vo. cloth, 4s.

II.
ON HEALTHY AND DISEASED STRUCTURE, AND THE TRUE PRINCIPLES OF TREATMENT FOR THE CURE OF DISEASE, ESPECIALLY CONSUMPTION AND SCROFULA, founded on MICROSCOPICAL ANALYSIS. 8vo. cloth, 12s.

DR. ALDIS:

AN INTRODUCTION TO HOSPITAL PRACTICE IN VARIOUS COMPLAINTS; with Remarks on their Pathology and Treatment. 8vo. cloth, 5s. 6d.

DR. SOMERVILLE SCOTT ALISON, M.D.EDIN., F.R.C.P.

THE PHYSICAL EXAMINATION OF THE CHEST IN PUL-MONARY CONSUMPTION, AND ITS INTERCURRENT DISEASES. With Engravings. 8vo. cloth, 12s.

DR. ALTHAUS, M.D., M.R.C.P.
ON EPILEPSY, HYSTERIA, AND ATAXY. Cr. 8vo. cloth, 4s.

THE ANATOMICAL REMEMBRANCER; OR, COMPLETE POCKET ANATOMIST. Sixth Edition, carefully Revised. 32mo. cloth, 3s. 6d.

DR. McCALL ANDERSON, M.D.
I.
PARASITIC AFFECTIONS OF THE SKIN. With Engravings. 8vo. cloth, 5s.
II.
ECZEMA. 8vo. cloth, 5s.
III.
PSORIASIS AND LEPRA. With Chromo-lithograph. 8vo. cloth, 5s.

DR. ANDREW ANDERSON, M.D.
TEN LECTURES INTRODUCTORY TO THE STUDY OF FEVER. Post 8vo. cloth, 5s.

DR. THOMAS ANDERSON, M.D.
HANDBOOK FOR YELLOW FEVER: ITS PATHOLOGY AND TREATMENT. To which is added a brief History of Cholera, and a method of Cure. Fcap. 8vo. cloth, 3s.

DR. ARLIDGE.
ON THE STATE OF LUNACY AND THE LEGAL PROVISION FOR THE INSANE; with Observations on the Construction and Organisation of Asylums. 8vo. cloth, 7s.

DR. ALEXANDER ARMSTRONG, R.N.
OBSERVATIONS ON NAVAL HYGIENE AND SCURVY. More particularly as the latter appeared during a Polar Voyage. 8vo. cloth, 5s.

MR. T. J. ASHTON.
I.
ON THE DISEASES, INJURIES, AND MALFORMATIONS OF THE RECTUM AND ANUS. Fourth Edition. 8vo. cloth, 8s.
II.
PROLAPSUS, FISTULA IN ANO, AND HÆMORRHOIDAL AFFECTIONS; their Pathology and Treatment. Second Edition. Post 8vo. cloth, 2s. 6d

MR. W. B. ASPINALL.
SAN REMO AS A WINTER RESIDENCE. With Coloured Plates. Foolscap 8vo. cloth, 4s. 6d.

MR. THOS. J. AUSTIN, M.R.C.S.ENG.
A PRACTICAL ACCOUNT OF GENERAL PARALYSIS: Its Mental and Physical Symptoms, Statistics, Causes, Seat, and Treatment. 8vo. cloth, 6s.

DR. THOMAS BALLARD, M.D.
A NEW AND RATIONAL EXPLANATION OF THE DISEASES PECULIAR TO INFANTS AND MOTHERS; with obvious Suggestions for their Prevention and Cure. Post 8vo. cloth, 4s. 6d.

MESSRS. CHURCHILL & SONS' PUBLICATIONS.

DR. BARCLAY.

I.
A MANUAL OF MEDICAL DIAGNOSIS. Second Edition. Foolscap 8vo. cloth, 8s. 6d.

II.
MEDICAL ERRORS.—Fallacies connected with the Application of the Inductive Method of Reasoning to the Science of Medicine. Post 8vo. cloth, 5s.

III.
GOUT AND RHEUMATISM IN RELATION TO DISEASE OF THE HEART. Post 8vo. cloth, 5s.

DR. T. HERBERT BARKER, M.D., F.R.S., & MR. ERNEST EDWARDS, B.A.

PHOTOGRAPHS OF EMINENT MEDICAL MEN, with brief Analytical Notices of their Works. Vol. I. (24 Portraits), 4to. cloth, 24s.

DR. W. G. BARKER, M.D.LOND.

ON DISEASES OF THE RESPIRATORY PASSAGES AND LUNGS, SPORADIC AND EPIDEMIC; their *Causes*, Pathology, Symptoms, and Treatment. Crown 8vo. cloth, 6s.

DR. BARLOW.

A MANUAL OF THE PRACTICE OF MEDICINE. Second Edition. Fcap. 8vo. cloth, 12s. 6d.

DR. BARNES.

THE PHYSIOLOGY AND TREATMENT OF PLACENTA PRÆVIA; being the Lettsomian Lectures on Midwifery for 1857. Post 8vo. cloth, 6s.

DR. BASCOME.

A HISTORY OF EPIDEMIC PESTILENCES, FROM THE EARLIEST AGES. 8vo. cloth, 8s.

DR. BASHAM.

ON DROPSY, AND ITS CONNECTION WITH DISEASES OF THE KIDNEYS, HEART, LUNGS AND LIVER. With 16 Plates. Third Edition. 8vo. cloth, 12s. 6d.

MR. H. F. BAXTER, M.R.C.S.L.

ON ORGANIC POLARITY; showing a Connexion to exist between Organic Forces and Ordinary Polar Forces. Crown 8vo. cloth, 5s.

MR. BATEMAN.

MAGNACOPIA: A Practical Library of Profitable Knowledge, communicating the general Minutiæ of Chemical and Pharmaceutic Routine, together with the generality of Secret Forms of Preparations. Third Edition. 18mo. 6s.

MR. LIONEL J. BEALE, M.R.C.S.

I.
THE LAWS OF HEALTH IN THEIR RELATIONS TO MIND AND BODY. A Series of Letters from an Old Practitioner to a Patient. Post 8vo. cloth, 7s. 6d.

II.
HEALTH AND DISEASE, IN CONNECTION WITH THE GENERAL PRINCIPLES OF HYGIENE. Fcap. 8vo., 2s. 6d.

MESSRS. CHURCHILL & SONS' PUBLICATIONS. 9

DR. BEALE, F.R.S.

I.
URINE, URINARY DEPOSITS, AND CALCULI: and on the Treatment of Urinary Diseases. Numerous Engravings. Second Edition, much Enlarged. Post 8vo. cloth, 8s. 6d.

II.
THE MICROSCOPE, IN ITS APPLICATION TO PRACTICAL MEDICINE. Third Edition. With 50 Plates. 8vo. cloth, 16s.

III.
ILLUSTRATIONS OF THE SALTS OF URINE, URINARY DEPOSITS, and CALCULI. 37 Plates, containing upwards of 170 Figures copied from Nature, with descriptive Letterpress. 8vo. cloth, 9s. 6d.

MR. BEASLEY.

I.
THE BOOK OF PRESCRIPTIONS; containing 3000 Prescriptions. Collected from the Practice of the most eminent Physicians and Surgeons, English and Foreign. Third Edition. 18mo. cloth, 6s.

II.
THE DRUGGIST'S GENERAL RECEIPT-BOOK; comprising a copious Veterinary Formulary and Table of Veterinary Materia Medica; Patent and Proprietary Medicines, Druggists' Nostrums, &c.; Perfumery, Skin Cosmetics, Hair Cosmetics, and Teeth Cosmetics; Beverages, Dietetic Articles, and Condiments; Trade Chemicals, Miscellaneous Preparations and Compounds used in the Arts, &c.; with useful Memoranda and Tables. Sixth Edition. 18mo. cloth, 6s.

III.
THE POCKET FORMULARY AND SYNOPSIS OF THE BRITISH AND FOREIGN PHARMACOPŒIAS; comprising standard and approved Formulæ for the Preparations and Compounds employed in Medical Practice. Eighth Edition, corrected and enlarged. 18mo. cloth, 6s.

DR. HENRY BENNET.

I.
A PRACTICAL TREATISE ON INFLAMMATION AND OTHER DISEASES OF THE UTERUS. Fourth Edition, revised, with Additions. 8vo. cloth, 16s.

II.
A REVIEW OF THE PRESENT STATE (1856) OF UTERINE PATHOLOGY. 8vo. cloth, 4s.

III.
WINTER IN THE SOUTH OF EUROPE; OR, MENTONE, THE RIVIERA, CORSICA, SICILY, AND BIARRITZ, AS WINTER CLIMATES. Third Edition, with numerous Plates, Maps, and Wood Engravings. Post 8vo. cloth, 10s. 6d.

PROFESSOR BENTLEY, F.L.S.

A MANUAL OF BOTANY. With nearly 1,200 Engravings on Wood. Fcap. 8vo. cloth, 12s. 6d.

DR. BERNAYS.

NOTES FOR STUDENTS IN CHEMISTRY; being a Syllabus compiled from the Manuals of Miller, Fownes, Berzelius, Gerhardt, Gorup-Besanez, &c. Fourth Edition. Fscap. 8vo. cloth, 3s.

MESSRS. CHURCHILL & SONS' PUBLICATIONS.

MR. HENRY HEATHER BIGG.
ORTHOPRAXY: the Mechanical Treatment of Deformities, Debilities, and Deficiencies of the Human Frame. With Engravings. Post 8vo. cloth, 10s.

DR. BILLING, F.R.S.
ON DISEASES OF THE LUNGS AND HEART. 8vo. cloth, 6s.

DR. S. B. BIRCH, M.D.
CONSTIPATED BOWELS: the Various Causes and the Rational Means of Cure. Second Edition. Post 8vo. cloth, 3s. 6d.

DR. GOLDING BIRD, F.R.S.
I.
URINARY DEPOSITS; THEIR DIAGNOSIS, PATHOLOGY, AND THERAPEUTICAL INDICATIONS. With Engravings. Fifth Edition. Edited by E. LLOYD BIRKETT, M.D. Post 8vo. cloth, 10s. 6d.

II.
ELEMENTS OF NATURAL PHILOSOPHY; being an Experimental Introduction to the Study of the Physical Sciences. With numerous Engravings. Fifth Edition. Edited by CHARLES BROOKE, M.B. Cantab., F.R.S. Fcap. 8vo. cloth, 12s. 6d.

MR. BISHOP, F.R.S.
I.
ON DEFORMITIES OF THE HUMAN BODY, their Pathology and Treatment. With Engravings on Wood. 8vo. cloth, 10s.

II.
ON ARTICULATE SOUNDS, AND ON THE CAUSES AND CURE OF IMPEDIMENTS OF SPEECH. 8vo. cloth, 4s.

MR. P. HINCKES BIRD, F.R.C.S.
PRACTICAL TREATISE ON THE DISEASES OF CHILDREN AND INFANTS AT THE BREAST. Translated from the French of M. BOUCHUT, with Notes and Additions. 8vo. cloth. 20s.

MR. BLAINE.
OUTLINES OF THE VETERINARY ART; OR, A TREATISE ON THE ANATOMY, PHYSIOLOGY, AND DISEASES OF THE HORSE, NEAT CATTLE, AND SHEEP. Seventh Edition. By Charles Steel, M.R.C.V.S.L. With Plates. 8vo. cloth, 18s.

MR. BLOXAM.
CHEMISTRY, INORGANIC AND ORGANIC; with Experiments and a Comparison of Equivalent and Molecular Formulæ. With 276 Engravings on Wood. 8vo. cloth, 16s.

DR. BOURGUIGNON.
ON THE CATTLE PLAGUE; OR, CONTAGIOUS TYPHUS IN HORNED CATTLE: its History, Origin, Description, and Treatment. Post 8vo. 5s.

MR. JOHN E. BOWMAN, & MR. C. L. BLOXAM.
I.
PRACTICAL CHEMISTRY, including Analysis. With numerous Illustrations on Wood. Fifth Edition. Foolscap 8vo. cloth, 6s. 6d.

II.
MEDICAL CHEMISTRY; with Illustrations on Wood. Fourth Edition, carefully revised. Fcap. 8vo. cloth, 6s. 6d.

MESSRS. CHURCHILL & SONS' PUBLICATIONS.   11

DR. JAMES BRIGHT.
ON DISEASES OF THE HEART, LUNGS, & AIR PASSAGES; with a Review of the several Climates recommended in these Affections. Third Edition. Post 8vo. cloth, 9s.

DR. BRINTON, F.R.S.
I.
THE DISEASES OF THE STOMACH, with an Introduction on its Anatomy and Physiology; being Lectures delivered at St. Thomas's Hospital. Second Edition. 8vo. cloth, 10s. 6d.
II.
INTESTINAL OBSTRUCTION. Edited by Dr. Buzzard. Post 8vo. cloth, 5s.

MR. BERNARD E. BRODHURST, F.R.C.S.
I.
CURVATURES OF THE SPINE: their Causes, Symptoms, Pathology, and Treatment. Second Edition. Roy. 8vo. cloth, with Engravings, 7s. 6d.
II.
ON THE NATURE AND TREATMENT OF CLUBFOOT AND ANALOGOUS DISTORTIONS involving the TIBIO-TARSAL ARTICULATION. With Engravings on Wood. 8vo. cloth, 4s. 6d.
III.
PRACTICAL OBSERVATIONS ON THE DISEASES OF THE JOINTS INVOLVING ANCHYLOSIS, and on the TREATMENT for the RESTORATION of MOTION. Third Edition, much enlarged, 8vo. cloth, 4s. 6d.

MR. THOMAS BRYANT, F.R.C.S.
I.
ON THE DISEASES AND INJURIES OF THE JOINTS. CLINICAL AND PATHOLOGICAL OBSERVATIONS. Post 8vo. cloth, 7s. 6d.
II.
THE SURGICAL DISEASES OF CHILDREN. The Lettsomian Lectures, delivered March, 1863. Post 8vo. cloth, 5s.
III.
CLINICAL SURGERY. Parts I. to VII. 8vo., 3s. 6d. each.

DR. BUCKLE, M.D., L.R.C.P.LOND.
VITAL AND ECONOMICAL STATISTICS OF THE HOSPITALS, INFIRMARIES, &c., OF ENGLAND AND WALES. Royal 8vo. 5s.

DR. JOHN CHARLES BUCKNILL, F.R.S., & DR. DANIEL H. TUKE.
A MANUAL OF PSYCHOLOGICAL MEDICINE: containing the History, Nosology, Description, Statistics, Diagnosis, Pathology, and Treatment of Insanity. Second Edition. 8vo. cloth, 15s.

DR. BUDD, F.R.S.
I.
ON DISEASES OF THE LIVER. Illustrated with Coloured Plates and Engravings on Wood. Third Edition. 8vo. cloth, 16s.
II.
ON THE ORGANIC DISEASES AND FUNCTIONAL DISORDERS OF THE STOMACH. 8vo. cloth, 9s.

MR. CALLENDER, F.R.C.S.

FEMORAL RUPTURE: Anatomy of the Parts concerned. With Plates. 8vo. cloth, 4s.

DR. JOHN M. CAMPLIN, F.L.S.

ON DIABETES, AND ITS SUCCESSFUL TREATMENT. Third Edition, by Dr. Glover. Fcap. 8vo. cloth, 3s. 6d.

MR. ROBERT B. CARTER, M.R.C.S.

I.
ON THE INFLUENCE OF EDUCATION AND TRAINING IN PREVENTING DISEASES OF THE NERVOUS SYSTEM. Fcap. 8vo., 6s.

II.
THE PATHOLOGY AND TREATMENT OF HYSTERIA. Post 8vo. cloth, 4s. 6d.

DR. CARPENTER, F.R.S.

I.
PRINCIPLES OF HUMAN PHYSIOLOGY. With numerous Illustrations on Steel and Wood. Sixth Edition. Edited by Mr. HENRY POWER. 8vo. cloth, 26s.

II.
A MANUAL OF PHYSIOLOGY. With 252 Illustrations on Steel and Wood. Fourth Edition. Fcap. 8vo. cloth, 12s. 6d.

III.
THE MICROSCOPE AND ITS REVELATIONS. With numerous Engravings on Steel and Wood. Third Edition. Fcap. 8vo. cloth, 12s. 6d.

MR. JOSEPH PEEL CATLOW, M.R.C.S.

ON THE PRINCIPLES OF ÆSTHETIC MEDICINE; or the Natural Use of Sensation and Desire in the Maintenance of Health and the Treatment of Disease. 8vo. cloth, 9s.

DR. CHAMBERS.

I.
LECTURES, CHIEFLY CLINICAL. Fourth Edition. 8vo. cloth, 14s.

II.
THE INDIGESTIONS OR DISEASES OF THE DIGESTIVE ORGANS FUNCTIONALLY TREATED. Second Edition. 8vo. cloth, 10s. 6d.

III.
SOME OF THE EFFECTS OF THE CLIMATE OF ITALY. Crown 8vo. cloth, 4s. 6d.

DR. CHANCE, M.B.

VIRCHOW'S CELLULAR PATHOLOGY, AS BASED UPON PHYSIOLOGICAL AND PATHOLOGICAL HISTOLOGY. With 144 Engravings on Wood. 8vo. cloth, 16s.

MR. H. T. CHAPMAN, F.R.C.S.

I.
THE TREATMENT OF OBSTINATE ULCERS AND CUTANEOUS ERUPTIONS OF THE LEG WITHOUT CONFINEMENT. Third Edition. Post 8vo. cloth, 3s. 6d.

II.
VARICOSE VEINS: their Nature, Consequences, and Treatment, Palliative and Curative. Second Edition. Post 8vo. cloth, 3s. 6d.

MESSRS. CHURCHILL & SONS' PUBLICATIONS.

MR. PYE HENRY CHAVASSE, F.R.C.S.

I.
ADVICE TO A MOTHER ON THE MANAGEMENT OF HER CHILDREN. Eighth Edition. Foolscap 8vo., 2s. 6d.

II.
ADVICE TO A WIFE ON THE MANAGEMENT OF HER OWN HEALTH. With an Introductory Chapter, especially addressed to a Young Wife. Seventh Edition. Fcap. 8vo., 2s. 6d.

MR. LE GROS CLARK, F.R.C.S.

OUTLINES OF SURGERY; being an Epitome of the Lectures on the Principles and the Practice of Surgery, delivered at St. Thomas's Hospital. Fcap. 8vo. cloth, 5s.

MR. JOHN CLAY, M.R.C.S.

KIWISCH ON DISEASES OF THE OVARIES: Translated, by permission, from the last German Edition of his Clinical Lectures on the Special Pathology and Treatment of the Diseases of Women. With Notes, and an Appendix on the Operation of Ovariotomy. Royal 12mo. cloth, 16s.

DR. COCKLE, M.D.

ON INTRA-THORACIC CANCER. 8vo. 6s. 6d.

MR. COLLIS, M.B.DUB., F.R.C.S.I.

THE DIAGNOSIS AND TREATMENT OF CANCER AND THE TUMOURS ANALOGOUS TO IT. With coloured Plates. 8vo. cloth, 14s.

DR. CONOLLY.

THE CONSTRUCTION AND GOVERNMENT OF LUNATIC ASYLUMS AND HOSPITALS FOR THE INSANE. With Plans. Post 8vo. cloth, 6s.

MR. COOLEY.

COMPREHENSIVE SUPPLEMENT TO THE PHARMACOPŒIAS.

THE CYCLOPÆDIA OF PRACTICAL RECEIPTS, PROCESSES, AND COLLATERAL INFORMATION IN THE ARTS, MANUFACTURES, PROFESSIONS, AND TRADES, INCLUDING MEDICINE, PHARMACY, AND DOMESTIC ECONOMY; designed as a General Book of Reference for the Manufacturer, Tradesman, Amateur, and Heads of Families. Fourth and greatly enlarged Edition, 8vo. cloth, 28s.

MR. W. WHITE COOPER.

I.
ON WOUNDS AND INJURIES OF THE EYE. Illustrated by 17 Coloured Figures and 41 Woodcuts. 8vo. cloth, 12s.

II.
ON NEAR SIGHT, AGED SIGHT, IMPAIRED VISION, AND THE MEANS OF ASSISTING SIGHT. With 31 Illustrations on Wood. Second Edition. Fcap. 8vo. cloth, 7s. 6d.

SIR ASTLEY COOPER, BART., F.R.S.

ON THE STRUCTURE AND DISEASES OF THE TESTIS.
With 24 Plates. Second Edition. Royal 4to., 20s.

MR. COOPER.

A DICTIONARY OF PRACTICAL SURGERY AND ENCYCLO-
PÆDIA OF SURGICAL SCIENCE. New Edition, brought down to the present time. By SAMUEL A. LANE, F.R.C.S., assisted by various eminent Surgeons. Vol. I., 8vo. cloth, £1. 5s.

MR. HOLMES COOTE, F.R.C.S.

A REPORT ON SOME IMPORTANT POINTS IN THE
TREATMENT OF SYPHILIS. 8vo. cloth, 5s.

DR. COTTON.

I.

ON CONSUMPTION: Its Nature, Symptoms, and Treatment. To which Essay was awarded the Fothergillian Gold Medal of the Medical Society of London. Second Edition. 8vo. cloth, 8s.

II.

PHTHISIS AND THE STETHOSCOPE; OR, THE PHYSICAL
SIGNS OF CONSUMPTION. Third Edition. Foolscap 8vo. cloth, 3s.

MR. COULSON.

I.

ON DISEASES OF THE BLADDER AND PROSTATE GLAND.
New Edition, revised. *In Preparation.*

II.

ON LITHOTRITY AND LITHOTOMY; with Engravings on Wood.
8vo. cloth, 8s.

MR. WILLIAM CRAIG, L.F.P.S., GLASGOW.

ON THE INFLUENCE OF VARIATIONS OF ELECTRIC
TENSION AS THE REMOTE CAUSE OF EPIDEMIC AND OTHER DISEASES. 8vo. cloth, 10s.

MR. CURLING, F.R.S.

I.

OBSERVATIONS ON DISEASES OF THE RECTUM. Third Edition. 8vo. cloth, 7s. 6d.

II.

A PRACTICAL TREATISE ON DISEASES OF THE TESTIS,
SPERMATIC CORD, AND SCROTUM. Third Edition, with Engravings. 8vo. cloth, 16s.

DR. DALRYMPLE, M.R.C.P., F.R.C.S.

THE CLIMATE OF EGYPT: METEOROLOGICAL AND MEDI-
CAL OBSERVATIONS, with Practical Hints for Invalid Travellers. Post 8vo. cloth, 4s.

MR. JOHN DALRYMPLE, F.R.S., F.R.C.S.

PATHOLOGY OF THE HUMAN EYE. Complete in Nine Fasciculi: imperial 4to., 20s. each; half-bound morocco, gilt tops, 9l. 15s.

DR. HERBERT DAVIES.
ON THE PHYSICAL DIAGNOSIS OF DISEASES OF THE
LUNGS AND HEART. Second Edition. Post 8vo. cloth, 8s.

DR. DAVEY.
I.
THE GANGLIONIC NERVOUS SYSTEM: its Structure, Functions, and Diseases. 8vo. cloth, 9s.
II.
ON THE NATURE AND PROXIMATE CAUSE OF IN-SANITY. Post 8vo. cloth, 3s.

DR. HENRY DAY, M.D., M.R.C.P.
CLINICAL HISTORIES; with Comments. 8vo. cloth, 7s. 6d.

MR. DIXON.
A GUIDE TO THE PRACTICAL STUDY OF DISEASES OF
THE EYE. Third Edition. Post 8vo. cloth, 9s.

DR. DOBELL.
I.
DEMONSTRATIONS OF DISEASES IN THE CHEST, AND
THEIR PHYSICAL DIAGNOSIS. With Coloured Plates. 8vo. cloth, 12s. 6d.
II.
LECTURES ON THE GERMS AND VESTIGES OF DISEASE,
and on the Prevention of the Invasion and Fatality of Disease by Periodical Examinations. 8vo. cloth, 6s. 6d.
III.
ON TUBERCULOSIS: ITS NATURE, CAUSE, AND TREAT-MENT; with Notes on Pancreatic Juice. Second Edition. Crown 8vo. cloth, 3s. 6d.
IV.
LECTURES ON WINTER COUGH (CATARRH, BRONCHITIS,
EMPHYSEMA, ASTHMA); with an Appendix on some Principles of Diet in Disease. Post 8vo. cloth, 5s. 6d.
V.
LECTURES ON THE TRUE FIRST STAGE OF CONSUMP-TION. Crown 8vo. cloth, 3s. 6d.

DR. TOOGOOD DOWNING.
NEURALGIA: its various Forms, Pathology, and Treatment. THE JACKSONIAN PRIZE ESSAY FOR 1850. 8vo. cloth, 10s. 6d.

DR. DRUITT, F.R.C.S.
THE SURGEON'S VADE-MECUM; with numerous Engravings on Wood. Ninth Edition. Foolscap 8vo. cloth, 12s. 6d.

MR. DUNN, F.R.C.S.
PSYCHOLOGY—PHYSIOLOGICAL, 4s.; MEDICAL, 3s.

SIR JAMES EYRE, M.D.
I.
THE STOMACH AND ITS DIFFICULTIES. Fifth Edition. Fcap. 8vo. cloth, 2s. 6d.
II.
PRACTICAL REMARKS ON SOME EXHAUSTING DIS-EASES. Second Edition. Post 8vo. cloth, 4s. 6d.

DR. FAYRER, M.D., F.R.C.S.
CLINICAL SURGERY IN INDIA. With Engravings. 8vo. cloth, 16s.

DR. FENWICK.
ON SCROFULA AND CONSUMPTION. Clergyman's Sore Throat, Catarrh, Croup, Bronchitis, Asthma. Fcap. 8vo., 2s. 6d.

SIR WILLIAM FERGUSSON, BART., F.R.S.
A SYSTEM OF PRACTICAL SURGERY; with numerous Illustrations on Wood. Fourth Edition. Fcap. 8vo. cloth, 12s. 6d.

SIR JOHN FIFE, F.R.C.S. AND MR. URQUHART.
MANUAL OF THE TURKISH BATH. Heat a Mode of Cure and a Source of Strength for Men and Animals. With Engravings. Post 8vo. cloth, 5s.

MR. FLOWER, F.R.S., F.R.C.S.
DIAGRAMS OF THE NERVES OF THE HUMAN BODY, exhibiting their Origin, Divisions, and Connexions, with their Distribution to the various Regions of the Cutaneous Surface, and to all the Muscles. Folio, containing Six Plates, 14s.

DR. BALTHAZAR FOSTER, M.D., M.R.C.P.
THE USE OF THE SPHYGMOGRAPH IN THE INVESTIGATION OF DISEASE. With Engravings. 8vo. cloth, 2s. 6d.

MR. FOWNES, PH.D., F.R.S.
I.
A MANUAL OF CHEMISTRY; with 187 Illustrations on Wood. Ninth Edition. Fcap. 8vo. cloth, 12s. 6d.
Edited by H. BENCE JONES, M.D., F.R.S., and A. W. HOFMANN, PH.D., F.R.S.

II.
CHEMISTRY, AS EXEMPLIFYING THE WISDOM AND BENEFICENCE OF GOD. Second Edition. Fcap. 8vo. cloth, 4s. 6d.

III.
INTRODUCTION TO QUALITATIVE ANALYSIS. Post 8vo. cloth, 2s.

DR. D. J. T. FRANCIS.
CHANGE OF CLIMATE; considered as a Remedy in Dyspeptic, Pulmonary, and other Chronic Affections; with an Account of the most Eligible Places of Residence for Invalids, at different Seasons of the Year. Post 8vo. cloth, 8s. 6d.

DR. W. FRAZER.
ELEMENTS OF MATERIA MEDICA; containing the Chemistry and Natural History of Drugs—their Effects, Doses, and Adulterations. Second Edition. 8vo. cloth, 10s. 6d.

C. REMIGIUS FRESENIUS.
A SYSTEM OF INSTRUCTION IN CHEMICAL ANALYSIS, Edited by LLOYD BULLOCK, F.C.S.
QUALITATIVE. Sixth Edition, with Coloured Plate illustrating Spectrum Analysis. 8vo. cloth, 10s. 6d.——QUANTITATIVE. Fourth Edition. 8vo. cloth, 18s.

MESSRS. CHURCHILL & SONS' PUBLICATIONS. 17

DR. FULLER.

I.
ON DISEASES OF THE LUNGS AND AIR PASSAGES.
Second Edition. 8vo. cloth, 12s. 6d.

II.
ON DISEASES OF THE HEART AND GREAT VESSELS.
8vo. cloth, 7s. 6d.

III.
ON RHEUMATISM, RHEUMATIC GOUT, AND SCIATICA:
their Pathology, Symptoms, and Treatment. Third Edition. 8vo. cloth, 12s. 6d.

MR. GALLOWAY.

I.
THE FIRST STEP IN CHEMISTRY. Third Edition. Fcap. 8vo. cloth, 5s.

II.
THE SECOND STEP IN CHEMISTRY; or, the Student's Guide to the Higher Branches of the Science. With Engravings. 8vo. cloth, 10s.

III.
A MANUAL OF QUALITATIVE ANALYSIS. Fourth Edition. Post 8vo. cloth, 6s. 6d.

IV.
CHEMICAL TABLES. On Five Large Sheets, for School and Lecture Rooms. Second Edition. 4s. 6d.

MR. J. SAMPSON GAMGEE.

HISTORY OF A SUCCESSFUL CASE OF AMPUTATION AT THE HIP-JOINT (the limb 48-in. in circumference, 99 pounds weight). With 4 Photographs. 4to cloth, 10s. 6d.

MR. F. J. GANT, F.R.C.S.

I.
THE PRINCIPLES OF SURGERY: Clinical, Medical, and Operative. With Engravings. 8vo. cloth, 18s.

II.
THE IRRITABLE BLADDER: its Causes and Curative Treatment. Second Edition, enlarged. Crown 8vo. cloth, 5s.

SIR DUNCAN GIBB, BART., M.D.

I.
ON DISEASES OF THE THROAT AND WINDPIPE, as reflected by the Laryngoscope. Second Edition. With 116 Engravings. Post 8vo cloth, 10s. 6d.

II.
THE LARYNGOSCOPE IN DISEASES OF THE THROAT, with a Chapter on RHINOSCOPY. Second Edition, enlarged, with Engravings. Crown 8vo., cloth, 5s.

MRS. GODFREY.

ON THE NATURE, PREVENTION, TREATMENT, AND CURE OF SPINAL CURVATURES and **DEFORMITIES** of the **CHEST** and **LIMBS,** without **ARTIFICIAL SUPPORTS** or any **MECHANICAL APPLIANCES.** Third Edition, Revised and Enlarged. 8vo. cloth 5s.

DR. GORDON, M.D. C.B.

I.
ARMY HYGIENE. 8vo. cloth, 20s.

II.
CHINA, FROM A MEDICAL POINT OF VIEW, IN 1860 AND 1861; With a Chapter on Nagasak as a Sanatarium. 8vo. cloth, 10s. 6d.

DR. GAIRDNER.

ON GOUT; its History, its Causes, and its Cure. Fourth Edition. Post 8vo. cloth, 8s. 6d.

DR. GRANVILLE, F.R.S.

I.
THE MINERAL SPRINGS OF VICHY: their Efficacy in the Treatment of Gout, Indigestion, Gravel, &c. 8vo. cloth, 3s.

II.
ON SUDDEN DEATH. Post 8vo., 2s. 6d.

DR. GRAVES M.D., F.R.S.

STUDIES IN PHYSIOLOGY AND MEDICINE. Edited by Dr. Stokes. With Portrait and Memoir. 8vo. cloth, 14s.

DR. S. C. GRIFFITH, M.D.

ON DERMATOLOGY AND THE TREATMENT OF SKIN DISEASES BY MEANS OF HERBS, IN PLACE OF ARSENIC AND MERCURY. Fcap. 8vo. cloth, 3s.

MR. GRIFFITHS.

CHEMISTRY OF THE FOUR SEASONS—Spring, Summer, Autumn, Winter. Illustrated with Engravings on Wood. Second Edition. Foolscap 8vo. cloth, 7s. 6d.

DR. GULLY.

THE SIMPLE TREATMENT OF DISEASE; deduced from the Methods of Expectancy and Revulsion. 18mo. cloth, 4s.

DR. GUY AND DR. JOHN HARLEY.

HOOPER'S PHYSICIAN'S VADE-MECUM; OR, MANUAL OF THE PRINCIPLES AND PRACTICE OF PHYSIC. Seventh Edition, considerably enlarged, and rewritten. Foolscap 8vo. cloth, 12s. 6d.

GUY'S HOSPITAL REPORTS. Third Series. Vols. I. to XII., 8vo., 7s. 6d. each.

DR. HABERSHON, F.R.C.P.

I.
ON DISEASES OF THE ABDOMEN, comprising those of the Stomach and other Parts of the Alimentary Canal, Œsophagus, Stomach, Cæcum, Intestines, and Peritoneum. Second Edition, with Plates. 8vo. cloth, 14s.

II.
ON THE INJURIOUS EFFECTS OF MERCURY IN THE TREATMENT OF DISEASE. Post 8vo. cloth, 3s. 6d.

DR. C. RADCLYFFE HALL.

TORQUAY IN ITS MEDICAL ASPECT AS A RESORT FOR PULMONARY INVALIDS. Post 8vo. cloth, 5s.

DR. MARSHALL HALL, F.R.S.

I.
PRONE AND POSTURAL RESPIRATION IN DROWNING AND OTHER FORMS OF APNŒA OR SUSPENDED RESPIRATION. Post 8vo. cloth. 5s.

II.
PRACTICAL OBSERVATIONS AND SUGGESTIONS IN MEDICINE. Second Series. Post 8vo. cloth, 8s. 6d.

MESSRS. CHURCHILL & SONS' PUBLICATIONS.

MR. HARDWICH.

A MANUAL OF PHOTOGRAPHIC CHEMISTRY. With Engravings. Seventh Edition. Foolscap 8vo. cloth, 7s. 6d.

DR. J. BOWER HARRISON, M.D., M.R.C.P.

I.
LETTERS TO A YOUNG PRACTITIONER ON THE DISEASES OF CHILDREN. Foolscap 8vo. cloth, 3s.

II.
ON THE CONTAMINATION OF WATER BY THE POISON OF LEAD, and its Effects on the Human Body.' Foolscap 8vo. cloth, 3s. 6d.

DR. HARTWIG.

I.
ON SEA BATHING AND SEA AIR. Second Edition. Fcap. 8vo., 2s. 6d.

II.
ON THE PHYSICAL EDUCATION OF CHILDREN.. Fcap. 8vo., 2s. 6d.

DR. A. H. HASSALL.

I.
THE URINE, IN HEALTH AND DISEASE; being an Explanation of the Composition of the Urine, and of the Pathology and Treatment of Urinary and Renal Disorders. Second Edition. With 79 Engravings (23 Coloured). Post 8vo. cloth, 12s. 6d.

II.
THE MICROSCOPIC ANATOMY OF THE HUMAN BODY, IN HEALTH AND DISEASE. Illustrated with Several Hundred Drawings in Colour. Two vols. 8vo. cloth, £1. 10s.

MR. ALFRED HAVILAND, M.R.C.S.

CLIMATE, WEATHER, AND DISEASE; being a Sketch of the Opinions of the most celebrated Ancient and Modern Writers with regard to the Influence of Climate and Weather in producing Disease. With Four coloured Engravings. 8vo. cloth, 7s.

DR. HEADLAND, M.D., F.R.C.P.

I.
ON THE ACTION OF MEDICINES IN THE SYSTEM. Fourth Edition. 8vo. cloth, 14s.

II.
A MEDICAL HANDBOOK; comprehending such Information on Medical and Sanitary Subjects as is desirable in Educated Persons. Second Thousand. Foolscap 8vo. cloth, 5s.

DR. HEALE.

I.
A TREATISE ON THE PHYSIOLOGICAL ANATOMY OF THE LUNGS. With Engravings. 8vo. cloth, 8s.

II.
A TREATISE ON VITAL CAUSES. 8vo. cloth, 9s.

MESSRS. CHURCHILL & SONS' PUBLICATIONS.

MR. CHRISTOPHER HEATH, F.R.C.S.

I.
PRACTICAL ANATOMY; a Manual of Dissections. With numerous Engravings. Fcap. 8vo. cloth, 10s. 6d.

II.
A MANUAL OF MINOR SURGERY AND BANDAGING, FOR THE USE OF HOUSE-SURGEONS, DRESSERS, AND JUNIOR PRACTITIONERS. With Illustrations. Third Edition. Fcap. 8vo. cloth, 5s.

MR. HIGGINBOTTOM, F.R.S., F.R.C.S.E.

A PRACTICAL ESSAY ON THE USE OF THE NITRATE OF SILVER IN THE TREATMENT OF INFLAMMATION, WOUNDS, AND ULCERS. Third Edition, 8vo. cloth, 6s.

DR. HINDS.

THE HARMONIES OF PHYSICAL SCIENCE IN RELATION TO THE HIGHER SENTIMENTS; with Observations on Medical Studies, and on the Moral and Scientific Relations of Medical Life. Post 8vo. cloth, 4s.

MR. J. A. HINGESTON, M.R.C.S.

TOPICS OF THE DAY, MEDICAL, SOCIAL, AND SCIENTIFIC. Crown 8vo. cloth, 7s. 6d.

DR. HODGES.

THE NATURE, PATHOLOGY, AND TREATMENT OF PUERPERAL CONVULSIONS. Crown 8vo. cloth, 3s.

DR. DECIMUS HODGSON.

THE PROSTATE GLAND, AND ITS ENLARGEMENT IN OLD AGE. With 12 Plates. Royal 8vo. cloth, 6s.

MR. JABEZ HOGG.

A MANUAL OF OPHTHALMOSCOPIC SURGERY; being a Practical Treatise on the Use of the Ophthalmoscope in Diseases of the Eye. Third Edition. With Coloured Plates. 8vo. cloth, 10s. 6d.

MR. LUTHER HOLDEN, F.R.C.S.

I.
HUMAN OSTEOLOGY; with Plates, showing the Attachments of the Muscles. Third Edition. 8vo. cloth, 16s.

II.
A MANUAL OF THE DISSECTION OF THE HUMAN BODY. With Engravings on Wood. Second Edition. 8vo. cloth, 16s.

MR BARNARD HOLT, F.R.C.S.

ON THE IMMEDIATE TREATMENT OF STRICTURE OF THE URETHRA. Second Edition, Enlarged. 8vo. cloth, 3s.

DR. W. CHARLES HOOD.
SUGGESTIONS FOR THE FUTURE PROVISION OF CRIMINAL LUNATICS. 8vo. cloth, 5s. 6d.

DR. P. HOOD.
THE SUCCESSFUL TREATMENT OF SCARLET FEVER; also, OBSERVATIONS ON THE PATHOLOGY AND TREATMENT OF CROWING INSPIRATIONS OF INFANTS. Post 8vo. cloth, 5s.

MR. JOHN HORSLEY.
A CATECHISM OF CHEMICAL PHILOSOPHY; being a Familiar Exposition of the Principles of Chemistry and Physics. With Engravings on Wood. Designed for the Use of Schools and Private Teachers. Post 8vo. cloth, 6s. 6d.

MR. LUKE HOWARD, F.R.S.
ESSAY ON THE MODIFICATIONS OF CLOUDS. Third Edition, by W. D. and E. HOWARD. With 6 Lithographic Plates, from Pictures by Kenyon. 4to. cloth, 10s. 6d.

DR. HAMILTON HOWE, M.D.
A THEORETICAL INQUIRY INTO THE PHYSICAL CAUSE OF EPIDEMIC DISEASES. Accompanied with Tables. 8vo. cloth, 7s.

DR. HUFELAND.
THE ART OF PROLONGING LIFE. Second Edition. Edited by ERASMUS WILSON, F.R.S. Foolscap 8vo., 2s. 6d.

MR. W. CURTIS HUGMAN, F.R.C.S.
ON HIP-JOINT DISEASE; with reference especially to Treatment by Mechanical Means for the Relief of Contraction and Deformity of the Affected Limb. With Plates. Re-issue, enlarged. 8vo. cloth, 3s. 6d.

MR. HULKE, F.R.C.S.
A PRACTICAL TREATISE ON THE USE OF THE OPHTHALMOSCOPE. Being the Jacksonian Prize Essay for 1859. Royal 8vo. cloth, 8s.

DR. HENRY HUNT.
ON HEARTBURN AND INDIGESTION. 8vo. cloth, 5s.

MR. G. Y. HUNTER, M.R.C.S.
BODY AND MIND: the Nervous System and its Derangements. Fcap. 8vo. cloth, 3s. 6d.

PROFESSOR HUXLEY, F.R.S.
LECTURES ON THE ELEMENTS OF COMPARATIVE ANATOMY.—ON CLASSIFICATON AND THE SKULL. With 111 Illustrations. 8vo. cloth, 10s. 6d.

MESSRS. CHURCHILL & SONS' PUBLICATIONS.

MR. JONATHAN HUTCHINSON, F.R.C.S.
A CLINICAL MEMOIR ON CERTAIN DISEASES OF THE EYE AND EAR, CONSEQUENT ON INHERITED SYPHILIS; with an appended Chapter of Commentaries on the Transmission of Syphilis from Parent to Offspring, and its more remote Consequences. With Plates and Woodcuts, 8vo. cloth, 9s.

DR. INMAN, M.R.C.P.
I.
ON MYALGIA: ITS NATURE, CAUSES, AND TREATMENT; being a Treatise on Painful and other Affections of the Muscular System. Second Edition. 8vo. cloth, 9s.

II.
FOUNDATION FOR A NEW THEORY AND PRACTICE OF MEDICINE. Second Edition. Crown 8vo. cloth, 10s.

DR. JAGO, M.D.OXON., A.B.CANTAB.
ENTOPTICS, WITH ITS USES IN PHYSIOLOGY AND MEDICINE. With 54 Engravings. Crown 8vo. cloth, 5s.

MR. J. H. JAMES, F.R.C.S.
PRACTICAL OBSERVATIONS ON THE OPERATIONS FOR STRANGULATED HERNIA. 8vo. cloth, 5s.

DR. PROSSER JAMES, M.D.
SORE-THROAT: ITS NATURE, VARIETIES, AND TREATMENT; including the Use of the LARYNGOSCOPE as an Aid to Diagnosis. Second Edition, with numerous Engravings. Post 8vo. cloth, 5s.

DR. HANDFIELD JONES, M.B., F.R.C.P.
CLINICAL OBSERVATIONS ON FUNCTIONAL NERVOUS DISORDERS. Post 8vo. cloth, 10s. 6d.

DR. H. BENCE JONES, M.D., F.R.S.
LECTURES ON SOME OF THE APPLICATIONS OF CHEMISTRY AND MECHANICS TO PATHOLOGY AND THERAPEUTICS. 8vo. cloth, 12s.

DR. HANDFIELD JONES, F.R.S., & DR. EDWARD H. SIEVEKING.
A MANUAL OF PATHOLOGICAL ANATOMY. Illustrated with numerous Engravings on Wood. Foolscap 8vo. cloth, 12s. 6d.

DR. JAMES JONES, M.D., M.R.C.P.
ON THE USE OF PERCHLORIDE OF IRON AND OTHER CHALYBEATE SALTS IN THE TREATMENT OF CONSUMPTION. Crown 8vo. cloth, 3s. 6d.

MR. WHARTON JONES, F.R.S.

I.
A MANUAL OF THE PRINCIPLES AND PRACTICE OF OPHTHALMIC MEDICINE AND SURGERY; with Nine Coloured Plates and 173 Wood Engravings. Third Edition, thoroughly revised. Foolscap 8vo. cloth, 12s. 6d.

II.
THE WISDOM AND BENEFICENCE OF THE ALMIGHTY, AS DISPLAYED IN THE SENSE OF VISION. Actonian Prize Essay. With Illustrations on Steel and Wood. Foolscap 8vo. cloth, 4s. 6d.

III.
DEFECTS OF SIGHT AND HEARING: their Nature, Causes, Prevention, and General Management. Second Edition, with Engravings. Fcap. 8vo. 2s. 6d.

IV.
A CATECHISM OF THE MEDICINE AND SURGERY OF THE EYE AND EAR. For the Clinical Use of Hospital Students. Fcap. 8vo. 2s. 6d.

V.
A CATECHISM OF THE PHYSIOLOGY AND PHILOSOPHY OF BODY, SENSE, AND MIND. For Use in Schools and Colleges. Fcap. 8vo., 2s. 6d.

MR. FURNEAUX JORDAN, M.R.C.S.

AN INTRODUCTION TO CLINICAL SURGERY; WITH A Method of Investigating and Reporting Surgical Cases. Fcap. 8vo. cloth, 5s.

MR. JUDD.

A PRACTICAL TREATISE ON URETHRITIS AND SYPHILIS: including Observations on the Power of the Menstruous Fluid, and of the Discharge from Leucorrhœa and Sores to produce Urethritis: with a variety of Examples, Experiments, Remedies, and Cures. 8vo. cloth, £1. 5s.

DR. LAENNEC.

A MANUAL OF AUSCULTATION AND PERCUSSION. Translated and Edited by J. B. Sharpe, M.R.C.S. 3s.

DR. LANE, M.A.

HYDROPATHY; OR, HYGIENIC MEDICINE. An Explanatory Essay. Second Edition. Post 8vo. cloth, 5s.

SIR WM. LAWRENCE, BART., F.R.S.

I.
LECTURES ON SURGERY. 8vo. cloth, 16s.

II.
A TREATISE ON RUPTURES. The Fifth Edition, considerably enlarged. 8vo. cloth, 16s.

DR. LEARED, M.R.C.P.

IMPERFECT DIGESTION: ITS CAUSES AND TREATMENT. Fourth Edition. Foolscap 8vo. cloth, 4s.

DR. EDWIN LEE.

I.
THE EFFECT OF CLIMATE ON TUBERCULOUS DISEASE, with Notices of the chief Foreign Places of Winter Resort. Small 8vo. cloth, 4s. 6d.

II.
THE WATERING PLACES OF ENGLAND, CONSIDERED with Reference to their Medical Topography. Fourth Edition. Fcap. 8vo. cloth, 7s. 6d.

III.
THE PRINCIPAL BATHS OF FRANCE. Fourth Edition. Fcap. 8vo. cloth, 3s. 6d.

IV.
THE BATHS OF GERMANY. Fourth Edition. Post 8vo. cloth, 7s.

V.
THE BATHS OF SWITZERLAND. 12mo. cloth, 3s. 6d.

VI.
HOMŒOPATHY AND HYDROPATHY IMPARTIALLY APPRECIATED. Fourth Edition. Post 8vo. cloth, 3s.

MR. HENRY LEE, F.R.C.S.

I.
ON SYPHILIS. Second Edition. With Coloured Plates. 8vo. cloth, 10s.

II.
ON DISEASES OF THE VEINS, HÆMORRHOIDAL TUMOURS, AND OTHER AFFECTIONS OF THE RECTUM. Second Edition. 8vo. cloth, 8s.

DR. ROBERT LEE, F.R.S.

I.
CONSULTATIONS IN MIDWIFERY. Foolscap 8vo. cloth, 4s. 6d.

II.
A TREATISE ON THE SPECULUM; with Three Hundred Cases. 8vo. cloth, 4s. 6d.

III.
CLINICAL REPORTS OF OVARIAN AND UTERINE DISEASES, with Commentaries. Foolscap 8vo. cloth, 6s. 6d.

IV.
CLINICAL MIDWIFERY: comprising the Histories of 545 Cases of Difficult, Preternatural, and Complicated Labour, with Commentaries. Second Edition. Foolscap 8vo. cloth, 5s.

DR. LEISHMAN, M.D., F.F.P.S.

THE MECHANISM OF PARTURITION: An Essay, Historical and Critical. With Engravings. 8vo. cloth, 5s.

MR. LISTON, F.R.S.

PRACTICAL SURGERY. Fourth Edition. 8vo. cloth, 22s.

MR. H. W. LOBB, L.S.A., M.R.C.S.E.

ON SOME OF THE MORE OBSCURE FORMS OF NERVOUS AFFECTIONS, THEIR PATHOLOGY AND TREATMENT. Re-issue, with the Chapter on Galvanism entirely Re-written. With Engravings. 8vo. cloth, 8s.

DR. LOGAN, M.D., M.R.C.P.LOND.

ON OBSTINATE DISEASES OF THE SKIN. Fcap. 8vo. cloth, 2s. 6d.

MESSRS. CHURCHILL & SONS' PUBLICATIONS. 25

LONDON HOSPITAL.
CLINICAL LECTURES AND REPORTS BY THE MEDICAL AND SURGICAL STAFF. With Illustrations. Vols. I. to III. 8vo. cloth, 7s. 6d.

LONDON MEDICAL SOCIETY OF OBSERVATION.
WHAT TO OBSERVE AT THE BED-SIDE, AND AFTER DEATH. Published by Authority. Second Edition. Foolscap 8vo. cloth, 4s. 6d.

MR. M'CLELLAND, F.L.S., F.G.S.
THE MEDICAL TOPOGRAPHY, OR CLIMATE AND SOILS, OF BENGAL AND THE N. W. PROVINCES. Post 8vo. cloth, 4s. 6d.

DR. MACLACHLAN, M.D., F.R.C.P.L.
THE DISEASES AND INFIRMITIES OF ADVANCED LIFE. 8vo. cloth, 16s.

DR. A. C. MACLEOD, M.R.C.P.LOND.
ACHOLIC DISEASES; comprising Jaundice, Diarrhœa, Dysentery, and Cholera. Post 8vo. cloth, 5s. 6d.

DR. GEORGE H. B. MACLEOD, F.R.C.S.E.
I.
OUTLINES OF SURGICAL DIAGNOSIS. 8vo. cloth, 12s. 6d.
II.
NOTES ON THE SURGERY OF THE CRIMEAN WAR; with REMARKS on GUN-SHOT WOUNDS. 8vo. cloth, 10s. 6d.

MR. JOSEPH MACLISE, F.R.C.S.
I.
SURGICAL ANATOMY. A Series of Dissections, illustrating the Principal Regions of the Human Body.
The Second Edition, imperial folio, cloth, £3. 12s.; half-morocco, £4. 4s.
II.
ON DISLOCATIONS AND FRACTURES. This Work is Uniform with the Author's "Surgical Anatomy;" each Fasciculus contains Four beautifully executed Lithographic Drawings. Imperial folio, cloth, £2. 10s.; half-morocco, £2. 17s.

MR. MACNAMARA.
ON DISEASES OF THE EYE; referring principally to those Affections requiring the aid of the Ophthalmoscope for their Diagnosis. With coloured plates. 8vo. cloth, 10s. 6d.

DR. MONICOLL, M.R.C.P.
A HAND-BOOK FOR SOUTHPORT, MEDICAL & GENERAL; with Copious Notices of the Natural History of the District. Second Edition. Post 8vo. cloth, 3s. 6d.

DR. MARCET, F.R.S.
I.
ON THE COMPOSITION OF FOOD, AND HOW IT IS ADULTERATED; with Practical Directions for its Analysis. 8vo. cloth, 6s. 6d.
II.
ON CHRONIC ALCOHOLIC INTOXICATION; with an INQUIRY INTO THE INFLUENCE OF THE ABUSE OF ALCOHOL AS A PREDISPOSING CAUSE OF DISEASE. Second Edition, much enlarged. Foolscap 8vo. cloth, 4s. 6d.

DR. J. MACPHERSON, M.D.
CHOLERA IN ITS HOME; with a Sketch of the Pathology and Treatment of the Disease. Crown vo. cloth, 5s.

DR. MARKHAM.
I.
DISEASES OF THE HEART: THEIR PATHOLOGY, DIAGNOSIS, AND TREATMENT. Second Edition. Post 8vo. cloth, 6s.

II.
SKODA ON AUSCULTATION AND PERCUSSION. Post 8vo. cloth, 6s.

III.
BLEEDING AND CHANGE IN TYPE OF DISEASES. Gulstonian Lectures for 1864. Crown 8vo. 2s. 6d.

SIR RANALD MARTIN, K.C.B., F.R.S.
INFLUENCE OF TROPICAL CLIMATES IN PRODUCING THE ACUTE ENDEMIC DISEASES OF EUROPEANS; including Practical Observations on their Chronic Sequelæ under the Influences of the Climate of Europe. Second Edition, much enlarged. 8vo. cloth, 20s.

DR. MASSY.
ON THE EXAMINATION OF RECRUITS; intended for the Use of Young Medical Officers on Entering the Army. 8vo. cloth, 5s.

MR. C. F. MAUNDER, F.R.C.S.
OPERATIVE SURGERY. With 158 Engravings. Post 8vo. 6s.

DR. MAYNE, M.D., LL.D.
I.
AN EXPOSITORY LEXICON OF THE TERMS, ANCIENT AND MODERN, IN MEDICAL AND GENERAL SCIENCE. 8vo. cloth, £2. 10s.

II.
A MEDICAL VOCABULARY; or, an Explanation of all Names, Synonymes, Terms, and Phrases used in Medicine and the relative branches of Medical Science. Second Edition. Fcap. 8vo. cloth, 8s. 6d.

DR. MERYON, M.D., F.R.C.P.
PATHOLOGICAL AND PRACTICAL RESEARCHES ON THE VARIOUS FORMS OF PARALYSIS. 8vo. cloth, 6s.

DR. MILLINGEN.
ON THE TREATMENT AND MANAGEMENT OF THE INSANE; with Considerations on Public and Private Lunatic Asylums. 18mo. cloth, 4s. 6d.

DR. W. J. MOORE, M.D.
I.
HEALTH IN THE TROPICS; or, Sanitary Art applied to Europeans in India. 8vo. cloth, 9s.

II.
A MANUAL OF THE DISEASES OF INDIA. Fcap. 8vo. cloth, 5s.

PROFESSOR MULDER, UTRECHT.
THE CHEMISTRY OF WINE. Edited by H. BENCE JONES, M.D., F.R.S. Fcap. 8vo. cloth, 6s.

DR. W. MURRAY, M.D., M.R.C.P.
EMOTIONAL DISORDERS OF THE SYMPATHETIC SYSTEM OF NERVES. Crown 8vo. cloth, 3s. 6d.

DR. MUSHET, M.B., M.R.C.P.
ON APOPLEXY, AND ALLIED AFFECTIONS OF THE BRAIN. 8vo. cloth, 7s.

MR. NAYLER, F.R.C.S.
ON THE DISEASES OF THE SKIN. With Plates. 8vo. cloth, 10s. 6d.

DR. BIRKBECK NEVINS.
THE PRESCRIBER'S ANALYSIS OF THE BRITISH PHARMACOPEIA of 1867. 32mo. cloth, 3s. 6d.

DR. THOS. NICHOLSON, M.D.
ON YELLOW FEVER; comprising the History of that Disease as it appeared in the Island of Antigua. Fcap. 8vo. cloth, 2s. 6d.

DR. NOAD, PH.D., F.R.S.
THE INDUCTION COIL, being a Popular Explanation of the Electrical Principles on which it is constructed. Second Edition. With Engravings. Fcap. 8vo. cloth, 3s.

DR. NOBLE.
THE HUMAN MIND IN ITS RELATIONS WITH THE BRAIN AND NERVOUS SYSTEM. Post 8vo. cloth, 4s. 6d.

MR. NUNNELEY, F.R.C.S.E.
I.
ON THE ORGANS OF VISION: THEIR ANATOMY AND PHYSIOLOGY. With Plates, 8vo. cloth, 15s.
II.
A TREATISE ON THE NATURE, CAUSES, AND TREATMENT OF ERYSIPELAS. 8vo. cloth, 10s. 6d.

DR. OPPERT, M.D.
HOSPITALS, INFIRMARIES, AND DISPENSARIES; their Construction, Interior Arrangement, and Management, with Descriptions of existing Institutions. With 58 Engravings. Royal 8vo. cloth, 10s. 6d.

MR. LANGSTON PARKER.
THE MODERN TREATMENT OF SYPHILITIC DISEASES, both Primary and Secondary; comprising the Treatment of Constitutional and Confirmed Syphilis, by a safe and successful Method. Fourth Edition, 8vo. cloth, 10s.

DR. PARKES, F.R.S., F.R.C.P.
I.
A MANUAL OF PRACTICAL HYGIENE; intended especially for the Medical Officers of the Army. With Plates and Woodcuts. 2nd Edition, 8vo. cloth, 16s.
II.
THE URINE: ITS COMPOSITION IN HEALTH AND DISEASE, AND UNDER THE ACTION OF REMEDIES. 8vo. cloth, 12s.

MESSRS. CHURCHILL & SONS' PUBLICATIONS.

DR. PARKIN, M.D., F.R.C.S.

I.
THE ANTIDOTAL TREATMENT AND PREVENTION OF THE EPIDEMIC CHOLERA. Third Edition. 8vo. cloth, 7s. 6d.

II.
THE CAUSATION AND PREVENTION OF DISEASE; with the Laws regulating the Extrication of Malaria from the Surface, and its Diffusion in the surrounding Air. 8vo. cloth, 5s.

MR. JAMES PART, F.R.C.S.

THE MEDICAL AND SURGICAL POCKET CASE BOOK, for the Registration of important Cases in Private Practice, and to assist the Student of Hospital Practice. Second Edition. 2s. 6d.

DR. PAVY, M.D., F.R.S., F.R.C.P.

DIABETES: RESEARCHES ON ITS NATURE AND TREATMENT. 8vo. cloth, 8s. 6d.

DR. PEACOCK, M.D., F.R.C.P.

I.
ON MALFORMATIONS OF THE HUMAN HEART. With Original Cases and Illustrations. Second Edition. With 8 Plates. 8vo. cloth, 10s.

II.
ON SOME OF THE CAUSES AND EFFECTS OF VALVULAR DISEASE OF THE HEART. With Engravings. 8vo. cloth, 5s.

DR. W. H. PEARSE, M.D.EDIN.

NOTES ON HEALTH IN CALCUTTA AND BRITISH EMIGRANT SHIPS, including Ventilation, Diet, and Disease. Fcap. 8vo. 2s.

DR. PEET, M.D., F.R.C.P.

THE PRINCIPLES AND PRACTICE OF MEDICINE; Designed chiefly for Students of Indian Medical Colleges. 8vo. cloth, 16s.

DR. PEREIRA, F.R.S.

SELECTA E PRÆSCRIPTIS. Fourteenth Edition. 24mo. cloth, 5s.

DR. PICKFORD.

HYGIENE; or, Health as Depending upon the Conditions of the Atmosphere, Food and Drinks, Motion and Rest, Sleep and Wakefulness, Secretions, Excretions, and Retentions, Mental Emotions, Clothing, Bathing, &c. Vol. I. 8vo. cloth, 9s.

PROFESSOR PIRRIE, F.R.S.E.

THE PRINCIPLES AND PRACTICE OF SURGERY. With numerous Engravings on Wood. Second Edition. 8vo. cloth, 24s.

PROFESSOR PIRRIE & DR. KEITH.

ACUPRESSURE: an excellent Method of arresting Surgical Hæmorrhage and of accelerating the healing of Wounds. With Engravings. 8vo. cloth, 5s.

PROFESSORS PLATTNER & MUSPRATT.

THE USE OF THE BLOWPIPE IN THE EXAMINATION OF MINERALS, ORES, AND OTHER METALLIC COMBINATIONS. Illustrated by numerous Engravings on Wood. Third Edition. 8vo. cloth, 10s. 6d.

MESSRS. CHURCHILL & SONS' PUBLICATIONS. 29

DR. HENRY F. A. PRATT, M.D., M.R.C.P.

I.
THE GENEALOGY OF CREATION, newly Translated from the Unpointed Hebrew Text of the Book of Genesis, showing the General Scientific Accuracy of the Cosmogony of Moses and the Philosophy of Creation. 8vo. cloth, 14s.

II.
ON ECCENTRIC AND CENTRIC FORCE: A New Theory of Projection. With Engravings. 8vo. cloth, 10s.

III.
ON ORBITAL MOTION: The Outlines of a System of Physical Astronomy. With Diagrams. 8vo. cloth, 7s. 6d.

IV.
ASTRONOMICAL INVESTIGATIONS. The Cosmical Relations of the Revolution of the Lunar Apsides. Oceanic Tides. With Engravings. 8vo. cloth, 5s.

V.
THE ORACLES OF GOD: An Attempt at a Re-interpretation. Part I. The Revealed Cosmos. 8vo. cloth, 10s.

THE PRESCRIBER'S PHARMACOPŒIA; containing all the Medicines in the British Pharmacopœia, arranged in Classes according to their Action, with their Composition and Doses. By a Practising Physician. Fifth Edition. 32mo. cloth, 2s. 6d.; roan rock (for the pocket), 3s. 6d.

DR. JOHN ROWLISON PRETTY.

AIDS DURING LABOUR, including the Administration of Chloroform, the Management of Placenta and Post-partum Hæmorrhage. Fcap. 8vo. cloth, 4s. 6d.

MR. P. C. PRICE, F.R.C.S.

AN ESSAY ON EXCISION OF THE KNEE-JOINT. With Coloured Plates. With Memoir of the Author and Notes by Henry Smith, F.R.C.S. Royal 8vo. cloth, 14s.

DR. PRIESTLEY.

LECTURES ON THE DEVELOPMENT OF THE GRAVID UTERUS. 8vo. cloth, 5s. 6d.

DR. RADCLIFFE, F.R.C.P.L.

LECTURES ON EPILEPTIC, SPASMODIC, NEURALGIC, AND PARALYTIC DISORDERS OF THE NERVOUS SYSTEM, delivered at the Royal College of Physicians in London. Post 8vo. cloth, 7s. 6d.

MR. RAINEY.

ON THE MODE OF FORMATION OF SHELLS OF ANIMALS, OF BONE, AND OF SEVERAL OTHER STRUCTURES, by a Process of Molecular Coalescence, Demonstrable in certain Artificially-formed Products. Fcap. 8vo. cloth, 4s. 6d.

DR. F. H. RAMSBOTHAM.

THE PRINCIPLES AND PRACTICE OF OBSTETRIC MEDICINE AND SURGERY. Illustrated with One Hundred and Twenty Plates on Steel and Wood; forming one thick handsome volume. Fourth Edition. 8vo. cloth, 22s.

DR. RAMSBOTHAM.
PRACTICAL OBSERVATIONS ON MIDWIFERY, with a Selection of Cases. Second Edition. 8vo. cloth, 12s.

PROFESSOR REDWOOD, PH.D.
A SUPPLEMENT TO THE PHARMACOPŒIA: A concise but comprehensive Dispensatory, and Manual of Facts and Formulæ, for the use of Practitioners in Medicine and Pharmacy. Third Edition. 8vo. cloth, 22s.

DR. DU BOIS REYMOND.
ANIMAL ELECTRICITY; Edited by H. BENCE JONES, M.D., F.R.S. With Fifty Engravings on Wood. Foolscap 8vo. cloth, 6s.

DR. REYNOLDS, M.D.LOND.
I.
EPILEPSY: ITS SYMPTOMS, TREATMENT, AND RELATION TO OTHER CHRONIC CONVULSIVE DISEASES. 8vo. cloth, 10s.
II.
THE DIAGNOSIS OF DISEASES OF THE BRAIN, SPINAL CORD, AND THEIR APPENDAGES. 8vo. cloth, 8s.

DR. B. W. RICHARDSON.
I.
ON THE CAUSE OF THE COAGULATION OF THE BLOOD. Being the ASTLEY COOPER PRIZE ESSAY for 1856. With a Practical Appendix. 8vo. cloth, 16s.
II.
THE HYGIENIC TREATMENT OF PULMONARY CONSUMPTION. 8vo. cloth, 5s. 6d.

DR. RITCHIE, M.D.
ON OVARIAN PHYSIOLOGY AND PATHOLOGY. With Engravings. 8vo. cloth, 6s.

DR. WILLIAM ROBERTS, M.D., F.R.C.P.
AN ESSAY ON WASTING PALSY; being a Systematic Treatise on the Disease hitherto described as ATROPHIE MUSCULAIRE PROGRESSIVE. With Four Plates. 8vo. cloth, 5s.

DR. ROUTH.
INFANT FEEDING, AND ITS INFLUENCE ON LIFE; Or, the Causes and Prevention of Infant Mortality. Second Edition. Fcap. 8vo. cloth, 6s.

DR. W. H. ROBERTSON.
I.
THE NATURE AND TREATMENT OF GOUT. 8vo. cloth, 10s. 6d.
II.
A TREATISE ON DIET AND REGIMEN. Fourth Edition. 2 vols. 12s. post 8vo. cloth.

DR. ROWE.
NERVOUS DISEASES, LIVER AND STOMACH COM-
PLAINTS, LOW SPIRITS, INDIGESTION, GOUT, ASTHMA, AND DIS-
ORDERS PRODUCED BY TROPICAL CLIMATES. With Cases. Sixteenth
Edition. Fcap. 8vo. 2s. 6d.

DR. ROYLE, F.R.S., AND DR. HEADLAND, M.D.
A MANUAL OF MATERIA MEDICA AND THERAPEUTICS.
With numerous Engravings on Wood. Fourth Edition. Fcap. 8vo. cloth, 12s. 6d.

DR. RYAN, M.D.
INFANTICIDE: ITS LAW, PREVALENCE, PREVENTION, AND
HISTORY. 8vo. cloth, 5s.

ST. BARTHOLOMEW'S HOSPITAL.
A DESCRIPTIVE CATALOGUE OF THE ANATOMICAL
MUSEUM. Vol. I. (1846), Vol. II. (1851), Vol. III. (1862), 8vo. cloth, 5s. each.

ST. GEORGE'S HOSPITAL REPORTS. Vol. I. 8vo. 7s. 6d.

MR. T. P. SALT, BIRMINGHAM.
I.
ON DEFORMITIES AND DEBILITIES OF THE LOWER
EXTREMITIES, AND THE MECHANICAL TREATMENT EMPLOYED
IN THE PROMOTION OF THEIR CURE. With numerous Plates. 8vo.
cloth, 15s.
II.
ON RUPTURE: ITS CAUSES, MANAGEMENT, AND CURE,
and the various Mechanical Contrivances employed for its Relief. With Engravings.
Post 8vo. cloth, 3s.

DR. SALTER, F.R.S.
ON ASTHMA: its Pathology, Causes, Consequences, and Treatment.
8vo. cloth, 10s.

DR. SANKEY, M.D.LOND.
LECTURES ON MENTAL DISEASES. 8vo. cloth, 8s.

DR. SANSOM M.D.LOND.
I.
CHLOROFORM: ITS ACTION AND ADMINISTRATION. A Hand-
book. With Engravings. Crown 8vo. cloth, 5s.
II.
THE ARREST AND PREVENTION OF CHOLERA; being a
Guide to the Antiseptic Treatment. Fcap. 8vo. cloth, 2s. 6d.

MR. SAVORY.
A COMPENDIUM OF DOMESTIC MEDICINE, AND COMPA-
NION TO THE MEDICINE CHEST; intended as a Source of Easy Reference for
Clergymen, and for Families residing at a Distance from Professional Assistance.
Seventh Edition. 12mo. cloth, 5s.

DR. SCHACHT.
THE MICROSCOPE, AND ITS APPLICATION TO VEGETABLE
ANATOMY AND PHYSIOLOGY. Edited by FREDERICK CURREY, M.A. Fcap.
8vo. cloth, 6s.

DR. SCORESBY-JACKSON, M.D., F.R.S.E.
MEDICAL CLIMATOLOGY; or, a Topographical and Meteorological
Description of the Localities resorted to in Winter and Summer by Invalids of various
classes both at Home and Abroad. With an Isothermal Chart. Post 8vo. cloth, 12s.

DR. SEMPLE.
ON COUGH: its Causes, Varieties, and Treatment. With some practical Remarks on the Use of the Stethoscope as an aid to Diagnosis. Post 8vo. cloth, 4s. 6d.

DR. SEYMOUR.
I.
ILLUSTRATIONS OF SOME OF THE PRINCIPAL DISEASES OF THE OVARIA: their Symptoms and Treatment; to which are prefixed Observations on the Structure and Functions of those parts in the Human Being and in Animals. On India paper. Folio, 16s.

II.
THE NATURE AND TREATMENT OF DROPSY; considered especially in reference to the Diseases of the Internal Organs of the Body, which most commonly produce it. 8vo. 5s.

DR. SHAPTER, M.D., F.R.C.P.
THE CLIMATE OF THE SOUTH OF DEVON, AND ITS INFLUENCE UPON HEALTH. Second Edition, with Maps. 8vo. cloth, 10s. 6d.

MR. SHAW, M.R.C.S.
THE MEDICAL REMEMBRANCER; OR, BOOK OF EMERGENCIES. Fifth Edition. Edited, with Additions, by JONATHAN HUTCHINSON, F.R.C.S. 32mo. cloth, 2s. 6d.

DR. SHEA, M.D., B.A.
A MANUAL OF ANIMAL PHYSIOLOGY. With an Appendix of Questions for the B.A. London and other Examinations. With Engravings. Foolscap 8vo. cloth, 5s. 6d.

DR. SHRIMPTON.
CHOLERA: ITS SEAT, NATURE, AND TREATMENT. With Engravings. 8vo. cloth, 4s. 6d.

DR. SIBSON, F.R.S.
MEDICAL ANATOMY. With coloured Plates. Imperial folio. Fasciculi I. to VI. 5s. each.

DR. E. H. SIEVEKING.
ON EPILEPSY AND EPILEPTIFORM SEIZURES: their Causes, Pathology, and Treatment. Second Edition. Post 8vo. cloth, 10s. 6d.

DR. SIMMS.
A WINTER IN PARIS: being a few Experiences and Observations of French Medical and Sanitary Matters. Fcap. 8vo. cloth, 4s.

MR. SINCLAIR AND DR JOHNSTON.
PRACTICAL MIDWIFERY: Comprising an Account of 13,748 Deliveries, which occurred in the Dublin Lying-in Hospital, during a period of Seven Years. 8vo. cloth, 10s.

DR. SIORDET, M.B.LOND., M.R.C.P.
MENTONE IN ITS MEDICAL ASPECT. Foolscap 8vo. cloth, 2s. 6d.

MR. ALFRED SMEE, F.R.S.
GENERAL DEBILITY AND DEFECTIVE NUTRITION; their Causes, Consequences, and Treatment. Second Edition. Fcap. 8vo. cloth, 3s. 6d.

DR. SMELLIE.
OBSTETRIC PLATES; being a Selection from the more Important and Practical Illustrations contained in the Original Work. With Anatomical and Practical Directions. 8vo. cloth, 5s.

MR. HENRY SMITH, F.R.C.S.
I.
ON STRICTURE OF THE URETHRA. 8vo. cloth, 7s. 6d.

II.
HÆMORRHOIDS AND PROLAPSUS OF THE RECTUM: Their Pathology and Treatment, with especial reference to the use of Nitric Acid. Third Edition. Fcap. 8vo. cloth, 3s.

III.
THE SURGERY OF THE RECTUM. Lettsomian Lectures. Fcap. 8vo. 2s. 6d.

DR. J. SMITH, M.D., F.R.C.S.EDIN.
HANDBOOK OF DENTAL ANATOMY AND SURGERY, FOR THE USE OF STUDENTS AND PRACTITIONERS. Fcap. 8vo. cloth, 3s. 6d.

DR. W. TYLER SMITH.
A MANUAL OF OBSTETRICS, THEORETICAL AND PRACTICAL. Illustrated with 186 Engravings. Fcap. 8vo. cloth, 12s. 6d.

DR. SNOW.
ON CHLOROFORM AND OTHER ANÆSTHETICS: THEIR ACTION AND ADMINISTRATION. Edited, with a Memoir of the Author, by Benjamin W. Richardson, M.D. 8vo. cloth, 10s. 6d.

MR. J. VOSE SOLOMON, F.R.C.S.
TENSION OF THE EYEBALL; GLAUCOMA: some Account of the Operations practised in the 19th Century. 8vo. cloth, 4s.

DR. STANHOPE TEMPLEMAN SPEER.
PATHOLOGICAL CHEMISTRY, IN ITS APPLICATION TO THE PRACTICE OF MEDICINE. Translated from the French of MM. BECQUEREL and RODIER. 8vo. cloth, reduced to 8s.

MR. BALMANNO SQUIRE, M.B.LOND.
I.
COLOURED PHOTOGRAPHS OF SKIN DISEASES. With Descriptive Letterpress. Series I. (12 Parts), 42s.; Series II. (6 Parts), 22s. 6d.; Series III. (6 Parts), 21s.

II.
A STUDENT'S ATLAS OF SKIN DISEASES. (A Series of Chromolithographs, with Descriptive Letterpress). With cloth portfolio, 25s.

MR. PETER SQUIRE.
I.
A COMPANION TO THE BRITISH PHARMACOPŒIA. Third Edition. 8vo. cloth, 8s. 6d.

II.
THE PHARMACOPŒIAS OF THIRTEEN OF THE LONDON HOSPITALS, arranged in Groups for easy Reference and Comparison. 18mo. cloth, 3s. 6d.

DR. STEGGALL.
STUDENTS' BOOKS FOR EXAMINATION.

I.
A MEDICAL MANUAL FOR APOTHECARIES' HALL AND OTHER MEDICAL BOARDS. Twelfth Edition. 12mo. cloth, 10s.

II.
A MANUAL FOR THE COLLEGE OF SURGEONS; intended for the Use of Candidates for Examination and Practitioners. Second Edition. 12mo. cloth, 10s.

III.
GREGORY'S CONSPECTUS MEDICINÆ THEORETICÆ. The First Part, containing the Original Text, with an Ordo Verborum, and Literal Translation. 12mo. cloth, 10s.

IV.
THE FIRST FOUR BOOKS OF CELSUS; containing the Text, Ordo Verborum, and Translation. Second Edition. 12mo. cloth, 8s.

V.
FIRST LINES FOR CHEMISTS AND DRUGGISTS PREPARING FOR EXAMINATION AT THE PHARMACEUTICAL SOCIETY. Second Edition. 18mo. cloth, 3s. 6d.

MR. STOWE, M.R.C.S.
A TOXICOLOGICAL CHART, exhibiting at one view the Symptoms, Treatment, and Mode of Detecting the various Poisons, Mineral, Vegetable, and Animal. To which are added, concise Directions for the Treatment of Suspended Animation. Twelfth Edition, revised. On Sheet, 2s.; mounted on Roller, 5s.

MR. FRANCIS SUTTON, F.C.S.
A SYSTEMATIC HANDBOOK OF VOLUMETRIC ANALYSIS; or, the Quantitative Estimation of Chemical Substances by Measure. With Engravings. Post 8vo. cloth, 7s. 6d.

DR. SWAYNE.
OBSTETRIC APHORISMS FOR THE USE OF STUDENTS COMMENCING MIDWIFERY PRACTICE. With Engravings on Wood. Fourth Edition. Fcap. 8vo. cloth, 3s. 6d.

MR. TAMPLIN, F.R.C.S.E.
LATERAL CURVATURE OF THE SPINE: its Causes, Nature, and Treatment. 8vo. cloth, 4s.

SIR ALEXANDER TAYLOR, M.D., F.R.S.E.
THE CLIMATE OF PAU; with a Description of the Watering Places of the Pyrenees, and of the Virtues of their respective Mineral Sources in Disease. Third Edition. Post 8vo. cloth, 7s.

DR. ALFRED S. TAYLOR, F.R.S.

I.
THE PRINCIPLES AND PRACTICE OF MEDICAL JURISPRUDENCE. With 176 Wood Engravings. 8vo. cloth, 28s.

II.
A MANUAL OF MEDICAL JURISPRUDENCE. Eighth Edition. With Engravings. Fcap. 8vo. cloth, 12s. 6d.

III.
ON POISONS, in relation to MEDICAL JURISPRUDENCE AND MEDICINE. Second Edition. Fcap. 8vo. cloth, 12s. 6d.

MR. TEALE.
ON AMPUTATION BY A LONG AND A SHORT RECTAN-GULAR FLAP. With Engravings on Wood. 8vo. cloth, 5s.

DR. THEOPHILUS THOMPSON, F.R.S.
CLINICAL LECTURES ON PULMONARY CONSUMPTION; with additional Chapters by E. SYMES THOMPSON, M.D. With Plates. 8vo. cloth, 7s. 6d.

DR. THOMAS.
THE MODERN PRACTICE OF PHYSIC; exhibiting the Symptoms, Causes, Morbid Appearances, and Treatment of the Diseases of all Climates. Eleventh Edition. Revised by ALGERNON FRAMPTON, M.D. 2 vols. 8vo. cloth, 28s.

MR. HENRY THOMPSON, F.R.C.S.
I.
STRICTURE OF THE URETHRA; its Pathology and Treatment. The Jacksonian Prize Essay for 1852. With Plates. Second Edition. 8vo. cloth, 10s.

II.
THE DISEASES OF THE PROSTATE; their Pathology and Treatment. Comprising a Dissertation "On the Healthy and Morbid Anatomy of the Prostate Gland;" being the Jacksonian Prize Essay for 1860. With Plates. Second Edition. 8vo. cloth, 10s.

III.
PRACTICAL LITHOTOMY AND LITHOTRITY; or, An Inquiry into the best Modes of removing Stone from the Bladder. With numerous Engravings, 8vo. cloth, 9s.

DR. THUDICHUM.
I.
A TREATISE ON THE PATHOLOGY OF THE URINE, Including a complete Guide to its Analysis. With Plates, 8vo. cloth, 14s.

II.
A TREATISE ON GALL STONES: their Chemistry, Pathology, and Treatment. With Coloured Plates. 8vo. cloth, 10s.

DR. TILT.
I.
ON UTERINE AND OVARIAN INFLAMMATION, AND ON THE PHYSIOLOGY AND DISEASES OF MENSTRUATION. Third Edition. 8vo. cloth, 12s.

II.
A HANDBOOK OF UTERINE THERAPEUTICS, AND OF MODERN PATHOLOGY OF DISEASES OF WOMEN. Second Edition. Post 8vo. cloth, 6s.

III.
THE CHANGE OF LIFE IN HEALTH AND DISEASE: a Practical Treatise on the Nervous and other Affections incidental to Women at the Decline of Life. Second Edition. 8vo. cloth, 6s.

DR. GODWIN TIMMS.
CONSUMPTION: its True Nature and Successful Treatment. Re-issue, enlarged. Crown 8vo. cloth, 10s.

MESSRS. CHURCHILL & SONS' PUBLICATIONS.

DR. ROBERT B. TODD, F.R.S.
I.
CLINICAL LECTURES ON THE PRACTICE OF MEDICINE.
New Edition, in one Volume, Edited by Dr. BEALE, 8vo. cloth, 18s.

II.
ON CERTAIN DISEASES OF THE URINARY ORGANS, AND ON DROPSIES. Fcap. 8vo. cloth, 6s.

MR. TOMES, F.R.S.
A MANUAL OF DENTAL SURGERY. With 208 Engravings on Wood. Fcap. 8vo. cloth, 12s. 6d.

DR. TURNBULL.
I.
AN INQUIRY INTO THE CURABILITY OF CONSUMPTION,
ITS PREVENTION, AND THE PROGRESS OF IMPROVEMENT IN THE TREATMENT. Third Edition. 8vo. cloth, 6s.

II.
A PRACTICAL TREATISE ON DISORDERS OF THE STOMACH
with FERMENTATION; and on the Causes and Treatment of Indigestion, &c. 8vo. cloth, 6s.

DR. TWEEDIE, F.R.S.
CONTINUED FEVERS: THEIR DISTINCTIVE CHARACTERS, PATHOLOGY, AND TREATMENT. With Coloured Plates. 8vo. cloth, 12s.

VESTIGES OF THE NATURAL HISTORY OF CREATION.
Eleventh Edition. Illustrated with 106 Engravings on Wood. 8vo. cloth, 7s. 6d.

DR. UNDERWOOD.
TREATISE ON THE DISEASES OF CHILDREN. Tenth Edition,
with Additions and Corrections by HENRY DAVIES, M.D. 8vo. cloth, 15s.

DR. UNGER.
BOTANICAL LETTERS. Translated by Dr. B. PAUL. Numerous Woodcuts. Post 8vo., 2s. 6d.

MR. WADE, F.R.C.S.
STRICTURE OF THE URETHRA, ITS COMPLICATIONS
AND EFFECTS; a Practical Treatise on the Nature and Treatment of those Affections. Fourth Edition. 8vo. cloth, 7s. 6d.

DR. WALKER, M.B.LOND.
ON DIPHTHERIA AND DIPHTHERITIC DISEASES. Fcap. 8vo. cloth, 3s.

DR. WALLER.
ELEMENTS OF PRACTICAL MIDWIFERY; or, Companion to the Lying-in Room. Fourth Edition, with Plates. Fcap. cloth, 4s. 6d.

MR. HAYNES WALTON, F.R.C.S.
SURGICAL DISEASES OF THE EYE. With Engravings on Wood. Second Edition. 8vo. cloth, 14s.

DR. WARING, M.D., M.R.C.P.LOND.
I.
A MANUAL OF PRACTICAL THERAPEUTICS. Second Edition, Revised and Enlarged. Fcap. 8vo. cloth, 12s. 6d.

II.
THE TROPICAL RESIDENT AT HOME. Letters addressed to Europeans returning from India and the Colonies on Subjects connected with their Health and General Welfare. Crown 8vo. cloth, 5s.

DR. WATERS, M.R.C.P.
I.
THE ANATOMY OF THE HUMAN LUNG. The Prize Essay to which the Fothergillian Gold Medal was awarded by the Medical Society of London. Post 8vo. cloth, 6s. 6d.

II.
RESEARCHES ON THE NATURE, PATHOLOGY, AND TREATMENT OF EMPHYSEMA OF THE LUNGS, AND ITS RELATIONS WITH OTHER DISEASES OF THE CHEST. With Engravings. 8vo. cloth, 5s.

DR. ALLAN WEBB, F.R.C.S.L.
THE SURGEON'S READY RULES FOR OPERATIONS IN SURGERY. Royal 8vo. cloth, 10s. 6d.

DR. WEBER.
A CLINICAL HAND-BOOK OF AUSCULTATION AND PERCUSSION. Translated by JOHN COCKLE, M.D. 5s.

MR. SOELBERG WELLS, M.D., M.R.C.S.
ON LONG, SHORT, AND WEAK SIGHT, and their Treatment by the Scientific Use of Spectacles. Second Edition. With Plates. 8vo. cloth, 6s.

MESSRS. CHURCHILL & SONS' PUBLICATIONS.

MR. T. SPENCER WELLS, F.R.C.S.

I.

DISEASES OF THE OVARIES: THEIR DIAGNOSIS AND TREATMENT. Vol. I. 8vo. cloth, 9s.

II.

SCALE OF MEDICINES WITH WHICH MERCHANT VESSELS ARE TO BE FURNISHED, by command of the Privy Council for Trade; With Observations on the Means of Preserving the Health of Seamen, &c. &c. Seventh Thousand. Fcap. 8vo. cloth, 3s. 6d.

DR. WEST.

LECTURES ON THE DISEASES OF WOMEN. Third Edition. 8vo. cloth, 16s.

DR. UVEDALE WEST.

ILLUSTRATIONS OF PUERPERAL DISEASES. Second Edition, enlarged. Post 8vo. cloth, 5s.

MR. WHEELER.

HAND-BOOK OF ANATOMY FOR STUDENTS OF THE FINE ARTS. With Engravings on Wood. Fcap. 8vo., 2s. 6d.

DR. WHITEHEAD, F.R.C.S.

ON THE TRANSMISSION FROM PARENT TO OFFSPRING OF SOME FORMS OF DISEASE, AND OF MORBID TAINTS AND TENDENCIES. Second Edition. 8vo. cloth, 10s. 6d.

DR. WILLIAMS, F.R.S.

PRINCIPLES OF MEDICINE: An Elementary View of the Causes, Nature, Treatment, Diagnosis, and Prognosis, of Disease. With brief Remarks on Hygienics, or the Preservation of Health. The Third Edition. 8vo. cloth, 15s.

THE WIFE'S DOMAIN: the YOUNG COUPLE—the MOTHER—the NURSE—the NURSLING. Post 8vo. cloth, 3s. 6d.

DR. J. HUME WILLIAMS.

UNSOUNDNESS OF MIND, IN ITS MEDICAL AND LEGAL CONSIDERATIONS. 8vo. cloth, 7s. 6d.

MESSRS. CHURCHILL & SONS' PUBLICATIONS.

MR. ERASMUS WILSON, F.R.S.

I.
THE ANATOMIST'S VADE-MECUM: A SYSTEM OF HUMAN ANATOMY. With numerous Illustrations on Wood. Eighth Edition. Foolscap 8vo. cloth, 12s. 6d.

II.
ON DISEASES OF THE SKIN: A SYSTEM OF CUTANEOUS MEDICINE. Sixth Edition. 8vo. cloth, 18s.
THE SAME WORK; illustrated with finely executed Engravings on Steel, accurately coloured. 8vo. cloth, 36s.

III.
HEALTHY SKIN: A Treatise on the Management of the Skin and Hair in relation to Health. Seventh Edition. Foolscap 8vo. 2s. 6d.

IV.
PORTRAITS OF DISEASES OF THE SKIN. Folio. Fasciculi I. to XII., completing the Work. 20s. each. The Entire Work, half morocco, £13.

V.
THE STUDENT'S BOOK OF CUTANEOUS MEDICINE AND DISEASES OF THE SKIN. Post 8vo. cloth, 8s. 6d.

VI.
ON SYPHILIS, CONSTITUTIONAL AND HEREDITARY; AND ON SYPHILITIC ERUPTIONS. With Four Coloured Plates. 8vo. cloth, 16s.

VII.
A THREE WEEKS' SCAMPER THROUGH THE SPAS OF GERMANY AND BELGIUM, with an Appendix on the Nature and Uses of Mineral Waters. Post 8vo. cloth, 6s. 6d.

VIII.
THE EASTERN OR TURKISH BATH: its History, Revival in Britain, and Application to the Purposes of Health. Foolscap 8vo., 2s.

DR. G. C. WITTSTEIN.

PRACTICAL PHARMACEUTICAL CHEMISTRY: An Explanation of Chemical and Pharmaceutical Processes, with the Methods of Testing the Purity of the Preparations, deduced from Original Experiments. Translated from the Second German Edition, by STEPHEN DARBY. 18mo. cloth, 6s.

DR. HENRY G. WRIGHT.

HEADACHES; their Causes and their Cure. Fourth Edition. Fcap. 8vo. 2s. 6d.

DR. YEARSLEY, M.D., M.R.C.S.

I.
DEAFNESS PRACTICALLY ILLUSTRATED; being an Exposition as to the Causes and Treatment of Diseases of the Ear. Sixth Edition. 8vo. cloth, 6s.

II.
ON THROAT AILMENTS, MORE ESPECIALLY IN THE ENLARGED TONSIL AND ELONGATED UVULA. Eighth Edition. 8vo. cloth, 5s.

CHURCHILL'S SERIES OF MANUALS.

Fcap. 8vo. cloth, 12s. 6d. each.

"We here give Mr. Churchill public thanks for the positive benefit conferred on the Medical Profession, by the series of beautiful and cheap Manuals which bear his imprint."— *British and Foreign Medical Review.*

AGGREGATE SALE, 150,000 COPIES.

ANATOMY. With numerous Engravings. Eighth Edition. By ERASMUS WILSON, F.R.C.S., F.R.S.

BOTANY. With numerous Engravings. By ROBERT BENTLEY, F.L.S., Professor of Botany, King's College, and to the Pharmaceutical Society.

CHEMISTRY. With numerous Engravings. Ninth Edition. By GEORGE FOWNES, F.R.S., H. BENCE JONES, M.D., F.R.S., and A. W. HOFMANN, F.R.S.

DENTAL SURGERY. With numerous Engravings. By JOHN TOMES, F.R.S.

MATERIA MEDICA. With numerous Engravings. Fourth Edition. By J. FORBES ROYLE, M.D., F.R.S., and FREDERICK W. HEADLAND, M.D., F.L.S.

MEDICAL JURISPRUDENCE. With numerous Engravings. Eighth Edition. By ALFRED SWAINE TAYLOR, M.D., F.R.S.

PRACTICE OF MEDICINE. Second Edition. By G. HILARO BARLOW, M.D., M.A.

The MICROSCOPE and its REVELATIONS. With numerous Plates and Engravings. Third Edition. By W. B. CARPENTER, M.D., F.R.S.

NATURAL PHILOSOPHY. With numerous Engravings. Sixth Edition. By CHARLES BROOKE, M.B., M.A., F.R.S. *Based on the Work of the late Dr. Golding Bird.*

OBSTETRICS. With numerous Engravings. By W. TYLER SMITH, M.D., F.R.C.P.

OPHTHALMIC MEDICINE and SURGERY. With coloured Plates and Engravings on Wood. Third Edition. By T. WHARTON JONES, F.R.C.S., F.R.S.

PATHOLOGICAL ANATOMY. With numerous Engravings. By C. HANDFIELD JONES, M.B., F.R.S., and E. H. SIEVEKING, M.D., F.R.C.P.

PHYSIOLOGY. With numerous Engravings. Fourth Edition. By WILLIAM B. CARPENTER, M.D., F.R.S.

POISONS. Second Edition. By ALFRED SWAINE TAYLOR, M.D., F.R.S.

PRACTICAL ANATOMY. With numerous Engravings. (10s. 6d.) By CHRISTOPHER HEATH, F.R.C.S.

PRACTICAL SURGERY. With numerous Engravings. Fourth Edition. By Sir WILLIAM FERGUSSON, Bart., F.R.C.S., F.R.S.

THERAPEUTICS. Second Edition. By E. J. WARING, M.D., M.R.C.P.